Lecture Notes in Physics

Editorial Board

AF801916

H. Araki
Research Institute for Mathematical Sciences
Kyoto University, Kitashirakawa
Sakyo-ku, Kyoto 606, Japan

E. Brézin
Ecole Normale Supérieure, Département de Physique
24, rue Lhomond, F-75231 Paris Cedex 05, France

J. Ehlers
Max-Planck-Institut für Physik und Astrophysik, Institut für Astrophysik
Karl-Schwarzschild-Strasse 1, W-8046 Garching, FRG

U. Frisch
Observatoire de Nice
B. P. 229, F-06304 Nice Cedex 4, France

K. Hepp
Institut für Theoretische Physik, ETH
Hönggerberg, CH-8093 Zürich, Switzerland

R. L. Jaffe
Massachusetts Institute of Technology, Department of Physics
Center for Theoretical Physics
Cambridge, MA 02139, USA

R. Kippenhahn
Rautenbreite 2, W-3400 Göttingen, FRG

H. A. Weidenmüller
Max-Planck-Institut für Kernphysik
Postfach 10 39 80, W-6900 Heidelberg, FRG

J. Wess
Lehrstuhl für Theoretische Physik
Theresienstrasse 37, W-8000 München 2, FRG

J. Zittartz
Institut für Theoretische Physik, Universität Köln
Zülpicher Strasse 77, W-5000 Köln 41, FRG

Managing Editor

W. Beiglböck
Assisted by Mrs. Sabine Landgraf
c/o Springer-Verlag, Physics Editorial Department V
Tiergartenstrasse 17, W-6900 Heidelberg, FRG

The Editorial Policy for Proceedings

The series Lecture Notes in Physics reports new developments in physical research and teaching – quickly, informally, and at a high level. The proceedings to be considered for publication in this series should be limited to only a few areas of research, and these should be closely related to each other. The contributions should be of a high standard and should avoid lengthy redraftings of papers already published or about to be published elsewhere. As a whole, the proceedings should aim for a balanced presentation of the theme of the conference including a description of the techniques used and enough motivation for a broad readership. It should not be assumed that the published proceedings must reflect the conference in its entirety. (A listing or abstracts of papers presented at the meeting but not included in the proceedings could be added as an appendix.)

When applying for publication in the series Lecture Notes in Physics the volume's editor(s) should submit sufficient material to enable the series editors and their referees to make a fairly accurate evaluation (e.g. a complete list of speakers and titles of papers to be presented and abstracts). If, based on this information, the proceedings are (tentatively) accepted, the volume's editor(s), whose name(s) will appear on the title pages, should select the papers suitable for publication and have them refereed (as for a journal) when appropriate. As a rule discussions will not be accepted. The series editors and Springer-Verlag will normally not interfere with the detailed editing except in fairly obvious cases or on technical matters.

Final acceptance is expressed by the series editor in charge, in consultation with Springer-Verlag only after receiving the complete manuscript. It might help to send a copy of the authors' manuscripts in advance to the editor in charge to discuss possible revisions with him. As a general rule, the series editor will confirm his tentative acceptance if the final manuscript corresponds to the original concept discussed, if the quality of the contribution meets the requirements of the series, and if the final size of the manuscript does not greatly exceed the number of pages originally agreed upon.

The manuscript should be forwarded to Springer-Verlag shortly after the meeting. In cases of extreme delay (more than six months after the conference) the series editors will check once more the timeliness of the papers. Therefore, the volume's editor(s) should establish strict deadlines, or collect the articles during the conference and have them revised on the spot. If a delay is unavoidable, one should encourage the authors to update their contributions if appropriate. The editors of proceedings are strongly advised to inform contributors about these points at an early stage.

The final manuscript should contain a table of contents and an informative introduction accessible also to readers not particularly familiar with the topic of the conference. The contributions should be in English. The volume's editor(s) should check the contributions for the correct use of language. At Springer-Verlag only the prefaces will be checked by a copy-editor for language and style. Grave linguistic or technical shortcomings may lead to the rejection of contributions by the series editors.

A conference report should not exceed a total of 500 pages. Keeping the size within this bound should be achieved by a stricter selection of articles and not by imposing an upper limit to the length of the individual papers.

Editors receive jointly 30 complimentary copies of their book. They are entitled to purchase further copies of their book at a reduced rate. As a rule no reprints of individual contributions can be supplied. No royalty is paid on Lecture Notes in Physics volumes. Commitment to publish is made by letter of interest rather than by signing a formal contract. Springer-Verlag secures the copyright for each volume.

The Production Process

The books are hardbound, and the publisher will select quality paper appropriate to the needs of the author(s). Publication time is about ten weeks. More than twenty years of experience guarantee authors the best possible service. To reach the goal of rapid publication at a low price the technique of photographic reproduction from a camera-ready manuscript was chosen. This process shifts the main responsibility for the technical quality considerably from the publisher to the authors. We therefore urge all authors and editors of proceedings to observe very carefully the essentials for the preparation of camera-ready manuscripts, which we will supply on request. This applies especially to the quality of figures and halftones submitted for publication. In addition, it might be useful to look at some of the volumes already published. As a special service, we offer free of charge LATEX and TEX macro packages to format the text according to Springer-Verlag's quality requirements. We strongly recommend that you make use of this offer, since the result will be a book of considerably improved technical quality. To avoid mistakes and time-consuming correspondence during the production period the conference editors should request special instructions from the publisher well before the beginning of the conference. Manuscripts not meeting the technical standard of the series will have to be returned for improvement.

For further information please contact Springer-Verlag, Physics Editorial Department V, Tiergartenstrasse 17, W-6900 Heidelberg, FRG

B. Baschek G. Klare J. Lequeux (Eds.)

New Aspects of Magellanic Cloud Research

Proceedings of the Second European Meeting on the Magellanic Clouds Organized by the Sonderforschungsbereich 328 „Evolution of Galaxies"
Held at Heidelberg, Germany, 15-17 June 1992

Springer-Verlag Berlin Heidelberg GmbH

Editors

Bodo Baschek
Institut für Theoretische Astrophysik
Im Neuenheimer Feld 561, W-6900 Heidelberg, Germany

Gerhard Klare
Landessternwarte Königstuhl
W-6900 Heidelberg, Germany

James Lequeux
Radioastronomie ENS
F-75231 Paris Cedex 05, France

ISBN 978-3-662-13944-8 ISBN 978-3-540-47556-9 (eBook)
DOI 10.1007/978-3-540-47556-9

This work is subject to copyright. All rights are reserved, whether the whole or part of the material is concerned, specifically the rights of translation, reprinting, re-use of illustrations, recitation, broadcasting, reproduction on microfilms or in any other way, and storage in data banks. Duplication of this publication or parts thereof is permitted only under the provisions of the German Copyright Law of September 9, 1965, in its current version, and permission for use must always be obtained from Springer-Verlag. Violations are liable for prosecution under the German Copyright Law.

© Springer-Verlag Berlin Heidelberg 1993

Originally published by Springer-Verlag Berlin Heidelberg New York in 1993.

Typesetting: Camera ready by author/editor using the $\mathrm{T_EX}$ macro package from Springer-Verlag
58/3140-543210 - Printed on acid-free paper

Preface

The Heidelberg Conference on New Aspects of Magellanic Cloud Research was held from 15 to 17 June 1992. This was the first meeting devoted to the Magellanic Clouds to take place in Heidelberg.

The idea behind this conference originated on the one hand from research carried out in Heidelberg on Magellanic Cloud topics within the framework of the Sonderforschungsbereich (SFB) No. 328 "Evolution of Galaxies" (Project B: Chemical Evolution of Galaxies), a special cooperative programme supported by the Deutsche Forschungsgemeinschaft. On the other hand the need for discussing problems of the Clouds in a wider international group of colleagues arose out of the activities within the ESO key programme 'Coordinated Investigations of Selected Regions in the Magellanic Clouds' and the ESO-Swedish SEST key programme 'CO Studies of the Magellanic Clouds'. In the series of the associated European Meetings on the Magellanic Clouds this Heidelberg conference, organized by the SFB 328, is the second one (after the Paris Colloquium 1989).

The main goal of this conference was to bring together astronomers working on different fields of Magellanic Cloud research to review recent progress, to discuss unsolved problems, and to coordinate investigations.

The Scientific Organizing Committee of the conference consisted of

J. Lequeux (France), Chair
B. Baschek (Germany)
M. S. Bessell (Australia)
K. S. de Boer (Germany)
J. Fried (Germany)
F. Spite (France)
B. Wolf (Germany),

and the Local Organizing Committee of

B. Baschek
G. Klare
M. Pakull
O. Stahl
R. Wehrse
B. Wolf.

After the welcoming address by the Speaker of the Heidelberg Sonderforschungsbereich, I. Appenzeller, and an introduction by B. Baschek, during the three days of the conference 32 invited speakers discussed the different aspects of present-day Magellanic Cloud research. Summary talks were given by K. S. de Boer and J. Lequeux. The oral contributions were supplemented by about 50 poster presentations. This book contains the proceedings of the invited oral contributions as well as the contributed posters, grouped according to the different topics.

The organizers express their gratitude to the Deutsche Forschungsgemeinschaft (SFB 328) for financial support which made this conference possible. Additional support by IBM Deutschland is also gratefully acknowledged.

We thank Ms. B. Farr for taking the fine photographs of participants of the conference during their talks, in the audience, during the coffee breaks, and at the barbecue party at the Landessternwarte on the Königstuhl on Tuesday evening. The pictures can be found throughout this book. We also wish to thank A. Jüttner for his valuable help in editing some of the contributions.

We are very indebted to Ms. E. Bär, Ms. B. Farr, Ms. B. Hoffmann, Ms. U. Kapczynski, and to many students for their help in the organization of the conference, at the registration desk, for projecting the slides, and invaluable further assistance. H. Radlinger and S. and T. Zinser are thanked for their help in organizing the barbecue party.

Heidelberg, Paris
September 1992

B. Baschek, G. Klare, J. Lequeux

Contents

Novae, Supernovae, Supernova Remnants

Stars, Stellar Evolution

Clusters, Chemical Evolution

Nomenclature

Summary Talks

Introduction

Bodo Baschek

Institut für Theoretische Astrophysik der Universität Heidelberg
Im Neuenheimer Feld 561, D-6900 Heidelberg, Germany

First I would like to welcome you all to this Magellanic Cloud meeting, also on behalf of our Heidelberg colleagues. While spectroscopists connect Heidelberg with the discovery of spectral analysis in 1859 by G. Kirchhoff and R. Bunsen, Magellanic Cloud researchers may wonder why this meeting is to take place here on a northern-hemisphere site. Thus let me give you a little information about the work on the Magellanic Clouds done in Heidelberg. Since 1987 all Heidelberg astronomical institutes cooperate in the frame of the Sonderforschungsbereich (Special Cooperative Programme) No. 328 "Evolution of Galaxies" which is financed by the Deutsche Forschungsgemeinschaft (DFG). Here within project B (Chemical Evolution of Galaxies) groups at the Landessternwarte and the Institut für Theoretische Astrophysik are engaged in Magellanic Cloud research with the main topics of interest being high-resolution spectroscopy and quantitative abundance analyses of B stars and modelling of the chemical evolution of irregular galaxies.

On this occasion, I would like to express our gratitude to the Deutsche Forschungsgemeinschaft for supporting our reseach and in particular for enabling us to hold this conference. Our thanks are also due to the referees of the DFG, of whom two are among us in the audience, for spending much time and effort on evaluating our proposals.

Our observations of the Clouds are coordinated within the ESO key programme 'Coordinated Investigation of Selected Regions in the Magellanic Clouds', furthermore we have a close collaboration with M.S. Bessell of Mount Stromlo and Siding Spring Observatories.

Of course, the geographic latitude of Heidelberg, $\varphi = 49.4°$, and the southern declination e.g. of the LMC around $-70°$ clearly forces the local astronomers to go to southern-hemisphere locations to carry out their observations. You may be surprised, however, to hear that this situation is to change in the (astronomically) near future. Due to the earth's precession at least the center and the northern parts of the LMC will be above the horizon in about 10,000 years from now, as H. Schwan of the Astronomisches Rechen-Institut kindly calculated for us. Realizing that only about 5,000 years ago observations took place at Stonehenge I, we may imagine the wonderful possibilities for observations we may have in Heidelberg in 10^4 years time. Now that I revealed our future plans to you, let us return to the more serious problems of present day Magellanic Cloud research.

Currently there are two ESO key programmes concerned with research on the Magellanic Clouds: (i) The *Coordinated Investigation of Selected Regions in the Magellanic Clouds* based upon the optical and ultraviolet spectral ranges concentrates on a total of six selected regions, four in the LMC and two in the SMC (de Boer et al. 1989, de Boer et al. 1990). (ii) The ESO–Swedish SEST key programme *CO Studies of the Magellanic Clouds* is based on observations with the Swedish–ESO Submillimetre Telescope SEST (Booth et al. 1987, Israel & Johansson 1989).

The last meetings devoted to the Magellanic Clouds were held in May 1989 in Paris and in July 1990 in Sydney. The proceedings of these conferences (de Boer et al. 1989, Haynes & Milne 1991), and the review article by Westerlund (1990) document the recent progress in this field.

Now after only one year's time about 100 astronomers meet again to discuss "new aspects of Magellanic Cloud research", showing that this topic indeed is a lively and active field of research. Why is it that this interest in our neighbouring galaxies continues so intensely? The motive for this may be the following: The Magellanic Clouds are *galaxies*, they are *near-by*, and they are *different* from the Milky Way Galaxy.

The different irregular types and the low metallicities compared to our Galaxy of type Sbc originate in a *different* evolution of the lower-mass Magellanic Clouds during which interactions, in particular with the Milky Way, play an important role. The blue "globular" clusters, unique to the Clouds, are a special attraction and may be a key to the understanding of the Clouds' evolution. Only recently a population of blue "globular" clusters has been observed in another galaxy, NGC 1275 = Per A, with the HST planetary camera (Holtzman et al. 1992).

The *proximity* of the Magellanic Clouds allows very detailed studies which in particular serve as fundamental calibration for more distant galaxies. New instruments and new observational techniques as well as new theoretical methods are usually directed and applied to the Clouds, e.g. the new instruments available in space such as the Hubble Space Telescope, the X-ray satellite ROSAT, and the Compton Gamma Ray Observatory. Incidentally, the central parts of the LMC were the first target of the "first light" observation by ROSAT performed on 16 June 1990, exactly two years before this conference (Trümper et al. 1991), and by kind permission of the ROSAT team this image was used on the poster inviting participation in this meeting. Concerning ground-based optical instruments we are on the threshold of observing the Clouds with the future class of "Very Large Telescopes".

As a consequence of these developments, *the study of the Magellanic Clouds has in many ways become more "galactic" than "extragalactic"* as Westerlund (1990) expressed it in his review article.

For the Introduction to this conference, I would like to present you with a list of as yet open questions, desiderata etc. by looking through the recent reviews and conference proceedings. Here, without going into details, is the compilation of catchwords:

Problems of Magellanic Cloud Research

— distances, interstellar extinction
— structure

extent in depth, fragmentation, tilt
halo / old disk
— kinematics
rotation
motion of gas complexes
— Magellanic Cloud System (LMC, SMC, Intercloud Region, Magellanic Stream)
interactions, "history", star bursts
— population structure
stellar and interstellar contents
completeness of brightest objects (LBV, PN, B[e] stars, ...)
giant H II regions (30 Dor)
star clusters, blue globulars, ages / metallicity
interstellar matter: molecules, dust, magnetic fields
— chemical composition
homogeneity of metal content
distribution of Wolf–Rayet subtypes
— evolution
metal-poor galaxies
star formation, star burst
variations of IMF of clusters
stellar evolution, CMDs, SN 1987A

Of course, from my personal research interest I was aware that our knowledge of e.g. abundance patterns and chemical evolution is far from being complete. But I was surprised by the extent of the list, and even began to wonder what really *is* known about the Clouds since we began to explore them with more and more details so that complexity is not masked by "unfavourable" angular resolution as for the more distant galaxies.

Clearly the long list of unsettled topics underlines the importance of exchanging ideas among the various fields of Magellanic Cloud research, as well as the necessity of coordinating observations and research. I wish that this Conference will fruitfully contribute to the progress in this field. Three days for solving all the open problems admittedly is a fairly short time. Nevertheless, friends of the Magellanic Clouds, let us start with the work.

References

Booth R.S., de Jonge M.J., Shaver P.A., 1987, ESO Messenger 48, 2
de Boer K.S., Azzopardi M., Baschek B. et al., 1989, ESO Messenger 57, 27
de Boer K.S., Spite F., Stasińska G. (eds.), 1989, Recent Developments of Magellanic Cloud Research — A European Colloquium, Observatoire de Paris, Paris
de Boer K.S., Spite F., François P. et al., 1990, ESO Messenger 66, 14
Haynes R., Milne D. (eds.), 1991, The Magellanic Clouds (IAU Symposium No. 148), Kluwer Academic, Dordrecht
Holtzman J.A., Faber S.M., Shaya E.J. et al., 1992, AJ 103, 691

Israel F.P., Johansson L.E.B., 1989, in: de Boer K.S., Spite F., Stasińska G. (eds.) Recent Developments of Magellanic Cloud Research — A European Colloquium, Observatoire de Paris, Paris, p. 11
Trümper J., Hasinger G., Aschenbach B. et al., 1991, Nature 349, 579
Westerlund B.E., 1990, A&AR 2, 29

Overviews, Structure, Populations

Welcoming address by Immo Appenzeller

Bodo Baschek

Roy S. Booth, Frank P. Israel, Robin E.S. Clegg, (Joachim Köppen), Volker Weidemann, Bodo Baschek, Ulrich Kinkel, Hildegard Domgörgen (left to right)

The Stellar Populations in the Magellanic Clouds. An Overview.

Bengt E. Westerlund

Astronomical Observatory, Uppsala University, Box 515, S- 75120 Uppsala, Sweden

Abstract: Recent investigations of the stellar populations in the Magellanic Clouds are reviewed in the light of the evolution and the present structure of the Clouds.

1. Introduction

The Magellanic Clouds continue to attract the interest of more and more astronomers and new results appear from observations over the whole spectrum, from X- rays to radio wavelengths. The amount of available information continues to increase from year to year. More detailed studies are carried out spectroscopically and photometrically of the brighter objects; higher spatial resolution and higher sensitivity is used in surveying the Clouds for molecules and dust; and the limiting magnitude is pushed further and further down in photometric studies so that the main sequence of the oldest population is now reached. The first results from the Hubble Space telescope have just appeared confirming and ameliorating the resolution of R136a in the 30 Doradus nebula into the long foreseen cluster of stars.

Nevertheless, the global properties of the Magellanic System seem to remain much as I summarized them two years ago (Westerlund 1990). The mass of Population II in the LMC may now be down to $\sim 2\%$, and the average depth of the SMC is more likely ≤ 10 kpc. The total mass of molecular gas in the SMC has been estimated to 3×10^7 $M_\odot$ (Rubio et al. 1991) leading to a ratio of molecular to atomic mass of roughly 7%. Most of the ^{12}CO emission arises from two large complexes, located in the SW and NE regions of the Bar, and closely associated with the most intense HI emission. In the LMC the total molecular mass was previously estimated to 1.4×10^8 $M_\odot$, nearly half of it from the complex south of 30 Dor.

Most problems dealt with, repeatedly, in the past 30 years are still in the running, not because they are unsolved but more because refined techniques raise hopes of much improved results able to satisfy demanding theoreticians. Some problems are, however, still rather disturbing. Most serious for stellar research is the difficulty to determine reddening corrections for individual clusters and field populations, at least with regard to the accuracy we wish to achieve. Observing time is seldom available for accurate photometry and spectroscopy of a sufficient number of stars in the region of interest. Therefore, mean values are frequently used and an unknown uncertainty factor introduced. This affects all attempts to fit theoretical isochrones to colour-magnitude diagrams (CMDs) for cluster and field- populations and, of course, also all temperature determinations of stars with the aid of colour indices. It seems, however, still too early to use, as is sometimes done, the fitting of cluster main sequences to theoretical isochrones to determine the reddening and the distance. In such cases other information which ought to be derived from the observations, e.g. age and metallicity, have to be assumed.

For the moment there seems, fortunately, to be less controversy regarding the distances of the Clouds than a couple of years ago, and the problem regarding their tilts appears also to be(temporarily?) set aside.

Unavoidably(?), spiral structure in both Clouds is still discussed on much the same basis as before. In the LMC the "octopus" with its head in 30 Doradus (cf. Westerlund 1991) is still alive even if it has shrunk a little. Before global models of the LMC are presented with 30 Dor as origin the contributions of the older populations must be considered (Cf. Laspias and Meaburn 1991) and the existence of the enormous gas and dust clouds south of 30 Dor will have to be explained. The Fujimoto, Noguchi (1990) model seems to be a very good beginning in that direction. We may also expect the new radio data showing conspicuous emission ridges and highly ordered magnetic fields in this region (Haynes et al. 1990) to contribute to an acceptable explanation of the structure in this part of the LMC.

2. Distance Moduli

Several new results have been presented. Capaccioli et al. (1990) have used the maximum magnitude versus rate of decline for novae to derive $(m-M)_0 = 18.70$ for the LMC. The properties of the SN 1987A circumstellar ring have given $(m-M)_0 = 18.50 \pm 0.13$ (Panagia et al.1991), and absorption line velocities of its expanding shell $(m-M)_0 = 18.47 \pm 0.18$ (Hanuschik and Schmidt- Kaler 1991). Arellano Ferro et al. (1991) have used luminous F supergiants to derive 19.33 ± 0.31 and 18.19 ± 0.22 for the SMC and the LMC, respectively. Supergiants are difficult to use and these data may have less weight than the others. There seems to be no reason to change the distance moduli of 18.5 and 18.9 mag given in my A&AR review (Westerlund 1990).

3. The Interstellar Reddening of the Clouds

The Galactic foreground colour excess towards the Magellanic Clouds has been investigated by Schwering and Israel (1991) on scales of 48 arcmin. They find that over the LMC surface the foreground reddening varies over a range in E(B- V) of 0.07 to 0.17 mag. For the SMC they find more constancy; the range is 0.07 to 0.09 mag. On smaller scales the amplitude may be much greater.

Bessell(1991) has summarized reddening determinations from photometry, stellar polarization, and HI column densities, and also used CaII K lines and UV extinction.

He finds lower values than Schwering and Israel (1991); in the SMC a foreground reddening of E(B-V) between 0.04 and 0.06 mag and in the LMC a more varied reddening between 0.04 and 0.09 mag. The average reddening within the SMC is probably about 0.06 mag but there are regions with reddenings up to 0.3 mag. The average reddening in the LMC is similar but the variations are larger with many regions with higher as well as lower than average reddening. This is confirmed by Greve et al.(1990) in their investigation of dust in emission nebulae in the LMC; the maximum value reported is E(B-V) = 1.1 mag. The galactic foreground reddening can also be surprisingly large in the outskirts of the Clouds. Thus, it amounts to E(B-V) = 0.18 ± 0.02 mag at NGC 1841, 15° from the centre of the LMC (Walker 1990).

The reddening is generally small in the Bridge region. Grondin et al. (1992) find E(B-V) between 0.0 and 0.10 mag. They keep, however, in most cases colour excess and distance as free parameters in fitting their CMDs to theoretical isochrones.

4. The Structure of the Clouds

The structure of the Clouds is to a high degree the result of their mutual interaction and their interaction with the Galaxy. The two Clouds have had many close passages. The SMC came within 2 - 5 kpc from the centre of the LMC about 0.2 Gyr ago. This collision led to the fragmentation of the SMC and had a severe effect on the LMC. Many of the SMC gas clouds passed undisturbed through the LMC but many collisions occurred. Fujimoto and Noguchi (1991) have shown how the hydrodynamical collisions led to the formation of larger clouds which moved in the common gravitational field in the LMC, but not necessarily in its equatorial plane. The largest is the CO and hydrogen rich cloud that now contains the 30 Dor nebula and the supergiant shell LMC SGS- 2 south of it. The collisions led also to a highly asymmetric distribution of the gas clouds in the LMC; many, perhaps also the large cloud just discussed, ended up outside the main plane of the LMC.

5. The Stellar Populations

Since our conference in Paris three years ago many investigations of the stellar populations in the Clouds have appeared. Many of them have dealt with clusters and the populations of their surrounding fields.

5.1 The Bridge Region

All attempts to detect stellar members of the Clouds between the SMC Wing and the LMC were unsuccessful for a long time. The Magellanic Bridge remained an HI feature (cf. Mathewson et al. 1979 for references and details). Until 1980, when Kunkel identified a group of young stars at $2^h30^m, -74^o$, the tip of the wing was at$2^h15^m, -74.5^o$ (Westerlund and Glaspey 1971). An extensive young stellar population has now been discovered by Irwin et al.(1990, IDK) and discussed in a series of papers (Cf. Grondin et al. 1992 for references). The series of associations found by IDK form an arc between the Clouds at about decl. -74°. Their ages are ≤ 0.1 Gyr; the youngest are suggested to be only 0.016 Gyr. A difference in distance modulus between two associations separated by only 17 arcmin amounts to $\approx 12 \pm 8$ kpc indicate a certain depth of the Bridge. The LMC halo population was found to extend to R.A.=3^h40^m, i.e. to about 9 degrees from its optical center. On the SMC side no halo population was found eastward of the SMC Wing tip.

5.2 The Small Magellanic Cloud

The extension of the major body of the SMC in depth is likely to be less than 10 kpc; it may be appreciably larger in the some outer parts (Cf. Westerlund 1990 for references and discussion). Recent investigations appear to confirm this structure: Gardiner and Hawkins (1991) have investigated the stellar populations in the northern and north- western outlying regions of the SMC by using the red-giant clump(RGC) stars, considered to be "halo" stars . They find the halo to extend further in the northern direction than in the western. Also in the NE the halo population drops sharply, beginning about 4 kpc from the centre. This asymmetry in the old population is probably caused by the interaction between our Galaxy and the Magellanic Clouds. It may be connected with the event that led to the formation of the Bridge and the Magellanic Stream.

The previously determined depth of the SMC halo in the NE is confirmed and found to exist even further westward; it is about 12 - 16 kpc. Most of the regions in the N and NW exhibit smaller depths, of the order of 4 to 9 kpc. In the NE a distinct structure may be projected in front of the main structure.

Reid and Mould's (1990) study of the asymptotic giant branch stars in the SMC seems to confirm that the depth along the line of sight is small in the main body; they suggest a depth of ≈ 5 kpc from the observed width of the giant branch. Their radial-velocity determinations confirm the bimodal distribution displayed by most young components in the SMC but previously not seen in the older objects. The investigation of a sample of carbon stars in the central parts of the SMC showed for instance no evidence of bimodality (Hardy et al. 1989).

The carbon star population in the SMC has been studied by Rebeirot et al.(1992) who have prepared a catalogue of 1707 objects. The searched area covers most of the main body of the SMC. The survey will be described further by Azzopardi in this volume.

5.3 The Large Magellanic Cloud

In my A&AR review I pointed out that the number of true globular clusters in the LMC and the SMC was sufficient to indicate that the initial star formation in both Clouds had been on a normal scale as compared with our Galaxy (cf. Westerlund 1990); they would then also represent a halo population. In a recent study of field RR Lyrae variables in the outer parts of the LMC Kinman et al.(1991) concluded that the LMC field halo of Population II stars makes up about 2% of the mass of the LMC, a similar percentage to that found for our Galactic halo. The efficiencies of the initial star-formation processes may then have been comparable in the two systems. Kinman et al. find also that the relative strengths of the halo to an "old-disk" population is within a factor of two of that which would be seen in our Galaxy at the solar distance from the center.

Hughes et al. (1992) have investigated the radial velocities of a sample of long-period variables (LPVs) in the LMC. A kinematic analysis applied to a wide range of LMC populations (HI gas, CO molecular clouds, PNs, clusters, CH stars) showed that the dynamics of the LMC is dominated by a single rotating disk, that the travers velocity of the LMC is $\approx$ 200 kms^{-1}, and that all major populations of the Bar have solid body rotation. The dispersion ages for the PN, clusters, CH stars and intermediate age LPVs are all similar at $\approx$ 4 Gyr.

The velocity distribution of the old LPVs indicates that the mass of the LMC is $\leq 6.2 \pm 1.5 \times 10^9 M_\odot$. The distribution of the population represented by the old LPVs, with an age of$\approx$ 10 Gyr, is highly flattened, with an axial ratio $\approx$ 0.3 to 0.5, and its mass is$\approx$ 2% of the LMC's total mass. This is the same value as derived by Kinman et al. for the oldest population,$\geq$ 14 Gyr. Thus, three old generations are now defined: the halo,$\geq$ 14 Gyr old and metal- poor, [Fe/H] ≤ -1.4, the flattened spheroidal,$\approx$ 10 Gyr, [Fe/H]≈ -1, and the intermediate- age disk population,$\approx$ 4 Gyr, and less metal poor.

Also the CH stars belong to two distinct groups according to Cowley and Hartwick (1991). The observed stars were found to form at least two distinct kinematical sub-groups, one associated with the LMC disk population and the other with the old-cluster population. (See, however, Feast 1993).

It is interesting to note that among the old globulars in the LMC for which individual distances are available NGC1466, H11, NGC2210 (See Westerlund 1990), NGC2257 (Walker 1989), and GLC0435-59 (Reticulum, Walker 1990) lie close to the plane defined by the SMC and the LMC when their distance moduli are 18.9 and 18.5 mag, respectively. The SMC old cluster NGC 121 follows the same pattern. This suggests that the Magellanic System, excluding the Stream, is flattened. Only NGC 1841, with a distance modulus of 18.19 mag (Walker 1990), falls (about 12 kpc) in front of this plane. Its membership of as well as it origin in the Magellanic System may be questioned. For its membership in the LMC speaks that its CMD is similar to that of NGC2257 and NGC2210 (but also to that of M92!) and the fact that a halo population, of RR Lyr stars, exists at that distance from the LMC.

In their continued analysis of the red giant stars in the LMC Bar West field Frogel and Blanco (1990) confirm (See Westerlund 1990, p.47 for previous studies) the existence of two separate generations. The new results confirm also two episodes of star formation with a quiet period inbetween. The red population is dominated by asymptotic giant branch stars (AGB) with properties similar to the intermediate age clusters, i.e. of ages$\approx$ a few Gyr. The metallicity of this intermediate age population is tentatively assigned a value of [Fe/H] $= -0.5 \pm 0.3$. About 20 % of the AGB population has properties similar to the young, $\leq$ 100 Myr, clusters. They find as Lundgren (1988) that the most luminous red giants are not carbon stars but of type M, probably MS. There is still a significant deficit of luminous AGB stars (M(bol) ≤ -6). Reid et al. (1990) searched for optically obscured AGB stars in the LMC by looking for IRAS sources with [12-25] colours consistent with their being either stellar photospheres or circumstellar dust shells. Of 63 such objects 24 had no obvious optical counterparts. The existence of a sufficient number of optically invisible high- luminosity AGB stars was ruled out; several hundred would be needed to explain the difference between the predicted numbers and those optically observed. The non- identified objects may be dusty AGB stars in a phase of enhanced mass loss. Their luminosities would be low, M(bol) ~ -5, and their phase of evolution triggered at low luminosity and thus explaining the deficit of luminous AGB stars just mentioned.

Sequential star formation has been discussed frequently on the scale of superassociations; it has mostly concerned SGS LMC-2 and LMC-4 (cf. Lortet and Testor 1988 and Westerlund 1990). Massey et al. (1989) found in the OB associations LH 117 and LH 118 three supergiants of lower mass than several of the unevolved members; they must have formed about 6 to 10 million years earlier. Parker et al. (1992) have investigated the stellar content of LH 9 and LH 10, two associations of quite different appearances. LH 10 is surrounded by nebulosity, LH 9 has hardly any. The colour excess E(B-V) is three times as large in LH 10 than in LH 9. All these features indicate differences in age with LH 10 being the younger. The authors suggest that the evolution of the massive stars in LH 9 triggered star formation in LH 10. The initial mass functions of the two associations have quite different slopes, -1.6 ± 0.1 for LH 9 and -1.1 ± 0.1 for LH 10, the latter being the flattest so far measured for an association in the LMC.

Two distinct supergiant branches have been found in the young LMC cluster NGC 2214 (Sagar et al. 1991). It is composted of two clumps, one compact and the other amorphous. The data derived by Sagar et al. demonstrate the presence of two stellar populations in the cluster with age differences of 75 to 150 Myr. The older population is spatially the more compact one. As the difference in age between the two populations is too large for a binary cluster the most likely explanation is a sequential star formation.

The luminosity functions (LF) of cluster and field populations are useful tools for investigation of the star formation history in various parts of the LMC. Linde, Lyngå, and I have secured deep CCD photometry in B,V of three faint clusters in each of two regions in the LMC, one in the NW and one in the SW. In each region we have determined the CMDs of the three clusters and of the field populations. The observed distributions are shown for two of our clusters and their surrounding fields in Fig.:s 1 and 2 (Distance modulus 18.5 mag, A(V) = 0.25 mag). The stars form well defined sequences in all four figures: a main sequence(MS), a red giant clump(RGC), less well defined in Fig.1, and in Fig.2 red giant (RG) and subgiant (RSG) branches (For details see Linde et al. 1993, Westerlund et al. 1993).

Fig.1a shows that HS96 is a young cluster, its age is estimated to 0.2 Gyr. The surrounding field (Fig.1b) has a relatively young population but also a large number of red giants around $M(V) \approx 0.5$ mag,$(B-V)_0 \approx 0.8$ mag. In the SW region neither the cluster nor the field has any blue stars more luminous than $M(V) \approx 1$ mag. The age of LW55 is about 1 Gyr. LW55 as well as its field has a pronounced RGC, it is particularly concentrated in the cluster CMD (Fig.2a) and is also bluer in the cluster than in the field. The red star populations are different in the HS96 and the LW55 fields.

The LFs of the three NW clusters are as well represented by the differential initial luminosity function (ILF, Sandage 1957) as can be expected from their small numbers of stars. There is, however, a fundamental difference: there is in all three clusters a sharp cutoff of the MS at about $M(V) \approx 2-2.5$ mag. This is at too bright a magnitude to be explained as incompleteness in our material and must be considered as a real

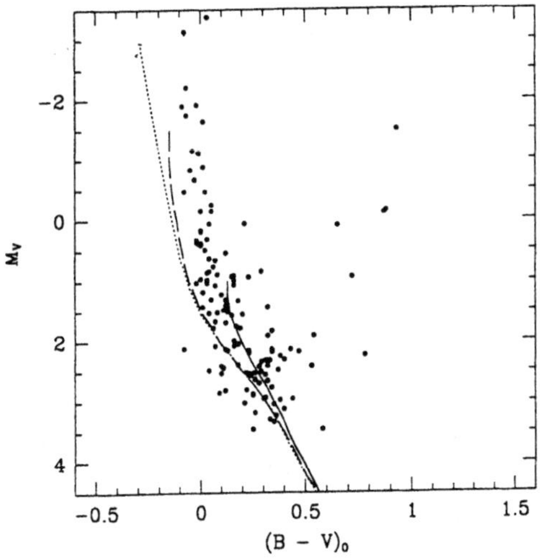

Fig. 1a. The colour-magnitude diagram for HS96 in the NW region. The ZAMS is drawn as well as the Pleiades and Hyades main sequences

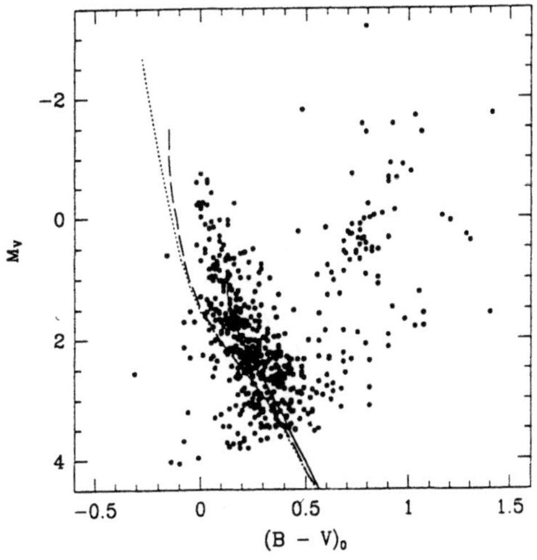

Fig. 1b. The colour-magnitude diagram for the field around HS96. Sequences as in Fig.1a

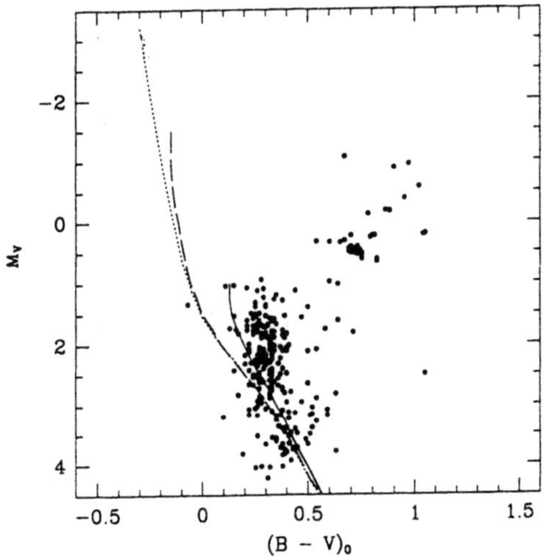

Fig. 2a. The colour-magnitude diagram for LW55 in the SW region. Sequences as in Fig.1a

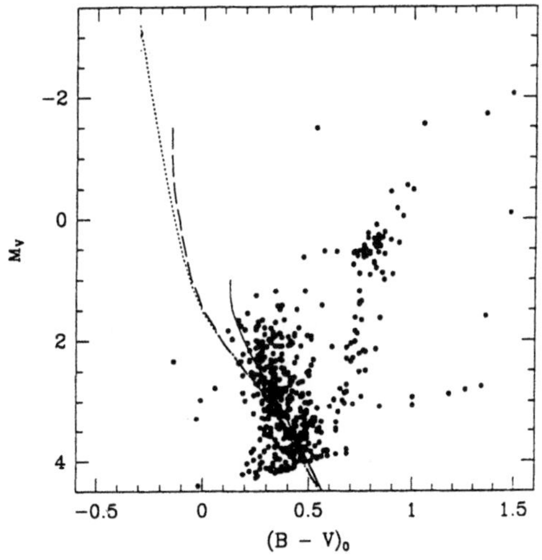

Fig. 2b. The colour-magnitude diagram for the field around LW55. Sequences as in Fig.1a. LW55 and HS96 occupy equal areas and so do the two fields

break in the mass function. It is important to find out if this is typical for the fainter open clusters in the LMC.

In the case of LW55 the normalized differential general luminosity function (GLF) fits not only the MS part well but also the red giant stars with $M(V) \leq 0.5$ mag when added to the top of the MS.

The NW field population is well represented by the GLF and agrees with most previously determined field LFs. In the SW field the LF has a different slope, showing a complete lack of bright MS stars. Similar LFs are seen in the Hodge (1987) field, also far out in the SW, and in the Ardeberg et al. (1985) more centrally situated field. All LFs show a break at about M(V) = 3 mag.

We have normalized the LF for M3 (Sandage 1957, with M(V) = 0.6 mag for the RR Lyr) to the red-star LF for our SW field. The agreement in shape between the two LFs is surprisingly good and confirms the existence of an old field population in the LMC. Also the horisontal branches are surprisingly equal in strength though the age difference, our field being the younger, shows up in their colours, blue in M3 and red in our field. The red-star population in the SW field is also similar to the one in the Draco dwarf spheroidal galaxy (Carney 1986).

Judging from the shape of the RSG part the LF of the NW field an old population of the same kind as in the SW field exists there too (See Fig.1b and 2b). At higher luminosities the appearance of a number of younger red giants is visible as a widening of the RGC. The same effect is seen in the Hardy et al. field (1984).

In an attempt to determine the star formation history of the LMC Mateo et al. (1990) interpreted their photometric data with the aid of synthetic CMDs. They found that the field CMDs and LFs could be modelled with a star formation rate that showed a burst about 7 ± 1 Gyr ago, the precise age depending on the use of overshoot models. Classical models would yield a burst age 2-3 Gyr younger, in better agreement with the age of this burst normally counted with.

We have attempted to separate the various generations in our fields by first eliminating from their CMDs the generations most easily identified in the observed data. Thus, we subtract the LW 55 cluster population from the SW field by normalizing the former to the latter in the RGC box. The result is shown in Table 1. The remaining stars identify two populations, one very old and discussed above, the other younger and less metal- poor.

Table 1. The distribution of stars in the SW field after subtraction of the LW55 cluster type population (per 10 arcmin^2). Isochrones for 3 Gyr, Z=0.017 (VandenBerg 1985), and 10 Gyr, Z=0.001, X=0.70 (Green et al. 1987, fulldrawn), are outlined

M(V) \ (B−V)o	−0.2 – 0.0	0.0 – 0.2	0.2 – 0.4	0.4 – 0.6	0.6 – 0.8	0.8 – 1.0	1.0 – 1.4	≥1.4
−2.5								2
−2.0								2
−1.5				2			4	
−1.0								
−0.5						6		
0.0						6		2
+0.5				3	28	29		
1.0				1	6	9		
1.5				2	11	2	2	
2.0		9	33	21	20	2		
2.5	2	5	85	34	14			
3.0	2	5	141	51	20	7	5	
3.5	2		68	112	7			
4.0		2	97	91	7			

We have also subtracted the SW field population from that in the NW field in order to obtain a more pure intermediate- age population. The result is shown in Table 2. It appears quite possible to determine the star formation history of the Magellanic Clouds directly from the observations by a stepwise elimination of generations,

Table 2. The distribution of stars in the NW field after subtraction of the SW field component (per 10 arcmin^2). Normalization has been done in the box $0.25 \leq M(V) \leq 1.25$, $0.61 \leq (B-V)o \leq 1.00$. The VandenBerg (1985) isochrone for 0.4 Gyr, Z=0.01 is outlined

M(V)	(B−V)o −0.2 – 0.0	0.0 – 0.2	0.2 – 0.4	0.4 – 0.6	0.6 – 0.8	0.8 – 1.0	1.0 – 1.4	≥1.4
−3.0					10			
−2.5								
−2.0				10				
−1.5					13	10		
−1.0						20	10	
−0.5	30	50			10	8		
0.0	40	90		13	50	10	13	
+0.5	30	160	15				20	
1.0		280	60	4			10	
1.5	40	505	90					
2.0	10	330	188					
2.5	3	178	473	86				
3.0	3	58	39					

beginning with an old well defined one, and using the proper boxes in the CMDs for the normalization procedure.

6. Conclusions

Recent results concerning the structure and the evolution of the Magellanic Clouds have been briefly summarized with the emphasis on the stellar field population. It appears now possible to identify with some degree of certainty sub- generations within the previously acknowledged major ones. Thus, in addition to the oldest generation with an age of $\geq$ 14 Gyr an old generation around 10 Gyr may be seen. The gap in active star formation, from 10 to 4 Gyr ago, remains. The intermediate- age generation is composed of sub- groups of ages of about 3 and $\leq$ 1 Gyr. The youngest generation, formed at a burst about 50 million years ago, has not been considered here.

References

Ardeberg, A., Linde, P., Lindgren, H., Lyngå, G. (1985): A&A 148, 263

Arellano Ferro, A., Mantegazza, L., Antonello, E. (1991): A&A 246, 341

Bencivenni, D., Brocato, E., Buonanno, R., Castellani V. (1991): AJ 102, 137

Bessell, M.S. (1991): A&A 242, L17

Capaccioli, M., Della Valle, M., D'Onofrio, M., Rosino, L. (1990): ApJ 360, 63

Carney, B.W. (1986): AJ 92, 23

Cowley, A.P., Hartwick, F.D.A. (1991): ApJ 373, 80

Feast, M.W. (1993): This conference

Frogel, J.A., Blanco, V.M. (1990): ApJ 365, 168

Fujimoto, M., Noguchi, M. (1990): PAS Japan 42, 505

Gardiner, L.T., Hawkins, M.R.S. (1991): MNRAS 251, 174

Green, E.M., Demarque, P., King, Ch.R. (1987): The Revised Yale Isochrones and Luminosity Functions, Yale Univ. Obs.

Greve, A., van Genderen, A.M., Laval, A. (1990): A&AS 85, 895

Grondin, L., Demers, S., Kunkel, W.E. (1992): AJ 103, 1234
Hanuschik, R.W., Schmidt- Kaler, Th. (1991): A&A 249, 36
Hardy, E., Buonanno, R., Corsi, C.E., Janes, K.A., Schommer, R.A. (1984): ApJ 278, 592
Hardy, E., Suntzeff, N.B., Azzopardi, M. (1989): ApJ 344, 210
Haynes, R.F., Klein, U., Wayte, S.R., Wielebinski, R., Murray, J.D., Bajaja, E., Meinert, D., Buczilowski, U.R., Harnett, J.I., Hunt, A.J., Wark, R., Sciacca, L. (1991): A&A 252, 475
Hodge, P.W. (1987): PASP 99,730
Hughes, Sh.M.G., Wood, P.R., Reid, N. (1992) preprint
Irwin, M.J., Demers, S., Kunkel, W.E. (1990): AJ 99, 191 (IDK)
Kinman, T.D., Stryker, L.L., Hesser, J.E., Graham, J.A., Walker, A.R., Hazen, M.L., Nemec, J.M. (1991): PASP 103, 1279
Kunkel, W.E. (1980): IAU Symp. 85, ed. J.E. Hesser, p.353
Laspias, V.N., Meaburn, J. (1991): MNRAS 249, 1P
Linde, P., Lyngå, G., Westerlund, B.E. (1993): to be submitted to A&AS
Lortet, M.-C., Testor, G. (1988): A&A 194, 11
Lundgren, K.(1988): A&A 200, 85
Massey, Ph., Garmany, C.D., Silkey, M., Degioia- Eastwood, K. (1989): AJ 97, 107
Mateo, M., Bertelli, G., Chiosi, C. (1990): 5. IAP Workshop: Astrophysical Ages and Dating Methods, p.225, eds. E. Vangioni- Flam, M. Cassé, J. Audouze, J. Tran Thanh Van
Mathewson, D.S., Ford, V.L., Schwarz, M.P., Murray,J.D. (1979): IAU Symp. 84, ed. W.B. Burton, p.547.
Panagia, N., Gilmozzi, R., Macchetto, F., Adorf, H.-M., Kirschner, R.P. (1991): ApJ 380, L 23
Parker, J.Wm., Garmany, C.D., Massey, Ph., Walborn, N.R. (1991): AJ 103, 1205
Rebeirot, E., Azzopardi, M., Westerlund, B.E. (1992): A&AS, in press
Reid, N., Mould, J. (1990): ApJ 360, 490
Reid, N., Tinney, Ch., Mould, J. (1990): ApJ 348, 98
Rubio, M., Garay, G., Montani, J., Thaddeus, P. (1991): ApJ 368, 173
Sagar, R., Richtler, T., de Boer, K.S. (1991): A&A 249, L5
Sandage, A. (1957): ApJ 125, 422
Schwering, P.B.W., Israel, F.P. (1991): A&A 246, 231
VandenBerg, D.A. (1985): ApJS 58, 711
Walker, A.R. (1989): AJ 98, 2086
Walker, A.R. (1990): AJ 100, 1532
Walker, A.R. (1992): AJ 103, 1166
Westerlund, B.E. (1990): A&AR 2, 29
Westerlund, B.E. (1991): IAU Symp. No. 148, eds. R. Haynes and D. Milne, p.15
Westerlund, B.E., Glaspey, J. (1971): A&A 10, 1
Westerlund, B.E., Linde, P., Lyngå, G. (1992): to be submitted to A&A

Kinematics in the Small Magellanic Cloud

D.Hatzidimitriou [1], and R.D.Cannon [2]

[1]Institute of Astronomy, University of Cambridge, UK

[2]Anglo-Australian Observatory, Australia

During the last decade several authors have undertaken extensive studies of the kinematics and structure of the SMC, which is generally thought to be affected by the dynamics of the triple interacting system consisting of the Galaxy and the Magellanic Clouds (MCs). In this paper we point out the importance of studying the kinematics of old rather than young populations when investigating the purely gravitational effects on the structure of the SMC of the previously mentioned tidal interaction. We also describe briefly the results obtained for such old stars in the NE 'corner' of the SMC.

Kinematics of young populations: The majority of existing kinematical studies in the SMC refer to 'Pop.I' objects (Torres and Carranza 1987; Mathewson *et al.* 1986, 1988; Martin *et al.* 1989). The picture drawn by these studies is complicated and often contradictory. One of the few unchallenged results is the existence of at least two main complexes with different velocities, with the lower velocity component located in front of the other.

Several factors can affect the kinematics and apparent structure of the gas (and the associated newly formed stars), such as the overall disk rotation, the large inclination of the Bar along the line of sight, the possible presence of several star-forming expanding shells, shocks that may have developed in the gas (in the case of collision, e.g. if the disks of the two MCs went through each other), gas falling back into the disk and finally the purely gravitational effects of the interaction. Unless all these effects are properly taken into account, a realistic model of the kinematics of young populations in the SMC will remain elusive.

Kinematics of old populations: Already existing stars at the time of a close passage between the MCs will respond in the same way to the gravitational forces as the gas clouds and will not suffer from most of the complicating factors described above.

It is obvious that the gravitational effects will be more evident in the outer less-strongly bound regions of the galaxy. Hence, studies directed towards evaluating these effects should be conducted in the outer regions of the SMC rather that in the central area where all of the extensive studies have been confined to date (Dopita *et al.* 1985 and Hardy *et al* 1989).

It should also be pointed out that studies of the radial velocity distributions alone (as in all of the studies of older stars to date) can be useful but cannot give a complete picture without information about the location of the stars along the line-of-sight.

New kinematical study: This study combines the three characteristics that optimise the detection and quantification of the effects of the gravitational interaction on the SMC, as discussed above: it considers relatively old stars, located in the outer parts of the SMC and it combines kinematical information with the location of the stars along the line-of-sight.

The sample includes 32 clump/horizontal branch stars located within a 40arcmin field in the NE outer regions of the SMC, and at an average projected distance of 3.3kpc from the optical centre of the SMC. The stars in their majority have ages larger that 1Gyr (Hatzidimitriou and Hawkins 1989 (HH89)). Their distances along the line-of-sight are estimated from their magnitudes (under the assumption that they are zero-age HB stars older than $\simeq 5 \times 10^8$yr; see HH89). Their velocities were derived using standard cross-correlation procedures from intermediate-dispersion spectra obtained with the multi-object spectroscopic facility Autofib at the 3.9m Anglo-Australian Telescope (AAT). The velocity accuracy achieved was $\simeq$10km/sec.

The mean velocity of the sample was 151 ± 6km/sec with an estimated true dispersion of 33 ± 4 km/sec. Most importantly, there is a strong linear correlation between line-of-sight distance and radial velocity for these stars (Spearman rank correlation coefficient $r = 0.84$; see Hatzidimitriou *et al* 1992 for details and figure), with a slope of 6km/sec/kpc, and with the more distant stars having the highest radial velocities. The timescale of the suggested motion is $\simeq 2 \times 10^8$yr, which coincides with the epoch of the recent encounter between the SMC and the LMC inferred by the tidal models of Fujimoto and Murai (1984).

Two more samples of clump/HB stars should be studied in the same way in a forthcoming observing run with the AAT (Sepetember 1992), one in the W outer regions of the SMC where there is no large line-of-sight depth among the older stars (HH89) and one further to the E of the present sample.

References

Dopita M.A.,Ford H.C.,Laurence C.J.,Webster B.L. (1985) Ap.J. **296**,390.

Fujimoto M.,Murai T. (1984) in "The Magellanic Clouds" ed. S.van den Bergh and K. de Boer [Reidel:Dordrecht] p115.

Hatzidimitriou D.,Hawkins M.R.S. (1989) MNRAS **241**,667.

Hatzidimitriou D.,Cannon R.D.,Hawkins M.R.S. (1992) MNRAS submitted.

Hardy E.,Suntzeff N.B.,Azzopardi M. (1989) Ap.J. **344**,210.

Martin N.,Maurice E.,Lequeux J. (1989) A.A. **215**,219.

Mathewson D.S.,Ford V.L.,Visvanathan N. (1986) Ap.J. **301**,664.

Mathewson D.S.,Ford V.L.,Visvanathan N. (1988) Ap.J. **333**,617.

Torres G.,Carranza G.J. (1987) MNRAS **226**, 513.

Star Formation History in the Keyprogram Regions C and E in the LMC

W. Seggewiss, D.J. Bomans, K.S. de Boer, T. Richtler,

A. Vallenari, J.-M. Will

Sternwarte der Universität Bonn, Auf dem Hügel 71, D-5300 Bonn 1

Abstract: CCD photometry of 4 fields in Region C (de Boer et al., 1989 and 1991) at the NW side of the LMC and of 10 fields in Region E at the NW edge of supershell LMC4 in the NE shows that the stellar population comes from a wide range of star formation epochs. We have reduced the data with DAOPHOT and we can thus produce colour-magnitude diagrams (CMDs) with error indications and with proper information on completeness of the data. Field stars are subtracted. The CMDs are used to derive ages and mass functions.

1 The Young Populous Cluster NGC 1818

In Region C the target of our investigations was the young populous cluster NGC 1818. From the CMD we derive an age of 40 Myr, when using stellar evolution models with overshoot included. The CMD of the cluster area is shown in Figure 1. The main sequence is used to construct the luminosity function. Assuming the mentioned age and with the help of the appropriate isochrone we find that the initial mass function has a slope of $x = 1.1 \pm 0.3$, compared to the Salpeter slope of $x = 1.35$. This result is in good agreement with the results of Sagar & Richtler (1991) for five young LMC clusters. Details will be published elsewhere (Will et al., in preparation).

2 The Young Association NGC 1948

The first interesting structure in Region E is the association NGC 1948. It is included in 4 of our CCD fields while a fifth contains just LMC field stars. Figure 2 shows the CMD of NGC 1948. We have analysed the association data, in order to derive the age of the association. The models used are without and with treatment of the overshoot. The effects of the presence of binary stars on the shape of the main sequence (data points becoming brighter by 0.75 mag) and of the effects of rotation on the appearance of a star (colour change of up to 0.5, brightness change of up to 0.3 mag for objects of the same age and mass) are taken into account. After all these considerations we

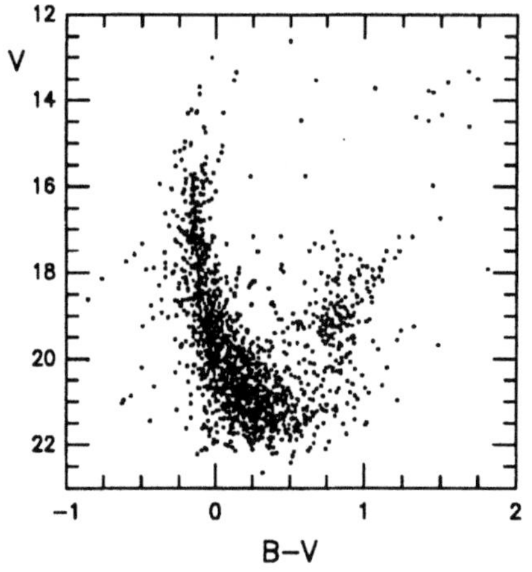

Fig. 1. CMD of NGC 1818

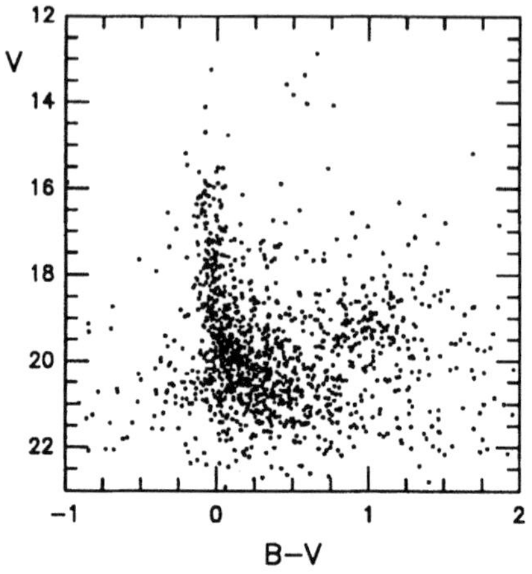

Fig. 2. CMD of NGC 1948

conclude that the association NGC 1948 has an age of 10 to 20 Myr (depending on the just mentioned CMD effects) or, when excluding overshoot, of 7 to 15 Myr (depending on CMD effects). Details will be published by Vallenari et al. (1992).

The young stellar associations at the edge of the supershell LMC4 are thought to be the product of star formation propagating radially outward from the centre of LMC4. The expansion is supposed to have started about 15 Myr ago, according to calculations based on the observed expansion velocities and suggest an age for NGC 1948 of 4 Myr only (Dopita et al., 1985). The ages we derived for NGC 1948 are similar to those set for the onset of star formation in the centre of LMC4. Thus the dynamical model is in clear conflict with the derived ages.

3 The Intermediate Age Cluster NGC 1978

In Region E the most conspicuous structure in the observed field is the large and very elliptical globular cluster NGC 1978. Our photometry allowed to derive a well defined CMD, with a turn-off at $V = 21$ mag which, at the distance of the Large Magellanic Cloud of $(m - M) = 18.5$ mag implies an age of about 3 Gyr. The details on this work will be published by Bomans et al. (in preparation).

References

Bertelli, G., Bressan, A., Fagotto, F., Chiosi, C., 1992b, preprint
de Boer, K.S., Azzopardi, M., et al., 1989, ESO Messenger 57, 27
de Boer, K.S., Spite, F., et al., 1991, ESO Messenger 66, 14
Dopita, M.A., Mathewson, D.S., Ford, V.L., 1985, ApJ 297, 599
Sagar, R., Richtler, T., 1991, A&A 250, 324
Vallenari, A., Bomans, D.J., de Boer, K.S., 1992, A&A submitted

New Coravel Radial Velocities, Structure and Kinematics of the Magellanic Clouds

Nicole Martin , Eric Maurice

Observatoire de Marseille, 2 Place Le Verrier, F - 13248 Marseille, France

Abstract: We have observed 49 late-type supergiants in the Small and 25 in the Large Magellanic Cloud. After these observations,accurate radial velocities are known for approximately 450 supergiants in the Small Magellanic Cloud and approximately also 450 in the Large Magellanic Cloud. Several of these stars belong to the bridge between the two Clouds.

A Coravel scanner is in use since 1981 at the Cassegrain focus of the Danish 1.54 m telescope at ESO, La Silla, in Chile. For details regarding the instrument and observing procedure we refer to Baranne et al. (1979). Since then, the supergiants in both Magellanic Clouds have been observed and accurate radial velocities have been published for 233 F to M type stars in and near the Small Magellanic Cloud (SMC) by Maurice et al. (1987) and for 404 supergiant stars of the same spectral types in the Large Magellanic Cloud (LMC) by Prévot et al. (1985). The standard errors for a single measurement range from 1.3 km s^{-1} for M-type stars to 2.0 km s^{-1} for F-type stars. More information concerning the measurements, their accuracy and their immediate interpretation in terms of structure and kinematics of these galaxies can be found in the two above mentionned references.

Since the publication of these results, new Coravel radial velocity determinations of late-type stars have been made in both Clouds. In the SMC, these measurements mostly concern stars in a list of 372 known and suspected late-type supergiants by Sanduleak (1989). In both Clouds the necessity of a better understanding of the structure, depth, and kinematics of these galaxies was the fundamental reason for these observations.

These new measurements confirm the results obtained previously in the SMC and bring a new light on the kinematics of the outer parts of the wing of the SMC. In the LMC, the new measurements concern stars in the so-called constellations 30 Dor and Shappley II. These results are presented in the proceedings of the present conference by Martin and Lortet. They will also be published in a forthcoming paper.

References

Baranne, A., Mayor, M. Poncet, J. L.: 1979, Vistas Astron. **23**, 279.

Maurice, E., Andersen, J., Ardeberg, A., Bardin, C., Imbert, C., Lindgren, H., Martin, N., Mayor, M., Nordström, B., Prévot, L., Rebeirot, E., Rousseau, J.: 1987, Astron. Astrophys. Suppl. Ser. **67**, 423.

Prévot, L., Andersen,J., Ardeberg, A., Benz, W., Imbert, M., Lindgren, H., Martin, N., Maurice, E., Mayor, M., Nordström, B., Rebeirot, E., Rousseau, J.: 1985, Astron. Astrophys. Suppl. Ser. **62**, 23.

Sanduleak, N.:1989, Astron. J. **98**, 825.

The Morphology of the Large Magellanic Cloud and its Surroundings

Theodor Schmidt-Kaler

Astronomisches Institut, Ruhr-Universität Bochum, Postfach 102148,
D-W-4630 Bochum 1, Germany

The spiral structure of our own Galaxy has been inferred from the geometrical distribution of individual tracers belonging to the extreme population I (HII-regions, young OB-stars and O-associations, supergiants, dark clouds, cepheids, WR-stars) and from the kinematics of the interstellar gas (HI, HII, CO). Using the same tracers the overall spiral structure of the LMC was inferred (Schmidt-Kaler 1977): a multi-armed asymmetric spiral of type Sc(-c+) III-IV with 30 Dor as nucleus.

The magnetic field pattern, as demonstrated by the optical polarizations of individual stars and of the optical surface intensity, appears generally oriented along the spiral arm fragments, centered on the central star R136 of 30 Dor, and in the south-west area gradually merging into the panmagellanic stream. Analysis of past star formation and of the widths of the spiral features led to the conclusion that the last burst of star formation occurred about $1 \cdot 10^7$ yrs ago and subsequent ejection led to the nearly radial pattern around 30 Dor. Thus, the LMC seems composed of an elliptical-barred galaxy with the optical bar, superimposed with an asymmetric Sc(-c+)III-IV spiral located somewhat above the plane of the elliptical disk galaxy. This picture has recently received strong support from quite independent new observations:

1. A high sensitivity radio survey at 1.4 GHz produced evidence for large-scale star-forming filaments in striking similarity to the 1977 picture (Feitzinger et al. 1987).
2. New surveys at 2.5, 4.75, 8.55 GHz (Haynes et al. 1991) confirm this.
3. The FIR emission confirms the radio continuum results. The cool 100 μm dust emission shows the same ridges as the 1.4 GHz radio map. Again 30 Dor dominates the structure, the optical bar is non-existing (Laspias, Meaburn 1991, Xu et al. 1992).
4. A new HI survey has been studied by means of Gaussian analysis of the profiles (Luks & Rohlfs 1992). Two global components are recognized: a disk component, extending over the whole LMC, and an asymmetrically distributed lower velocity component, roughly coinciding with the two main optical spiral arms, representing about 20% of the neutral hydrogen mass. This component must be located about 400 pc above the disk.
5. The magnetic field has been determined from the polarization of the nonthermal radiation at 1.4 and 2.5 GHz (Haynes et al. 1992). It is in general agreement with the spiral picture. Furthermore, direct wide-angle photographs in the optical

domain (obtained by W. Schlosser 1985) show the overall filamentary structure connected with the LMC B-field.

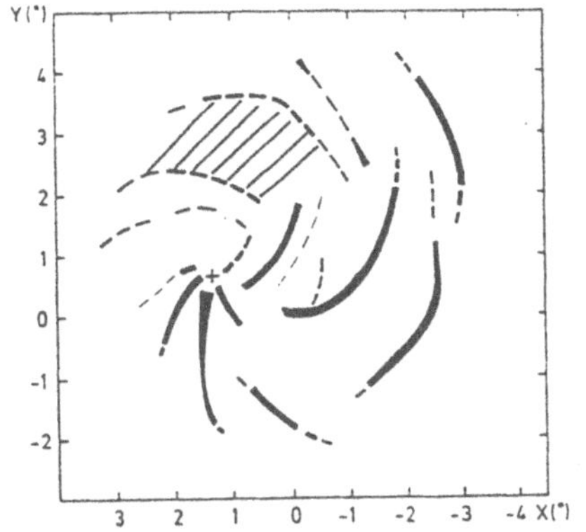

Fig. 1. The overall spiral structure of the LMC. The various features are schematically indicated according to their strength: less certain features are given by broken lines. 30 Dor is given by the cross (Schmidt-Kaler 1977).

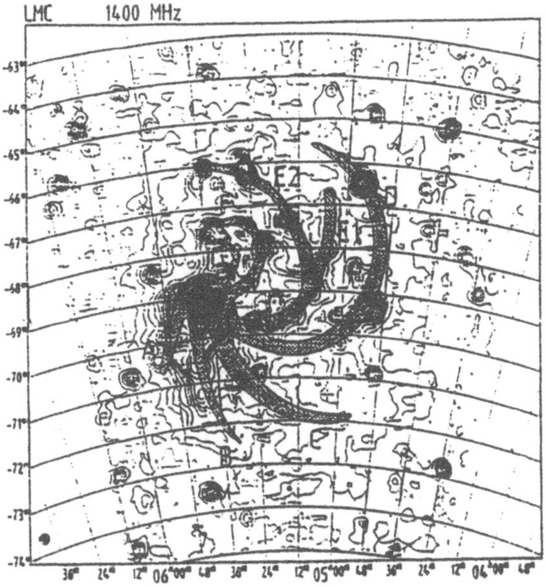

Fig. 2. Star forming filaments as delineated by the ridges of radio continuum (Feitzinger et al. 1987, Haynes et al. 1992)

The Bochum photographic UBVR surface photometry (Kimeswenger et al. 1992) shows a slightly curved narrow bridge at $l = 273°$ almost perpendicular to the galactic plane, connecting the Galaxy and the LMC. The bridge is of reddish colour $B - V = 1.0$, and shows up clearly in the $(B - V)$ colour map, but not in the intensity map.

A detailed paper will be given elsewhere.

References

Feitzinger, J.V. et al. (1987): Vistas Astr. 30, 243
Haynes, R.F. et al. (1991): A&A 252, 475
Haynes, R.F. et al. (1992): preprint
Kimeswenger, S. et al. (1992): A&AS in press
Laspias, V.N., Meaburn, J. (1991): MNRAS 249, 1p
Luks, Th., Rohlfs, K. (1992): A&A in press
Schmidt-Kaler, Th. (1977): A&A 54, 771
Xu, C. et al. (1992): A&A 257, 47

The Distance to the Large Magellanic Cloud from Observations of SN1987A

Theodor Schmidt-Kaler

Astronomisches Institut, Ruhr-Universität Bochum, Postfach 102148, D-W-4630 Bochum 1, Germany

The distance to the LMC, although of basic importance to astrophysics, is still not known very well. The latest review (van den Bergh 1989) quotes m_o-M=18.45±0.1 or 49.0±2.2 kpc as result of many methods applied to thousands of objects. Here we present a distance dermination which attains comparable accuracy using only one object, SN1987A. The radius of the photosphere of SN1987A, calculated from luminosity L_{bol} and temperature T_{eff}, is distance-dependent. Spectroscopic radii, calculated from observed Doppler shifts, are distance-independent. Comparison of r_{ph} with r_{sp} yields the distance.

1 The Baade-Wesselink Method

The nine weakest lines display general agreement with $v_{ph}(t)$ and yield $d = 52.3 \pm 1.5$ kpc (Hanuschik & Schmidt-Kaler 1991). To account for the time delay by the initial acceleration this is diminished by 0.8 kpc. But even those weakest lines originate slightly above the photosphere. Increasingly stronger lines originate at increasingly higher levels with correspondingly higher expansion velocities: $\Delta = \log r_{sp}/r_{ph} = -0.01 + 0.11 \log W_\lambda$, suggesting a reduction of r_{sp} by 2% or 1.2 kpc.

The basic assumptions in deriving the distance by Baade's method were

(1) the envelope is spherically symmetric

(2) the flux of the expanding envelope can be approximated by a black-body.

The second effect is influenced by flux dilution due to scattering-dominated opacity (decreasing $\varepsilon = \pi F/\sigma T_{eff}{}^4$), and atmospheric extension (increasing ε). From atmospheric models we adopt $\epsilon = 0.8 - 1.0$, or a distance correction of 0.95.

The asymmetry effect makes the radiating surface deviate from $4\pi r_{ph}^2$, the flux distribution non-uniform over the surface, and the net radiation polarized. From calculations of Shapiro and Sutherland (1982) the observed continuum polarization requires an axis ratio $\mu = 0.77$ for an oblate, and $\mu = 0.70$ for a prolate spheroid, correspondingly $L/L_S = 1.10$ and 0.88 (if viewed equator-on). The pole-on case can be discarded because spheroids viewed pole-on appear in projection as circular disks. The distance correction then is $(L/L_S)^{-1/2} = 0.94$ if prolate, 1.05 if oblate. Physical intuition led to adopt the prolate case, but meanwhile this has been proved by studies of the explosion process (Steinmetz & Höflich 1992). The final result is $d = 44.8 \pm 2.0$ kpc.

2 The Light Parallax of the Circumstellar Ring and the Expansion Parallax of SN 1987A

Another accurate distance dermination is possible due to the light parallax of the circumstellar ring (Panagia et al. 1991). Rediscussion, accounting for the finite thickness D of the ring and for some retardation of the first light t_0, with newly determined $t_0 = 78\overset{d}{.}5 \pm 1.5$, $t_{\max} = 403\overset{d}{.}5 \pm 3$, yields $d = 49.1 \pm 2.9$ kpc.

Practically all of the hydrogen mass is ejected by day 80. The velocity of the Hα absorption trough around that day is $v_\alpha = 5450 \pm 37$ km/s, agreeing with the mean velocity of the strong lines. An independent estimate of the expansion velocity of the nebular disk is from the full width of the Hα emission line in the nebular stage, e.g. at day 201 about 10600 km/s, corresponding to $v_{exp} = 5300$ km/s, in later phases nearly constant at 10000 km/s.

Taking $v = 5400 \pm 100$ (total m.e.) km/s, and the angular diameter $162 \pm 17mas$ of the expansion disk as measured on day 1278, the distance is $d = 49.2kpc \pm 5.2kpc$.

3 The Position of SN1987A relative to the LMC–Centre

The east side of the LMC is closer to us, with p.a.=55°±17°for the closest part determined from cepheids. The p.a. of the minor axis of the best-fitting ellipse, 78°±7°(Schmidt-Kaler&Gochermann 1992) agrees well with that, and is confirmed by the HI distribution and kinematics (Luks & Rohlfs 1992). With i=38°the disk plane in the direction of 30 Dor is 0.75 kpc nearer to us. Both effects add up to 1.1±0.15 kpc for 30 Dor, and slightly less (0.9 kpc) for SN1987A since it appears to be (with $E_{B-V} = 0\overset{m}{.}16$) at almost the same depth as most of the 30 Dor stars.

Giving weights 6, 3 and 1, resp. to the three methods, the final distance estimate for the LMC from SN1987A is 47.4 ± 1.5 kpc or m_0-M=18.35±0.07 m.e.

References

Hanuschik R.W., Schmidt-Kaler Th. (1991): A&A 249, 36
Luks Th., Rohlfs K. (1992): A&A in press
Schmidt-Kaler Th., Gochermann J. (1992): PASP in press
Steinmetz M., Höflich P. (1992): A&A 257, 641
van den Bergh S. (1989): A&AR 1, 111

CO in the Large Magellanic Cloud - a SEST key project

R. S. Booth

Onsala Space Observatory, Chalmers University of Technology
S-43992, Onsala, Sweden

Abstract: Results from the SEST key project to study carbon monoxide in the Large Magellanic Cloud (LMC) are reviewed. A survey of the general properties of the CO emission in LMC molecular clouds has revealed that the lines are weaker and narrower than equivalent lines in the Galaxy. Attempts to determine the conversion factor, X, between CO emission and molecular hydrogen column density in two major CO complexes show that, taken at face value, it varies between 1.5 and 3 times the 'standard' Galactic value. However, the significance of these values of X are discussed in terms of the physical conditions in the LMC. Finally, some more detailed results on CO isotopically substituted species in the Doradus region are presented.

1 Introduction

One of the prime motivations for the construction of SEST, the Swedish-ESO Submillimetre Telescope was the study of the Magellanic Clouds (MC) at millimetre wavelengths with high resolution. Thus it was not surprising that, among the the responses to the first call for proposals, there were several to study the CO distribution in the 'clouds'. The ESO and Swedish programme committees therefore recommended that the interested groups should work together on a 'Key Project'[1] to study the LMC and SMC. Drs. Frank Israel and Lars E.B. Johansson were invited to coordinate the observations in terms of the ESO and Swedish communities respectively.

There are several important reasons to study the MC with SEST. Its beam width at the frequency of the lowest rotational transition of CO is 43 arcsec giving a linear resolution of 10 pc at the distance of the LMC. Thus we have, for the first time, the opportunity for detailed study of molecular clouds in an external galaxy, which

[1] The co-investigators in the Magellanic Clouds Key Project: F.P. Israel (Leiden), L.E.B. Johansson (Onsala), J. Lequeux (Meudon), R.S. Booth (Onsala), L.-Å. Nyman (Onsala/SEST), P. Crane (ESO, Garching), M. Rubio (Univ. de Chile), Th. de Graauw (Groningen), M.L. Kutner, (Rensselaer Poly., Troy, NY), R. Gredel (ESO, Chile), F. Boulanger (Meudon), G. Garray (Univ. de Chile), B. Westerlund (Uppsala), A. Greve (IRAM).

are in the size range typical of those in the Galaxy (50 pc). It is of great interest to investigate the physical properties of the molecular gas in the MC since they are so different from the Galaxy. Such dwarf irregular galaxies, with significantly lower metallicities, a higher gas to dust ratio and a more hostile radiation environment than the Galaxy may be expected to contain less molecular gas and hence may yield important clues about the survival of giant molecular clouds. Above all they should provide a strict test of the constancy, from galaxy to galaxy, of the conversion factor between CO intensity and molecular hydrogen column density. Progress on the SEST MC key programmes has been reported at several meetings; see for example Israel and de Graauw (1991), Johansson (1991) and Rubio (1991). This paper reviews the LMC observations to date; in a companion paper Dr. Monica Rubio reports on SEST observations of the SMC. More detailed results will soon be published by the key project teams in a series of papers. Paper 1, by Israel et al. (1992), will be referred to here.

2 Earlier work

Early observations of molecules in the MC have been reviewed by Israel (1984). Following the detection of CO by Huggins et al. (1975), Israel et al. (1982, 1986) made a limited survey in the CO (2-1) line at 230 GHz using the ESO 3.6 m telescope equipped with a millimetre line receiver, and concluded that CO was weaker in the MC than in the Galaxy. A complete CO map of the LMC (linear resolution, 200 pc) was made by Cohen et al. (1988) using the Columbia 1.2 m telescope. Their results show CO to be associated with all major star formation regions but they also find a large CO cloud to the south of 30 Dor, where there is no apparent star formation. The CO map is very similar to the warm dust distribution as derived from IRAS data by Schwering and Israel (1989).

3 SEST observations of the LMC - observing strategy

The first observations of the LMC with SEST[2] were carried out in the ^{12}CO (1-0) rotational transition at a frequency of 115 GHz. All observations were performed using the frequency switching technique with a frequency offset of 15 kHz (40 km s^{-1}) and the spectra were recorded with an acousto-optical spectrometer, (frequency resolution 0.2 km s^{-1}). The integration time per point was typically 4 min giving a 3 sigma detection limit of about 0.15 K. Follow up observations in ^{13}CO have been made in some positions and for a few clouds the (2-1) transition has been observed.

A complete, fully sampled map of the CO in the LMC will take nearly 10 years of full time observations, thus the observing strategy adopted was to study selected regions in the first instance. Because of the high degree of similarity between the IRAS map of Schwering and Israel and the CO distribution mapped by Cohen et al., we chose to observe 100 positions of peak IR emission from the IRAS survey. In addition, we mapped in detail (20″ grid spacing) the regions near 30 Dor and

[2] see Booth et al., 1989, for more details of the telescope and observing system

the large CO complex , N11. Finally, to effect an unbiased set of observations, we observed two strips, N-S and E-W, passing through the 30 Dor region. The constant right ascension (RA) strip had a 20″ spacing in declination; CO emission was detected in 100 positions, representing some 15 different clouds. The constant declination strip had a spacing in RA of 40″; CO emission was detected in 10 positions, probably representing 4 different clouds. The observations so far are summarised in Table 1.

Table 1. Summary of LMC observations

IRAS peaks	100 positions observed in ^{12}CO and ^{13}CO
E-W strip through α=5^{h} 18^{m}, δ=-69° 30′	500 positions, 40″ spacing. 4 different CO clouds detected
N-S strip through α=5^{h} 40.5^{m}, δ=-69° 52′	500 positions, 20″ spacing. 15 different CO clouds detected
Mapping N-S strip clouds	3000 positions
30 Dor region map	2500 positions
N11 map	1000 positions

4 The relationship between CO and infrared emission

The observations of the IRAS peak emission regions may be summarised as follows: CO is detected in 90% of the IRAS regions but some strong IRAS sources do not have CO. A plot of far-infrared intensity v. integrated CO emission, I_{CO}, (paper 1) shows that while there is a general correlation between the FIR luminosity and I_{CO}, there is a large scatter. However, much of the scatter is attributed to the large difference in the CO and IRAS beams (even the 12 mm IRAS beam covers a solid angle nearly 20 times the SEST beam at 115 GHz) and the fact that the IRAS clouds were sampled in only one position. Cool IRAS sources, for which $F_{60}/F_{100} < 0.6$, are relatively weak in both CO and far-infrared, whereas sources with strong CO emission and high FIR are hot. Israel et al. (1992) also show that strong CO sources are associated with low ratios, F_{12}/F_{25}, i.e. they are anticorrelated with near-infrared colours which are characteristic of 'cirrus'.

5 The CO maps

5.1 The N11 complex

The HII region complex, N11 (see Israel and de Graauw, 1991) consists of some 21 separate CO clouds whose distribution follows the infrared emission as mapped by IRAS. A group of clouds to the south of the region appear to delineate a shell of molecular gas surrounding a rich cluster of stars. Israel and de Graauw (1991) have pointed out that the CO map shows a remarkable resemblance to the Orion molecular cloud complex, although the latter is about half the size. Both regions show clumps of molecular gas distributed like beads on a string, except that the Orion map contains diffuse emission between the clumps, not seen in the N11 map, perhaps because of sensitivity limitations.

5.2 The region near 30 Doradus

Mapping of the 30 Dor region was begun by the Swedish team whilst commissioning SEST and maps have been published, e.g. by Booth and Johansson (1991). 23 clouds were identified in these observations having diameters in the range 5 to 30 pc. Fig 1 shows maps of the CO integrated intensity in the 30 Dor and N159 regions together with a map of the thermal emission as measured in the continuum at 5 GHz by McGee et al. (1982). The lower resolution data of Cohen et al. (1987) has shown that while the CO delineates the large HII region complex to the south of 30 Dor, it is relatively weak around the main strong HII region, 30 Dor itself, and becomes stronger south of N159, where there are no other indicators of star formation. The recent SEST data reveals three small fairly luminous CO clouds within a few minutes of the main HII region, 30 Dor but confirms the paucity of CO in this area. All of the detected CO clouds are weaker by a factor of 2 or more than the relatively luminous clouds associated with N159. Although the data is incomplete, there are indications of a relatively strong CO cloud in the direction of the radio continuum hole in the map of McGee et al.

Further south, the SEST map of the N159 complex shows weak CO associated with N158C and N160A; there are also a pair of luminous CO clouds associated with N159. Another luminous CO cloud is found to the south of N159 where the radio continuum has cut off. The IRAS maps also show much weaker infrared emission south of N159 than in the 30 Dor region as a whole (Schwering and Israel, 1989). The three strongest IRAS point sources in the CO survey area are indicated by filled triangles; any other IRAS object is weaker by a factor of 10.

The region south of 30 Dor has attracted considerable attention in the literature on the LMC as a possible example of sequential star formation.(see e.g. Israel, 1984,). Deharveng and Caplan (1992) have identified a group of recently formed stars associated with N159 and Haynes et al (1991), using the radio continuum to CO intensity ratio as an indicator of star formation efficiency (SFE), find that SFE is high in 30 Dor and south down to N159, but that it decreases abruptly further south. Thus we may speculate that the molecular gas south of N159 may indicate a region of the LMC which is currently poised for a burst of star formation.

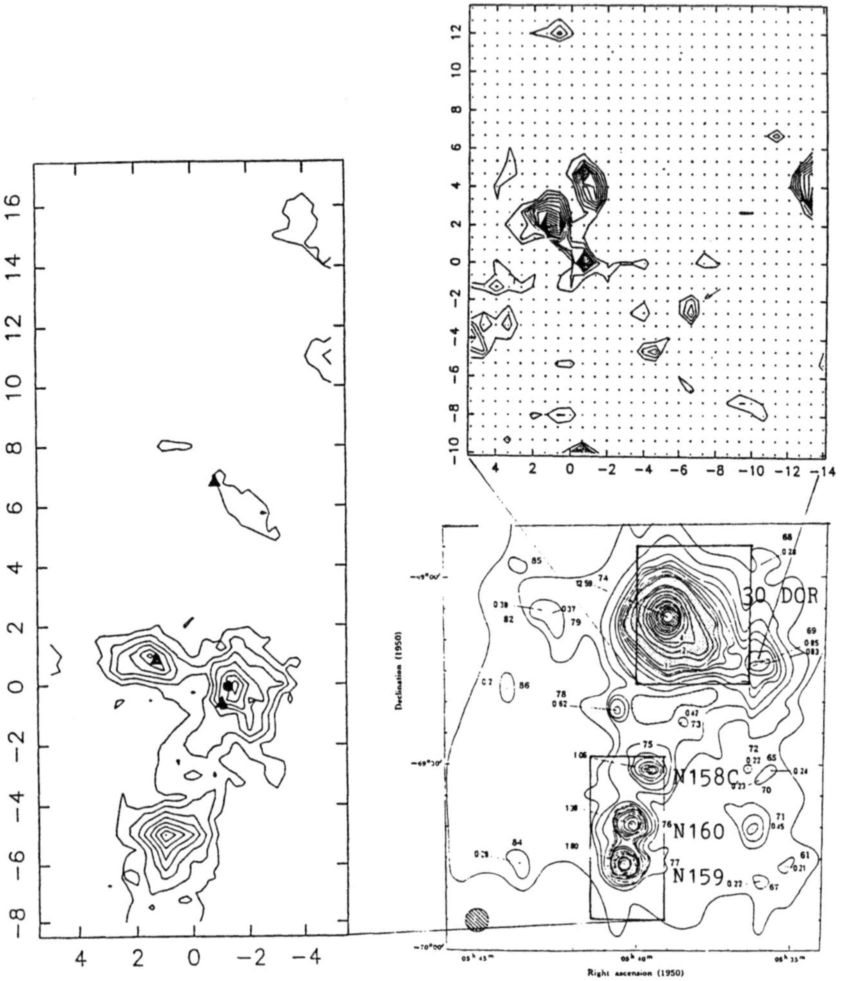

Fig. 1. Maps of the CO in those parts of the 30 Doradus complex identified on the 5 GHz continuum map of McGee et al. (1982). Note that the contour intervals on the two CO maps are not the same. The centres (0,0) of the CO maps are (1950 coords.):
30 Dor: α=5^{h} 39^{m} 0.66^{s}, δ=-69° 08′, N159: α=5^{h} 40^{m} 18.2^{s}, δ=-69° 47′

6 General properties of the CO in the LMC

6.1 Weaker emission in the MC

In paper 1, the general properties of the CO (1-0) in the MC are discussed in relation to average Galactic CO clouds. The data for the Galaxy were taken from surveys and positions such that the linear resolution was the same as for SEST on the MC. Thus, an average ^{12}CO (1-0) intensity for the Galaxy is found to be 30 K km s^{-1}. The main result of paper 1 is that the CO lines in the MC are less intense and narrower than in the Galaxy. In the LMC the velocity integrated ^{12}CO and ^{13}CO intensities are lower than those in the Galaxy by at least a factor of three. In the SMC they are an order of magnitude less than in the Galaxy.

6.2 Isotopic ratios

^{13}CO emission was measured for 27 clouds in the LMC. The ratio of the main beam brightness temperatures, $T(^{12}CO)/T(^{13}CO)$ lies in the range 8 to 14 with a mean value of about 10. In the Galaxy this ratio is nearer 5. In the LMC the $^{12}CO/^{13}CO$ line width ratio varies between 1.1 and 2 with a mean of about 1.25 and the integrated CO intensity ratio, $I(^{12}CO)/I(^{13}CO)$ ranges between 10 and 20 with a mean of 12.5±0.5. The stronger sources favour the smaller ratios. Compared to the Galaxy, where a typical ratio is more like 5 or 6, these values are again high.

6.3 Metallicity related effects and the abundance of CO

One explanation of the lower intensities of CO in the LMC relative to the Galaxy may be the result of the lower metallicity in the LMC. The CO weakness in both MC is of the same order of magnitude as the underabundance of oxygen and especially carbon and dust relative to the Galaxy. The oxygen abundance ratio, Galaxy : LMC is 1 : 0.5 and for carbon it is 1 : 0.3 (Dufour, 1984). For the dust-to-gas abundance, expressed as $E(B\text{-}V)/N_{HI}$, the ratio is 1 : 0.25 (Koorneef, 1982; Bouchet et al., 1985). Another explanation, also related to the lower metallicity and the consequent higher gas-to-dust ratio in the LMC relative to that in the solar neighbourhood may be the effect of the high UV radiation field on the CO gas. Photodissociation of the CO in the LMC will result in smaller CO clouds (greater threshold depth for CO self shielding) than in the Galaxy and hence a smaller beam filling factor. This effect will also increase the $^{12}CO/^{13}CO$ intensity ratio since the photodissociation rate is greater for ^{13}CO than ^{12}CO (van Dishoeck and Black, 1988).

7 Detailed observations near 30 Dor

7.1 The physical state of the molecular gas

The ^{12}CO lines in the Galaxy are heavily saturated and are therefore broad; $^{12}CO/^{13}CO$ line width ratios lie in the range 1.5 - 2. The narrower lines in the LMC and corresponding lower line width ratios may indicate smaller CO optical depths in the LMC. Johansson (1991) has discussed the optical depth and excitation conditions of the CO clouds in the LMC with reference to the four main clouds in the 30 Dor and N159 regions. These clouds have been observed in the two lowest rotational lines of ^{12}CO and ^{13}CO (Johansson et al., in prep.) and the data has been analysed using the standard large velocity gradient (LVG) approximation. The solutions give values of the ^{12}CO (1-0) optical depth close to unity, kinetic temperatures in the range 10-40 K (the upper limit being constrained to be no greater than the dust temperature determined by Schwering, 1988 from IRAS data) and isotopic ratios, $^{12}CO/^{13}CO$ in the range 10 to 30 (mean 17±3). Column densities of ^{12}CO are in the range $10^{16.5}$ to $10^{17.4}$, significantly lower than for Galactic molecular clouds of similar sizes.

7.2 Isotopic abundances

In the main N159 CO cloud, $C^{18}O$ has been detected in the (1-0) and (2-1) transitions (Johansson et al, in prep.). This implies a $^{13}CO/C^{18}O$ abundance ratio of 20±5 and a corresponding $^{12}CO/C^{18}O$ abundance ratio of 340±150. The ratios $^{12}CO/^{13}CO$ and $^{12}CO/C^{18}O$ are comparable with the isotopic ratios $^{12}CO/^{13}CO$ and $^{16}O/^{18}O$ as derived by Langer et al. (1984) for the Galactic centre, by modelling to account for chemical fractionation. However such comparisons must be treated with caution because of fractionation and also isotope selective photodissociation effects in the LMC. See Booth and de Graauw (1991) and references therein, for discussion.

8 The CO - molecular hydrogen relation in the LMC

8.1 CO as a tracer of molecular hydrogen

Molecular hydrogen is a major constituent of the interstellar medium of galaxies but because it cannot, in general, be observed directly, other species, notably CO must be used to trace the molecular gas. CO is most important since it is very abundant and its lowest rotational transitions are easily excited. The problem now is to convert the measured CO line intensity data, $I(^{12}CO)$, into molecular hydrogen column densities.

The approach normally taken is to assume that the conversion factor is constant from galaxy to galaxy and that it is equal to average value found for Galactic CO clouds This is based on a number of assumptions (Young and Scoville, 1982, Maloney and Black, 1988):

1. The ensemble average of the CO brightness temperatures and mean densities is constant across the galactic disc, in the sense that $T_b/(n_{H_2})^{0.5}$ = const.
2. The peak antenna temperature is a measure of beam filling factor and hence the number of clouds in the beam
3. All clouds are in approximate virial equilibrium in the sense that CO line widths are directly related to cloud mass and hence to the cloud area-averaged column density.

Implicit in these assumptions is a basic assumption that the CO and the molecular hydrogen clouds are co-extensive. However, referring to our earlier discussion on photodissociation rates we note that the data of van Dishoek and Black (1988) shows that molecular hydrogen is much better at self-shielding than CO and that it would not be at all surprising if the CO does not trace the true H_2 distribution in low metallicity systems like the MC, where CO may already be underabundant. This point has been argued by Maloney and Black (1988) and by Elmegreen (1989).

Solomon et al. (1987), from a study of 273 Galactic CO clouds identified in their Massachusetts- Stony Brook CO Galactic Plane Survey, have shown that there is a relationship between CO velocity dispersion, Δv and cloud size, s, of the form $\Delta v \propto s^{0.5}$, which may be explained if the clouds are in virial equilibrium and have constant surface density. They also find that the virial masses, M_v of the clouds are related to the CO luminosity, L_{CO} in units of K km s^{-1} pc^2 as $M_v \propto (L_{CO})^{0.8}$. The conversion factor, X, normally defined as the ratio of the molecular hydrogen column density to CO integrated intensity $(N(H_2)/I_{CO})$ is then determined since

$L_{CO} = AI_{CO}$ and the cloud mass, $M = AN(H_2)m_{H_2}$, where A is the area of the cloud projected on to the line of sight.

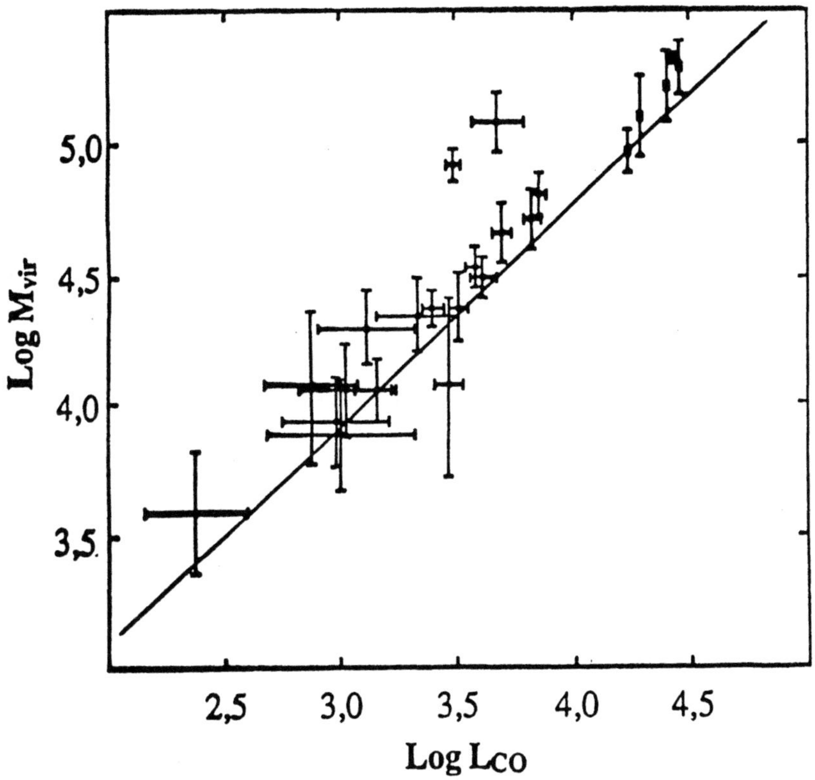

Fig. 2. A plot of virial mass, $M_V(M_0)$ v. CO luminosity, L_{CO} in units of (K km s^{-1} pc^2) for the CO clouds in the vicinity of 30 Doradus. The Galctic relationship (Solomon et al. 1987) is shown as the solid line.

8.2 Results for the LMC

Johansson (1991) has performed a similar analysis on the 23 CO clouds identified in the 30 Dor region. The resulting plot between M_V and L_{CO} is shown in Fig 2. where the Galactic relation (Solomon et al, 1987) is also plotted as a solid line. He finds that the value of X for these clouds is greater than the Galactic value by less than a factor of two. Israel and de Graauw (1991) have conducted a similar analysis including the 20 clouds in the N11 region and find a value of X which is three times the Galactic value. These are surprising results in view of the smaller CO optical depths in the LMC and the discussion above on self shielding from UV radiation. However, if the CO clouds are indeed of smaller extent than the H_2 clouds because of their greater susceptibility to UV dissociation, then the true virial (molecular hydrogen) masses will be greater than those calculated from the measured sizes of the CO clouds and

X will be larger. Clearly, more work, both observational and theoretical is required before the problem is satisfactorily resolved.

9 Future work

This summary of current progress in the SEST key project to investigate the CO properties of the LMC shows that the results are both interesting and tantalising. While there are clear differences in the masses, sizes and probably optical depths of the LMC CO clouds, relative to the Galaxy, the apparent conversion between CO integrated intensity and molecular hydrogen column density is not very different, that is is we assume the CO and molecular hydrogen are co-extensive. More work on the physical conditions in the LMC molecular clouds is needed and the new SEST 345 GHz receiver will enable us to measure the (3-2) rotational transition of CO and its isotopically substituted species. Such observations, together with more data on the lower transitions will help us refine the excitation analysis and derive more accurate physical parameters for the molecular clouds.

More CO complexes will be mapped in the next year or so and comparison with e.g. radio continuum and HI data should help to throw more light on some of the interesting features of the LMC like the possible molecular shell in N11 and the postulated sequential star formation south of 30 Dor.

References

Booth, R.S., Delgado, G., Hagström, M., Johansson, L.E.B., Murphy, D.C., Olberg, M., Whyborn, N.D., Greve, A., Hansson, B., Linström, C-O., Rydberg, A. (1989) Astron. Astrophys., 216, 315.

Booth, R.S , de Graauw, Th. (1991) in IAU Symp. 148, The Magellanic Clouds, eds. R. Haynes and D.E. Milne, (Kluwer: Dordrecht), p. 415.

Booth, R.S., Johansson, L.E.B., (1991) ibid, p.157.

Bouchet, P., Lequeux, J., Maurice, E., Prévot, L., Prévot-Burnichon, M-L. (1985) Astron. Astrophys. 149, 330.

Cohen, R.S., Dame, T.M., Garay, G., Montani, J., Rubio, M., Thaddeus, P. (1988) Astrophys. J., 331, L95.

van Dishoec k, E.F., Black, J.H..(1988) in Molecular Clouds in the Milky Way and Extenal Galaxies eds. R.L. Dickman, R.L. Snell, J.S. Young

Dufour, R.J. (1984) in IAU Symp. 108, Structure and Evolution of the Magellanic Clouds, ed. S. van den Bergh and K.S. de Boer, (Reidel, Dordrecht). p. 353.

Deharveng, L, Caplan, J. (1992) Astron. Astrophys. (in press)

Elmegreen, B.G. (1989) Astrophys. J., 338, 178

Haynes, R.F., Klein, U., Wayte, S.R., Wielebinski, R., Murray, J.D., Bajaja, E., Meinert, D., Buczilowski, U.R., Harnett, J.I., Hunt, A.J., Wark, R., Sciacca, L. (1991) Astron. Astrophys. 252, 475.

Huggins, P.J., Gillespie, A.R., Phillips, T.G., Gardner, F., Knowles, S. (1975) Mon. Not R. Astr. Soc., 173, 69P.

Israel, F.P., (1984) in IAU Symp. 108, Structure and Evolution of the Magellanic Clouds, eds. S. van den Bergh and K.S. de Boer (Reidel: Dordrecht), p.397.

Israel, F.P., de Graauw, Th., Lidholm, S., van de Stadt, H., de Vries, C.P. (1982) Astrophys. J., 262, 100.

Israel, F.P., de Graauw, Th., van de Stadt, H., de Vries, C.P. (1986) Astrophys. J., 303, 186.

Israel, F.P., de Graauw, Th. (1991) in IAU Symp. 148, The Magellanic Clouds, eds. R. Haynes and D.E. Milne, (Kluwer: Dordrecht), p. 45.

Israel, F.P., Johansson, L.E.B., Lequeux, J., Booth, R.S., Nyman, L-Å, Crane, P., Rubio, M., de Graauw, Th., Kutner, M., Gredel, R., Boulanger, F., Garay, G., Westerlund, B. (1992) Astron. Astrophys. (in press)

Johansson, L.E.B., (1991) in IAU Symp. 146, Dynamics of Galaxies and their Molecular Cloud Distributions, eds. F. Casoli and F. Combes. (Kluwer: Dordrecht), p 1.

Koorneef, J. (1982) Astron. Astrophys., 107, 247

Langer, W.D., Graedel, T.E., Frerking, M.A., Armentrout, P.B. (1984) Astrophys. J., 277, 581.

Maloney, P., Black., J.H. (1988) Astrophys. J., 325 389.

Rubio, M. (1991) in IAU Symp. 146, Dynamics of Galaxies and their Molecular Cloud Distributions, eds. F. Casoli and F. Combes. (Kluwer: Dordrecht), p 9.

Schwering , P.B.W., Israel, F.P. (1989) Astron. Astrophys. Suppl., 79. 79.

Solomon, P.M. Rivolo, A.R., Barrett, J., Yahil, A. (1987) Ap.J., 319, 730

Young, J.S., Scoville, N.Z. (1982) Astrophys. J., 258, 467.

Overview of CO SEST observations: Small Magellanic Cloud

Mónica Rubio
Departamento de Astronomía
Universidad de Chile

Summary:

We present the results of observations of the $^{12}CO(1-0)$ emission line from the Small Magellanic Cloud done with the 15m SEST telescope. We have fully mapped two areas in the SW region of the SMC bar and have identified several molecular clouds. Molecular clouds in the SMC show different properties than that of Galactic molecular clouds. They follow the same linewidth-size relation ($\Delta V \ R^{1/2}$) as Galactic clouds for a large range of radii, but they are underluminous in CO. At the smallest scales we can resolve (10 pc), the SMC CO clouds are less luminouds in CO by a factor of 2, while at large scales they are less luminous by a factor of 20. The physical properties derived from two clouds where $^{12}CO(2-1)$, $^{13}C0(1-0)$, and $^{13}C0(2-1)$ observations were done indicate that the CO clouds are clumpy, with a higher kinetic temperature and a smaller filling factor than that of Galactic CO clouds. These properties can be explained as a consequence of a higher photodissociation rate of CO. Assuming virial equilibrium for the CO structures we derive a preliminary estime of the conversion factor to derive total mass of gas from the CO luminosity. This factor is larger that the canonical value adopted for our Galaxy and depends on the size of the molecular cloud.

1. Introduction

The knowledge of the distribution and properties of molecular gas in external galaxies is based on the interpretation of the emission of the CO molecule and its relation to the unobservable H_2 molecule. In our Galaxy, as in the disks of other spiral galaxies, the interstellar medium contains about 50% of the total mass of gas in the form of massive molecular clouds to which almost all massive star formation is associated(Combes 1992, Young and Scoville 1992). Irregular galaxies show a strong defficiency in CO emission and this has lead to the conclusion that these galaxies may contain a small amount of molecular gas based on the critical assumption that the factor to derive molecular mass from the CO lominosity is the same as that found in our Galaxy.

The Magellanic Clouds are the nearest irregular galaxies and therefore they are the best candidates to study in detail the distribution and properties of molecular gas. Moreover, we can study in detail the molecular gas in an environment different to that found in our Galaxy, specially in the Small Magellanic Cloud (SMC) where these differences are more extreme. This study should enable us to determine if the properties of molecular clouds in the SMC are different to that of Galactic molecular clouds and how does the conversion factor between CO luminosity and mass depend on ambient conditions.

2. Observations

Observations of the $^{12}CO(1\text{-}0)$ emission line from the SMC have been done as part of the ESO-Swedish SEST Key Programme: CO in the Magellanic Clouds (see Israel et al 1992; for the LMC: Johansson and Booth, these proceedings). The observations have been done with the 15m SEST telescope at La Silla Observatory, Chile, with an angular resolution of 43" at 115 Ghz, corresponding to a linear resolution at the SMC of 12pc. As backend we used an acousto optical spectrometer with a total velocity coverage of 230 km/s and a velocity resolution of 0.11 km/s. We observed in frequency switching mode, with a sensitivity of about 0.1 K r.m.s. noise level per channel obtained after an integration time of 10 min per position.

We have mapped with full resolution, two regions in the SW part of the Bar. The mapping was done in a 20" x 20" (6.1x6.1 pc) grid and covered an area of 5' x 8' for SMCB-1 and 5' x 10' for SMCB-2, corresponding to the regions SW-1 and SW-2 of Rubio et al. 1991. respectively (Rubio et al 1992).

3. Results

We present the CO integated velocity maps of SMCB-1 and SMCB-2 in Figure 1. The CO emission is found associated to HII regions, dark clouds and cold IRAS sources. It has only been detected at a velocity of Vlsr(CO) =120-130 km/s, the low velocity component of the HI gas in this area. This velocity component has been associated to the nearer component. The CO emission shows a complex spatial and velocity distribution with structures at all scales and velocity gradients. We have identified about a dozen molecular clouds by studying the

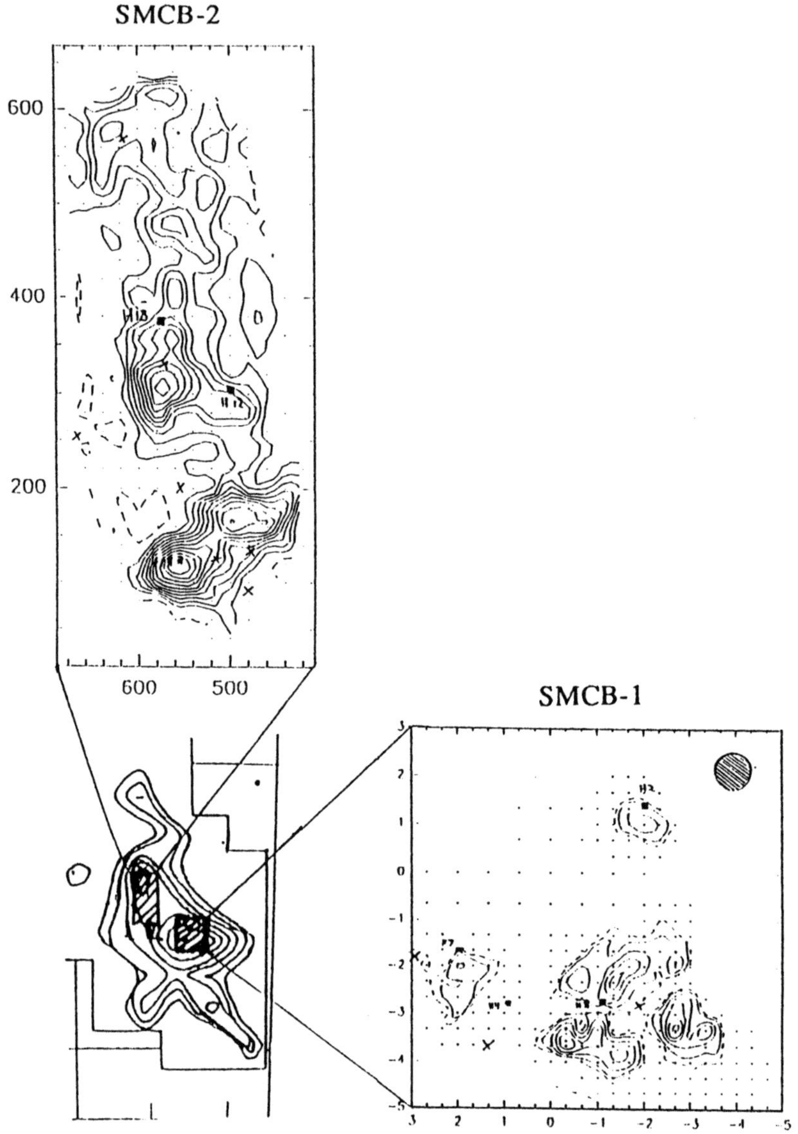

Fig.1. CO integrated velocity maps of SMCB-1 and SMCB-2.

velocity distribution in channel maps of 2km/s velocity integration intervals. The observed parameters of the molecular clouds are given in Table 1.

We have derived the properties of the molecular clouds identified and we have found that the SMC CO clouds follow the same size-linewidth relation that Galactic molecular clouds. A least-square fit to ΔV and R gives that $\log \Delta V = 0.5 \log R + 0.04$, which is similar to that found for molecular clouds in our Galaxy(i.e. Solomon at al 1987). This relation holds for a large range of radii as can be seen in Figure 2a where we have combined these data to the SMC CO complexes (Rubio et al. 1991) and compared to Galactic data of various authors. (For a complete description of this comparison please refer to Rubio, Lequeux and Boulanger 1992). However, a comparison between the CO luminosity and the CO linewidth, and between the CO luminosity and virial mass, shows that the SMC CO clouds are underluminous in CO with respect to Galactic molecular clouds (Figure 2b). At the smallest scale that we can resolve, 10 pc, the SMC CO clouds are 2 times less luminous in CO than molecular clouds from the outer Galaxy having the same CO linewidth. But at the largest scales, 150 pc, this factor is larger and the SMC complexes are 10 to 20 times less luminous than Galactic molecular complexes. Thus, the SMC molecular clouds show less and less CO emission as the CO linewidth increases or the virial mass is larger when compared to Galactic molecular clouds.

Table 1: $^{12}CO(1-0)$ properties of a few SMC clouds

Name	α(1950)	δ(1950)	T_{mb} K	V km/s	ΔV km/s	L_{CO} K km/s pc^2	r pc	M_{vir} ($M_\odot$)
SMC-B1#1	0h 43m 42.4s	-73°35'10"	1.38	122.5	3.1	1.21 10^3	13.8	2.4 10^4
SMC-B1#2	0 44 29.6	-73 38 30	0.71	125.6	4.9	1.85 10^3	16.2	7.4 10^4
SMC-B1#3	0 44 34.3	-73 38 50	0.61	115.2	3.2	1.00 10^3	12.9	2.5 10^4
LIRS 36	0 44 50.5	-73 22 33	2.37	125.9	3.8	2.80 10^3	18.6	5.1 10^4
SMC-B2#1	0 45 59.2	-73 33 10	0.89	121.1	3.7	1.89 10^3	15.4	4.0 10^4
SMC-B2#2	0 46 03.9	-73 33 10	0.54	122.0	4.7	1.10 10^3	10.9	4.5 10^4
SMC-B2#3	0 46 22.6	-73 31 10	0.97	121.2	4.6	3.79 10^3	19.8	8.0 10^4
SMC-B2#4	0 46 22.7	-73 32 10	0.72	139.0	2.9	1.07 10^3	11.4	1.8 10^4
SMC-B2#5	0 46 27.7	-73 28 10	0.53	119.1	4.2	1.38 10^3	15.0	5.0 10^4
SMC-B2#6	0 46 28.1	-73 34 10	1.24	119.6	4.5	2.36 10^3	15.0	5.8 10^4
LIRS 49	0 46 32.9	-73 21 50	1.88	113.6	6.2	6.41 10^3	16.7	1.2 10^5
N 88	1 22 54.1	-73 24 33	0.77	147.9	3.7	0.54 10^3	*	

* unresolved

Table 2. LIRS 49, LIRS 36

Line	LIRS 49 T_{mb} K	LIRS 36 T_{mb} K
$^{12}CO(1-0)$	1.99	2.32
$^{12}CO(2-1)/^{12}CO(1-0)$	1.35	1.2
$^{13}CO(2-1)/^{13}CO(1-0)$	1.71	1.3
$^{12}CO(1-0)/^{13}CO(1-0)$	16.2	10.5

Table 2 summarizes the line ratios observed for two resolved CO clouds, LIRS49 and LIRS36, in which observations of the $^{12}CO(2-1)$, $^{13}CO(1-0)$, and $^{13}CO(2-1)$ emission lines were made. These clouds are associated to HII regions comparable to the Orion nebula. We have derived the physical properties of these clouds assuming uniform physical conditions following the results of the LVG model for Orion from Castets et al 1990. The observed brightness temperature of a few degress implies that the CO emmision must arise from clumps having smaller filling factors than the CO clumps in Galactic molecular clouds if the ^{12}CO emission is most probably optically thick. The $^{12}CO(1-0)/^{13}CO(1-0)$ line intensity ratio of 16 and 11 for LIRS49 and LIRS36 respectively suggests that the $^{12}CO(1-0)$ line is moderately optically thick, if the $^{12}CO/^{13}CO$ ratio is at least equal to the solar value. The $^{12}CO(2-1)/^{12}CO(1-0)$ ratio of 1.3 indicate that ^{12}CO is warmer than TK=10 K and not very subthermally excited, and that the density of molecular hydrogen is at least 10^3 mol. cm^{-3}. Finally the $^{13}(2-1)/^{13}(1-0)$ ratio of 1.7 for LIRS49 implies kinetic temperatures larger than about 13K and densities of a few 10^3 mol cm^{-3} for the ^{13}CO region. Although, the asumptions made are probably unrealistic since the spatial distribution of ^{13}CO is not the same as that of ^{12}CO (i.e. inside the clumps), the kinetic and excitation temperature are not uniform, and we do not know what the $^{12}CO/^{13}CO$ ratio is for the SMC, the LVG analysis should give correct orders of magnitudes for the temperatures, densities and optical depth. We conclude that these clouds are warm and clumpy.

The properties exhibit by the SMC molecular clouds can be explained as a consequence of a higher photodissociation rate of CO due to the combined effect of the low C and O abundance, the higher gas to dust ratio, and the stronger UV field in the SMC. The lack of dust plays against the formation of H_2 molecule and allows deeper penetration of the stronger UV field. The lower abundances imply that less CO will be

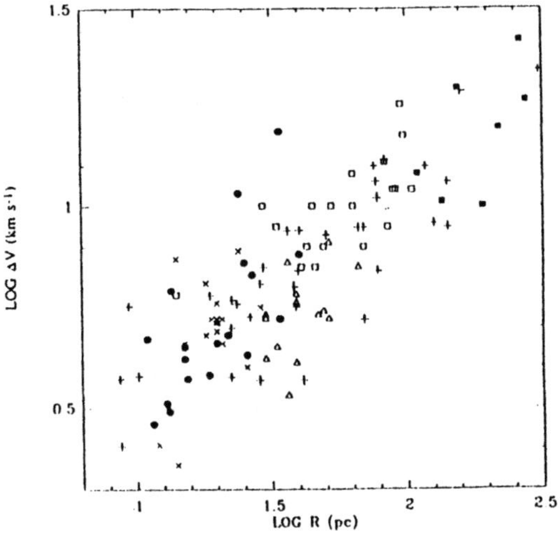

Fig. 2 a) Linewidth-size relationship for SMC molecular clouds. Filled circles represent SMC data from SEST observations (this work) and filled squares are from Rubio et al. 1991. For comparison Galactic data has been also plotted as described in Rubio, Lequeux, and Boulanger, 1992.

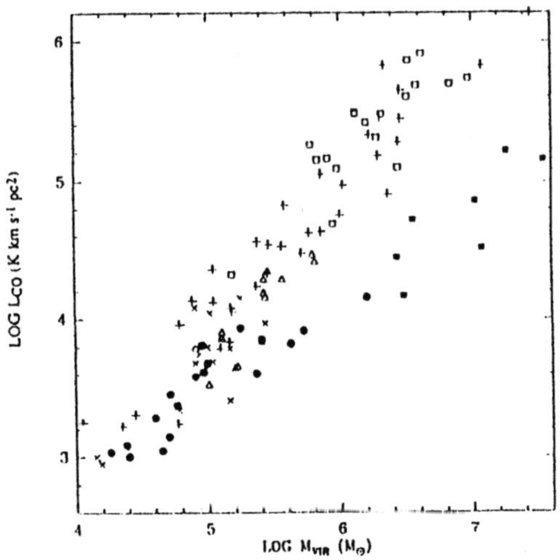

Fig. 2 b) CO luminosity, LCO, as a function of virial mass, Mvir, for SMC molecular clouds. See caption a) for details.

formed. Since the photodissociation of both CO and H_2 is a non linear phenomena in which as you decrease the amount of H_2 or CO you increase the destruction of the molecules, the photodissocition process in the SMC is more efficient. Thus, in the SMC, CO can only survive in the dense clumps, while it is photodissociated in the diffuse and interclump medium. It might be even possible that the molecular hydrogen is also destroyed and that the large structures in the SMC contain mainly atomic gas.

Assuming that the interstellar structures in the SMC are in virial equilibrium we can determine a preliminary estimate of the conversion factor to derive the total mass of gas from the CO luminosity. The total gas of mass is therefore the sum of the molecular plus the atomic hydrogen mass. The numerical value of the SMC conversion factor, X_{SMC}, depends on the way the various parameters are defined and therefore it should not be cosidered too literally. We found that X_{SMC} obtained from the M_{vir}/L_{co} ratio is larger than the canonical value adopted for our Galaxy and depends on the size of the molecular clouds. From this work we obtain $X_{SMC} = 9 \times 10^{20} (R/10\ pc)^{0.7}$ mol. cm^{-2} (K km/s)$^{-1}$. This result can explain why is has been so difficult to detect CO emission from other irregulars as they have been observed with poor linear resolution.

The molecular mass of the SMC of 3×10^7 Mo, determined from the complete $^{12}CO(1-0)$ survey (Rubio et al 1991), is thus an upper limit. The amount of molecular gas mass compared to the atomic gas mass in the SMC is negligable.

We would like to aknowledge support from the Local Organizing Committee and from FONDECYT (Chile) through grant 1093-90 .

References

Castets, A., Duvert G., Dutrey A., Bally, J., Langer, W.D., Wilson, R.W., 1990 A&A, 234 469

Combes, F., 1992, Ann.Rev.A&A, 29 125

Israel, F.P., Johansson, L.E.B., Lequeux, J.Booth, R.,S., Nyman, L.A., Crane, P., Rubio, M., de Graauw, Th., Kutner, M., Gredel, R., Boulanger, F., Garay, G., Westerlund., 1992, A&A in press (Paper I)

Johansson, L. E. B., Booth, R.S., 1992, these proceedings

Rubio, M., Garay, G., Montani, J., Thaddeus, P. 1991, ApJ, 368 173

Rubio, M., Lequeux, J., Boulanger, F., Booth, R.,S., de Graauw, Th., Israel, F.P., Johansson, L.E.B., Kutner, M., Nyman, L.A., 1992, A&A in press (Paper II)

Rubio, M., Lequeux., J., Boulanger, F. 1992, A&A, in press (Paper III)

Solomon. P. M., Rivolo, A.R., Barrett,J., Yahil,A., 1987, ApJ, 319, 730

Young, J., S., Scoville, N.Z., 1992, Ann.Rev.A&A, 29 yyy;

Ionized Carbon [CII] in the Magellanic Clouds

F.P. Israel [1], P.R. Maloney [2]

[1]Sterrewacht Leiden, Postbus 9513, 2300 RA Leiden, The Netherlands
[2]Astrophysics Branch, NASA Ames Research Center, Moffet Field, CA 94035, USA

Abstract: We present preliminary results of recently obtained observations of the important cooling line of singly ionized carbon in the Magellanic Clouds. Three LMC and six HII region/CO cloud complexes have been detected and mapped. Comparison shows that the strength and distribution of [CII] regions is well-correlated with those of dust complexes emitting far-infrared radiation, but not with molecular cloud complexes traced by CO emission. The results confirm that photo-dissociation processes are of relatively greater importance in Magellanic Cloud star formation regions than in Galactic regions.

1. [CII]: A Major Cooling Line of Photo-Dissociation Regions.

The Magellanic Clouds are rich in gas and young, luminous stars. The relatively low metallicity and dust content of the Clouds create an environment in which neutral gas clouds are provided with relatively little shielding against UV radiation longwards of the Lyman limit. This UV radiation, when impinging on such clouds, will ionize or dissociate molecular and atomic species with ionization or dissociation potentials less than the hydrogen ionization potential of 13.6 eV. In doing so, it creates photo-dissociation regions enveloping the clouds. Nothwithstanding the low carbon abundances in the Magellanic Clouds, carbon with its ionization potential of 11.3 eV is still the most abundant atom that can be ionized by UV radiation longwards of the Lyman limit. The ^{2}P, J=3/2 - 1/2 fine structure line of [CII] occurs at a wavelength of 157.7 microns and has a critical density for excitation of order 3×10^3 cm^{-3}; it is a major cooling line for photo-dissociation regions.

Because of its wavelength, observations of [CII] cannot be carried out from the ground, but must be conducted from airborne platforms such as the NASA Kuiper Airborne Observatory (KAO). With the use of the KAO, [CII] has been detected towards a wide variety of Galactic sources, as well as several external galaxies (Stacey et al. 1991a). Galactic observations (e.g. Stutzki et al. 1988; Howe et al. 1991; Stacey et al. 1992) shows that Galactic [CII] emission arises from warm (100-300 K) and dense

($> 10^4$ cm^{-3}) regions at the surface of cooler molecular clouds. The observed distribution and intensities can be modelled by clumpy or filamentary molecular clouds with high density contrast, allowing UV radiation to penetrate deeply into the clouds. Such models also explain the observed distribution of far-infrared emission from heated dust in these clouds, so that one would expect a good correlation between [CII] and far-infrared continuum emission (Howe et al. 1991). In Orion, the mass of the photo-dissociation region is of order 3 per cent of the total mass of the Orion molecular cloud (Stacey et al. 1992). [CII] is widespread in the Galaxy, as shown by the COBE results presented by Wright et al. 1991.

The [CII] line measurements of about 20 external galaxies (Stacey et al. 1991a) show that it to be among the brightest single emission lines from galaxies, representing between 10^{-3} and 10^{-2} of the total far-infrared luminosity of these galaxies. Extragalactic [CII] emission is correlated nonlinearly with far-infrared continuum emission and its intensity traces local UV radiation field strengths. In distant external galaxies, [CII] emission is also well-correlated with ^{12}CO, J=1-0 line intensities, indicating that at least in starburst galaxies the latter may not be a reliable tracer of total molecular mass.

2. [CII]: Readily Detectable in Magellanic Cloud Objects.

The Magellanic Clouds present an excellent opportunity to further study the relationship between [CII] emission on the one hand and CO and far infrared continuum emission on the other hand. Linear resolutions and fields of view are of order 10-100 times larger than in the case of Galactic observations, and of order 50-150 times smaller than in the case of extragalactic observations conducted to date. Thus, they allow observations of significantly larger areas than in the Galaxy, without the loss of structural information encountered in observations of much more distant galaxies.

In addition, the low metallicity and strong radiation fields of the Magellanic Clouds suggest that photo-dissociation processes will be relatively important, so that significant [CII] emission is expected. In both Clouds, we selected sources with high far-infrared intensities (Schwering & Israel, 1990), for which preferably also J=1-0 ^{12}CO observations were available from the ESO-Swedish SEST Key Programme (see the contributions by Booth and Rubio elsewhere in these proceedings. We did not observe 30 Doradus in the LMC, as this has already been done by A. Poglitsch et al. (private communication; see also Stacey et al. 1991a). We note that Boreiko & Betz 1991 searched for [CII] emission from 17 positions in the LMC, out of which 10 yielded detections. They did not attempt mapping, but instead obtained velocity-resolved spectra of the line. As we concentrated on mapping velocity-integrated [CII] emission, their work is complementary to ours.

For our observations, we used the Far-Infrared Fabry-Perot Interferometer (FIFI) array of the Max Planck Institut fur Physik und Astrophysik, Institut for Extraterrestrische Physik and the University of California at Berkely (Geis, 1991). The FIFI array consists of 5×5 detectors separated by 40 arcsec. At a wavelength of 158 micron, the resolution on the 92cm telescope of the NASA Kuiper Airborne Observatory (KAO) is 55 arcsec. The observations were conducted in "stare" mode during KAO flights from New Zealand in 1991 and 1992. In such a mode, the line spectrum is not scanned, but the line is made to fall within a single spectral resolution element. In addition, spatial chopping was performed to positions about 8 arcmin from the target position. The chop direction was always chosen such as to minimize the chance of emission in the off-set position. On the LMC, we integrated each array position for 5 minutes, yielding an r.m.s. sensitivity of 1.4×10^{-5}ergs^{-1}cm^{-2}sr^{-1}, while on the SMC we integrated for 10 minutes to a sensitivity of 0.9×10^{-5}ergs^{-1}cm^{-2}sr^{-1}.

[CII] emission was readily detected from all sources. In Table 1 we summarize the results.

Table 1. Preliminary Results of [CII] Mapping Magellanic Cloud Objects.

Object	IR #	Map Area	FIR Strength	Peak [CII]	[CII] Source
		($'^2$)	(10^{-12}Wm^{-2})	(10^{-14}Wm^{-2})	Size ($'$)
LMC N11	217	170	14.49	1.59±0.06	Complex
N159	1518	100	21.32	2.72±0.07	4.2×2.7
N160	1503		31.23	2.66±0.07	3.0×2.0
SMC N22	45	10	3.43	0.52±0.06	1.3×2.1
N27	49	30	3.56	0.84±0.05	1.8×2.0
N66	131	45	9.55	0.49±0.05	6.4×3.0
N83	199	65	2.70	0.47±0.07	2.2×2.2
N84	202		2.15	0.41±0.07	2.7×0.6
N88	215	10	2.37	0.45±0.04	0.8×1.3

The first two columns give the Henize name of the associated HII region and the IRAS identification number in the catalog by Schwering and Israel 1990. The third column gives the area included by the maps, the fourth column the far-infrared strength of the associated IRAS source, derived from the 60 and 100 micron fluxes listed by Schwering and Israel 1990. The fifth and sixth columns give observed [CII] peak fluxes with their errors and the approximate FWHM size of the [CII] emitting area corrected for finite beamwidth, respectively. Note that LMC-N159/N160 and SMC N83/N84 are included in single maps. All objects, with the exception of LMC-N11 were fully mapped. N11 is a very large and complex source, covering an area of about 45 arcmin across of which we mapped the central ten per cent centered on the brightest far-infrared emission (see Figure 1 in Israel & de Graauw 1991). The peak [CII] intensities, averaged over the beam, are somewhat lower than those obtained for galactic sources such as W3 and Orion B (Howe et al. 1991) which probably reflects structure on scales smaller than 15 pc, and thus unresolved by us. At the same time, these peak intensities are gnerally higher than those obtained by Stacey et al. 1991a towards external galaxy nuclei.

3. Structure of [CII] Emission Regions in the Magellanic Clouds.

With a resolution of 55 arcsec (corresponding to a linear resolution of about 15 pc on the Magellanic Clouds), SMC sources N22, N27 and N88 show little structure in [CII]. All other objects show often more than one clear peak embedded in a large halo of extended, low surface brightness emission of about 15-20 per cent of the observed peak surface brightness In addition, it appears that large areas in the Magellanic Clouds radiate diffusely in [CII] at levels of about $10^{-10} Wm^{-2}sr^{-1}$, but this is hard to determine quantitatively because of our spatial chopping. Positions of peak [CII] emission are found close to infrared peaks and optical/radio HII region positions but not coincident with the latter. The relation with CO emission is more complex. First, several [CII] peaks are found close to but displaced from CO clouds. Second, [CII] emission is seen towards positions where no CO has been found. Third, some CO clouds have no or at best a weak counterpart in [CII]. In general, at the observed resolutions of order 15 pc, integrated [CII] intensities correlate well but not precisely with far-infrared continuum brightness, and rather poorly with CO luminosities. We will illustrate this with a few specific examples.

4. The N159/N160 Complex in the LMC.

In Figure 1 (center) we show the [CII] map of N159 and N160. Dots mark the positions sampled. N160 (at -1′, +7′) has a bright peak surrounded by a much larger region of lower surface brightness. South of N160 we find at (0′, +1′) the N159 complex which consists of three major [CII] peaks, again surrounded by weaker emission. The westernmost peak at (-1′, 0′) coincides with the very bright nebula N159A. In the N159/N160 map all four peaks are approximately equally bright. Figure 1 (left) shows a map of the J=1-0 ^{12}CO emission, integrated over a velocity range of 218 to 261 km s^{-1} (cf. Johansson, 1991; Booth, these proceedings). The map has a resolution of 40 arcsec which is close to the resolution of the [CII] map. The CO cloud associated with N160 has a shape and dimensions roughly corresponding to the [CII] contours, but the [CII] peak is displaced from the CO maximum in a northeasterly direction by about one [CII] beam (15 pc). In N159, the eastern [CII] peak is displaced by a similar amount in a northwesterly direction.

The western peak (N159A) is displaced by a somewhat smaller amount towards the northeast and the central [CII] peak has no associated CO emission at all, but falls rather at the edge of the lowest CO contour connecting the two CO maxima In addition to these significant positional shifts, the ratio of [CII] to CO emission is obviously variable: all [CII] peaks are roughly equally strong, whereas the N160 CO cloud is much less bright than the two N159 CO clouds. Moreover, whereas the central [CII] peak in N159 has no clear CO counterpart, the rather strong CO cloud south of N159 has virtually no [CII] emission, even though it was fully sampled in our [CII] map.

Finally, Figure 1 (right) shows the distribution of 843 MHz radio continuum emission, obtained with the MOST in Australia (kindly provided by B.Y. Mills, see also

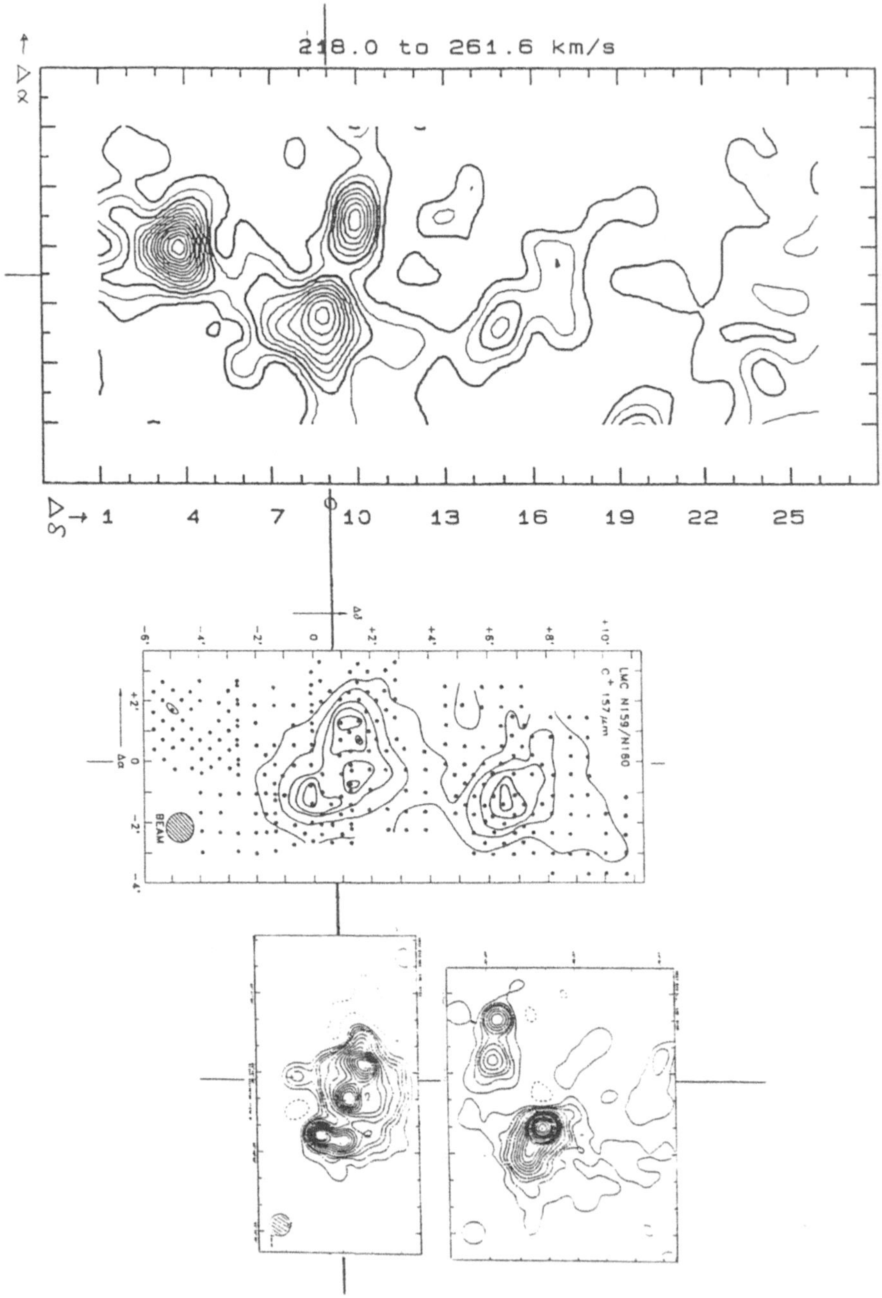

Fig. 1. Maps of N159 (Infrared Source #1518; bottom) and N160 (IR Source #1503; top) made with similar resolutions in [CII] (center), J=1-0 ^{12}CO (left) and 843 MHz radio continuum (right). The maps are on the same scale, and are lined up in declination.

Mills and Turtle 1984). The emission is mostly thermal, and therefore traces the distribution of ionized hydrogen, excited by early-type stars. As such, the Figure indicates the location and distribution of the photon-producing stars that are not only ionizing hydrogen, but also heating dust, dissociating molecules and ionizing carbon. The bright N160 radio core is displaced by another 15 pc from the [CII] peak again in northeasterly direction. In contrast, the eastern and western N159 radio peaks are virtually coincident with the [CII] peaks. The central radio peak falls between the two CO clouds and is about 20 pc to the southeast of the central [CII] peak. At the position of the southern CO cloud, no radio source is known.

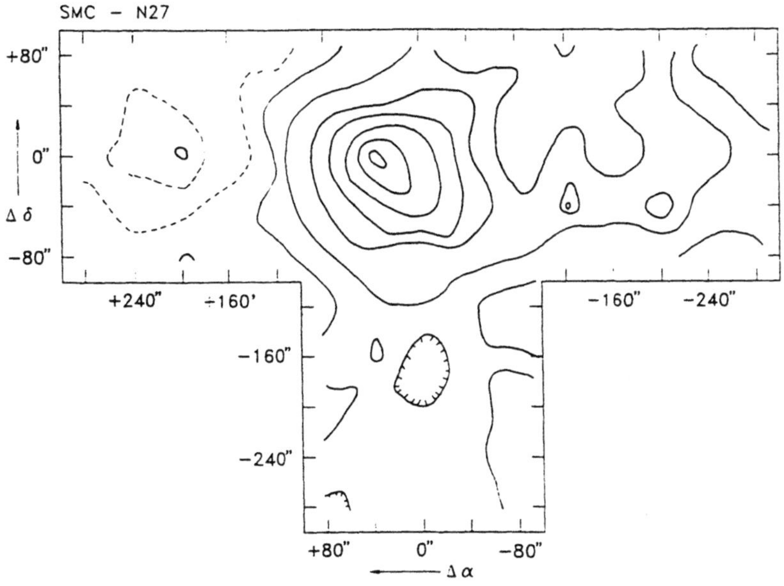

Fig. 2. [CII] Map of N27 in the SMC (Infrared Source #49).

The results are consistent with the existence CO/HII/[CII] complexes, where the maxima of [CII] emission occur between the exciting star cluster and the associated cold molecular cloud, surrounded by much larger envelopes of diffuse [CII]. The N160 complex appears to be seen largely edge-on, whereas the N159 complex should be mostly face-on. An image-sharpened IRAS 60 micron image with a resolution of about one arcmin, kindly provided by R. Bontekoe (Estec) shows clear infrared peaks roughly coincident with the eastern and western [CII] peaks in N159. The central [CII] peak coincides with a weak infrared extension. Likewise, the N160 [CII] peak coincides with a pronounced unresolved infrared source. However, the N160 infrared source is significantly stronger than the N159 infrared sources (see also Table 1), emphasizing the good, but not perfect correlation between far-infrared continuum and [CII] emission.

5. The SMC-N27 and N66 Complexes.

In the SMC, the strongest CO detection to date was obtained towards N27 (Israel et al. 1992), while N66 is the strongest IRAS infrared source; N66 is also the first-ranked HII region complex in the SMC. The [CII] map of N27 shows a featureless, resolved source with low surface brightness extensions to the south and to the west (Figure 2). The IRAS source is unresolved and coincides with the [CII] peak. The CO source is extended with a diameter of about 25 pc (Rubio et al. 1992) and peaks about 10 pc west of the [CII] peak. The radio continuum source (Mills & Turtle 1984) is close to the CO peak. The radio source has a weaker extension to the east, coinciding with the [CII] peak. Again, the situation is consistent with a molecular cloud complex in the proces of being eroded by radiation from the exciting stars of N27.

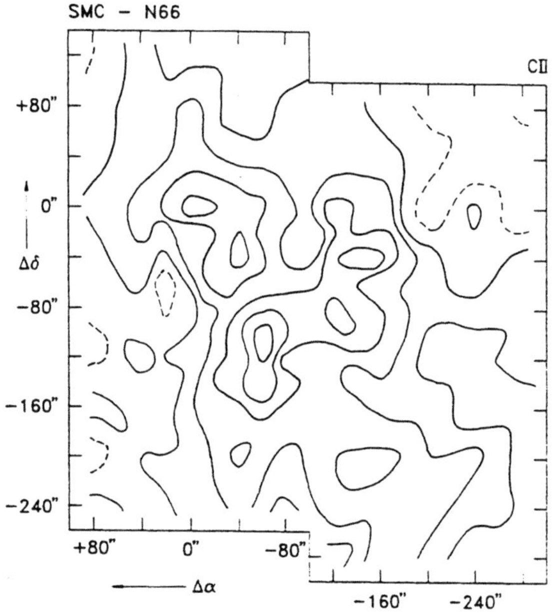

Fig. 3. [CII] map of SMC N66 (Infrared Source #131), the firstranked HII region complex in the Small Magellanic Cloud

The N66 map (Figure 3) shows a rather extended [CII] source with several peaks. However, the peak intensities are relatively low, and the contrast between peaks and surrounding diffuse emission is likewise low. N66 is associated with a very strong infrared source which peaks close to the [CII] peak; the infrared source is extended and has a size of about 3×3 arcmin (Schwering & Israel 1990). Unpublished SEST CO observations show a small and relatively weak CO source in the northeast of the [CII] complex; the CO source is much smaller than the [CII] emitting region. It appears that the strong UV radiation produced by the N66 ionizing star cluster has by now destroyed most of its parent molecular cloud complex - at least as far as the CO tracer is concerned. The extended, relatively diffuse [CII] cloud should then be considered as the dissociated and ionized remnant of this CO cloud complex. We also note that N66 is associated with a strong local HI maximum (at X = -0.1, Y = +0.5, V_{LSR} = +151 km s^{-1}) in the position-velocity maps published by Bajaja & Loiseau 1982. Radio and H-alpha observations of N66 have been published by Ye et al. 1991. In particular their Hα maps (their Figure 3 and Plate 2) show that the distribution of ionized hydrogen is very similar to that of ionized carbon.

In particular, the southeast to north-west "bar"-like feature in the center of the [CII] map (centered at -120$''$, -90$''$) corresponds to a similar structure in Hα, while a local [CII] maximum at (-140$''$, -190$''$) corresponds to what appears to be a pronounced ionization front in the Hα images Although this suggests that HII and [CII] occur in the same volume in N66, further analysis is required to substantiate this speculation.

6. Conclusion: Ionized Carbon in the Magellanic Clouds.

Although the recently obtained [CII] measurements of several Magellanic Cloud objects await further analysis and careful modelling, some tentave conclusions are apparent. Ionized carbon is readily detectable in the Magellanic Clouds, suggesting vigorous photo-dissociation of molecular (CO) material by intense radiation fields under conditions of relatively poor shielding. The enhanced photodissociation, at least with respect to Galactic objects, tends to increase the dimensions of [CII] emitting photodissociation regions at the expense of molecular clouds. Because the [CII] line has a low optical depth (for instance, approximately unity in Orion, Stacey et al. 1991b), its intensity to a large extent reflects column density. Dust is underabundant in the Magellanic Clouds and [CII] column densities are mostly limited by dust absorption (penetration depth corresponding to A_V of about 4, Tielens & Hollenbach 1985). Thus, in the Magellanic Clouds [CII] intensities comparable to those in Galactic sources are reached only in much larger regions, as is indeed observed. This may, in turn, explain the apparently poorer relation between [CII] and CO emission which extends to the presence of [CII] where no CO is seen, vice versa. This confirms a similar conclusion by Boreiko & Betz 1991, see also the brief discussion on 30 Doradus by Stacey et al. 1991a.

The relation between [CII] and far-infrared continuum is much better. Generally, [CII] and far-infrared peaks coincide suggesting a close coupling between ionized carbon and warm dust which can be understood if both primarily trace the same radiation field. As noted before, the correlation is good but not perfect. For the six SMC objects observed, we find a ratio of integrated [CII] strength to total far-infrared strength (defined by the FIR parameter) [CII]/FIR = 9.9×10^{-3}. The mean deviation is 2.1×10^{-3}, and actual values range from 4.6×10^{-3} for N88 to 17.1×10^{-3} for N83. Thus, in the SMC sources the [CII] line intensity is of order 1% of the total infrared luminosity. This is almost identical to the upper limit of the same ratio found in extragalactic nuclei.

References

Bajaja, E., Loiseau, N.: 1982 A&AS 48, 71

Boreiko, R.T., Betz, A.L.: 1991 ApJL 380, L27

Geis, N.: 1991 Ph.D. Thesis Ludwig-Maxilians-Universitat, Munchen (FRG), MPE Report 231

Howe, J.E., Jaffe, D.T., Genzel. R., Stacey, G.J.: 1991 ApJ 373, 158

Israel, F.P., de Graauw, Th.: 1991 in 'The Magellanic Clouds, IAU Symposium 148, eds. R. Haynes & D. Milne, (Dordrecht: Kluwer), p. 45

Israel, F.P., Johansson, L.E.B., Lequeux, J., Booth, R.S., Nyman, L.-A., Crane, P., Rubio, M., de Graauw, Th., Kutner, M.L., Gredel, R., Boulanger, F., Garay, G., Westerlund, B.: 1992, A&A in press.

Johansson, L.E.B.: 1991 in: 'Dynamics of Galaxies and their Molecular Cloud Distributions', IAU Symposium 146, eds. F. Combes & F. Casoli (Dordrecht: Kluwer), p. 1

Mills, B.Y., Turtle, A.J., 1984 in: Structure and Evolution of the Magellanic Clouds', IAU Symposium 108, eds. S. van den Bergh & K.S. de Boer (Dordrecht: Reidel), p. 283

Rubio, M., Lequeux, J., Boulanger, F., Booth, R.S., Garay, G., de Graauw, Th., Israel, F.P., Johansson, L.E.B., Kutner, M.L., Nyman, L.-A.: 1992 A&A in press.

Schwering, P.B.W., Israel, F.P.: 1990, Atlas and Catalogue of Infrared Sources in the Magellanic Clouds (Dordrecht: Kluwer)

Stacey, G.J., Geis, N., Genzel, R., Lugten, J.B., Poglitsch, A., Sternberg, A., Townes, C.H.: 1991a ApJ 373, 423

Stacey, G.J., Townes, C.H., Poglitsch, A., Madden, S.C., Jackson, J.M., Herrmann, F., Genzel, R., Geis, N.: 1991b ApJL 382, L37

Stacey, G.J., Jaffe, D.T., Geis, N., Genzel, R., Harris, A.I., Poglitsch, A., Stutzki, J., Townes, C.H.: 1992 ApJ submitted

Stutzki, J., Stacey, G.J., Genzel, R., Harris, A.I., Jaffe, D.T., Lugten, J.B.: 1988 ApJ 332, 379

Tielens, A.G.G.M., Hollenbach, D.: 1985 ApJ 291, 722

Wright, E.L. and 21 others, 1991 ApJ 381, 200

Ye, T., Turtle, A.J., Kennicutt, R.C.: 1991 MNRAS 249, 722

The Magellanic Clouds as seen by IRAS

L. Vigroux

The essential contents of this Review have been published:

M. Sauvage, T. X. Thuan, L. Vigroux, 1990, Astron. Astrophys. **237**, 296

DEEP INFRARED IMAGES of the SMALL MAGELLANIC CLOUD and COMPARISON with the DISTRIBUTION of ULTRAVIOLET EMISSION

K.Okumura [1], F.Viallefond [1], M.Viton [2], W.Rice [3]

[1]Observatoire de Paris Section de Meudon, France.
[2]Laboratoire d'Astronomie Spatiale, Marseille, France.
[3]IPAC, Jet Propultion Laboratory, Pasadena, and NOAO, Tucson, U.S.A.

Abstract: Deep infrared images of the Small Magellanic Cloud have been produced and compared to an ultraviolet image at an angular resolution of 8′. There is a strong correlation between the far infrared and the ultraviolet emission but the dispersion in this correlation is unrelated to infrared colors. Comparing with the results for the nearby spiral M33 at different radial distances, it is suggested that the population of dust grains in the Small Magellanic Cloud has very different properties.

1 Observations

The Additional Observations (AOs) obtained with the Infrared Astronomical Satellite (IRAS) in the direction of the Small Magellanic Cloud (SMC) have been processed to improve the image quality (removal of the striping, relative calibration in brightness between the different AOs, calibration of the positions of the AOs between each other) in order to detect, in particular, regions with low surface brightness outside the bar. We produced low resolution (8′ angular resolution), high brightness sensitivity images of the SMC region in the four IRAS bands (Viallcfond, Okumura and Rice 1988). Using the Super Skyflux images of the survey data recently released by IPAC, we compared them with the AOs images and combine them to improve the signal to noise ratio. The brightness sensitivity of 1.6×10^{-2} MJy sr^{-1} is achieved at both 12 μm and 25 μm. In the 60 μm and 100 μm bands there is emission over the entire imaged field because of confusion with foreground Galactic cirrus; the measured fluctuation in brightness due to these cirrus is, at this resolution, at levels of 4.0×10^{-2} MJy and 2.0×10^{-1} MJy (rms) in these two bands respectively.
We also processed far ultraviolet (UV) images obtained with the Very Wide Field Camera aboard of Spacelab 1 (experiment 1ES022). These UV images were calibrated

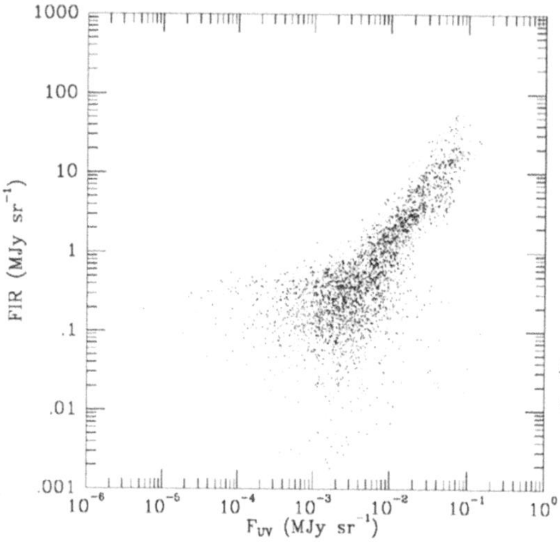

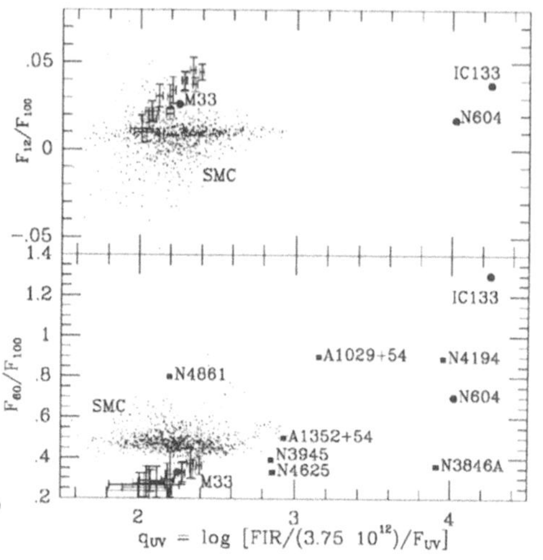

Fig. 1 and 3. *1 (left)*: Comparison between the FIR and the UV surface brightness in the SMC; there are 4 points per resolution element. *3 (right)* : 12 to 100 and 60 to 100 μm flux density ratios as a function q_{UV} for the SMC (4 points per resolution element), for M33 at various radial distances, for the M33 HII regions NGC604 and IC133 and for a few galaxies.

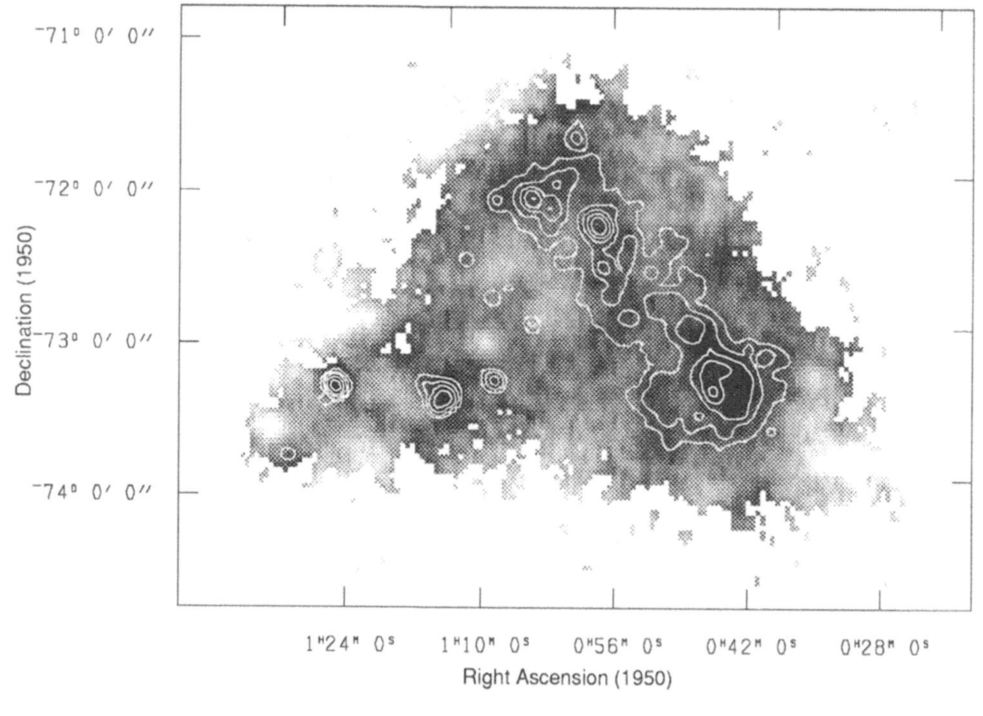

Fig. 2. Distribution of q_{UV} (gray scale) and of the ridge of emission at 25 μm (contours) over the face of the SMC. Contour values are 0.13, 0.37, 0.9, 2.0, 4.6 and 10.6 MJy sr^{-1}.

relative to field stars observed with the TD1 satellite. The ultraviolet image at 1950 Å, also with a resolution of 8′, has been projected onto the grid used for the IRAS images to allow direct comparisons at a linear scale corresponding to about 150 pc. The measured brightness sensitivity for this UV image is 1.7 Jy sr^{-1}.

2 Results

The general comparison of the surface brightness between FIR and UV is plotted in Fig. 1. The correlation is tight with no apparent saturation effects at the brightest levels; notice also that the slope is significantly larger than 1. Following Helou et al. (1985) for the radio-FIR comparison, we also define a parameter q_{UV} expressing the FIR to UV ratio. q_{UV} shows a remarkable spatial distribution with maxima related to the distribution of the active star forming regions and giant molecular complexes. This is clearly illustrated in Fig. 2 where q_{UV} (gray scale) is compared with the distribution of the emission at 25 μm, the infrared band in which the HII regions may contribute the most (Rice et al. 1990). The infrared colors observed in the SMC (Fig. 3) have been compared with those observed in the nearby spiral M33 at different galactocentric distances. The two bright M33 HII regions NGC 604 and IC 133 (Rice et al. 1990) and a few other galaxies are also included in the comparison. The range of q_{UV} values observed on large scale in M33 (from 2 at large radial distances to 2.4 in the central part) is smaller compared to the range observed in the SMC. Contrarily to the SMC where the metallicity is presumably small, in the case of M33 the spatial variation of q_{UV} and of the infrared colors are thought to be associated mainly to two effects, the radial variation of the intensity of the radiation field and the radial variation of the metallicity. These two effects acting simultaneously make the interpretation difficult. The outer part of M33 has a similar color index between 12 and 100 μm compared to the SMC. These regions are both characterized by low metallicities and low intensities for the radiation field. By contrast, the color index between 60 and 100 μm for these regions is very different and in this case it is the central part of M33, with metallicity of about the solar value, which behaves as the SMC! This suggests that the properties of the population of dust grains in the SMC are very different, even when compared to the outermost regions of M33.

References

Helou G. Soifer, B.T. and Rowan-Robinson, M. 1985. ApJ, **298**, L7

Viallefond, F.,Okumura, K. and Rice W. 1989 in *Recent Developments of Magellanic Cloud Research* A European Colloquium, Eds K.S. de Boer, F. Spite and G. Stasinska, p 143

Rice, W., Boulanger, F., Viallefond, F., Soifer, B.T., Freedman, W.L. 1990. ApJ, **358**, 418.

Magellanic Cloud Research With ISO

Stephen C. Russell

Dublin Institute for Advanced Studies, School of Cosmic Physics, 5 Merrion Square, Dublin 2, Republic of Ireland

Abstract: With the high sensitivity and spectral resolution of the Infrared Space Observatory (ISO) there are great opportunities for future research of the mid- and far- infrared properties of the Magellanic Clouds (MCs). This paper reviews suggested ISO central programme proposals for MC research. It is hoped, herewith, to stimulate thought, preparation, and coordination of ISO open time proposals for MC research.

Introduction

In what follows I simply list all proposals I know of that have been suggested in writing for the ISO central programme that involve observations of the Magellanic Clouds. The name of instrument team (or mission scientist, or country) is followed by the first author in brackets, then the title of the proposal. After that, I give a brief outline of the aim of the proposal. Further details are given in Tables 1 and 2 at the end.

Proposals

1. CAM (C. Bonoli) - Star Forming Regions in Nearby Galaxies
   AIM - To study the distribution of dust and PAHs in the vicinity of active star formation as traced by supergiant HII regions in nearby galaxies.
2. CAM (F. Boulanger) - A Coordinated Study of Nearby Normal Galaxies
   AIM - To study, in a coordinated way (with other ISO instruments and ground based observations), the distribution of dust and PAHs in nearby galaxies.
3. LWS (D. Pequignot) - Far Infrared Spectroscopy of Nuclearly Processed Material in Young Supernova Remnants
   AIM - To detect and establish characteristics of cold nuclearly processed material expected to be present in the inner shocked regions of young supernova remnants.
4. PHOT (D. Lemke) - Interstellar Dust in Galaxies
   AIM - To investigate the life cycle of dust throughout the universe, and in particular, the contribution of PAH emission/absorption to the mid-infrared continuum in the Magellanic Clouds.

5. PHOT (H. Hippelein) - Star Formation in Disk HII Regions
   AIM - To study the far infrared (FIR) properties of star forming regions, and their dependence on the physical characteristics of the regions.
6. PHOT (H. Hippelein) - Heating of the Interstellar Medium in Spirals: Mapping of Nearby Galaxies
   AIM - To study the spatial distribution of dust in different archetypical spiral and irregular galaxies.
7. PHOT (R. Tuffs) - Observations of Supernova Remnants with ISO
   AIM - To study the FIR properties, and statistical distribution of luminosities of an unbiased sample of supernova remnants (SNRs) as derived from the MCs.
8. SWS (J. van der Hulst) - Spectral Study of Extragalactic HII regions: Abundances and Physical Conditions
   AIM - To probe the physical conditions of star forming regions and surrounding interstellar medium in nearby galaxies.
9. SWS (R. Genzel) - Evolution of the Interstellar Medium in Galactic Nuclei and the Connection between Star Formation and AGN's
   AIM - To use 30 Dor (LMC) as one of several local 'prototypical template' sources in the investigation of the nature, physical conditions and evolution of the nuclear regions of luminous Seyfert's, AGN/QSO's and IRAS galaxies.
10. SWS (S. Drapatz) - Physical and Chemical Conditions in Dense, Warm Interstellar Gas
    AIM - Specifically, to explore the characteristics of X-ray excited molecular gas around SN1987A.
11. Japan (T. Tsuji) - Systematic Studies on Mass Loss and Its Effect on the Late Stage of Stellar Evolution
    AIM - To study mass loss in cool evolved stars by performing photometry and spectrophotometry on a sample of stars drawn from the Magellanic Clouds.
12. Mission Scientists (E. Oliva) - ISO Observations of Very Bright Supernova Remnants
    AIM - To obtain template spectra of shock excited gas and gain new information on physical conditions inside a few very high surface brightness SNRs.

Table 1. Summary of LMC Observations

Target	Instrument[a]	Proposal	Observation	No. of Fields	λ (μm)	S/N	Time (min)
30 Dor	CAM	1	multi-filter images	10	2.5-17	>>10	120
	CAM-CVF	1	central region	1	5-17	>30	24
	CAM	2	2o long strip map[b]	60	4.0-17	20	198
	PHT-P,C	5	mapping 20'x20'	>1	25-160	>5	28
	SWS/LWS	8	complete grating scan	1	2.4-180	>5	150
	SWS/LWS	8	10 grating postions	>1	selected	>5	>100
	P-S/S/L	9	complete grating scan	1	2.4-180	>5	244
N4A	SWS/LWS	8	10 grating postions	>1	selected	>5	100
N11	CAM	2	250 arcmin2 map[c]	40	4.0-17	20	132
	PHT-C	6	mapping 7'x6'	65-262	60-160	?	20

	SWS/LWS	8	10 grating postions	1	selected	>5	100
N49	LWS	3	full grating scan	1	45-180	?	7
	C/P-S/S/L	12	complete coverage	1-64	2.4-180	various	180
N49B	LWS	3	full grating scan	1	45-180	?	7
N63A	LWS	3	full grating scan	1	45-180	?	7
	C/P-S/S/L	12	complete coverage	1-64	2.4-180	various	180
N79A	SWS/LWS	8	10 grating postions	>1	selected	>5	100
N103B	LWS	3	full grating scan	1	45-180	?	7
N132D	LWS	3	full grating scan	1	45-180	?	7
N157	SWS/LWS	8	10 grating postions	>1	selected	>5	100
N157B	LWS	3	full grating scan	1	45-180	?	7
N159	SWS/LWS	8	10 grating postions	>1	selected	>5	100
	PHT-C	6	mapping 8'x10'	108-432	60-160	?	32
0505-67.9	LWS	3	full grating scan	1	45-180	?	7
0509-67.5	LWS	3	full grating scan	1	45-180	?	7
0519-69.0	LWS	3	full grating scan	1	45-180	?	7
0540-69.3	LWS	3	full grating scan	1	45-180	?	7
0548-70.4	LWS	3	full grating scan	1	45-180	?	7
SN1987A	LWS	3	full grating scan	1	45-180	?	7
	SWS/LWS	10	selected line scans	1	3.6-180	6-30	42.2

a P-S/S/L = PHT-S/SWS/LWS
C/P-S/S/L = CAM/PHT-S/SWS/LWS
b strip map includes N159 as well
c includes SN1987A

Table 2. Summary of SMC Observations

Target	Instrument*a*	Prop-osal	Observation	No. of Fields	λ (μm)	S/N	Time (min)
LIRS36,49	PHT-C	6	mapping 4'x4'	21-86	60-160	?	6
N12	SWS/LWS	8	10 grating postions	1	selected	>5	100
N27	SWS/LWS	8	10 grating postions	1	selected	>5	100
N66	CAM	1	multi-filter images	4	2.5-17	20	150
	CAM-CVF	1	Full CVF-LW plus LW1	1	5-17	>15	60
	SWS/LWS	8	10 grating postions	1	selected	>5	100
N81	SWS/LWS	8	10 grating postions	1	selected	>5	100
SW bar	CAM	2	complete 120 arcmin2	20	4.0-17	15	264
1E0102-7219	LWS	3	full grating scan	1	45-180	?	7

Plus the mapping of all MC supernova remnants (33 in the LMC and 11 in the SMC) with CAM, PHT-P and PHT-C (proposal 7); PAH emission/absorption in the MCs (proposal 4); and spectrophotometry of luminous AGB stars (15) and OH/IR stars (4) from the LMC as well as ir-photometry of cool stars from clusters in the MCs (proposal 11).

ROSAT Observations of the Magellanic Clouds

Wolfgang Pietsch [1], Peter Kahabka [1]

[1]Max Planck Institut für extraterrestrische Physik,
D-8046 Garching, Fed. Rep. of Germany

Abstract: The Magellanic Clouds have been covered with the X-ray astronomy satellite ROSAT (Röntgen Observatory SATellite) during the all sky survey and during pointed observations. A few hundred sources have been detected in the fields. Super-soft sources are detected as a new class of X-ray emitters, that may solve the missing millisecond pulsar progenitor problem. Transient super-soft as well as hard X-ray binary sources are discovered or rediscovered in both clouds. Extended sources are candidates for new supernova remnants and some of them already have been identified with supernova remnants. Large scale diffuse emission is clearly found in the LMC and extends beyond the supershell optical boundaries.

1 Introduction

Before the observations with the imaging X-ray telescope aboard the ***Einstein*** satellite only a few persistent (LMC X−1, X−2, X−3, X−4, SMC X−1) and some transient (LMC X−5, X−6, A0538−66, X0544−665, SMC X−2, X−3) very bright X-ray binaries have been detected by the collimated detector systems of the UHURU, SAS−3, ***Copernicus***, ***Ariel***−V and HEAO−1 satellites. The pointed surveys with the imaging proportional counter (IPC) and high resolution imager (HRI) as focal plane detectors of the X-ray telescope aboard the ***Einstein*** satellite in the energy range of 0.15 − 3.5 keV were sensitive enough to detect supernova remnants, extremely soft sources and diffuse emission in the Magellanic Clouds (MCs). More than 100 sources have been reported in the LMC area (Long, Helfand, and Grabelsky 1981, Wang et al. 1991) and more than 70 in the SMC area (Seward and Mitchell 1981, Inoue et al. 1983, Bruhweiler et al. 1987, Wang and Wu, 1992). EXOSAT observations added only a few additional sources (Jones et al. 1985, Pakull et al. 1985, Pietsch et al. 1989). A substantial number of the X-ray sources detected in the fields of the MCs have been identified with foreground stars or background active galaxies or clusters of galaxies. See Pakull 1990, Wang et al. 1991, Wang and Wu 1992 for reviews on the X-ray view of the MC.

We report on observations of the LMC with the Röntgen Observatory SATellite (ROSAT, Trümper 1983) during the calibration and verification phase (Trümper et al. 1991) and also on deep observations of the LMC during the ROSAT All Sky Survey (RASS). Deep ROSAT observations of the northern part of the LMC containing the X-ray binary LMC X–4 are covered by a separate paper in this issue (Dennerl et al. 1992). For the SMC we report on deep pointed ROSAT observations. The RASS observations of the SMC are also covered by a separate paper in this issue (Kahabka and Pietsch 1992).

The ROSAT X-ray telescope consists of 4 nested Wolter type I mirrors (Aschenbach 1988). The ROSAT design reduces mirror scattering to a minimum. This leads to high-contrast images that are essentially free of non-X-ray background and allow to directly map the diffuse sky background. For all the observations a position sensitive proportional counter (PSPC, Pfeffermann et al. 1986) was the focal instrument in use. The PSPC covers the energy range 0.1 – 2.4 keV and has a field of view of 57 arcmin radius. The on axis angular resolution is limited by the PSPC to 25" and degrades for off axis angles larger than 20 arcsec due to the telescope blur, the energy resolution is 45% FWHM at 1 keV. During the RASS ROSAT scanned the sky with constant speed on great circles perpendicular to the sun direction covering all the sky within half a year starting July 30, 1990. The angular resolution degrades to about 1 arcmin, analysing all the photons accumulated for a source at different PSPC off axis angles during the RASS. The typical exposure in the ecliptic plane was 400 sec adding up to 50,000 sec at the ecliptic poles. Due to the position close to the south ecliptic pole the LMC region on average got a much deeper exposure than the SMC during the RASS.

2 "First Light" Observations of the LMC

The first target observed by ROSAT starting from 16 June 1990 was the central region of the LMC around the bright X-ray binary LMC X–1. During the following days (17–28 June) this region and its surroundings were viewed by a mosaic of observations covering an area of $\sim$ 8 square degrees, with exposure times ranging from 1,000 to 20,000 seconds. Trümper et al. (1991) report on a search for discrete sources, diffuse emission and soft X-rays from SN1987A in these data.

45 X-ray sources have been detected in the field, 30 of which were already known from the *Einstein* LMC survey (Long, Helfand, and Grabelsky 1981, Wang et al. 1990). Five of these sources seem to be extended in the ROSAT analysis. Foreground stars and supernova remnants (SNRs) can clearly be identified by their hardness ratio (soft versus hard). CAL 83, one of the two known low-mass X-ray binaries (LMXBs) in the LMC (Smale et al. 1988) has an unusually soft spectrum and emits essentially no X-rays above 0.5 keV.

For the 15 newly detected X-ray sources in the ROSAT image candidates have been looked for on ESO Schmidt plates of the LMC. Six of the new sources appear to be extended in X-rays. One is identified with the optical emission nebula N158D, some of the others are associated with diffuse optical emission. Three of the pointlike objects are identified with bright ($<$ 9 mag) stars and also fall in the same X-ray hardness range as the *Einstein* identified stars. 50 arcsec of the position of one of the sources, there is a 10 mag star as the possible identification, one position coincides with a Harvard variable star of 16 – 17 mag. The remaining four sources have no clear identification and may be either objects in the LMC or more distant active galactic nuclei; their error circle typically contains several blue objects fainter than 14 – 17 mag.

The LMC area covered by the ROSAT observations contains 14 of the 27 SNRs discovered during the ***Einstein*** survey of the LMC. All 14 SNRs are detected by ROSAT and additional 10 extended ROSAT sources remain as SNR candidates. Optical follow up observations of these objects using H_α and [OIII] yield 7 new SNR identifications (Aschenbach, private communication).

The most remarkable new source in the list is RX J05278–6954, a bright source with a very soft X-ray spectrum, that must have brightened since the LMC observations by ***Einstein*** by more than a factor of ten. The spectrum and luminosity of the source is very similar to CAL 83. The ROSAT data of these two supersoft sources have been further investigated by Greiner, Hasinger, and Kahabka (1991). Their X-ray spectra resemble black-body spectra with a temperature of 20 – 40 eV and N_H values that clearly indicate a LMC membership. They radiate at or even above the Eddington limit for a $1.4M_\odot$ object at the distance of the LMC. Additional sources with similar soft spectra and luminosity have been found in LMC and SMC by ***Einstein*** (Long et al. 1981, Wang et al. 1991, Wang and Wu 1992) and ROSAT (see below) and in ROSAT pointed observations of M31 (Hasinger, private communication) and in the ROSAT galactic plane survey (Motch, private communication). Several models may explain the nature and the soft spectra of these sources. There may be a compact object with supercritical accretion. The extremely soft spectrum could be produced by Compton scattering in a cocoon of surrounding material (Ross 1979). An other explanation may be accreting white dwarfs with steady nuclear burning on their surface, a kind of persistent nova (Van den Heuvel et al. 1992). This new class of super-soft sources in external galaxies might solve the millisecond pulsar birthrate problem (Kulkarni and Narayan 1988) by providing accretion powered compact sources that before ROSAT have not been detected in our galaxy but may be progenitors to millisecond pulsars.

Due to the high sensitivity of ROSAT the highly structured diffuse X-ray emission of the LMC could be directly seen in the first images. The ***Einstein*** LMC survey already revealed this diffuse component, but due to the lower intrinsic background, lower mirror scattering, as well as the higher spatial and spectral resolution of the ROSAT PSPC, the LMC is imaged in unprecedented detail. The brightest regions are east of LMC X–1 and in the star-forming region 30 Doradus. The structure of the diffuse emission of 30 Doradus is strikingly similar to the optical picture. Below 0.4 keV LMC radiation is strongly absorbed by the galactic foreground N_H and only the flat soft diffuse forground radiation of our own galaxy is visible. A coarse temperature map of the diffuse LMC radiation is shown in figure 1 (Fig. 3 of Trümper et al. 1991). The bulk of the diffuse emission has a temperature of $(3 - 7) \times 10^6$K with maxima occurring to the northeast of LMC X–1 and in 30 Dor. The diffuse emission shows structure down to the angular resolution of the instrument. In the 30 Doradus area the raw ROSAT PSPC image is very similar to the maximum-entropy map deconvolved from the ***Einstein*** observations (Wang and Helfand 1991). There is also a strong correlation between diffuse features in the ROSAT image and ESO colour images of the LMC in visible light. Bright knots in the X-ray map correspond to H II emission in the optical. The bright X-ray region to the northeast of LMC X–1 has been attributed to the blow-out of the hot supergiant bubble LMC–2. The detailed structure of the

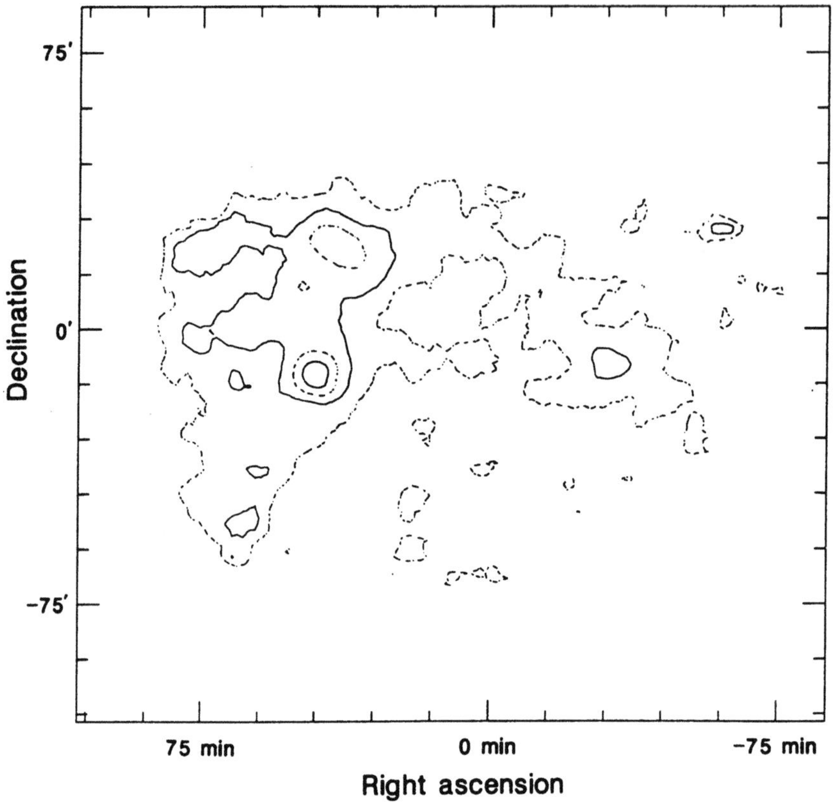

Fig. 1. Temperature map of the diffuse LMC emission. Contours correspond to temperatures of 4.5, 5.8, 6.7 and 10 million K. (Fig. 3 from Trümper et al. 1991)

X-ray emission however suggests that the diffuse emission extends much further than the actual superbubble.

From the region around supernova 1987A (SN1987A) hard X-rays have been detected by the HEXE experiment aboard the ***MIR-Kvant*** observatory (Sunyaev et al. 1987) and the ***Ginga*** satellite (Dotani et al. 1987) starting in August 1987. Soft X-rays below 2.5 keV could not be detected by rocket experiments (Aschenbach et al. 1987, Burrows, Nousek, and Garmire 1987). Also during the ROSAT observations no positive signal above background was detected to an upper limit of 2.7 μCrab. SN1987A is also not detected during the RASS yielding similar upper limits about 5 month after the pointed observations. These upper limits however due not yet severely constrain models the propose a soft X-ray origin from a central pulsar or accreting neutron star. The depth for photoelectric absorption in the expanding envelope is too big. The expected soft X-ray emission from the shock-wave-heated supernova plasma is still about two orders of magnitude below the ROSAT upper limit. For a more detailed discussion see Trümper et al. 1991.

3 All Sky Survey View of the LMC

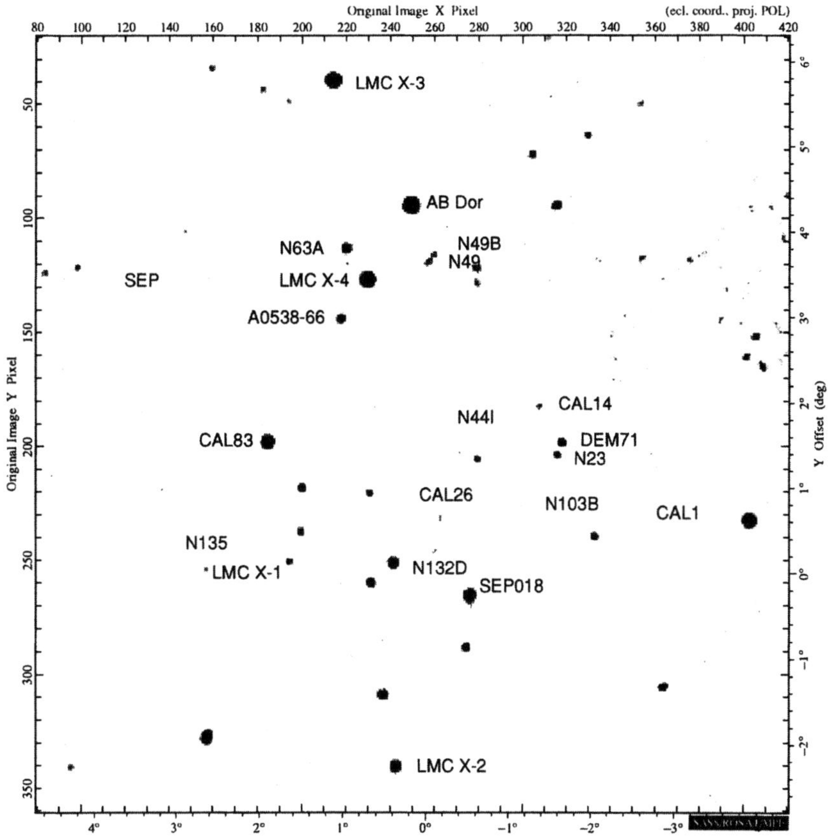

Fig. 2. ROSAT all sky survey image of the LMC in the energy range 0.1 – 0.4 keV. Some source identifications and the position of the south ecliptic pole (SEP) are indicated.

The exposure corrected RASS images of the LMC in the energy range 0.1 – 0.4 keV and 0.4 – 2.4 keV (fig. 2, 3) demonstrate the wealth of information about discrete sources and diffuse emission that is available to us in the LMC region for detailed time and spectral variability studies. The closer to the south ecliptic pole (SEP) the longer the LMC area was exposed during the survey. As explained above regions of ~2 degree diameter at the ecliptic poles were covered every day during the survey. Some well known bright LMC sources, X-ray binaries and SNRs are indicated as well as the bright RS CVn star AB Dor and the SEP region (fig. 2). The Be type recurrent transient A0538–66 (Skinner 1981) that was inactive during ***Einstein*** observations carried out 1979 to 1981 (Wang et al. 1991) and EXOSAT observations in 1983 and 1984 (Pakull et al. 1985) was clearly active during the RASS. Some of the bright sources have been analysed in detail for time variability and spectra. The source SEP018 is a transient super-soft source that turned on during the RASS (Schaeidt, Hasinger, and Trümper 1991, Schaeidt, private communication). In the soft RASS image no structured diffuse emission is visible for the reason discussed already for the

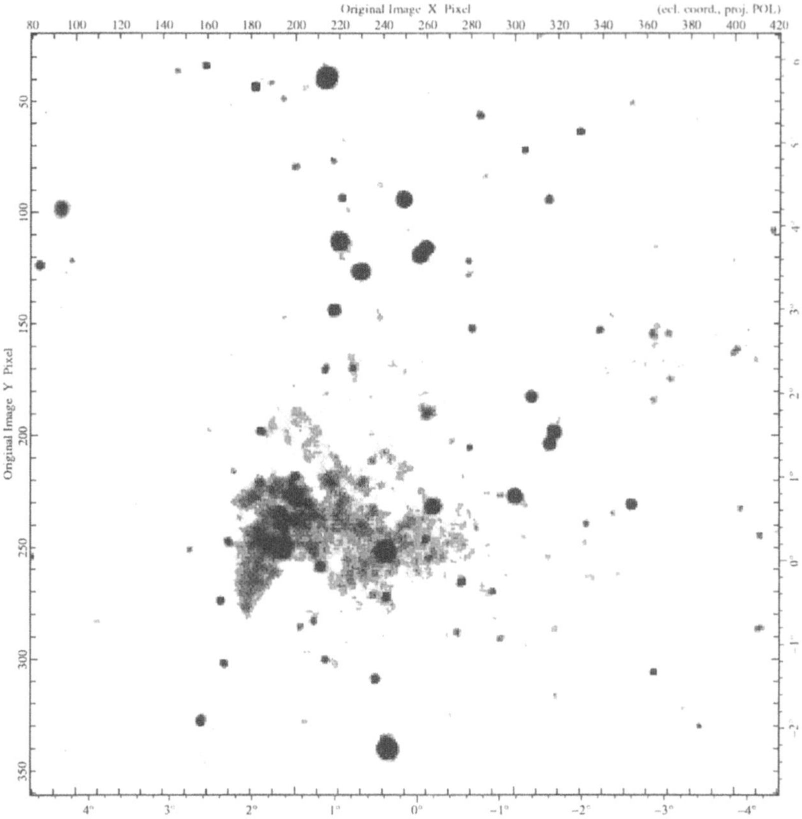

Fig. 3. **ROSAT all sky survey image of the LMC in the energy range 0.4 – 2.4 keV. Besides the hard sources structured diffuse emission is clearly visible.**

first light images; the RASS hard image shows the diffuse structures more clearly and extending further out than the area covered by the first light images (fig. 2, 3).

We analysed the RASS data of the LMC in a field of ~ 13.5 deg× 13.5 deg centered at $\alpha_{2000} = 5^h20^m$, $\delta_{2000} = -69^o00'$ for point sources. The exposure over the field varies from ~350 seconds in badly exposed areas to ~20,000 seconds at the SEP. The average exposure time in the field is 2,000 seconds. Altogether 508 sources can be identified in the field merging detections in the 0.1 – 2.4, 0.1 – 0.5, and 0.5 – 2.4 keV band. 55 of the 114 sources reported by Wang et al. 1991 are in this detection list. ***Einstein*** sources missed in the RASS are either faint sources from deep ***Einstein*** pointings or are in crowded areas that have not been resolved (like in the 30 Dor region). All bright ***Einstein*** sources are also detected in the RASS. For the ROSAT first light observation 20 of the 45 sources are not identified in the RASS since they were too faint or located in the 30 Dor confused region. A plot of the 114 ***Einstein*** sources, the 45 ROSAT first light detected sources, and the RASS sources contained in the same area is given in figure 4. In the overlapping area of the observations the soft transient SEP018 and the Be type transient A0538–66 are only detected in the RASS. The EXOSAT LMC transient EXO053109–6609.2 was not active during the RASS. Identifications of the newly detected sources is in progress.

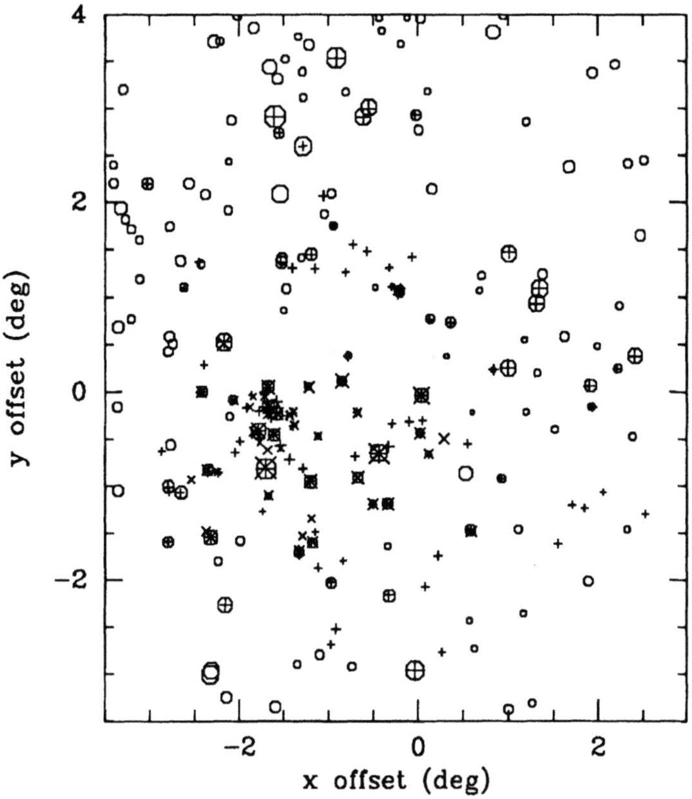

Fig. 4. Tangential projection of the RASS detected sources (o), the *Einstein* sources (+) and the ROSAT first light (×) sources for the overlapping area. North is up and East to the left; 0,0 correspond to $\alpha_{2000} = 5^h20^m$, $\delta_{2000} = -69°00'$. The size of the symbols is coded according to the logarithm of the count rates.

4 Pointed Observations of the SMC

Observations of the SMC were carried out during the RASS from October to November 1990 (Kahabka and Pietsch 1992, this conference) and during 3 pointed observations from 8 October to 3 November 1991. The field centers of the three pointed observations are

$\alpha_{2000} = 0^{\mathrm{h}}58^{\mathrm{m}}$, $\delta_{2000} = -72°17'$ for field A,
$\alpha_{2000} = 0^{\mathrm{h}}50^{\mathrm{m}}$, $\delta_{2000} = -73°14'$ for field B and
$\alpha_{2000} = 1^{\mathrm{h}}13^{\mathrm{m}}$, $\delta_{2000} = -72°49'$ for field C.

The exposure times are 17, 1.4, and 22 ksec respectively. Region B which turned out to be of specific interest was only covered with a much shorter observing time than originally planned.

A total of 99 sources were detected in an area of 8.5 deg^2 in the broad (0.1–2.4keV), hard (0.4–2.4keV) and soft (0.1–0.4keV) energy band (56 sources in the broad energy band, see figure 5). 38 sources (17 in the broad energy band) could be identified with ***Einstein*** detections (Wang and Wu 1991), one source is identified with the X-ray Be type transient SMC X-2. Identified sources are listed in table 1. Many of the faint unidentified sources are expected to be not associated with the SMC but background AGN.

Table 1. Identified sources detected in three ROSAT pointings A, B and C of the SMC (see figure 5). RASS source numbers (from Kahabka and Pietsch 1992) are given in case of a detection in the survey data.

Source No	Soure name	RASS source number
1	1E0056.8-7154; PN 67	9
3	1E0102.2-7219; SNR 0102-723	10
7	RX J0101.1-7207	-
9	1E0057.6-7228; SNR 0057-724; NGC 346; N 66	-
22	SNR 0103-726	14
28	RX J0049.1-7251	-
29	SNR 0045-734	-
33	SNR 0049-736	18
34	RX J048.3-7332; SMC 3	21
35	SMC X-2	-
55	HD 8191	19
56	SMC X-1	20

The two luminous but transient SMC X-ray sources, SMC X-2 and SMC X-3 were discovered by SAS 3 (Clark et al. 1978) during an outburst. The two transients have not been discovered by X-ray satellites since and also were not detected in the survey. These 2σ upper limit count rates for SMC X-2 and SMC X-3 are 7.4×10^{-3} and 9.84×10^{-3} cts/sec respectively. SMC X-2 was observed in outburst in the pointed ROSAT observation of field B (source number 35 in figure 5). A power law fit to the spectrum (fig. 6) yields a photon index of 2 and an hydrogen column density of 6×10^{21} cm^{-2}. This absorbing column density is a factor of 20 larger than the galactic foreground hydrogen column density in this direction. It has to be accounted for by absorption within the Be system and within the SMC. The ROSAT spectrum translates to a luminosity of $6. \times 10^{36}$ erg sec^{-1} in the SAS-3 energy band, a factor of 10 below the reported 1977 outburst. Another hard transient source was detected by ROSAT in field B (source number 28 in figure 5). This source was not detected by the ***Einstein*** satellite and during the RASS.

Two super-soft sources are visible in pointing A and B, an additional one in the RASS (Kahabka and Pietsch 1992). The source in pointing A (source number 1 in figure 5) is correlated with the planetary nebula PN 67 and was also seen in ***Einstein*** observations (Wang and Wu 1991, Wang 1992). From hardness ratio arguments out of the ***Einstein*** data they deduce a temperature of $T_{eff} \sim 3 \times 10^5$K. A blackbody fit to the ROSAT PSPC spectrum (fig. 6) yields a temperature of $T_{eff} = 1.4 \times 10^5$K. The absorbing column density of $N_H > 5 \times 10^{20}$ clearly puts that source into the SMC. The source in pointing B (source number 34 in figure 5) is a new ROSAT detection.

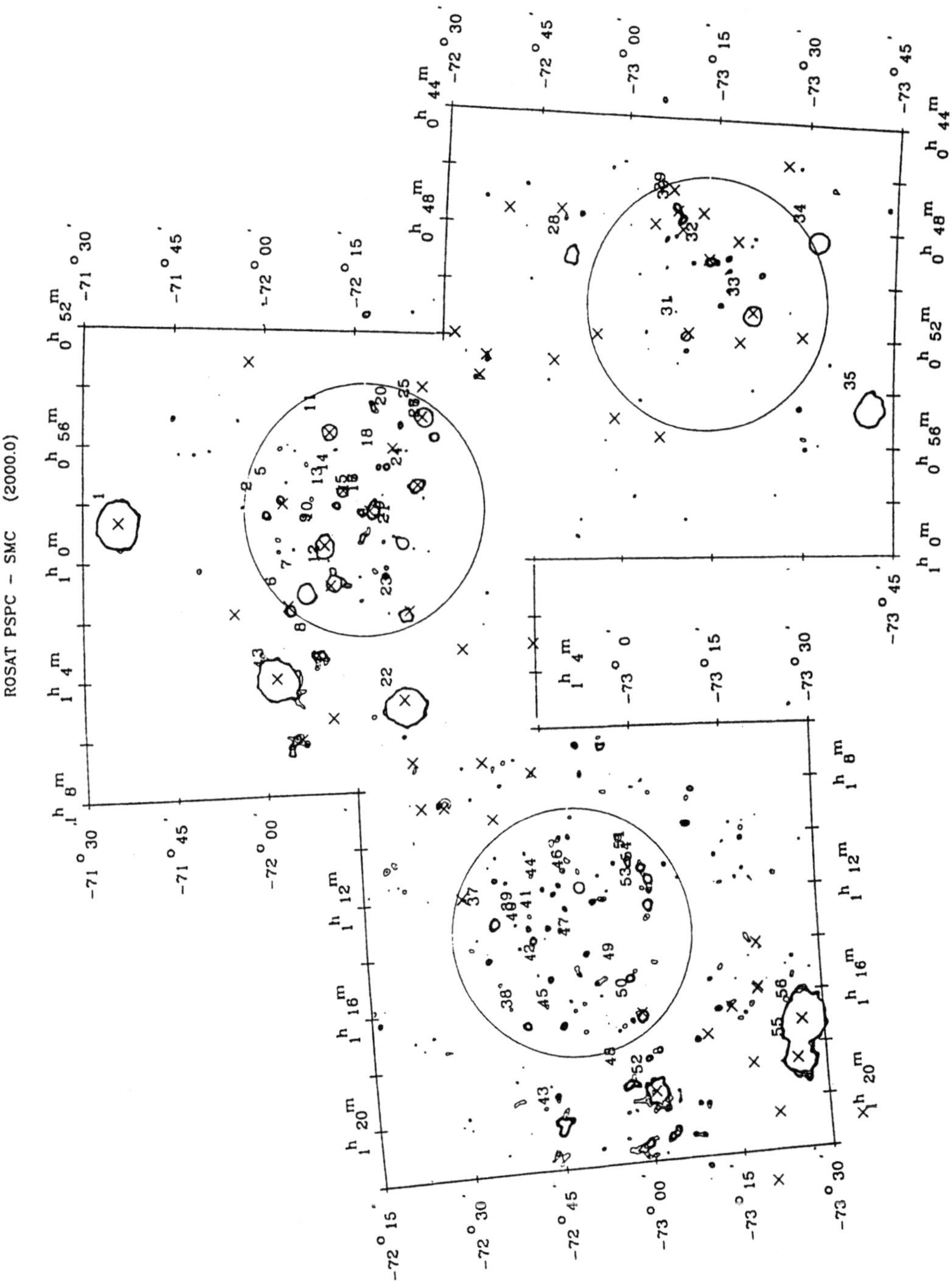

Fig. 5. ROSAT image of the three SMC pointings. Source numbers are indicated. Crosses give *Einstein* detection (Wang and Wu 1991). Big circles mark the inner 40 arcmin of the PSPC.

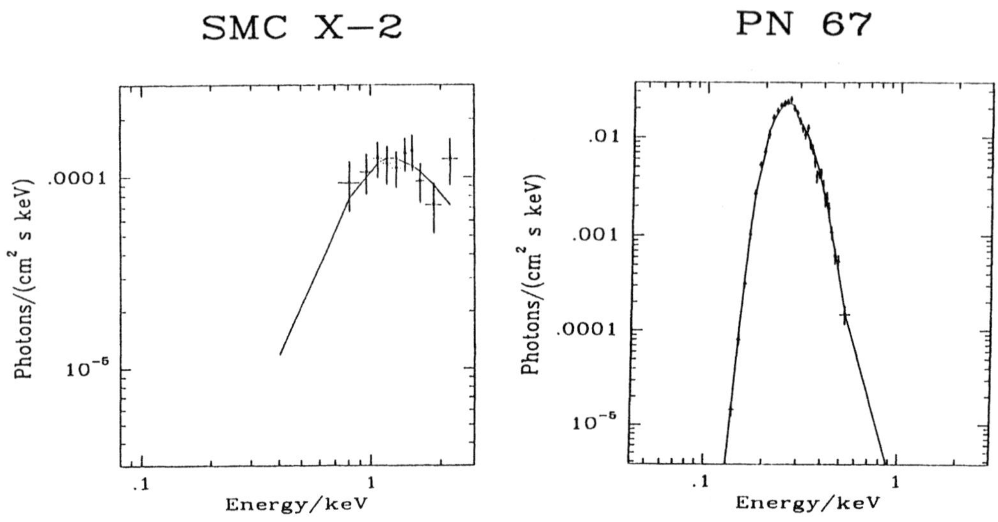

Fig. 6. Spectra of the Be type transient SMC X-2 (left panel) and the super-soft source in the planetary nebula PN 67 (right panel).

It could be identified with the symbiotic star SMC3 in the SMC cluster NGC 269 (Kahabka and Pietsch 1992, Morgan, private communication, Morgan 1992). This source was not detected by the ***Einstein*** satellite. The optical star SMC3 underwent an outburst in 1981 and is still active (Morgan 1992).

From ***Einstein*** observations 11 SNRs were identified in the SMC. Besides the prominent SNRs already seen in the RASS (Kahabka and Pietsch 1992) there were further SNRs and SNR candidates seen in the three pointed observations. Ye and Turtle (1991) detected 2 SNRs within the giant H II region N 66 in the radio band. Both of these are ROSAT sources. Within N 66 a third extended source is discovered by ROSAT (source number 7 in figure 5). Further SNR identifications are in progress.

5 Summary

ROSAT detected more than 500 sources in the LMC and more than 100 in the SMC area. The analysis of the data and the identification of sources detected is in progress. During the RASS and also during the SMC pointed observations several of the known bright X-ray transients in SMC and LMC were in outburst. In addition two new super-soft transients were detected in the LMC and one hard and one super-soft transient in the SMC. We now count four super-soft sources in the LMC and three in the SMC and include transient sources in the source category. The new class of super-soft sources has a similar number of members in the LMC and SMC as the class of classical high and low mass X-ray binaries and may be important as progenitors for millisecond pulsars. Several new SNRs are identified or proposed as SNRs by their X-ray extend. The diffuse emission detected by ***Einstein*** in the LMC is confirmed with much higher significance and resolution in the ROSAT pointed and RASS observations. Detailed correlations with observations in other wavelength is in progress. The RASS and first pointed ROSAT observations of the MCs have already indicated the potential of the data gathered. There are further ROSAT MC observations already taken or scheduled.

Acknowledgements

We would like to thank many colleagues at the Max Planck Institut für Extraterrestrische Physik who worked on ROSAT hardware and software. The help of K. Dennerl, J. Engelhauser, and W. Voges preparing the LMC survey image is acknowledged.

References

Aschenbach, B. (1988): Appl. Optics **27** 1404

Aschenbach, B., Briel, U.G., Pfeffermann, E., Bräuninger, H., Hippmann, H., and Trümper, J. (1987): Nature **330** 232

Bruhweiler, F.C., Klinglesmith III, D.A., Gull, T.R., and Sofia, S. (1987): Ap.J. **317** 152

Burrows, D., Nousek, J., and Garmire, G. (1987): IAU Circ. 4494

Clark, G., Doxsey, Li, F., Jernigan, J.G., and van Paradijs, J. (1978): Ap.J. **221** L37

Dennerl, K., Kürster, M., Pietsch, W., and Voges, W. (1992): *this volume*

Dotani, T. et al. (1987): Nature **330** 230

Greiner, J., Hasinger, G., and Kahabka, P. (1991): A.A. **246** L17

Inoue, H., Koyama, K., and Tanaka, Y. (1983): *Supernova Remnants and their X-Ray Emission*, ed. J. Danziger and P. Gorenstein, p. 535

Jones, L.R., Pye, J.P., McHardy, I.M., and Fairall, A.P. (1985): Sp.Sci.Rev. **40** 693

Kahabka, P., and Pietsch, W. (1992): *this volume*

Kulkarni, S.R., and Narayan, R. (1988): Ap.J. **335** 755

Long, K.S., Helfand, D.J., and Grabelsky, D.A. (1981): Ap.J. **248** 925

Morgan, D.H. (1992): "Symbiotic stars in the Magellanic Clouds", Edinburgh Astronomy Preprint Nr. 10

Pakull, M. (1989): *Recent Developments of Magellanic Cloud Research*, ed. K.S. Boer, F. Spite and G. Stasinska (Paris: Observatoire de Paris), p. 183

Pakull, M., Brunner, H., Pietsch, W., Staubert, A., Beuermann, K., van der Klis, M., and Bonnet-Bidaud, J.M. (1985): Sp.Sci.Rev. **40** 379

Pfeffermann, E., Briel, U.G., Hippmann, H., Kettenring, G., Metzner, G., Predehl, P., Reger, G., Stephan, K.H., Zombeck, M.V., Chappell, J., and Murray, S.S. (1988): Proc. SPIE **733** 519

Pietsch, W., Dennerl, K., and Rosso, C. (1989): Proc. 23rd ESLAB Symp. on Two-Topics in X-ray Astronomy, Bologna, ESA SP-296, 573

Ross, R.R. (1979): Ap.J. **233** 334

Schaeidt, S., Hasinger, G., and Trümper, J. (1991): *X-ray Emission from Active Galactic Nuclei and the Cosmic X-ray Background*, ed. W. Brinkmann and J. Trümper (MPE Garching), MPE report 235, p. 191

Seward, F.D., and Mitchell, M. (1981): Ap.J. **243** 736

Smale, A.P., Corbet, R.H.D., Charles, P.A., Ilovaisky, S.A., Mason, K.O., Motch, C., Mukai, K., Naylor, T., van der Klis, M., and van Paradijs, J. (1988): M.N.R.A.S. **233** 51

Sunyaev, R. et al. (1987): Nature **330** 227

Trümper, J. (1983): Adv.Sp.Res. **2** 241

Trümper, J., Hasinger, G., Aschenbach, B., Bräuninger, H., Briel, U.G., Burkert, W., Fink, H., Pfeffermann, E., Pietsch, W., Predehl, P., Schmitt, J.H.M.M., Voges, W., Zimmermann, U., and Beuermann, K. (1991): Nature **349** 579

van den Heuvel, E.P.J., Bhattacharya, B., Nomoto, K., and Rappaport, S.A. (1992): A.A., in press

Wang, Q., and Helfand, D.J. (1991): Ap.J. **370** 541

Wang, Q., Hamilton, T., Helfand, D.J., and Wu, X. (1991): Ap.J. **374** 475

Wang, Q. and Wu, X. (1992): Ap.J.Suppl. **78** 391

Ye, T., and Turtle, A.J. (1991): M.N.R.A.S. **249** 722

ROSAT Survey view of the SMC

Peter Kahabka, Wolfgang Pietsch [1]

[1]Max Planck Institut für extraterrestrische Physik, D-8046 Garching, Fed. Rep. of Germany

Abstract: We analysed a 8 × 8 degree field centered at the body of the Small Magellanic Cloud observed during ROSAT all-sky-survey from October to November 1990. We detected more than 40 sources, 15 could be identified with Einstein detections and further 9 with Simbad catalogue entries. We detected a new supersoft source in the SMC cluster NGC 269 which could be identified with the symbiotic star SMC3. In the body of the SMC we found a clear enhancement in flux at energies above 0.5 keV, which is at least partially explained by point sources detected during deep pointed observations.

1 Sources detected in a field centered on the SMC

A field of 8 × 8 deg was analysed centered at the body of the SMC at $\alpha(2000) = 1^h\ 0^m\ 0.0^s$, $\delta(2000) = -73^d\ 0^m\ 0^s$. 41 sources were detected in the energy band 0.1 - 2.4 keV. 7 sources are also covered by 3 ROSAT pointed observations performed from 8 October to 3 November 1991. 6 of these sources were already seen by the Einstein satellite, that is the pulsator SMC X-1, the supernova remnants SNR 0102-723, SNR 0103-726 and SNR 0049-73.6, a supersoft X-ray source related with the planetary nebula PN67, and the galactic foreground star HD 8191 (Seward and Mitchell, 1981, Inoue et al. 1983, Jones et al. 1985, Bruhweiler et al. 1987, Wang and Wu, 1991). Besides that a new supersoft X-ray source was detected which was also seen in one of the pointed observations. This source lies in the SMC cluster NGC 269 and could be identified with the symbiotic star SMC3 (Morgan 1992). The first supernova remnant detected in the SMC, SNR 0045-734, N 19 (Mathewson, 1972) was also seen during the survey but was detected with a lower existence likelihood than the detection threshold in the automatic maximum likelihood detection procedure. A 8th source, a supersoft source, probably a new type of LMXBs in the SMC detected by Einstein satellite (Wang and Wu 1991), but not yet identified, was also detected in the survey. For the two transient SMC X-ray sources, SMC X-2 and SMC X-3 (Clark et al. 1978) only 2σ upper limit count rates of 7.4×10^{-3} and 9.8×10^{-3} cts/sec respectively could be deduced. For the 7 sources the count rates, and in case of a SMC membership the luminosities for an absorbing column density of $N_H = 3. \times 10^{20}$ cm^{-2} and for a blackbody spectrum with a temperature of 10^7 K are listed in table 1.

Table 1. For 7 sources in the body and wing of the SMC detected during the survey count rates and luminosities $L_x(0.1 - 2.4$ keV) are given with 90% confidence errors given in brackets for an absorbing hydrogen column density of $N_H = 3. \times 10^{20}$ cm^{-2} and a blackbody spectrum with a temperature of 10^7 K. This spectrum is not valid for the three supersoft sources PN 67, SMC3 and the supersoft LMXB. For further 10 stars, one Abell cluster, the globular cluster 47 Tuc, one spiral galaxy and two radio quasars only count rates are given.

Source No	Identification	cts/s	$L_x(10^{37}$erg/sec)
9	PN 67	0.292 (63)	0.216 (47)
10	SNR 0102-723	2.02 (15)	1.49 (11)
13	supersoft LMXB	0.333 (56)	0.246 (42)
14	SNR 0103-726	0.240 (56)	0.178 (41)
18	SNR 0049-736	0.0531 (25)	0.0393 (19)
19	star, HD 8191	0.0537 (277)	
20	SMC X-1	0.595 (74)	0.440 (55)
21	symbiotic, SMC3	0.206 (42)	0.152 (31)
1	cluster Abell 2789	0.130 (49)	
2	star, HD 8519	0.076 (42)	
3	star, HR 83	0.069 (32)	
7	star, HD 9611	0.053 (30)	
8	star, HD 5028	0.113 (44)	
12	NGC 104 (47 Tuc)	0.162 (40)	
22	star, HD 2133	0.052 (24)	
23	star, HD 5303	2.41 (13)	
26	star, HD 987	0.359 (51)	
30	spiral galaxy ESO 29-4	0.837 (73)	
32	quasar, PKS 0101-760	0.018 (16)	
37	star, CPD-76	0.233 (42)	
39	star, HD 8813	0.039 (25)	
40	quasar, PKS 0202-765	0.091 (27)	

Comparison with the pointed data shows that the survey is complete down to a luminosity of $4. \times 10^{35}$ erg/sec in the region of the body of the SMC. The weakest source detected in the survey has a luminosity of $1. \times 10^{35}$ erg/sec. For comparison, the Einstein survey was complete to a luminosity of 10^{36} erg/sec and the weakest source detected had a luminosity of $3. \times 10^{35}$ erg/sec (Seward and Mitchell, 1981). Besides the 8 X-ray sources discussed above there are further 33 detections. 15 sources coincide with Einstein detections (Wang et al. 1991). 14 sources coincide with entries of the Simbad catalogue within a search radius of less than 20 arcsec, and 4 sources of less than 90 arcsec. These are 13 galactic or SMC stars, one cluster of galaxies, one spiral galaxy of type Sc, the globular cluster 47 Tuc, and two radio quasars. For 15 sources no identification was found in the Simbad catalogue.

2 Extended emission in the body and wing of the SMC

Extended emission was detected in the hard band image (0.5 – 2.4keV) after removal of the sources detected in the survey. The background corrected count rate of the extended emission is 0.69 cts/sec. This corresponds to a luminosity of $5. \times 10^{36}$ erg/sec in case of a SNR-like spectrum. This emission is fully accounted for by point sources with a luminosity below $4. \times 10^{35}$ erg/sec detected in the deep pointed observations of these regions. These sources add up to a luminosity of $\sim 6. \times 10^{36}$ erg/sec.

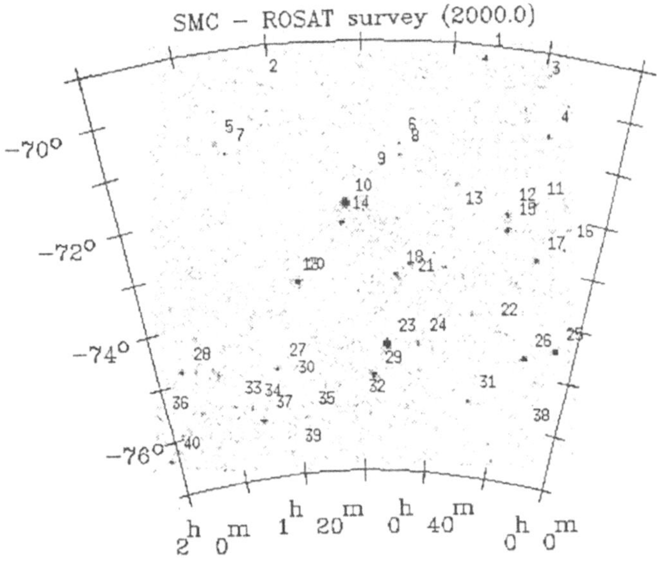

Fig. 1. Image (0.5 - 2.4 keV) of the 8 x 8 degree field centered at the body of the SMC. The image is background subtracted. The most prominent source (10) in the body of the SMC is the brightest SNR 0102-723. There are three supersoft sources in the SMC, the planetary nebula PN 67 (9), a possible CAL 83 type LMXB (13) and the symbiotic star SMC3 (21).

References

Bruhweiler, F.C., Klinglesmith III, D.A., Gull, T.R. and Sofia, S., (1987): Ap.J. **317** 152

Clark, G., Doxsey, Li, F., Jernigan, J.G., and van Paradijs, J. (1978): Ap.J. **221** L37

Inoue, H., Koyama, K., and Tanaka, Y. (1983): *Supernova Remnants and their X-Ray Emission*, ed. J. Danziger and P. Gorenstein pp. 535-540

Jones, L.R., Pye, J.P., Hardy, I.M. and Fairall, A.P. (1985): Sp.Sci.Rev, **40** 693

Mathewson, D.S., and Clarke, J.N. (1972): Ap.J. **178** L105

Morgan, D.H. 1992: "Symbiotic stars in the Magellanic Clouds", Edinburgh Astronomy Preprint Nr. 10.

Seward, F.D., and Mitchell, M. (1981): Ap.J. **243** 736

Wang, Q. and Wu, X. (1991): Ap.J. to be published

Deep ROSAT observations of the northern part of the LMC containing the X-ray binary LMC X-4

Konrad Dennerl, Martin Kürster,

Wolfgang Pietsch, Wolfgang Voges

Max–Planck–Institut für extraterrestrische Physik
W-8046 Garching bei München, Fed. Rep. of Germany

Abstract: From 29 October to 3 November 1991 we performed a 46 000 sec ROSAT PSPC observation of the northern part of the Large Magellanic Cloud containing the X–ray binary LMC X–4. Here we present first results of this pointed observation on LMC X–4. We also show the light curve of LMC X–4 obtained during the ROSAT all-sky survey which allowed its 30-day modulation to be followed for more than 50 days. In particular, these observations indicate that the length of the 30-day cycle of LMC X–4 is decreasing with time.

1 Introduction

The northern part of the LMC contains a wealth of interesting X–ray objects in a small area of the sky and offers an ideal opportunity for an imaging satellite like ROSAT to observe several objects simultaneously, an opportunity which we utilized in our pointed observation. Furthermore, its proximity to the south ecliptic pole allowed the soft X–ray emission of this region to be monitored over several weeks during the ROSAT all-sky survey (an X–ray image of the LMC can be found in the contribution by W. Pietsch, these proceedings). Here we summarize only our first results on LMC X–4, an accreting $1.4\,M_\odot$ neutron star in the LMC which orbits a mass donating $15\,M_\odot$ O star in 1.4 days. The deceleration of this matter near the neutron star causes very intense X–ray emission ($L_X = 10^{38} - 10^{39}$ erg/s). Besides short-term modulations due to the 13.5 sec rotation of the neutron star, eclipses by the optical star, dips caused by blobs of matter in the line of sight, and flares triggered by instabilities in the accretion flow, the observed X–ray flux is furthermore modulated on a time-scale of about 30 days, a phenomenon which has stimulated many models to explain its origin (e.g. Dennerl 1991).

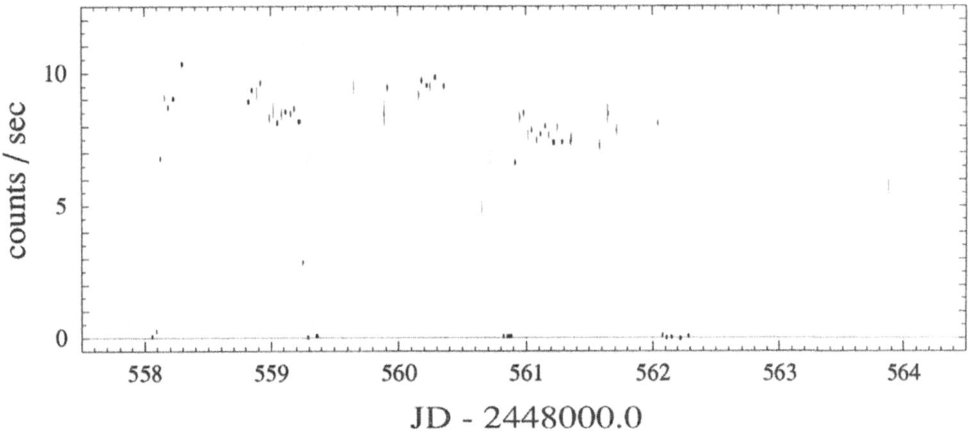

Fig. 1. Rosat light curve of LMC X–4 for the pointed observation (E = 0.07–2.4 keV). Shaded areas mark the predicted eclipse interval in hard X–rays as derived from EXOSAT measurements. There is clear evidence for an increased eclipse duration in soft X–rays. Also evident is a gradual decrease in intensity indicating the end of the ON-state.

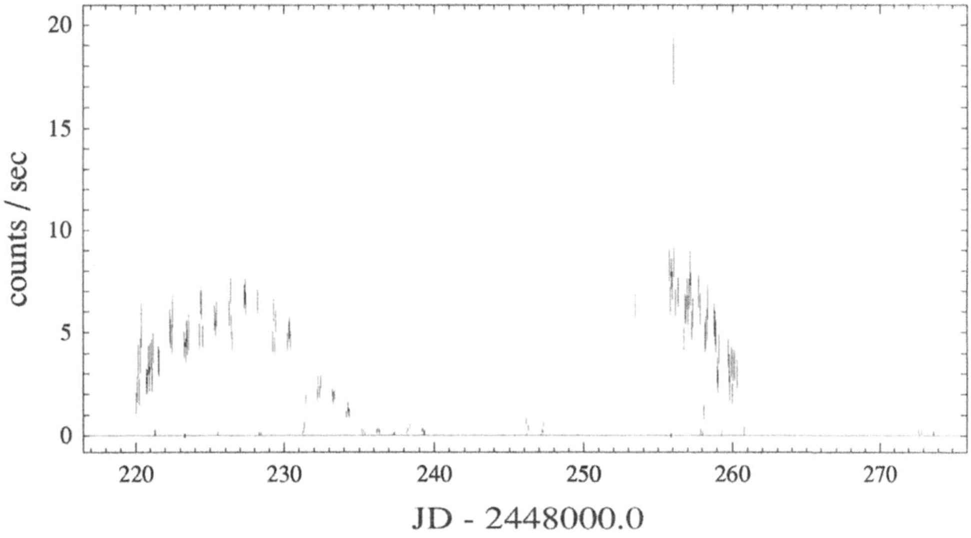

Fig. 2. Rosat survey light curve of LMC X–4, showing two consecutive 30-day cycles with several eclipses and one flare. This light curve is preliminary; it does not contain all survey data nor have the count rates been corrected for partial obscurations by the PSPC window support structure.

2 The duration of the X–ray eclipse of LMC X–4

The pointed ROSAT observation shows the soft X–ray flux to be significantly attenuated before and after the eclipse, resulting in an apparent increase of the eclipse duration (Fig. 1). This effect, which is also present in the survey data, confirms a tendency already indicated by the EXOSAT observations (Dennerl 1991).

3 The length of the 30-day cycle of LMC X–4

From the ROSAT all-sky survey which covered two 30-day cycles (Fig. 2) and the pointed ON-state observation (Fig. 1) we estimated a value of 30.24 ± 0.08 d for the average length of the 30-day cycle during 1991. By combining the ROSAT data with the EXOSAT data (Dennerl 1991) we derived a mean period of 30.25 ± 0.03 d for 1983–1991, which is significantly less than the value of 30.48 ± 0.06 d obtained for 1977–1978 (Lang et al. 1981). From Tab. 1, which summarizes similar measurements, it is evident that the ROSAT observations, extending the baseline to 14 years (171 cycles), show for the first time a clear indication that the length of the 30-day cycle is systematically decreasing. This result provides important constraints on models of the 30-day cycle, especially on those which assume that the epochs of the individual cycles are subject to some random fluctuation in phase but ultimately controlled by an underlying stable clock (e.g. a precessing neutron star). If subsequent measurements show this trend to continue, these models will have to be ruled out as soon as the required phase shifts will become comparable to 30 days.

Table 1. Measurements of the 30-day cycle of LMC X–4

Instrument	Time	P_{30} [d]	Reference
HEAO 1	1977–1978	30.48 ± 0.06	Lang et al. 1981
HEAO 1 / EXOSAT	1977–1983	30.42 ± 0.03	Pakull et al. 1985
ESO (opt. phot.)	1977–1988	30.36 ± 0.02	Heemskerk, van Paradijs 1989
EXOSAT	1983–1986	30.35 ± 0.05	Dennerl 1991
EXOSAT / ROSAT	1983–1991	30.25 ± 0.03	(this article)

References

Dennerl, K. (1991): "Eine vergleichende Untersuchung der Röntgendoppelsternsysteme LMC X–4 und Her X–1", PhD Thesis, Universität München, and MPE Report **232**.

Heemskerk, M.H.M., van Paradijs, J. (1989): Astron. Astrophys. **223**, 154.

Lang, F.L., Levine, A.M., Bautz, M., Hauskins, S., Howe, S., Primini, F.A., Lewin, W.H.G., Baity, W.A., Knight, F.K., Rothschild, R.E., Petterson, J.A. (1981): ApJ **246**, L21.

Pakull, M., Brunner, H., Pietsch, W., Staubert, A., Beuermann, K., van der Klis, M., Bonnet-Bidaud, J.M. (1985): Space Science Reviews **40**, 379.

H alpha Survey of the Small Magellanic Cloud

Etienne le Coarer [1], Margarita Rosado [2], Yvon Georgelin [1], Annick Viale [1]

[1]Observatoire de Marseille 2,pl Le Verrier F-13248 Marseille France
[2]Universidad Nacional Autonoma de Mexico, Instituto Astronomia. Apartado Postal 70-264, Ciudad Universitaria, CP-04510 Mexico,D.F.

Abstract: In 1990 we began an $H\alpha$ survey of the Milky Way and the Magellanic clouds at ESO La Silla with a 36 cm telescope. We discuss briefly the specialized instrumentation used for this survey (telescope, focal reducer, scanning Fabry-Perot interferometer, photon counting system). We present some aspects about the data reduction in order to give show the quality of our data, and finally images of the SMC showing both the intensity of HII regions and the velocity of the regions.

1 Introduction

The SMC and the LMC are the nearest external galaxies. Their position in a region of the sky of low reddening, their angular diameters, and the fact that both clouds are galaxies which have different properties (rotation, metallicity,etc.) compared with our Galaxy, makes them very attractive in the study of objects of a given specific class such as HII regions, SNRs, PNe, wind bubbles and superbubbles. On the other hand, the SMC reveals a quite complicated structure. Studies in HI, yellow stars, PNe, diffuse ionized hydrogen reveal that several velocity components are present along extended regions of the SMC. This complicated kinematical structure makes difficult to establish a rotational pattern for the SMC and consequently, to have a clear understanding of the properties of this galaxy; this difficulty has important consequences in the derivation of the mass of the SMC from its rotation curve. Furthermore, it is important to know the depth of the SMC, a rather controversial point, because if the SMC has a depth of about 40 kpc, the idea of studying a class of objects in the SMC and then, at the same distance, is non sense. We have studied several bubbles , HII regions and SNRs in the LMC by means of a scanning FP interferometer. This allowed us to know the kinematics of these objects with a large angular resolution (2".6/px) but with a modest angular field of view (8'). However, the limitations in the field of view make difficult the possibility of a photometric calibration of our observations because in order to calibrate properly the $H\alpha$ flux of a given region it is necessary to observe, in the same field, some other region already calibrated in its $H\alpha$ flux.

On the other hand, we wish to study specific classes of objects in the MCs, such as SNRs, wind bubbles, HII regions, superbubbles and PNe , to compare their properties with those of the objects of the same class located in other galaxies(in particular the Milky Way) and to study the interactions of these objects with the parent galaxy and then, to understand the energy exchanges between stars and gas in a galaxy.Thus, we require of a whole set of homogeneous observations which allow us both to have high sensitivity and to cover important regions in order to have a large number of objects in each galaxy. In addition, we wish to discriminate the kinematical components of the specific object from those of the extended galactic ones(as those discovered in the SMC and also in the LMC) and to study the kinematics of the diffuse hydrogen which permeates the SMC. These reasons motivate us to undertake a survey of the MCs with an equipment which insures high sensitivity to monochromatic H alpha emission, a large field of view and a spectral resolution comparable to our previous studies. In order to accomplish this we have used a small diameter telescope equipped with a focal reducer, a scanning FP interferometer and a photon counting camera. With this equipment we have covered the whole H alpha emission of the SMC and some selected regions of the LMC. The purpose of this paper is to report our first results derived from the observations of the SMC. In section 2 we will discuss about the equipment, the observations and the data reductions; in section 3 we will discuss the limiting sensitivity of our survey and the problem of the spurious sky lines emission ; in section 4 we will present maps of the $H\alpha$ emission of the SMC, built from several individual fields and presented in a mosaic form; in section 5 we will present the determinations of radial velocities of the DEM emission regions.

2 description of the instrument

For our Hα Survey we built a dedicated telescope and developed high performance software. The telescope of small diameter (36 cm) provides a large field of view (38'x38') on our detector. Because we observe extended sources, the collected flux is the same for a 0.36 m as for a 3.6 m diameter telescope. The loss in collecting area S is exactly compensated by the increase in the observed angular field Ω according to the throughput conservation principle ($S\Omega$ = constant). The limiting detection is thus the same for both large and small telescopes and in fact is most often set by our ability to subtract the parasitic nightskylines (OH 655.36nm , geocoronal $H\alpha$ 656.28nm , OH 656.88nm). Our 9" pixel size and wide field will enable us to observe the ionized hydrogen in the whole Galaxy in a manner directly comparable with radio observations.

This telescope is equipped with a focal reducer, a scanning Fabry-Prot interferometer and a Photon Counting Camera.The concept is similar to our CIGALE instrument (Boulesteix et al. 1984) already used successfully on several large telescopes. The main characteristics are given in Table 1. The telescope and its focal reducer were built at Marseille Observatory. Two scanning interferometers are avaible In these devices the spacing of the plates can be adjusted through a special electronics controller that drives piezo-electric spacers. The Image Photon Counting System - based on a photon counting camera manufactured by Thomson CSF (France) - allows us to analyse the photon events frame by frame with a time resolution of 1/50 s. The

Telescope
- Ritchey-Chretien type (for a large field)
- Primary mirror : 36 cm diameter
- Final aperture ratio F/D = 3.3
- Spatial resolution 9"/px
- Field of view 38' x 38'

Detector
- Photon counting camera composed of a microchannel plate intensifier
- electrostatically focused coupled to an SIT camera through optic fibres
- Number of pixels : 256 x 256
- Resolution : 1 px = 52 mm
- Dynamic range : 2 ev/px/h to 3000 ev/px/h
- Dark noise : 2 ev/px/h
- Time resolution 1/50 s for each frame
- Acquisition of the addresses of the centered events

Computer
- processor 68020, 16 MHz, 22 Moctets of memory
- color display 1024 x 768
- 120 Moctets disk
- 1/2" tape 4 density
- 1/4" streamer
- optocoupled parallel interface
- color printer
- video A/D converter Themis

Interference filters
-Wavelength Mainly Hα. Also available: [NII], [SII], Hb, [OIII]
- Bandpass 1.0 nm FWHM for each filter

Interferometers - 2 scanning Fabry-Perot. At Hα wavelength characteristics are:

- interference order for Ha	p = 796	p = 2604
- free spectral range	.82nm = 376 km/s	.25nm = 115 km/s
- spectral resolution	0.068nm = 31 km/s	0.025nm = 11.5 km/s
- typical scanning	24 steps	24 steps
- sampling step	0.034nm = 16 km/s	0.010nm = 5 km/s
- velocity accuracy	2 to 3 km/s	1 km/s

Typical exposure time
- 2h = 60 scanning sequences
- 1 scanning sequence = 5s per scanning step x 24 steps

Detection limit - 10-9 $Jm^{-2}\ s^{-1}\ sr^{-1}$ with a S/N ratio between 1 and 2 for a 15m exposure time

Fig. 1. Main characteristics

events are centered with special detection electronics. The huge amount of data (originally 240 Moctets/night when we recorded all the addresses of the detected events) and the systematic nature of this survey led us to develop a new acquisition system. High performance software allows the astronomer to alter the observational sequence (field identification, S/N ratio estimate, interferometer stability checking) and to have complete access to the reduced data while observing, for example to deconvolve and decompose profiles or to measure radial velocities in selected areas of the observed field. For each typical observation we obtain the following informations (all of which are accessible in real time on the TV monitor during the observation):

- 24 calibration interferograms from which we derive one phase-map giving the wavelength origin at each pixel . Instrumental profile and flat-field are also obtained from this series of calibration rings.
- 24 nebular interferograms of the observed field. From the calibration phase-map we can compute 24 $\lambda - maps$ analogous to the radio maps, to produce one data cube (x,y, λ). from these two cubes we derive:
- 65536 detailed $H\alpha$ profiles (one profile for each of the 256 x 256 pixels).
- one velocity map derived from the Doppler shift of the profile inside each pixel. This velocity map takes into account only the position of the barycenter of the profile and thus, disregards other velocity components that could be present in the profile.
- one monochromatic image computed by an algorithm which integrates the flux inside the emission line profile above from a fixed threshold.
- A continuum image of the field computed by integrating the flux of the profile below a fixed threshold.

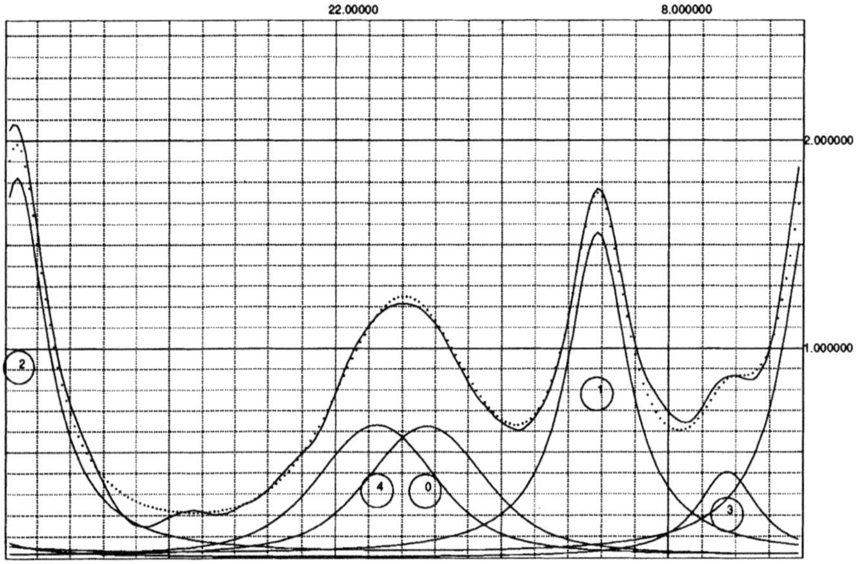

Fig. 2. Example of decomposition of profil seen in a empty field. Profiles 1,2 and 3 are nightsky emission lines and 4 and 0 are broad componants L at $155 kms^{-1}$ and H at $176 kms^{-1}$ with $5 erg.cm^{-6}.pc$ intensity

Most often we study rather large parts of the field of view chosen according to morphological and kinematical continuity criteria. Within such areas, containing typically 100 to 100 pixels, we add the fluxes found in the individual pixels so that the resulting Signal to Noise ratio is sufficient to allow detailed analysis of the profile (see figure 2). One can decompose the obtained velocity profiles in their gaussian components and deconvolve them from the instrumental function; one can correct the flat field, the continuum(obtained as mentioned above), one can substract the line sky emission, which allow us to show an emission morphology under the sky background.

3 Line Sky Parasitic Lines and Limiting Emission Measure

A shifted $H\alpha$ filter (lambda c= 656.9nm ; $\delta\lambda$=1nm) is more or less transparent to a series of night-sky emission lines which ultimately set the limit of detection of the nebular emission. Within this range we detect, in order of importance, geocoronal $H\alpha$ at 656.278nm , an OH nightglow line at 657.725nm and eventually, another OH line at 656.872nm. Figure 2 shows a typical profile obtained from our radial velocity data on a window of 10 x 10 px in one of the fields; the window is located in a region where no much nebulosity is seen in order to appreciate better the night-sky lines. As we can see, the radial velocities of the SMC nebulae are well located relative to the positions of the night-sky lines for this particular FP interferometer. This implies that, in general, night sky lines are easily discriminated and substracted from nebular emission for most of the velocity range of the SMC nebulosities. Extreme velocity nebulae are less easy to discriminate but, the fact that we can built a pattern of the night sky emission over the whole field help a lot in the sustraction of the night sky lines for these nebulae. Thus, in conclusion, we can detect easily most of the nebular emission even if it has intensities below the night sky value. Nebular emission at extreme velocities can be studied only if its intensity is comparable with that of the night sky lines. One of the results of this work is the improvement of the determination of the limiting $H\alpha$ surface brightness quoted in Amram et al.(1991) and Le Coarer et al.(1992). Indeed, under the asumption that the limiting surface brightness corresponds to that of the OH 657.7nm line, we obtain, from three independent methods , a value for the limiting Hα surface brightness to be detected of : $S(H\alpha)_{lim} = (1.1 \pm 0.3)10^{-15} ergcm^{-2}s^{-1}sr^{-1}$ corresponding to a limiting emission measure $EM_{lim} = (11.3 \pm 2.6)cm^{-6}pc$ (for a T=10^4 K nebula), reached in an exposure time of 8280 s, which confirms our previous estimate.

4 Results

4.1 Velocity field

For each profile a velocity can be determined by the barycenter, but this does not take into account the multiple components profiles. For a general view of the velocity field we have taken the row determination as we use to do in faraway galaxies, this velocity field is shown in the figure 3b. At first glance, the SMC is like a rotating bar, but the gradient is not continue in the bar and a cut along the bar shows some discontinuities already mentioned by Martin et al (1989) as four independent components (VL ,L ,H, VH)

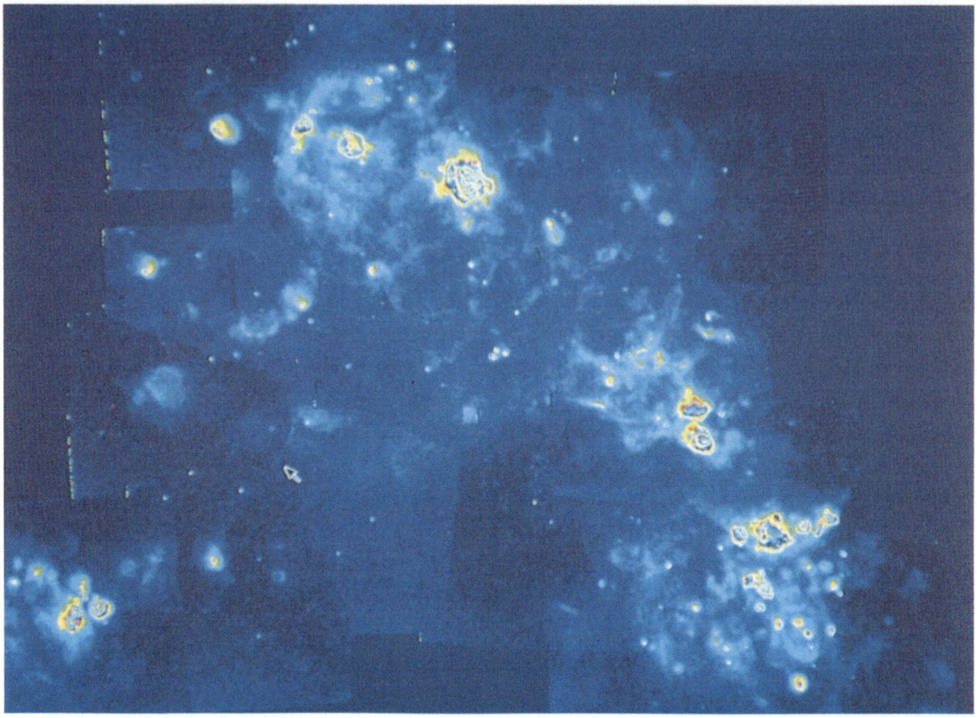

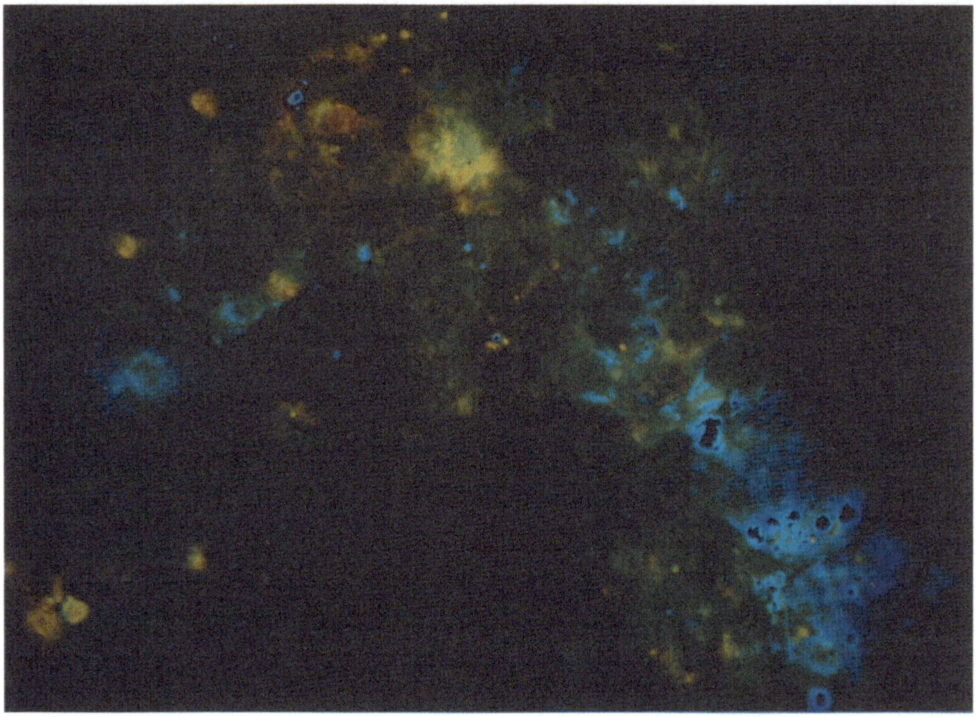

Fig. 3a. Monochromatic map of the SMC in the Hα line. The intensities are coded with a rainbow lookup table

Fig. 3b. Velocity field of the SMC, the color gives an idea of the value of the velocity, and the HII regions are shown by the intensity of the colour. dark blue= 110 kms^{-1}, light blue=140kms^{-1}, green=155kms^{-1} yellow=170 kms^{-1},red=190kms^{-1} white=205kms^{-1}

4.2 Monochromatique Emission

The monochromatic map of the SMC shows HII regions as Davies et al (1976). The limit of detection is 5 time better and some additionnal structure can be seen outside of the bar, these are well correlated with those seen in IRAS images in $60\mu m$ band, which can be explained by a heating process by the UV emission. A very diffuse emission can be detected over the whole SMC with a multi–component broad line structure. An example of this emission is given in figure 2. The figure 3a shows the image of the SMC given by mosaiquing the monochromatic map of 20 elementary fields. each fields has been expose 2 hours. The overlapping of fields gives a photometric continuity all over the SMC, and an absolute calibration is done comparing with absolute fluxes measured by Caplan and Deharveng (private communication).

4.3 HII Region Catalogue

We have completed the HII regions Catalog of Davies et al (1976), with an estimation of velocity, broadening value, absolute flux calibration, and spectral decomposition. Each region was mapped as the further example for DEM 35,37,38,43 association. Figure 4 shows a monochromatic image of this region, in figure 5a the we have the whole velocities map written over an isophot contour plot of the objets, and figure 5b gives a representation of these velocities over the same plot. A paper in preparation will present all results for this analysis.

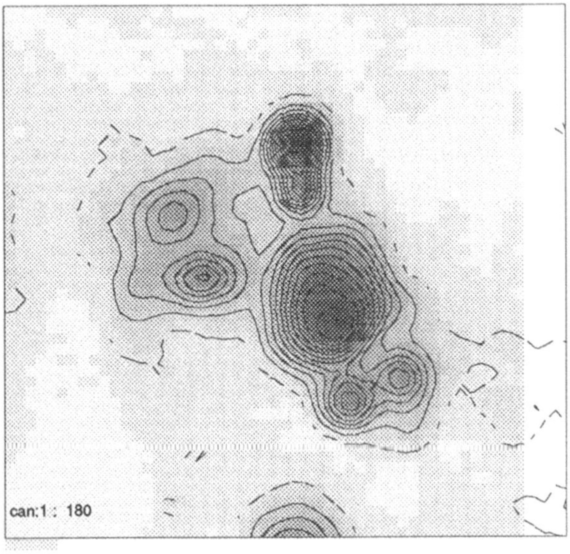

Fig. 4. monochromatic image of DEM 43 in Hα line

Fig. 5a. velocity field values in kms^{-1} over the monochromatic coutour

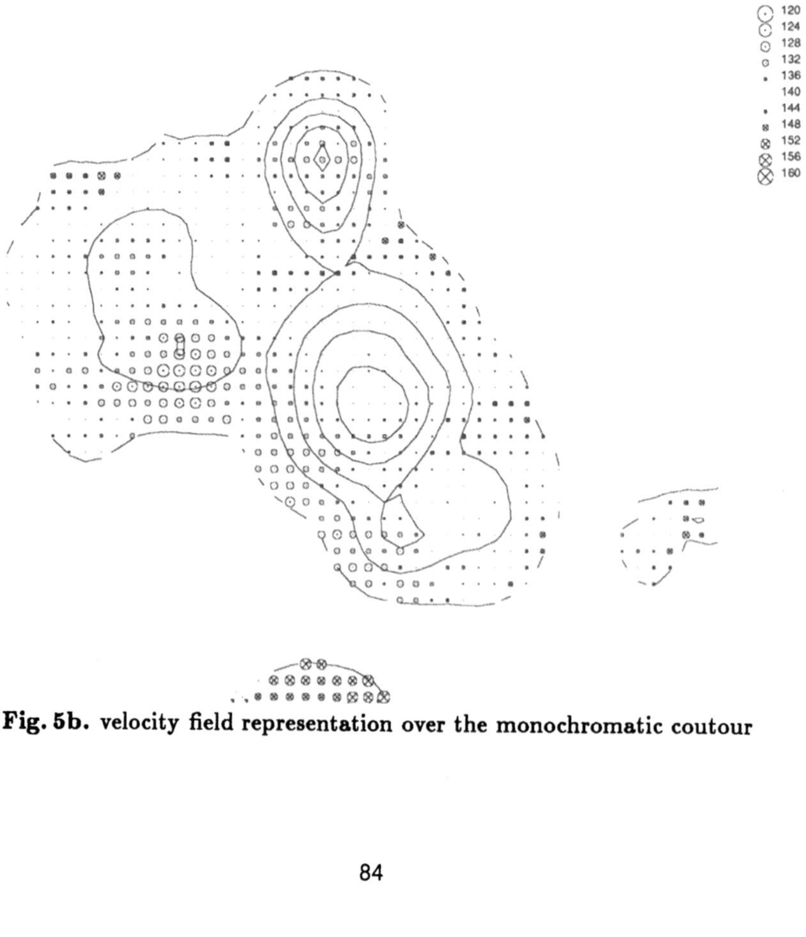

Fig. 5b. velocity field representation over the monochromatic coutour

5 Conclusion

We have presented very preliminary results of observations done on the SMC. definitive results will be publish in further paper as the HII region catalogue. Same observations for the LMC are in progress.

References

Amram ,P., Boulesteix, J., Georgelin, Y.P., Georgelin, Y.M., Laval, A., Le Coarer, E., Marcelin, M., Rosado, M.(1991): The Messenger

Boulesteix,J., Georgelin, Y.P., Marcelin, M., Monnet, G.(1984): SPIE Conf. Instrum. Astron. **445** 37

Davies, R.D., Eliott, K.H., Meaburn, J.,(1976): MemRAS**81** 89

Le Coarer, E., Amram ,P., Boulesteix, J., Georgelin, Y.M., Georgelin, Y.P., Marcelin, Joulie, P. , Urios, J., (1992):A&A bf 257 389

Martin, N., Maurice, E., Lequeux, J.(1989): A&A **215** 219

Survey of Carbon Stars and Emission Objects

Marc Azzopardi

Observatoire de Marseille, F-13004 Marseille, France

Abstract: The main purpose of this paper is to review the most extensive surveys for carbon stars and Hα emission-line objects of the Magellanic Clouds, with special emphasis on recent work.

1 Introduction

Because of their relative proximity, the Magellanic Clouds provide rather complete samples of the bright and most typical objects, thus are very suitable for detailed studies of populations of stars and nebulae. Since all sets of objects are located at the same relative distances with, generally, comparable low reddenings, they are unrivalled sites to assess the constraints to stellar evolution models.

Schmidt telescopes and astrographs equipped with objective-prisms have allowed systematic surveys (medium to low spectroscopic resolution) of the brightest objects exhibiting typical spectral features such as deep absorption- or emission-lines (see Azzopardi 1984, and references quoted therein), and molecular absorption bands. Among others, these are mainly blue and red supergiants, red giants, as well as emission-line stars (Be and Wolf-Rayet stars, planetary nebulae, etc.). However, those surveys were limited to the most luminous objects $M_V \lesssim -3$.

More recently, the use of large telescopes of the 4-m class, equipped with very low dispersion "nonobjective" gratings introduced by Hoag and Schroeder (1970) and improved by Bowen and Vaughan (1973) provided new opportunities for prospecting fainter Magellanic Cloud objects. For instance, the so called "grism" (grating on prism) devices allowed rather complete surveys of red giant stars, namely M-, S-, and carbon-type stars in both Clouds.

2 Field-Carbon Star Population of the MCs

Carbon stars have been searched for, both in the field and in intermediate-age clusters of the Magellanic Clouds. Since no recent work has been published on the carbon star detection in clusters, the paper by Azzopardi (1989) summarizing the main results on that field is still up to date, thus the present rewiew will be restricted to surveys of field carbon stars in the Magellanic Clouds.

Carbon stars as well as M giant stars are relatively cool and intrinsically bright objects lying on the asymptotic giant branch (AGB). Since carbon stars have C/O > 1, all oxygen present in the envelope and the atmosphere of the star is bound in carbon monoxide (CO) and the excess carbon forms molecules like C_2, CN, CH, etc.. Consequently, the carbon star surveys are made possible thanks to the strong molecular absorption bands present in their spectrum, both in the near-infrared spectral region through the CN-bands at 7945, 8125 and 8320 Å, and in the blue-green spectral domain via the pronounced Swan C_2-bands at 4737, 5165 and 5636 Å. Due to the lack of luminosity criteria, the membership of these carbon stars in the Magellanic Clouds is based on the very low probability of finding such kinds of faint objects at these relatively high galactic latitudes.

Early surveys of the field carbon stars were carried out in the Magellanic Cloud with Schmidt telescopes equipped with very low resolution objective-prisms; they were limited to stars brighter than $V \sim 16.5$, close to the typical magnitude of the carbon stars in the Large Magellanic Cloud (LMC), and completeness was not assured. Moreover, this kind of survey was not deep enough to allow the detection of carbon stars in the Small Magellanic Cloud (SMC).

The first survey for carbon stars in the Magellanic Clouds was carried out by Westerlund with the Uppsala Schmidt telescope at Mount Stromlo Observatory in the late 50s. The near-infrared objective-prism technique (2100 Å/mm dispersion at the atmospheric A-band; limiting magnitude $I \sim 13.5$) allowed him to identify a large number of red supergiant and giant stars in the LMC. Preliminary results by Westerlund (1960) give notice of the discovery of several hundred probable carbon stars but one had to await the paper by Westerlund and associates (1978) to get the identifications of 302 of these objects.

In the mean time, innovating the carbon star detection through their blue-green Swan C_2-bands, Sanduleak and Philip (1977) found 474 carbon stars in the LMC using the Curtis-Schmidt telescope at Cerro Tololo Inter-American Observatory (CTIO) and an objective-prism giving a dispersion of 1360 Å/mm at Hγ (plate limit $V \sim 16$). Although this detection missed the faintest, reddest objects, it, conversely, provided about 400 stars that escaped the former infrared survey, clearly showing the great efficiency of this novel carbon star survey technique.

Later on, Magellanic Cloud field carbon stars have also been searched for using transmission gratings (grisms) at the prime focus of large telescopes. With a limiting magnitude of about $V = 20$, these surveys insure carbon star detection in both Clouds with a reasonable degree of completeness and make statistical studies possible.

The near-infrared grism survey of selected regions at the prime focus of the CTIO 4-m telescope by Blanco, McCarthy and Blanco (1980), and Blanco and McCarthy (1983) , yielded a rather complete sampling of carbon stars in the Magellanic Clouds.

Very low spectroscopic observations of 52 circular fields of 0.12 deg^2 each (2350 Å/mm dispersion) in the LMC, and 28 circular fields of 0.12 deg^2 each plus 9 square fields of 0.38 deg^2 each (2350 Å/mm and 1700 Å/mm dispersion, respectively) in the SMC, resulted in the detection of about a thousand carbon star candidates in the LMC and 860 ones in the SMC. Identification of 849 cool-carbon LMC stars have been more recently published by Blanco and McCarthy (1990). From the integration of the surface densities of the carbon stars found in the different sample regions, Blanco and McCarthy (1983) estimated the total number of carbon stars to be 11 000 in the LMC and 2900 in the SMC.

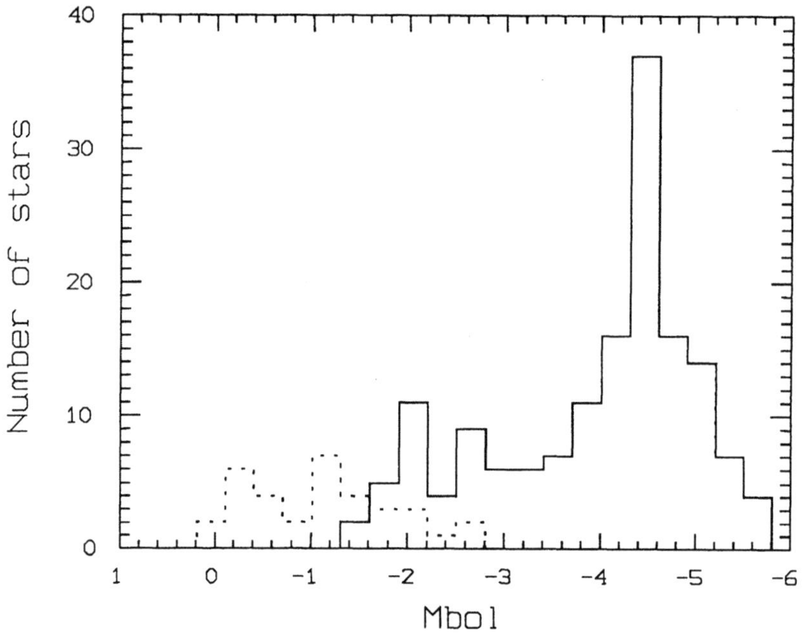

Fig. 1. Histograms (0.3 magnitude bin) showing the M_{bol} distribution of 155 SMC carbon stars with infrared photometry (full line) and of the 34 galactic carbon stars (dotted line)

Using a grism providing 2200 Å/mm dispersion in the blue-green spectral domain at the prime focus of the European Southern Observatory (ESO) 3.6-m telescope and the method pioneered by Sanduleak and Philip (1977), Rebeirot, Azzopardi and Westerlund (1992) found 1707 carbon stars in the main body of the SMC. In order to minimize the number of the overlaps and be able to survey even the most crowded parts of the SMC bar, they limited the length of the spectra to about 1000 Å. In that selected spectral range (4350–5300 Å) the pronounced C_2-band at 5165 Å makes easy the identification of the carbon stars (see Breysacher and Lequeux (1983) for more details on the observing technique they used). Unfortunately, the 13 partially overlapped grism fields covering the main body of the SMC do not extend enough towards its southern and eastern regions (see Fig. 1 in the paper by Azzopardi 1989). However, taking into account their surface distribution, Azzopardi and Rebeirot (1991) estimated the total number of the SMC carbon stars within the 75 carbon star isopleth of Blanco and McCarthy (1983) to be 2100; this number is somewhat larger than the estimate (1950 stars) made by the latter authors.

With the help of the spectrophotometry technique perfected by Westerlund, Azzopardi and Breysacher (1986), Rebeirot et al. (1992) measured, on their grism plates, a magnitude $m_{(5220)}$, a colour equivalent $m_{(4850)} - m_{(5220)}$ and the strength (equivalent width and depth) of the C_2-band at 5165 Å for most of the SMC carbon stars they identified. These quantities were used to discriminate the C_2-poor and the C_2-rich stars, and to recognize the J-type stars as well as a group of faint and relatively blue carbon star candidates. Subsequent *JHK* photometry and medium-resolution spectroscopy by Westerlund and associates (1992) resulted in the confirmation of the carbon star nature of the latter. Using a distance modulus of $(m - M)_0 = 18.9$, a mean colour excess $< E(B - V) >= 0.03$, and the relation by Wood et al. (1983) to compute the bolometric corrections $BC_K = 0.55 + 2.65(J - K)_0 - 0.67(J - K)_0^2$, they found a set of carbon stars with luminosities ($-2.8 \lesssim M_{bol} \lesssim -1.5$). Among these SMC carbon stars, those which are the least luminous ever found in a nearby galaxy, overlap with the most luminous ones of the galactic bulge. Although not representative of the luminosity function of the SMC carbon stars, the M_{bol} histogram built with 155 carbon stars – for which bolometric corrections BC_K have been calculated thanks to the available *J*,*K* photometry – is displayed in Fig. 1, together with the one that has been obtained for the 34 galactic bulge carbon stars presently known, using the data of Westerlund et al. (1991a) and the free-distance modulus $(m - M)_0 = 14.5$.

Table 1. Census of field carbon stars in local group galaxies

Galaxy	nb of C stars	total nb	references
LMC	849	11000	(1) (2)
SMC	1707	3100	(3) (4)
Fornax	77		(5)
Leo I	19		(6)
Leo II	12		(5) (6)
Carina	11		(6)
Sculptor	8		(6)
Draco	4		(6)
Ursa Minor	1		(6)
Sextans	0		(7)
Galactic bulge	34		(8)
Galactic halo	30		(9)

References to Table 1: (1) Blanco and McCarthy 1983; (2) Blanco and McCarthy 1990; (3) Rebeirot, Azzopardi and Westerlund 1992; (4) Azzopardi and Rebeirot 1991; (5) Azzopardi, Lequeux and Muratorio 1993; (6) Azzopardi, Lequeux and Westerlund 1986; (7) Irwin, Bunclark, Bridgeland and McMahon 1990; (8) Azzopardi, Lequeux and Westerlund 1991; (9) Green 1992 and references therein.

Comparison of the results (see the note added in proof by Westerlund et al. 1986, and the discussion by Blanco and McCarthy 1990) by the two observing techniques led to the conclusion that, obviously, near-infrared grism surveys favour the detection of the redder cool-carbon stars with strong near-infrared CN-bands, while conversely, the blue-green grism survey technique is better adapted for the identification of the bluer and C_2-rich ones. Thus, the two survey techniques together provide a high degree of completeness.

A review on carbon stars in dwarf spheroidal galaxies as been set out by Azzopardi and Lequeux (1992). Table 1 gives an up-to-date census of field carbon stars in those galaxies as well as for the galactic bulge and halo, and shows clearly that the Magellanic Clouds are, by far, the nearby galaxies containing the largest number of carbon stars, and consequently are ideal systems to study the evolution of this kind of object. Conversely, carbon stars are very useful objects to better understand chemical evolution, large-scale structure and kinematics (see for instance the paper by Hardy, Suntzeff and Azzopardi 1989) of the Magellanic Clouds.

3 Hα Emission-Line Object Population of the MCs

Hα emission-line surveys led to the identification of a large variety of objects. These are mainly OBe stars but also, among others, more interesting objects such as Of stars, P Cygni objects, Be supergiants with emission-lines of [FeII], [NiII], [OI], etc., symbiotic stars and planetary nebulae. Here is a quick overview of the most comprehensive surveys for Hα emission-line objects of the Magellanic Clouds.

Ω 3.1 Hα Emission-Line Stars and Nebulae

The first extensive survey of the Magellanic Clouds for Hα emission-line objects was made by Henize (1956) who found on objective-prism plates taken with the Mount Wilson 10-inch camera 172 emission-line stars and 221 nebulae in the LMC, as well as 65 emission-line stars and 117 nebulae in the SMC. Most of those nebulae were confirmed, later on, by the deep study of the Magellanic Cloud nebular complexes that was carried out by Davies, Elliott and Meaburn (1976). Then, deeper objective-prism surveys of the Magellanic Clouds with the ADH Schmidt telescope by Lindsay (1961, 1963), Lindsay and Mullan (1963), and Andrews and Lindsay (1964) resulted in the identification of 913 and 593 Hα emission-line objects in the LMC and the SMC, respectively. Also, an objective-prism survey with the CTIO Curtis-Schmidt telescope by Bohannan and Epps (1974) yielded 625 stellar-like objects in the LMC with Hα line in emission.

More recently, a new survey for Hα emission-line stars and small nebulae of the SMC has been carried out by Meyssonnier and Azzopardi (1991, 1993) using the CTIO Curtis-Schmidt telescope equipped with a 10° objective-prism providing 420 Å/mm dispersion at Hα and a 110 Å bandwith Hα + [NII] interference filter. Figure 2 shows a part of a plate. Exposures from 30-minutes to 4-hours on hypersensitized IIIa–F plates allowed them to identify 1898 Hα emission-line objects, almost quadrupling the number of those found by Lindsay (1961) in the same area. Among these objects, 1742 are stars, 65 are planetary nebulae or very low excitation objects (VLE), and 45 are compact HII regions. The surface distribution of the SMC emission-line stars has previously been displayed by Meyssonnier and Azzopardi (1991) in their Fig. 1.

Cross-identifications of the stars found by this survey with those identified by Rebeirot et al. (1992), resulted in the identification of some carbon stars showing Hα in emission, which are presumably SMC symbiotic stars. The nature of three of them has been confirmed by medium resolution spectroscopy (Westerlund et al. 1991b). Taking into account the three ones discovered by Morgan (1992) and those listed by Allen (1984), the total number of confirmed possible SMC symbiotic stars is at present 9, five of them having carbon star donors.

Ω 3.2 Planetary Nebulae

The planetary Nebula (PN) systems of the Magellanic Clouds are of special interest because they provide sets of objects that make possible the study of the late stages of stellar evolution, also providing important clues about the enrichment of the interstellar medium. Consequently, several extensive surveys for planetary nebulae in the Magellanic Clouds have been carried out for a long time.

Early surveys for planetary nebulae in the LMC (Westerlund and Rodgers 1959; Lindsay and Mullan 1963), as well as in the SMC (Lindsay 1961), resulted in the identification of several probable PN candidates in both Clouds. Subsequent observations by Westerlund and Smith (1964) confirmed the nature of 42 LMC PN listed by Lindsay and Mullan, while Henize and Westerlund (1963), using higher resolution direct photographs, concluded that the most luminous SMC PN candidates found by Lindsay were compact HII regions.

Later on, Sanduleak, MacConnell and Philip (1978) carried out a systematic objective-prism survey (1360 Å/mm and 580 Å/mm dispersion at Hγ, 420 Å/mm dispersion at Hα) for planetary nebulae with the CTIO Curtis-Schmidt telescope. Looking for objects showing up strongly due to the emission either from $\lambda\lambda$5007, 4959 [OIII] or Hα and λ6584 [NII], they found 102 and 28 PN candidates in the LMC and in the SMC, respectively. Then, the very deep survey by Jacoby (1980) of small areas of the Magellanic Cloud's bars, using the narrow-band photometry technique at the prime focus of the CTIO 4-m telescope, "on" and "off" the [OIII] (5007 Å) and Hα lines, resulted in the detection of 41 PN candidates (33 new) in the LMC and 27 ones (19 new) in the SMC. However, subsequent spectroscopic observations by Boroson and Liebert (1989) confirmed the nature of 19 PN in the LMC and 11 PN in the SMC only among the new objects detected by Jacoby.

More recently, a new objective-prism survey with the CTIO Curtis-Schmidt telescope allowed Sanduleak (1984) to detect 25 additional PN candidates (no finding charts) in the central region of the LMC. Also, the deep objective-prism surveys (800 Å/mm dispersion near Hγ) by Morgan and Good (1985, 1992) with the UK 1.2-m Schmidt telescope (UKST) resulted in the discovery of 86 new PN candidates in the LMC and 10 new planetary nebulae in the SMC. Finally, the objective-prism survey by Meyssonnier and Azzopardi (unpublished) quoted above, led to the identification of 18 new faint PN candidates in the SMC, the nature of 15 of them being already confirmed by subsequent medium resolution slit spectroscopy.

Taking into account the 30 additional candidate PN found by Morgan (1993) in the outer regions of the LMC and the 15 new planetary nebulae discovered in the central

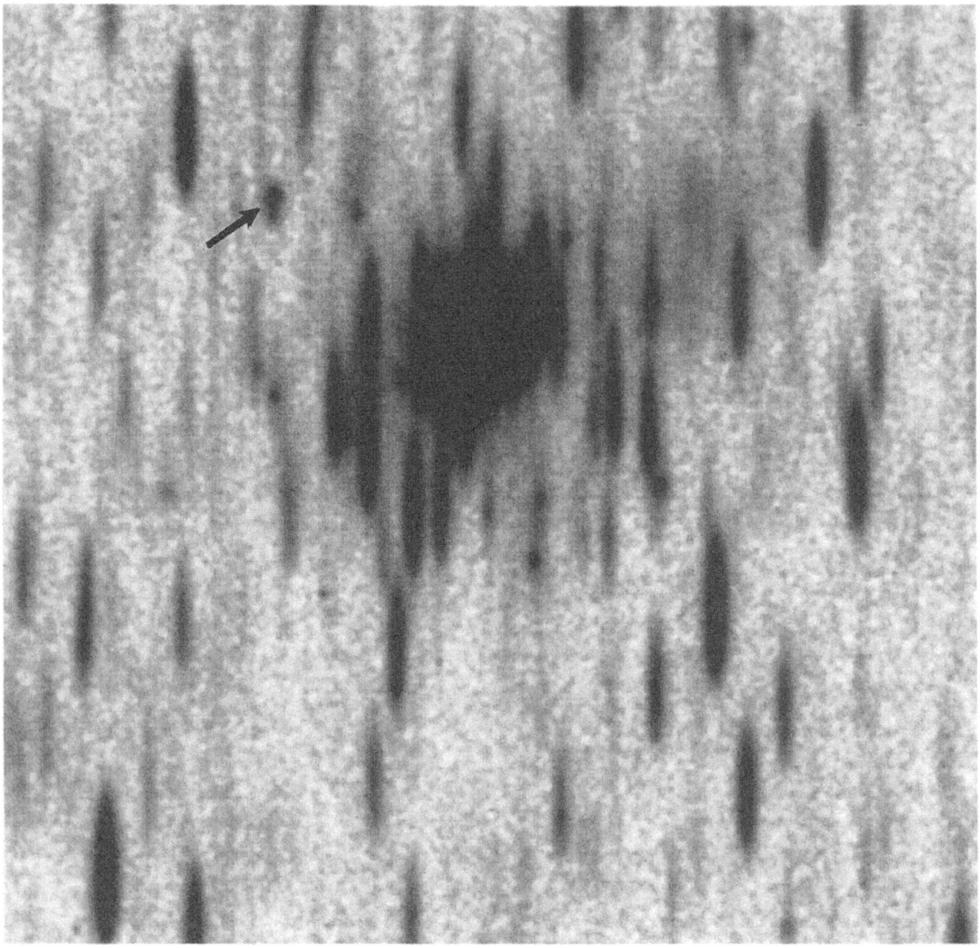

Fig. 2. Hα emission-line objects in the SMC globular cluster NGC 330 and its vicinity. Scan of a part of a CTIO Curtis-Schmidt objective-prism plate (forming gas baked IIIa–F emulsion, 4-hour exposure, 420 Å/mm dispersion at Hα and 110 Å bandwidth interference filter). The spectrum, marked with an arrow, showing both Hα and N[II] 6548–6583 Å in emission corresponds to the planetary nebula L 305 = SMP 21 (Lindsay 1961; Sanduleak et al. 1978)

regions of the SMC by Meyssonnier and Azzopardi (1991), the present number of detected PN/VLE the Magellanic Clouds is about 240 for the LMC and 75 for the SMC. These numbers have to be compared with the estimates by Boroson and Liebert (1989) of the total number of planetary nebulae in the LMC (643) and in the SMC (116) from an extrapolation according to Jacoby's formula (1980). Deeper surveys are needed to improve the planetary nebula statistics in the Magellanic Clouds and to better derive their spatial distribution as well as their luminosity function.

4 New Survey Technique

From the experience we gained from low and very low dispersion objective-prism and transmission grating spectroscopy in combination with intermediate and broad band-pass filter, we are now using the ESO Faint Object Spectrograph and Camera (EFOSC) in the field spectroscopy mode (slitless) at the Cassegrain focus of the ESO 3.6-m telescope in order to carry out very deep surveys in selected areas of the Magellanic Clouds. According to the type of object to be detected we are using various dispersions in combination with selected spectral ranges. When performing very low-resolution spectroscopic surveys (dispersions $\gtrsim$ 500 Å/mm), we are now using prisms instead of grisms in order to avoid the disturbing images corresponding to the different grating orders, mainly the "zero" order. Concurrently, we are also developing in the Munich Image Data Analysis System (MIDAS) environment (see Muratorio and Azzopardi, 1992a,b), a semi-automatic procedure to select objects that can be identified through either their emission-lines or molecular absorption bands. These are mainly Be and Wolf-Rayet stars, planetary nebulae and compact HII regions as well as cool giant stars such as carbon stars.

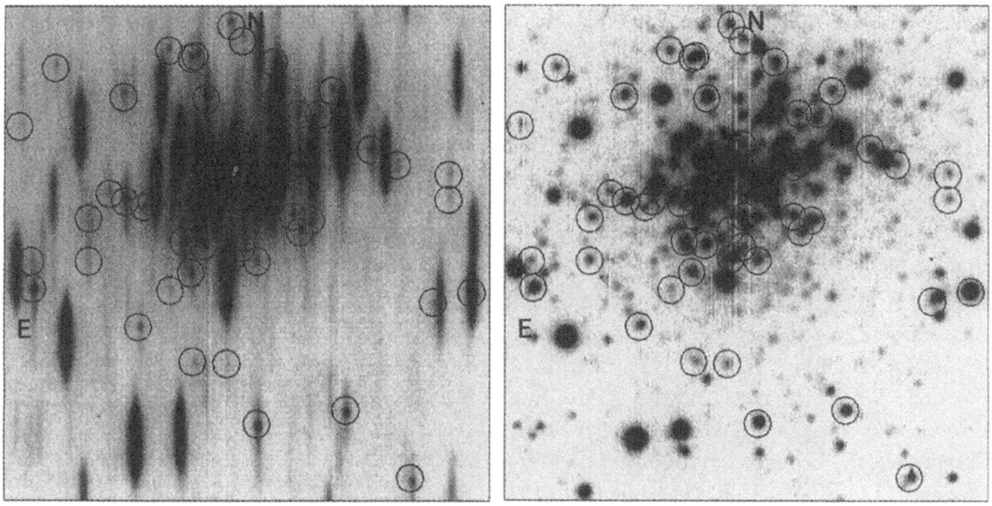

Fig. 3. Hα emission-line objects in the SMC globular cluster NGC 330. *Left:* Slitless spectroscopy with EFOSC (900 seconde exposure) through the red 300 grism (270 Å/mm dispersion) and an Hα + N[II] interference filter (6569 Å/100 Å); Encircled emission-line objects show up clearly due to the Hα line. *Right:* Finding chart (Hα + N[II] interference filter, 30 seconde exposure) for the Hα emission-line stars identified on the left-hand image.

Although the field of view of the EFOSC CCD camera is presently reduced to few ($\sim$ 20) square arcminutes, this survey technique is very efficient to survey very faint carbon stars or planetary nebulae. However, the latter is particularly well suitable for the search for typical objects in very crowded small fields such as some Magellanic

Cloud OB associations or globular clusters. For instance, this survey technique alowed us to identify, free from doubt, 15 Hα emission-line stars in the SMC globular cluster NGC 376 (Muratorio and Azzopardi, 1992b) as well as about 60 Hα emission-line stars in the youg SMC globular cluster NGC 330 (Fig. 3), while only 10 were previously listed by Feast (1972). The comparison of Fig. 2 with Fig. 3 is obviously self explanatory ! We plan to use this technique to survey faint carbon stars and planetary nebulae in nearby galaxies.

References

Allen, D.A. (1984): Proc. Astron. Soc. Austr. 5, 369

Andrews, A.D., Lindsay, E.M. (1964): Irish Astron. J. 6, 241

Azzopardi, M. (1984): "Narrow Spectral Range Objective-Prism Technique Applied to a Search for Small Magellanic Cloud Members", in Proc. IAU Coll. 78, ed. by M. Capaccioli (D. Reidel Publishing Company, Dordrecht), pp. 351–354

Azzopardi, M. (1989): "Surveys of Stars in Advanced Stages", in *Recent Developments of Magellanic Cloud Research* , ed. by K.S. de Boer, F. Spite and G. Stasinska (A European Colloquium, Paris), pp. 57–64

Azzopardi, M., Lequeux, J. (1992): "Surveying Carbon Stars in the Dwarf Spheroidal Galaxies", in Proc. IAU Symp. 149, ed. by B. Barbuy and A. Renzini (Kluwer Academic Publishers, Dordrecht), pp. 201–205

Azzopardi, M., Lequeux, J., Muratorio, G. (1993): Astron. Astrophys. Suppl. in preparation

Azzopardi, M., Lequeux, J., Westerlund, B.E. (1986): Astron. Astrophys. 161, 232

Azzopardi, M., Lequeux, J., Westerlund, B.E. (1991): Astron. Astrophys. Suppl. 88, 265

Azzopardi, M., Rebeirot, E. (1991): "Large-Scale Structure and Kinematics of the Magellanic Clouds from Carbon Star Studies", in Proc. IAU Symp. 148, ed. by R. Haynes and D. Milnes (Kluwer Academic Publishers, Dordrecht), pp. 71–76

Blanco, V.M., McCarthy, M.F. (1983): Astron. J. 88, 1442

Blanco, V.M., McCarthy, M.F. (1990): Astron. J. 100, 674

Blanco, V.M., McCarthy, M.F., Blanco, B.M. (1980): Astrophys. J. 242, 938

Bohannan, B., Epps, H.W. (1974): Astron. Astrophys. Suppl. 18, 47

Boroson, T.A., Liebert, J. (1989): Astrophys. J. 339, 844

Bowen, I.S., Vaughan, A.H.,Jr. (1973): Publ. Astron. Soc. Pacific 85, 175

Breysacher, J., Lequeux, J. (1983): The Messenger 33, 21

Davies, R.D., Elliott, K.H., Meaburn, J. (1976): Mem. Roy. Astron. Soc. 81, 89

Feast, M.W. (1972): Monthly Notices Roy. Astron. Soc. 159, 113

Green, P.J. (1992): Ph.D. Thesis, University of Washington

Hardy, E., Suntzeff, N.B., Azzopardi, M. (1989): Astrophys. J. 344, 210

Henize, K.G. (1956): Astrophys. J. Suppl. 2, 315

Henize, K.G., Westerlund, B.E. (1963): Astrophys. J. 137, 747

Hoag, A.A., Schroeder, D.J. (1970): Publ. Astron. Soc. Pacific 82, 1141

Irwin, M.J., Bunclarck, P.S., Bridgeland, M.T., McMahon, R.G. (1990): Monthly Notices Roy. Astron. Soc. 244, 16P

Jacoby, G.H. (1980): Astrophys. J. Suppl. 42, 1

Lindsay, E.M. (1961): Astron. J. 66, 169

Lindsay, E.M. (1963): Irish Astron. J. 6, 127

Lindsay, E.M., Mullan, D.J. (1963): Irish Astron. J. 6, 51

Meyssonnier, N., Azzopardi (1991): "A New Survey for Hα Emission-Line Stars and Related Objects in the SMC", in Proc. IAU Symp. 148, ed. by R. Haynes and D. Milnes (Kluwer Academic Publishers, Dordrecht), pp. 196–197

Meyssonnier, N., Azzopardi (1993): Astron. Astrophys. Suppl. in preparation

Morgan, D.H. (1992): Monthly Notices Roy. Astron. Soc. in press

Morgan, D.H. (1993): private communication

Morgan, D.H., Good,, A.R. (1985): Monthly Notices Roy. Astron. Soc. 213, 491

Morgan, D.H., Good,, A.R. (1992): Astron. Astrophys. Suppl. 92, 571

Muratorio, G., Azzopardi, M. (1992a): "Reduction of Slitless Spectroscopy Data", in Proc. 4th ESO/ST–ECF Data Analysis Workshop, ed. by P.J. Grosbol (European Southern Observatory, Garching), in press

Muratorio, G., Azzopardi, M. (1992b): these proceedings

Rebeirot, E., Azzopardi, M., Westerlund, B.E. (1992): Astron. Astrophys. Suppl. in press

Sanduleak, N. (1984): "A More Complete Listing of Planetary Nebulae", in Proc. IAU Symp. 108, ed. by S. van den Bergh and K.S. de Boer (D. Reidel Publishing Company, Dordrecht), pp. 231–232

Sanduleak, N., Philip, A.G.D. (1977): Publ. Warner and Swasey Obs. 2, 105

Sanduleak, N., MacConnell, D.J., Philip, A.G.D. (1978): Publ. Astron. Soc. Pacific 90, 621

Westerlund, B.E. (1960): Uppsala Astron. Obs. Ann. 4, No.7

Westerlund, B.E., Rodgers, A.W. (1959): The Observatory 79, 132

Westerlund, B.E., Smith, L.F. (1964): Monthly Notices Roy. Astron. Soc. 127, 449

Westerlund, B.E., Azzopardi, M., Breysacher, J. (1986): Astron. Astrophys. Suppl. 65, 79

Westerlund, B.E., Azzopardi, M., Breysacher, J., Rebeirot, E. (1991b): Astron. Astrophys. Suppl. 91, 425

Westerlund, B.E., Azzopardi, M., Breysacher, J., Rebeirot, E. (1992): Astron. Astrophys. Letter in press

Westerlund, B.E., Lequeux, J., Azzopardi, M., Rebeirot, E. (1991a): Astron. Astrophys. 244, 367

Westerlund, B.E., Olander, N., Richer, H.B., Crabtree, D.R. (1978): Astron. Astrophys. Suppl. 31, 61

Wood, P.R., Bessel, M.S., Fox, M.W. (1983): Astrophys. J. 272, 99

Detection of Faint Hα Emission-Line Objects and Carbon Stars in the Magellanic Clouds

Gerard Muratorio, Marc Azzopardi

Observatoire de Marseille, F-13004 Marseille, France

Observing and Reduction Techniques

CCD observations in the slitless spectroscopy mode were carried out with the ESO Faint Object Spectrograph and Camera (EFOSC) to survey Hα emission-line objects. For this purpose, direct images were secured through an interference filter (6563 Å/120 Å) while spectroscopy frames were obtained using a grism (270 Å/mm dispersion) and the same filter in order to reduce the length of the spectra and keep the number of overlaps as low as possible. The same technique was also used for a carbon star search; data were obtained with a Gunn R for direct imaging, and through a prism (1500 Å/mm dispersion) and a broad-band interference filter (4900 Å/1000 Å) for spectroscopy. More details and references on the observing technique are given in the paper by Azzopardi (these proceedings).

The semi-automatic procedure we used to reduce the spectroscopic data has been previously described by Muratorio and Azzopardi (1992). Briefly, the reduction technique involves in computing correlation factors between various normalized reference spectra and each background corrected normalized spectrum of the spectroscopy frame. The position of each spectrum is previously determined automatically by means of the direct frame. The selected spectra are sorted in order of the value of the smallest correlation factor. Then the shapes of the spectra are visually checked in order to avoid spurious objects. The main advantage of this technique is that the basic preselection of the objects – which is very time consuming on the conventional way – is done through impersonal criteria, while the final choice is nevertheless made visually so that "what we get is what we see".

References

Muratorio, G., Azzopardi, M. (1992): 4th ESO/ST-ECF Data Analysis Workshop, in press

Sanduleak, N., MacConnell, D.J., Philip, A.G.D. (1978): Publ. Astron. Soc. Pacific 90, 621

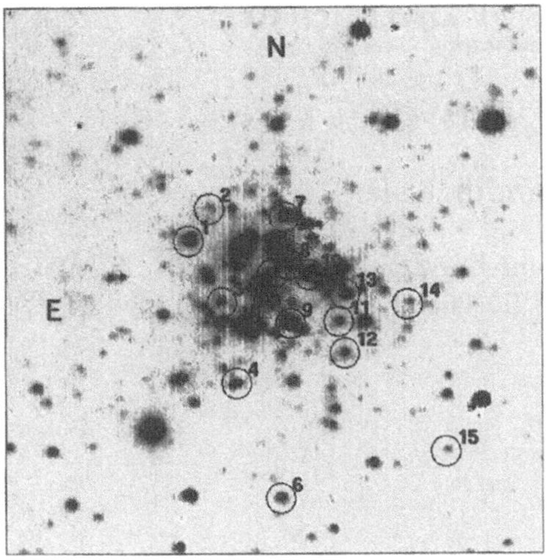

Fig. 1a. Hα emission-line stars identified in the centre of the SMC globular cluster NGC 376

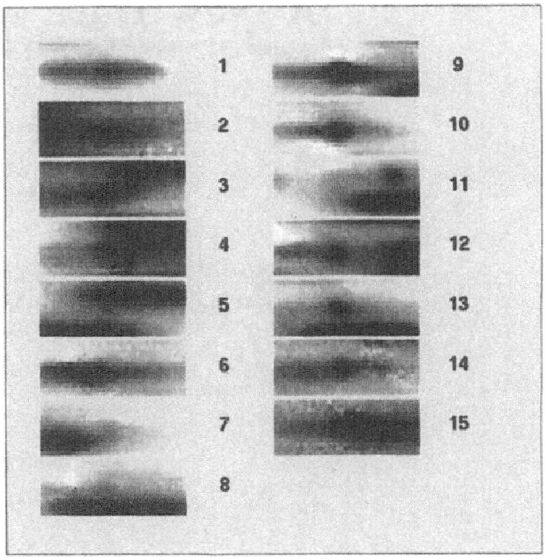

Fig. 1b. Corresponding individual spectra (270 Å/mm dispersion) extracted from the slitless spectroscopy frame of NGC 376

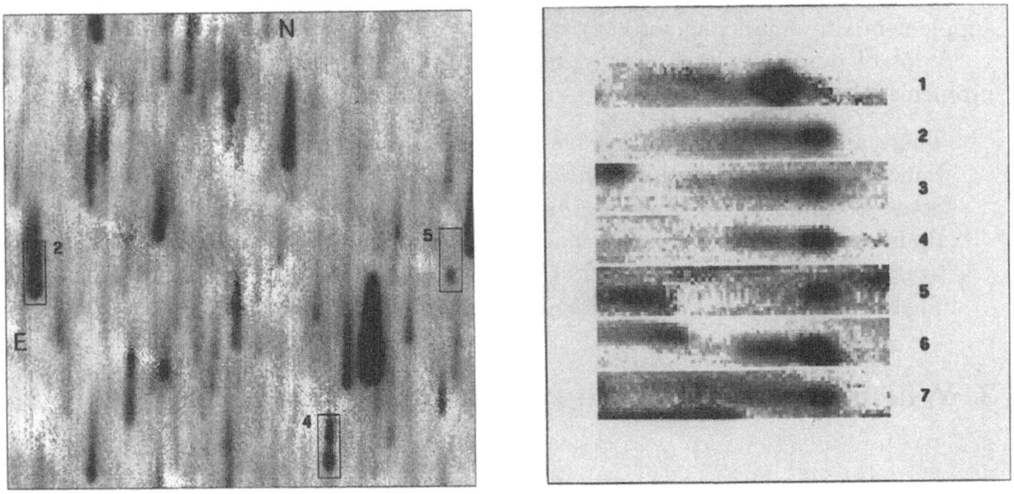

Fig. 2. *Left*: Part of a slitless spectroscopy frame in the SMC (prism giving 1500 Å/mm dispersion). *Right*: Individual spectra extracted from the left-hand image. Object No.1 is the planetary nebula SMP 10 (Sanduleak et al. 1978) exhibiting Hβ and [O III] in emission. The other ones are carbon stars; the depression in the continuum is due to the Swan C_2 -band at 5165 Å

A New Data Base of LMC Stars

Josef Gochermann, Michael O. Oestreicher
and Theodor Schmidt-Kaler

Astronomisches Institut der Ruhr-Universität Bochum,
Postfach 102148, D-W-4630 Bochum 1, Germany

1. Why a new data base ?

Complete data samples of stars in the general area of the Magellanic Clouds are necessary for HRDs useful for understanding the evolution of stars in the MCs and to investigate the galactic foreground. The systematic listing of MC members and galactic foreground stars has been initiated by Fehrenbach & Duflot (1970) and was supplemented by numerous authors.

A lot of new data have been published during the last 20 years but there are problems to combine them :

- There are many stars which have never been observed individually to get photometries, spectral types, luminosities etc. (see Table 1 for detailed listing)
- Some of the published data are uncertain (e.g. photometries in the catalogue of Rousseau et al. (1978) have in part been calculated from various measurements by simply taking the mean even if the data are different up to 0.5 mag)
- There are only a few cross-references for stars in different catalogues.

2. Which informations are stored in the data base ?

In a first step we try to get as much information as possible about stars down to 15-16 mag in the *LMC*. At present informations on more than 2600 galactic foreground stars, more than 2500 members and 173 stars with uncertain membership are stored (Table 1). Including our new UBV photometries of 960 foreground, 233 member and 69 uncertain stars good photometry now exists for 61% of the catalogued stars. In addition many unreliable old measurements were supplemented by new observations. A further campaign is planned for November 1992–February 1993.

Additionally the following informations will be stored: E_{B-V}, E_{U-B}, Stromgren $uvbyH\beta$, M_V, $T_{\rm eff}$, radial velocity, proper motion, distance, and special qualities.

The interstellar extinction split up into foreground and internal extinction will be determined for all stars, as well as their physical parameters.

Table 1. Statistics of some entries in the LMC data base (June 1992)

	Foreground No.	%	Member No.	%	Uncertain No.	%	Total No.	%
Number of Stars	2612	49	2542	48	173	3	5327	100
HDE Number	1881	72	305	12	81	47	2267	43
Hodge&Write entry	2365	91	1798	71	171	97	4334	81
Other Identifier	2549	97	2517	99	167	97	5233	98
α, δ	2586	99	2517	99	168	97	5271	99
V_{phot}	1855	71	2237	88	104	60	4196	79
Spectral Type	340	13	2339	92	126	73	2805	53
Luminosity Class	287	11	1347	53	55	32	1689	32
V	1330	51	1779	70	151	87	3260	61
B-V	1313	50	1778	70	151	87	3242	61
U-B	1275	49	1769	70	148	86	3192	60

3. How to use the data base

At present we are collecting *published* and *unpublished* data of stars in the MC or in their direction. **All colleagues are asked for supplements of the data sample.**

The data base is running on the system LARS (MS-DOS) on a normal 386 PC. The data sample will be published

1. in **digital** format (disk):
   - Version 1 (raw data) ≈ early 1993
   - Version 2 (with calculated supplements) ≈ end of 1993
2. as **printed catalogue** ≈ end of 1993

Until the catalogue is published colleagues who are interested in informations about *individual* stars may send their requests to J.G. Please indicate your interest for the complete catalogue in the same way.

Acknowledgements

This project is partly supported by the Deutsche Forschungsgemeinschaft grant Schm 160/39-1.

References

Fehrenbach Ch., Duflot M. (1970): A&ASS 1

Rousseau J., Martin N., Prévot L., Rebeirot R., Robin A., Brunet J.P. (1978): A&AS 31, 243

A correlation between Hα-extinction-excess and Hα surface brightness

Cong Xu

Max-Planck-Institut für Kernphysik, Postfach 103980, D-6900 Heidelberg

In the procedure of extinction correction for a Hα map of the LMC (Xu et al. 1992), we found a significant correlation between the Hα-extinction-excess $E(A_{H\alpha})$, defined by the difference between the measured extinction and the extinction estimated from the column density of neutral atomic and molecular gas, and the Hα surface brightness $B_{H\alpha}$.

The average foreground reddenning due to Galactic dust $E(B-V)_G$ and the average neutral-gas-to-reddenning ratio of the LMC $N(HI+H_2)/E(B-V)$ are determined through a linear regression of the column density of neutral atomic and molecular gas versus reddenning for a sample of stars which are in the vicinity of HII regions (Caplan and Deharveng 1986). Each point in **Fig.1** corresponds to a pair of averaged values, over a small region ($\sim 20'$ in size) centered on a HII region, of $E(B-V)_{star}$ and of $N(HI+H_2)$. Usually there are more than 10 stars in each of these small regions. It is assumed that half of the neutral atomic and molecular gas detected along the line of sight is between us and the sources. The linear correlation coefficient between the two quantities is **r** = **0.80**. The linear regression (the solid line) gives:

$$E(B-V)_G = 0.08 \pm 0.03 \quad \text{mag} \tag{1}$$

$$\frac{N(HI+H_2)}{E(B-V)} = 1.8(\pm 0.3)\, 10^{22} \qquad \text{cm}^{-2}\text{mag}^{-1} \tag{2}$$

Figure 2 plots Hα-extinction-excess $E(A_{H\alpha})$ versus the extinction-corrected Hα surface brightness $B_{H\alpha}$ (in $10^{-10}\text{erg cm}^{-2}\text{s}^{-1}\text{arcmin}^{-2}$) for a sample of 19 HII regions, also taken from Caplan and Deharveng (1986). By definition, the Hα-extinction-excess is

$$E(A_{H\alpha}) = A_{H\alpha} - 2.4 \times \left(0.08 + \frac{N(HI+H_2)}{1.8\, 10^{22}}\right), \tag{3}$$

where $A_{H\alpha}$ is the extinction estimated from thermal radio data. The linear correlation coefficient between the logarithms of $E(A_{H\alpha})$ and of $B_{H\alpha}$ is **r** = **0.71**. The linear regression (the solid line) gives:

$$\log E(A_{H\alpha}) = -0.29(\pm 0.13) + 0.55(\pm 0.20) \times \log\left(\frac{B_{H\alpha}}{10^{-10}\,\text{erg cm}^{-2}\text{s}^{-1}\text{arcmin}^{-2}}\right). \tag{4}$$

Interpretation: The following effects may be responsible to the Hα-extinction-excess: (1) extinction due to dust associated with ionized gas; (2) satuation of HI line; (3) underestimate of the column density of clouds of neutral atomic and molecular gas due to smearing effect of large beams. The correlation shown in Fig.2 indicates

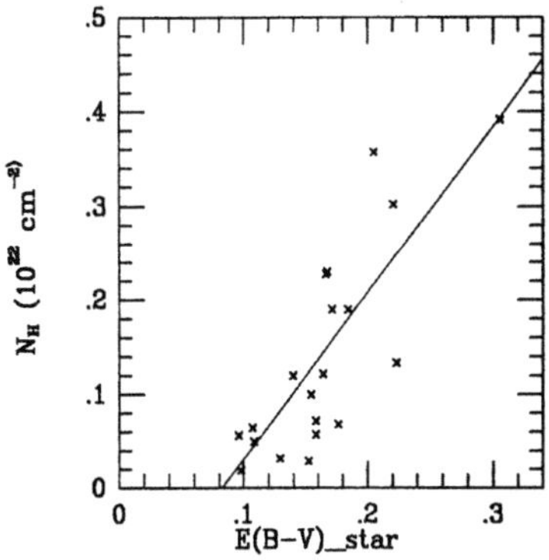

Fig. 1. Column density of neutral atomic and molecular gas versus reddenning for a sample of stars which are in the vicinity of HII regions (Caplan and Deharveng, 1986).

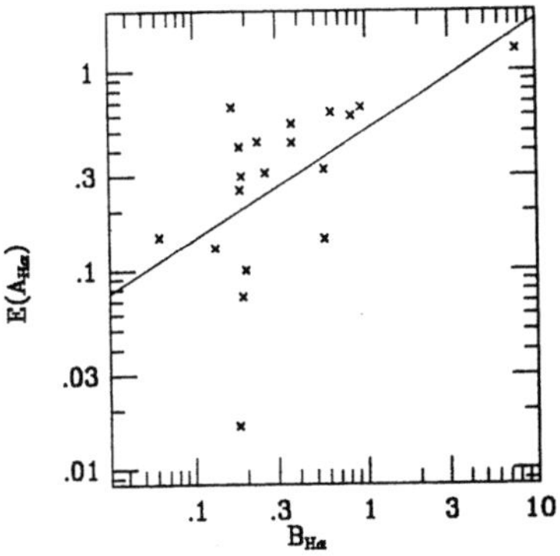

Fig. 2. Correlation between Hα-extinction -excess $E(A_{H\alpha})$ and extinction-corrected Hα surface brightness $B_{H\alpha}$ (in 10^{-10} erg $cm^{-2}s^{-1}arcmin^{-2}$).

that the extinction due to dust associated with ionized gas is the major effect. Assuming $N_e \simeq N(H^+)$ and $T_e \simeq 10^4$K, the relation between the $E(A_{H\alpha})$ and $B_{H\alpha}$ found in Fig.2 (i.e. EQ(4)) can be roughly approximated by:

$$E(A_{H\alpha}) \sim 0.004 \times EM^{0.5} , \tag{5}$$

where EM is the Hα emission-measure in cm^{-6} pc ($B_{H\alpha} = 10^{-10}$ erg $cm^{-2}s^{-1}arcmin^{-2}$ corresponding to EM $\simeq 1.4\ 10^4$). Accordingly, the ionized-gas-to-$A_{H\alpha}$ ratio in the LMC is estimated:

$$\frac{N(H^+)}{E(A_{H\alpha})} \sim 7\ 10^{21} \times k^{-1}\sqrt{\frac{D}{80pc}} \qquad (cm^{-2}mag^{-1}), \tag{6}$$

where $k = \sqrt{\langle n(H^+)^2\rangle}/\langle n(H^+)\rangle$ is the clumping factor ($k \gtrsim 1$), D ~ 80pc (corresponding to ~ 5′) is the typical size of the giant HII regions in Caplan and Deharveng sample. This is very close to the corresponding ratio for the neutual atomic and molecular gas in the LMC (assuming $A_{H\alpha}/E(B-V) = 2.4$) :

$$\frac{N(HI + H_2)}{A_{H\alpha,HI+H2}} = 7.5(\pm 1.3)\ 10^{21} \qquad (cm^{-2}mag^{-1}). \tag{7}$$

References

Caplan, J., Deharveng, L (1986): A&A **155** 297

Xu, C., Klein, U., Meinert, D., Wielebinski, R., Haynes, R.F. (1992): A&A **257** 47

Interstellar Matter, Planetary Nebulae

Monique and François Spite, Étienne Le Coarer

Klaas S. de Boer, (Kristen Rohlfs), (Ulrich Klein), Michael W. Feast, Stephen C. Russell

Recent Studies of Gas and Dust in the Magellanic System

Ulrich Mebold, Uwe Herbstmeier

Radioastronomisches Institut der Universität Bonn, Auf dem Hügel 71,
D-5300 Bonn 1, Fed. Rep. of Germany

Abstract: A brief summary of previous work emphazises the evidence for an origin of the Magellanic Stream (MS) by ram-pressure sweeping of gas from the Magellanic Clouds (MCs) and the Inter Cloud Region (ICR) by coronal galactic gas. Further we present results of recent investigations of the physical conditions of the diffuse gas and dust in the halo of the MCs and the MS. H I emission and absorption studies, carried out with the Parkes 64-m telescope and the Australia Telescope (AT) indicate that the spin temperature of the H I gas in the halo of the MCs and in the MS is warmer than that in the outer parts of our Galaxy. An unsuccessful search for IRAS 100 μm micron emission in the H I halo of the SMC and in a number of fields in the MS indicates a rater low intensity of the ambient radiation field. Detection of IR emission in an IVC and an HVC associated with the MS supports ram pressure models for the origin of the MS.

1. Introduction

The Large and the Small Magellanic Cloud (LMC and SMC) are extremely gas rich galaxies which are connected by a common gaseous halo and a mainly gaseous bridge called the Inter Cloud Region (ICR), here. Starting at the ICR and the SMC a long gaseous tail, called the Magellanic Stream (MS), extends over $\sim 100°$ on the sky (Fig. 1). The best known review of the H I distribution in this complex system, called the Magellanic System (MSY) here, has been compiled by Mathewson & Ford (1984). A more up-to-date review of the MSY can be found in Wayte (1990), a PhD Thesis which should be made more readily available to a larger community of astronomers.

After a brief summary of previous work this review presents recent results on the physical conditions of the diffuse gas and dust in the halo of the Clouds and the MS. In Sect. 4 we discuss results of a search with the Australia Telescope (AT) for 21-cm absorption lines towards radio continuum sources in or behind the MSY. In Sect. 5 we report about recent attempts to find infrared (IR) 100 μm emission in the H I-brightest parts of the Magellanic Stream. Figure 1 shows – superimposed on the H I outline of the MSY – the radio continuum sources which have been searched for H I absorption lines and the areas in the halo of the Clouds and in the MS which have been searched for IR 100 μm emission.

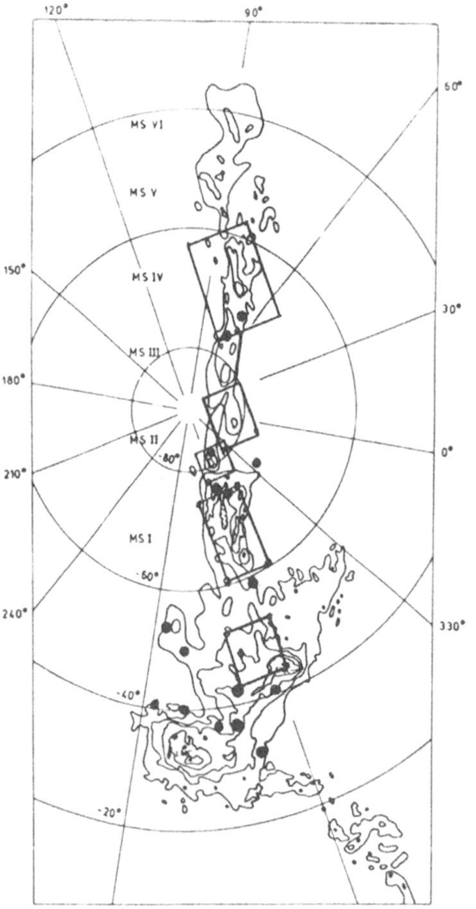

Fig. 1. H I outline of the Magellanic System in l,b. Indicated are the radio continuum sources (•) used for H I absorption studies and the fields which have been searched for IR emission.

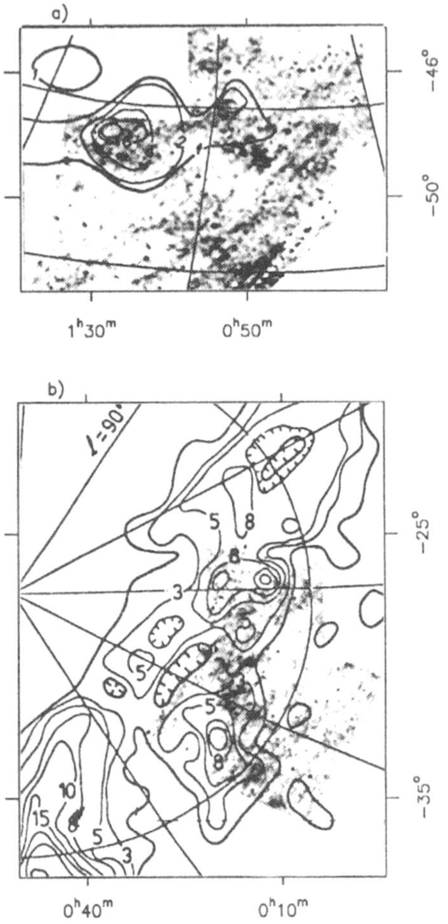

Fig. 2. IR 100 μm emission (grey-scale) superimposed onto H I features associated with the Magellanic Stream in α, δ (50): a) IVC 295−69−55, b) HVC 350−80−145

2. H I Surveys

A basic collection of references to previous H I work on the MSY is summarized in Table 1 for the LMC, the SMC, and the ICR and in Table 2 for the MS. Tables 1 and 2 also give some information on the size of the observed field, the grid, the telescope beam size, the velocity resolution ΔV, and the senitivity ΔT. Also included are the survey of Luks & Rohlfs (1992) of the LMC and that of the SMC (Rohlfs et al. in prep.) which is fully sampled and sensitive enough to be usefull for zero-spacing corrections of H I aperture synthesis observations. This is of interest because the Australia Telescope (AT) (Whiteoak 1989) has opened up the possibility for a synthesis of H I emission features with an angular resolution that allows detailed comparison with optical images.

Table 1. Basic references to H I observations of the MSY

	LMC	SMC	ICR
Author	McGee & Milton (1966)	Hindman (1967)	Mathewson et al. (1974)
HPBW	14.5'	14.5'	50'
Grid in α, δ	2^m, 12'	2^m, 6'	$?^m$, 2.5/5°
Field: $\alpha(50)$	04^h30^m to 06^h20^m	00^h00^m to 01^h30^m	00^h00^m to 24^h00^m
$\delta(50)$	-64° to -78°	-70° to -76°	-37°30′ to -90°
ΔV	$7\,km\,s^{-1}$	$7\,km\,s^{-1}$	$7\,km\,s^{-1}$
ΔT	1 K	1 K	0.3 K
Author	Luks & Rohlfs (1992)	Rohlfs et al.	McGee & Newton (1986)
HPBW	15'	15'	15'
Grid in l, b	0.2°, 0.2°	7', 7'	1°, 1°
Field:	8.2° * 8.2°	5.0° * 5.0°	$\sim 01^h30^m$ to $\sim 05^h30^m$ -68° to -77°
ΔV	$0.2\,km\,s^{-1}$	$0.2\,km\,s^{-1}$	$4.1\,km\,s^{-1}$
ΔT	0.05 K	0.05 K	0.1 K

Table 2. References to H I observations of the Magellanic Stream

Author	Field	Grid	Beam	ΔV	ΔT
Wannier & Wrixon (1972)	$\sim 22^h00^m$ to 01^h00^m -34° to 10°	2.0° * 2.0°	2°	$16.0\,km\,s^{-1}$	0.06 K
Mathewson et al. (1974)	00^h00^m to 24^h00^m -37.5° to -90°	2.5/5.0°	50'	$7.0\,km\,s^{-1}$	0.3 K
Haynes (1979)	23^h30^m to 02^h00^m -45° to -19°	~1.0°	22'	$4.2\,km\,s^{-1}$	~0.1 K
Cohen (1982)	23^h00^m to 00^h20^m -20° to 0°	2.0^m * 1.0°	~33'	$7.3\,km\,s^{-1}$	~0.03 K
Morras (1983)	00^h00^m to 02^h00^m -60° to -45°	4.0^m * 1.0°	34'	$2.0\,km\,s^{-1}$	0.2 K
Morras (1985)	23^h20^m to 02^h00^m -44° to -20°	4.0^m * 1.0°	34'	$2.0\,km\,s^{-1}$	0.2 K
Wayte (1990)	80° to 104° -56° to -38°	0.5° * 0.5°	15'	$4.1\,km\,s^{-1}$	0.02 K

The fundamental work on the LMC was carried out with the 64-m telescope in Parkes by McGee & Milton (1966), that on the SMC by Hindman (1967). The 18-m reflector in Parkes has been used for a survey of the ICR and the MS by Mathewson et al. (1974). Large but incompletely sampled data sets were obtained for the ICR with the 64-m reflector at Parkes by Mathewsons et al. (1979, 1984) and by McGee & Newton (1986). The H I surface density distributions of the LMC from McGee & Milton (1966) and of the SMC by Hindman (1967) is integrated in the synopsis of the LMC/SMC/ICR system presented by Mathewson et al. (1984). Rohlfs et al. (1984, 1991) observed the inner $6.2° * 6.2°$ of the LMC with the Parkes 64-m reflector. This survey has been extended to $8.2°*8.2°$ and is presented and discussed by Luks & Rohlfs (1992). Observations for a fully sampled survey of the SMC have been completed in 1991 by Rohlfs et al. (see Table 1).

The MS – as far as it is accessible from the northern sky – was discovered by Wannier & Wrixon (1972). The connection of the MS with the Magellanic Clouds has been established by Mathewson et al. (1974). Subsequently numerous investigation have been carried out in order to unveil the nature of the MS (see Table 2).

3. Distribution and Kinematics

3.1. The LMC and the SMC

From the recent observations of Rohlfs et al. (1991) Luks & Rohlfs (1992) identify a fairly regularly rotating H I disk and a low-velocity gas layer which they call the L-component. L-component and disk are receeding from each other with the L-component beeing in front (cf. however Sect. 4.1). Steep gradients of the H I and the radio continuum intensity on the eastern and northern side of the LMC indicate ram-pressure interaction with a gaseous galactic corona or an intergalactic medium.

More obvious than in the LMC is a bi-modal velocity structure in the SMC. It is seen in the H I line profiles, the velocity distribution of stars and H I regions, and even in the distances of stars. The SMC consists of two parts, the Mini-SMC and the SMC-remnant and may be an unbound system (cf. Mathewson & Ford 1984). The separation which started $\sim 2\,10^8$ years ago during a collision of the LMC and the SMC is probably accompanied by the release of large amounts of gas which can easily be swept out by a gaseous galactic corona. This would explain why the MS connects up to the SMC but not to the LMC (see below).

3.2. The ICR

Much of the ICR, which has $\sim 30\,\%$ of the gas of the MSY, appears to be at velocities in excess of $\sim 100\,\mathrm{km\,s^{-1}}$ with respect to the galactocentric standard of rest (GSR) and is therefore unbound to the MC's (Wayte 1991). This implies a) that the bridge is a rather young feature ($\sim 10^8$ years), b) that the Clouds have been experiencing a loss of large amounts of gas which probably started in the encounter $\sim 2\,10^8$ years ago, and c) that the two Clouds have experienced such a collision only once before. The bi-modal velocity distribution which is characteristic of the gas in the SMC and the LMC has also been found for the ICR. Here the two H I-line components are of about equal strength and separated by about $40\,\mathrm{km\,s^{-1}}$. There is a steep gradient of the H I column density at the leading edge of the ICR and of the Clouds (see above) compared to the fuzzy appearence at the downstream side, i.e. the side facing the MS. The fuzz is most clearly illustrated by "fingers" pointing in the direction of the MS.

All of this provides strong evidence for ram-pressure sweeping of gas from the ICR and the Clouds by coronal galactic gas. A sharp drop of about $70\,\mathrm{km\,s^{-1}}$ in velocity between ICR and MS indicates that the adjacent part of the MS is not just an extension of the fingers but that it has been produced by an interaction with gas that is not bound to the Clouds.

A consistent scenario is as follows: On its orbit through the outer galactic halo the rotating LMC-SMC system is experiencing the ram pressure of the probably inhomogeneous galactic corona. Because of its orientation in space and the sense of rotation of this system (cf. Fig. 5 of Mathewson et al. 1979) the gas experiences a larger ram-pressure on the far side of the system compared to the near side. This differential ram pressure tends to reduce the rotation speed at the far side and to boost it up on the near side of the system. Thus the gas at the far side becomes more tightly bound to the system. Similarly, the velocity of the gas at the near side might be boosted so much that it becomes unbound and shows up at rather larger radii and radial velocities when it shows up again on the side facing the MS.

This scenario explains why and how gas is lost at the extreme positive velocities on the down-stream side of the LMC/SMC/ICR system. Also, the sharp drop of about $70\,\mathrm{km\,s^{-1}}$ in velociy between ICR and MS is explained. Sweeping the outer layers of the Clouds has an advantage over the sweeping mechanism decribed by Wayte (1990) in that it provides a much larger sweeping cross section compared to the sweeping of the region around the libration point between the LMC and the SMC. It can also be used to predict the velocities of the gas swept out from different parts of the Clouds and the ICR by considering the systemic velocities and the rotation amplitudes of the LMC, the SMC, and the ICR. This leads to a natural explanation of the bi- or multi-furcation of the MS and of the velocity gradient perpendicular to the main axis of the MS. The side of the MS extending from the LMC consistently has a more positive radial velocity than that extending from the SMC. It is even conceivable that the change of the sign of that gradient at $\delta(50) = -30°$ is related to the encounter of the LMC and the SMC about $2\,10^8$ years ago.

3.3. The MS

The MS is a band of H I emission which starts close to the Clouds and extends over about 100° on the sky passing close to the south galactic pole (SGP). It is similar to filaments observed in interacting galaxies but has no stellar counterpart. It is widely accepted that the MS consists of material that has been pulled out of the Clouds and the ICR by one or another mechnism. Various models have been proposed in the past which have been reviewed and listed in decreasing order of compatibility with the observations by Wayte (1990): ram pressure models for a multi-phase coronal gas, a diffuse coronal gas, or a cloudy coronal gas, tidal models, turbulent wake models, and a primordial model.

The observational characteristics which have to be reproduced by these models are (cf. Wayte 1990):

1. The MS lies along a small circle about 7° from and parallel to a great circle passing through the SGP and cutting the galactic plane at $l = 280°$.

2. The MS can be subdivided into six main concentrations called MS I to MS VI (Fig. 1). But the H I surface density is essentially continuous decreasing with increasing distance from the Clouds, i.e. proceeding from MS I to MS VI.
3. The quoted concentrations consists of a number of filaments whose long axes are parallel to the general direction of the MS. These filaments are frequently linked into horse shoe shaped structures and make up for a bi-furcated structure for most of the length of the Stream.
4. The MS shows galactocentric radial velocities which vary from close to zero at MS I to about $-200\,\mathrm{km\,s^{-1}}$ at MS VI.
5. There is material alongside the MS which has more extreme velocities than the central ridges and which becomes less frequent towards the tip of the MS.
6. The emission lines of the MS have a rather large velocity width except for a single cloud close to the tip.
7. The MS is gaseous with no stellar component detected.
8. A discontinuity in velocity of $\sim$70 $\mathrm{km\,s^{-1}}$ is apparent between the MS and the LMC/SMC/ICR system.
9. The MS connects up to the SMC and the ICR but not to the LMC.
10. A velocity gradient shows up perpendicular to the Stream which, if viewed downstream, is positive towards the left of the MS and changes sign at $\delta(50) = -60°$.

Wayte (1990) presents ram-pressure models and argues that a multi-phase ram pressure model does reproduce most of these requirements. He does not consider items 8 to 10. The modification of the sweeping mechanism of the SMC and the ICR proposed above opens up the possibility to model the velocity discontinuity between MS and Clouds, the velocity gradient perpendicular to the Stream, and the bi- or multi-furcation of the MS.

The key problem of the ram pressure models is that relatively large densities of the halo gas in and close to the orbit of the Clouds are required. A gaseous polar ring consisting of H I high velocity clouds (HVCs) and a more diffuse, probably ionized component in an orbit retrograde to that of the Clouds (Mathewson et al. 1987, Haud 1988, Wayte 1990) can provide the required ram pressure if this ring is at a suitable distance. In fact, HVCs or intermediate velocity clouds (IVCs) which appear superimposed onto the MS are found alongside the MS (see 5. above) linked to it by H I features connecting them in the velocity dimension (cf. Fig. 8 of Mathewson et al. 1987, Wayte 1990). This is indicative of inelastic cloud collision predicted by the ram pressure hypothesis. We find that some of these IVCs or HVCs do have different IR properties from the rest of the gas in the MS giving further support to this hypothesis (see Sect. 5).

4. Gas Temperature in the Halo and in the Stream

The gas and the dust in the outer parts of the MSY can be used as a probe for the radiation field and the density. The spin temperature of the H I gas is the result of a heating-cooling balance where the input by the ambient radiation field is balanced by the loss from thermally excited heavy elements and ions. Observational estimates of this temperature can be used to derive the radiation field, if the element abundances and the density of the gas are known, and vice versa.

Radhakrishnan et al. (1972) present an H I absorption spectrum towards 30 Doradus which is very complex and has detectable optical depths $\tau \leq 0.6$ in the whole velocity range for which H I emission ($T_B < 60$ K) is found. This imlies that 30 Doradus is located at the far side of the gas distribution of the LMC (see above). These spectra indicate spin temperatures similar to galactic conditions.

A survey of H I absorption and emission lines has been carried out with the Australia Telescope Compact Array (ATCA) and the Parkes 64-m telescope towards extragalactic radio continuum background sources by Mebold et al. (1991). An initial source list was selected from the Parkes Catalog (Otrupcek & Wright 1991). It included sources with an 11-cm flux density > 2 Jy located within the N(H I) = 1 10^{19} cm^{-2} H I column density boundary of the MSY (see Fig. 1). The requirement T_B(H I) > 1 K for the 64-m telescope in the velocity domain of the MSY reduced the list to 20 sources. A search for unresolved compact cores with the ATCA finally selected 9 sources which were subsequently searched for H I absorption lines. No absorption was detected in any of the sight lines listed in Table 3. Details like optical depth limits $\Delta\tau$, are given in Mebold et al. (1991) for the strongest 5 radio sources of the survey. The other results will be published in due course.

From $\Delta\tau$ and T_B(H I) lower limits for spin temperature of the H I gas can be estimated from $T_{SP} = T_B(\text{H I}) / \Delta\tau$. Values as large as $T_{SP} > 600$ K have been found for the strongest of the observed radio sources, 0407-752. The temperature limits derived from this survey have been compared by Mebold et al. (1991) with the average spin temperatures for high latitude sources in the Galaxy and those derived for M 31 by Dickey & Brinks (1988). Mebold et al. (1991) find $T_{SP}(\text{MSY}) \approx 3 \cdot T_{SP}(\text{Galaxy})$ if only low brightness ($T_B < 3$ K) sight lines are used for our Galaxy. Since Dickey & Brinks (1988) find $T_{SP}(\text{Galaxy}) \approx T_{SP}(\text{M 31})$ the spin temperatures in the halo of the Clouds and in the MS appear larger than those in the Galaxy and in M 31.

A lack of cold clouds in the H I halo of the MSY may be caused by a comparatively large ambient radiation field or by a low cooling rate due to low gas density and/or low element abundances. Assuming abundances similar to those in the Magellanic Clouds, the intensity of the radiation field and the density are the critical parameters. The source 0252-712 is located behind the ICR while 0409-752 is located in the H I halo of the LMC. Obviously, the radiation field of the Clouds is not very diluted and/or the gas density is rather low. The source 0103-452 is located far from the Clouds. But here $\Delta\tau$ is only small enough to exclude the presence of really cold clouds ($T_{SP} < 20$ K). Whether this implies heating by an intergalactic radiation field or else can be explained by a low cooling efficiency – due to low gas density – in a gas having a rather high initial temperature after the release from the Clouds, has yet to be investigated.

5. IR Emission of the Magellanic Stream

The dust in the halo of the Magellanic Clouds and in the MS can be used as a probe for the ambient radiation field. Assuming that the gas-to-dust ratio and the dust composition in the halo of the MSY and in the MS is similar to that in the Clouds and using the fact that no stars have yet been detected in the MS, any IR radiation from the gas concentrations in the Stream far away from the Clouds will give us information on the strength of the intergalactic radiation field. IR emission of the H I gas in the ICR and the halo of the Clouds can be used to check on the dilution of the radiation field in these regions.

Fong et al. (1987) have compared the IRAS 100 μm flux densities in the region around the south galactic pole including part of MS II with H I column densities in the same area. Despite the presence of large amounts of gas in the MS, they detect no associated IRAS 100 μm flux density, indicating that the dust in the Stream is probably much cooler than that in the Clouds or the Galaxy.

We have extended the search for IRAS 100 μm emission to H I filaments located in the halo of the Clouds in MS I, MS II, MS III, and in MS IV using the GEISHA* analysis system (Wesselius et al. 1988). The iterative de-striping routine of GEISHA has been used to produce high sensitivity maps of regions of high H I column densities and high column density contrast. These regions are indicated in Fig. 1 by rectangles (see Table 3). The H I data used to derive the numbers in Table 3 are from Morras (1983 and 1985). The IR and the H I data refer to pixels of a diameter of about 10'. We present a preliminary evaluation here.

The intensity, I_{100}, in the 100 μm IRAS band for two of the regions studied is shown in Fig. 2. The zero-level of the intensity distribution is referred to a local reference plane. The regions displayed in Fig. 2 demonstrate that we typically find a) areas of high H I column densities, N(H I), of the MS and with no counter part in I_{100} and b) areas of local maxima of I_{100} with no N(H I)-counter part of the MS. This is also evident from Table 3. The areas MS IV peak, MS II A, and MS I A which are areas of relatively high MS column densities do not show any significant IR emission. In fact, MS II A and MS I A are the areas with the highest H I column densities in the MS; the very low limit of their I_{100}/N(H I) ratio indicates very low dust temperatures for a normal gas-to-dust ratio. In contrast to this, the areas MS III A, MS III HVC, and MS I IVC show significant I_{100} emission features which are not coincident with hydrogen emission features of the MS. There is a striking coincidence, however, of the latter feature, which we call MS I IVC, with IVC 295–69–50 (Morras 1983, Fig. 1b). Similarly, MS III HVC is located at the western slope of an H I feature in the MS (Morras 1985, Fig. 2). But checking more closely, one finds that the IR feature MS III HVC is coincident with the HVC 350–80–145 (Haynes 1979, Fig. 2; Morras 1985, Fig. 3a). As with MS I IVC the IR emission feature MS III HVC turns out to be coincident with an anomalous velocity cloud which appears superimposed onto the MS. HVC 350–80–145 is infact the first HVC which is discovered to have IR emission.

The I_{100}/N(H I) ratios in Table 3 are to be compared with typical values in the Galaxy (0.5-1 10^{-20} MJy ster^{-1} cm^2) and in the LMC (0.9 10^{-20} MJy ster^{-1} cm^2) or the SMC (0.06 10^{-20} MJy ster^{-1} cm^2) (Schwering 1988). So, the IVC, the HVC, and the features M III A, M IV A and B have I_{100}/N(H I) ratios which are normal compared to the LMC or the Galaxy.

The identification of the IR features in Tale 3 with the H I features mentioned above is, however, still ambigous. The problem is best illustrated by the IR features MS IV A and B in Table 3. Here we have again a coincidence with the slopes – rather

* GEISHA was developed by the Laboratory for Space Research at Groningen, The Netherlands. This development was partly funded by the University of Groningen, the Space Science Department of ESTeC and the Air Force Geophysics Laboratory (USA).

than the peak – of an H I-feature in the MS. The IR emission of the H I peak – i.e. of MS IV peak – is relatively weak. Inspection of the H I data of the Berkeley Survey (Heiles & Habing 1974) shows that the IR features MS IV A and B may be associated with H I features in our Galaxy rather than in the MS. Higher angular resolution H I data are required for a more definite conclusion. But also the identifications for MS I IVC and for MS III HVC require a cross check against a chance coincidence with a galactic H I feature. So, H I observations with the Parkes 64-m telescope are needed to verifiy the identification suggested above for the IR features in Table 3.

Table 3. Regions in the MS searched for IR emission

Name		$\alpha(50)$	$\delta(50)$	I_{100}	$N(\mathrm{H\,I})$	$I_{100}/N(\mathrm{H\,I})$
		(^{h} ^{m})	(°)	(MJy sr^{-1})	(10^{20} cm^{-2})	(10^{-20} MJy ster^{-1} cm^2)
MS IV	A	23 40	–12.5	0.76	0.8	0.95
	B	23 36	–12.0	0.41	0.5	0.82
	peak	23 40	–11.5	< 0.20	1.3	< 0.15
MS III	A	00 16	–32.0	0.50	0.4	1.25
	HVC	00 14	–34.0	0.26	0.3	0.87
MS II	A	00 42	–38.0	< 0.18	3.0	< 0.06
MS I	A	00 56	–50.0	< 0.15	2.5	< 0.06
	IVC	01 08	–48.5	0.32	0.6	0.53

In summary, no IR emission could be detected for the MS. This confirms and extends the results of Fong et al. (1987) and is consistent with a low-intensity ambient radiation field in the MS which is expected for most models of its origin. Results for the H I halo of the Clouds will be published in due course.

We did, however, find IR emission in some of the anomalous velocity features associated with the MS. Their IR radiation indicates that their dust is heated – probably by the galactic radiation field. So they are probably closer to the Galaxy than the MS. As the anomalous velocity features appear to have a velocity-connection to the MS (Wayte 1990) they have probably been produced by inelastic collisions between the gas in the Clouds and denser clouds in the Corona of the Galaxy. After losing orbital momentum they will sink through the galactic halo towards the Galaxy. So, the discovery of IR emission in anomalous velocity features associated with the MS provides additional evidence for its origin by ram pressure sweeping of the Clouds by a multi-phase coronal gas.

6. Conclusion

For the discussion in the last two sections we have assumed that the Magellanic Stream has been produced by ram pessure sweeping of the Magellanic Clouds by coronal gas of the Galaxy and that the dust and the chemical composition are therefore similar. The presence of heavy elements is confirmed by the discovery of Ca II, C II, C IV, O I, Mg II, Si II, and Fe II absorption lines in the Stream towards the background galaxy

Fairall 9 by Songaila (1981) and Penston (1982). But the quality of the data does not permit a quantitative determination of abundances.

The co-existence of cold dust and high spin temperature atomic hydrogen in the same volume of gas is an apparent contradiction. This can, however, be resolved if one assumes that the density of the gas in the MS is so low that the cooling time since its release from the Clouds some 10^8 years ago is rather long. The gas may still be warm enough that it can not be detected in H I absorption. Since the cooling time of the dust is much shorter, the lack of IR emission indicates a weak intergalactic radiation field inspite of the rather large H I spin temperatures found.

References

Cohen R.J., 1982, MNRAS 199, 281

Dickey J.M., Brinks E., 1988, MNRAS 233, 781

Fong R., Jones L.R., Shanks T., et al., 1987, MNRAS 224, 1059

Haud U., 1988, A&A 198, 125

Haynes M.P., 1979, AJ 84, 1173

Heiles C., Habing H.J., 1974, A&AS 14, 1

Hindman J.V., 1967, Aust. J. Phys. 20, 147

Luks T., Rohlfs K., 1992, A&A in press

Mathewson D.S., Cleary M.N., Murray J.D., 1974, ApJ 190, 291

Mathewson D.S., Ford V.L., 1984, in: Structure and Evolution of the Magellanic Clouds, IAU Symp. 108, (eds.: van den Bergh S., de Boer K.S.), Reidel Dordrecht, p. 125

Mathewson D.S., Ford V.L., Schwarz M.P., Murray J.D., 1979, in: The Large-Scale Characteristics of the Galaxy, IAU Coll. 84 (edt.: Burton W.B.), Reidel Dordrecht, p. 547

Mathewson D.S., Wayte S.R., Ford V.L., Ruan K., 1987, Proc. Astr. Soc. Australia 7, 19

McGee R.X., Milton J.A., 1966, Aust. J. Phys. 19, 343

McGee R.X., Newton L.M., 1986, Proc, ASA 6, 471

Mebold U., Greisen E.W., Wilson W., et al., 1991, A&A 251, L1

Morras R., 1983, AJ 88, 62

Morras R., 1985, AJ 90, 1801

Otrupcek R., Wright A.E., 1991, Proc. Astron. Soc. Australia in press

Penston M.V., 1982, The Observatory 102, 174

Radhakrishnan V., Brooks J.W., Goss W.M., Murray J.D., Schwarz U.J., 1972, ApJS 24, 1

Rohlfs K., Kreitschmann J., Siegman B.C., Feitzinger J.V., 1984, A&A 137, 343

Rohlfs K., Luks T., 1991, in: The Magellanic Clouds, IAU Symp. 148 (eds.: Haynes R., Milne D.), Kluwer Academic Publishers, p. 63

Schwering P., 1988, PhD Thesis, University of Groningen, The Netherlands

Songaila N., 1981, ApJ 243, L19

Wannier P., Wrixon G.T., 1972, ApJ 172, L119

Wayte S.R., 1990, PhD Thesis, Australian National University, Canberra

Wesselius P., Bontekoe T.R., de Jonge A.R.W., Kester D.J.M., 1988, in: Astronomy from large data bases, ESO Conf. 28 (eds.: Murthagh F., Heck A.), Garching, p. 85

Whiteoak J.B., 1989, Asia-Pacific Phys. News 4, 20

Structure and Kinematics of the ISM in the LMC from HI 21-cm line emission

Kristen Rohlfs and Thomas Luks

Astronomisches Institut der Ruhr-Universität Bochum, 4630 Bochum, Fed. Rep. of Germany

1 Line surveys

Investigations of the structure and kinematics of the ISM in the LMC have a long history dating back to the first low resolution 21-cm line survey by Kerr et al. (1954). A large 21-cm line survey made with the then new Parkes 61-m telescope by McGee and Milton (1964, 1966a, 1966b) has for many years been the main source of information on the H I gas distribution and kinematics. Using the same telescope but more sensitive receivers Rohlfs et al. (1984) have published a new survey of the central parts of the LMC giving an improved velocity resolution of $\Delta \mathrm{RV} = 1.65\,\mathrm{km\,s^{-1}}$ and $\sigma_T \approx 0.07\,\mathrm{K}$ compared to $\Delta \mathrm{RV} = 7.5\,\mathrm{km\,s^{-1}}$ and $\sigma_T = 1.75\,\mathrm{K}$ of the McGee survey. In 1989 the extent of this survey was about doubled so that it now covers a field of $8\overset{\circ}{.}2 \times 8\overset{\circ}{.}2$ with a raster of $0\overset{\circ}{.}2$ in both directions, forming a slightly undersampled data set (antenna beamwidth $0\overset{\circ}{.}25$).

Already McGee and Milton (1964) noted that the line emission is double peaked or complex in some parts of the LMC, and they concluded that possibly *the presence of an "arm" of H I extending outwards from the LMC plane* could be inferred from this. In his review on galaxies of the magellanic type Feitzinger (1980) too remarks that there must exist gas well above the principal plane of the system in the NE quadrant.

If a line profile is observed to be double- or multiple-peaked this means that there exist gas bodies on the line-of-sight with different mean velocities. And due to the short relaxation time for gas in the ISM (at most a few 1000 years) these bodies or sheets must be situated at different locations in space. The column density of each sheet then can be determined if its contribution to the total line emission at that position can be determined in some way.

During the last years we have made several attempts for such a separation, Fig. 6 in Rohlfs et al. (1984) shows some of the results, but when explicit values for both the amount of gas and the radial velocity of the gas for the separate sheets were to be determined, all automatic methods making use of statistical parameters were not successful. Only when we modeled the observed line profile as a superposition of several independent Gaussian components, a quantitative analysis was possible.

The procedure used for the introduction of extra components was rather restrictive so that we feel rather confident about the reality of the components given by us, we might, however, have missed some if the conditions for their visibility are unfavourable. An analysis of the list of the given components shows that a component will be recognized provided the gas causing it has

1. $N_{HI} \geq 3 \times 10^{19}$ cm^{-2},
2. $|\Delta RV| > 10 - 15$ km s^{-1} from the main body of gas.

Everywhere at most three components were sufficient to reproduce the measured line profiles.

A rather small number of gas sheets that are continuous in both the X- and the Y-coordinate and form contiguous regions could be outlined that contain basically all components. This was done by arranging them into position-velocity diagrams for cuts in the X- and Y-direction. Each component then could be classified in a rather unambiguous way. In this way two large-scale sheets were detected, sheet D (for disk) extending over the whole face of the LMC and sheet L (for lower radial velocity) extending over 23% of the sampled area. 9% of the measured gas belongs to features different from both D and L and is not considered further here.

total H I content of LMC	$(3.1 \pm 0.6) \times 10^8 M_\odot$
D component	$(2.2 \pm 0.4) \times 10^8 M_\odot$
L component	$(0.6 \pm 0.1) \times 10^8 M_\odot$

2 The disk

The contour map of the disk column density is quite similar to that of the total hydrogen column density, as is comprehensible considering that it contains 71% of the total gas content. There is no obvious similarity of its distribution to that of other known constitutents, in particular there is no sign of any influence of the bar (Fig. 1). The star formation region Shapley III shows up as a large void in the gas distribution, but the separate gas components showing the expansion of this feature as discussed by Dopita et al. (1985) are not obviously present here. This might be due to the fact, that most of these expansion features are weaker than those considered here.

Differences between the distribution of the total H I and that of the D feature are detected in the large gas cloud south of 30 Dor which is stronger in the southern part in the total H I and near $(X, Y) = (-2°, 1.°5)$ where there is a gas void in D similar to Shapley III which, however, is not present in the total H I .

When using multi-peaked or skew emission line profiles it is always a problem which measure to use for the radial velocity in the analysis: the velocity of the peak or that of the "center of gravity". For Gaussian components no such problem exists. In general the radial velocity field for the disk obtained in this way is quite similar to that of the mean radial velocity field of the total H I and it resembles that of a differentially rotating disk. But since the isolines for the heliocentric velocity ranges 265 – 290 km s^{-1}, where the steepest velocity gradients with position are found, has the shape of a large reversed S, this has always been interpreted as indicating a large-scale warp. If, however, the large angular extent of the LMC ($> 8°$) is properly taken

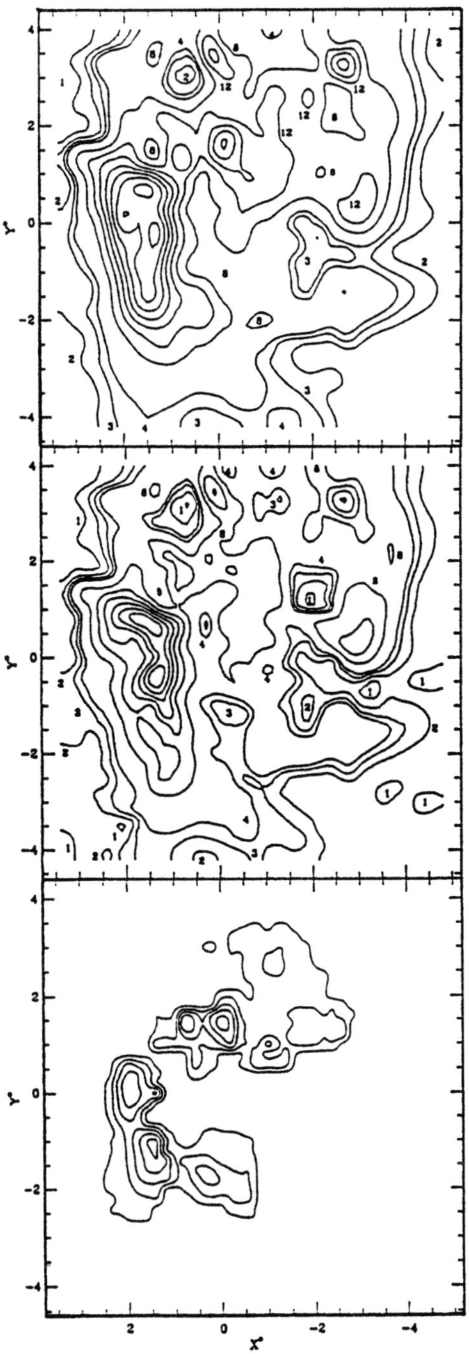

Fig. 1. Contour map of the H I column density. **(a)** Total H I , contour levels are $N_{HI} = (1, 2, 3, 4, 8, 12, 16, 20, 24, 28, 32) \times 10^{20}\ cm^{-2}$. **(b)** Disk component D, contour levels same as in (a). **(c)** L-component, contour levels: $N_{HI} = (2, 6, 10, 14, 18) \times 10^{20}\ cm^{-2}$.

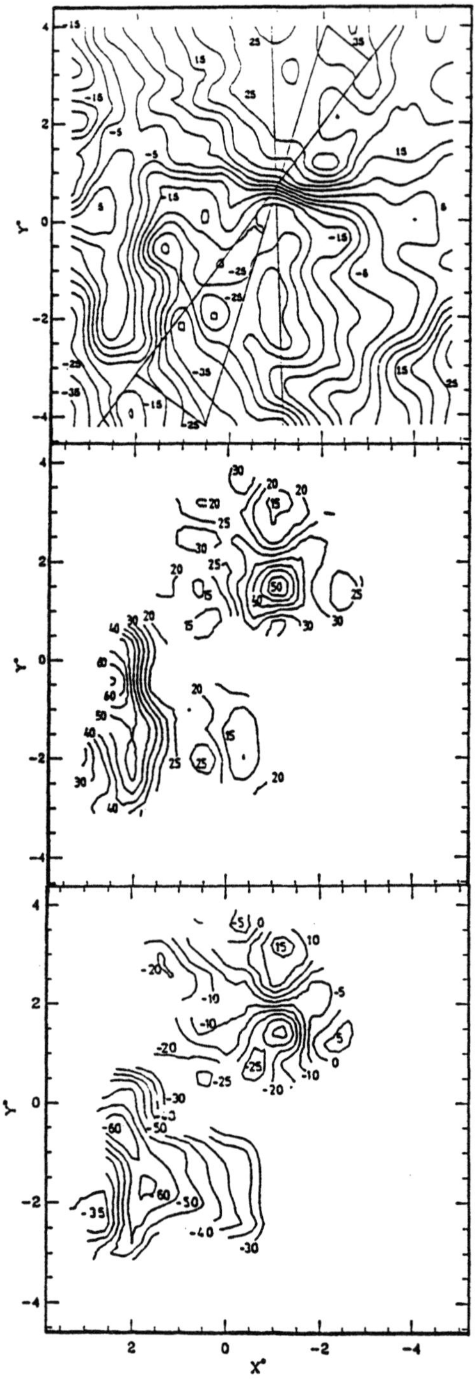

Fig. 2. (**a**) Magellanocentric radial velocity field of the disk component. The nodal line for $\vartheta_0 = 162°$ with a $\pm 20°$ sector is marked. (**b**) Radial velocity difference between the D and the L-component. (**c**) Magellanocentric radial velocity field of the L-component.

into account, we see that the (potentially large and unobservable) tangential velocity of this system is strongly influencing shape and orientation of the velocity diagram when it is transformed from the observed heliocentric field into the magellanocentric system (Fig. 2a).

Using the orbital parameters that Murai and Fujimoto (1980) dervied for the revolution of the Magellanic clouds about the Galaxy a velocity field without indications of a warp results. It is almost like a textbook example for a flat differentially rotating disk. The resulting rotation curve is without irregularities and bumps and it is symmetric in its N and S branch to within a few $\mathrm{km\,s^{-1}}$. For the kinematic line-of-nodes a position angle of $\vartheta_0 = 162°$ is found, quite similar to the geometric angle $\vartheta_0 = 168°$. Adopting a geometric inclination $i = 33°$ an observed turnover of the rotation curve at $r_1 \approx 3.5\,\mathrm{kpc}$ results in $M_{\mathrm{Kepler}} = (3.8 \pm 1.1) \times 10^9\,M_\odot$ (Fig.3).

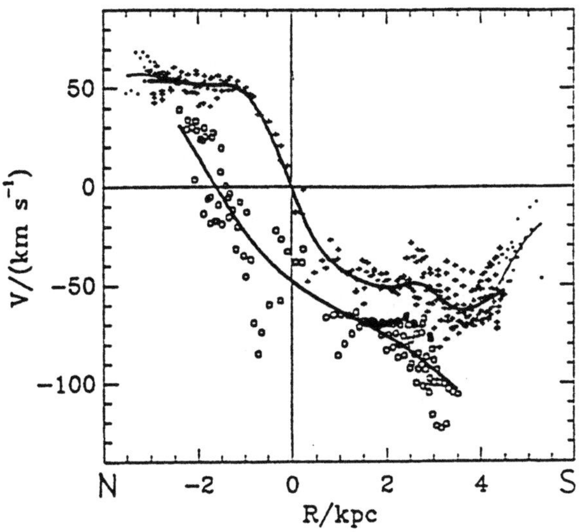

Fig. 3. Rotation curve for the D-component (+) and the L-component (o) using data from a ±20° sector.

3 The L-component and 30 Dor

This component consists of two large lobelike features, each extending over more than 2.5 kpc. The southern lobe contains part of the large cloud south of 30 Dor that is so clearly perceptible both here and in the disk. H I gas belonging to the L-component extends further to the W where it becomes indistinguishable from that of the disk because their radial velocity differ only by 10 - 15 $\mathrm{km\,s^{-1}}$. Therefore, it is well possible that the "tongue" visible in this region of the disk in reality belongs to L.

The kinematics of L is shown in two diagrams. Figure 2b presents the radial velocity difference between D and L, and Fig. 2c shows the magellanocentric radial velocity field of L computed from the measured heliocentric field using the same procedures and parameters as for the disk. The true intrinsic radial velocity field might well be different from that shown here, because it depends on both the unknown inclination i and position angle ϑ_0 of the line-of-nodes for the L-component. Here we adopted quite arbitrarily the same values as for the disk, and the resulting diagram has a reasonably regular appearance, and it can be interpreted showing a sheet of gas in differential rotation. Even a kind of rotation curve resulting in a mean angular rotation of $\Omega = 22\,\mathrm{km\,s^{-1}\,kpc^{-1}}$ could be derived, but one should not put too much significance into this. In the northern lobe of L there is a strong perturbation of the velocity field near $(X, Y) = (2°, -3° < Y < -0\overset{\circ}{.}5)$ the interpretation of which is still open.

A point that is still rather unclear is which other population tracers of the LMC are members of L. Concerning the constellation Shapley III the situation seems clear: the D-component shows a large void at the very position of this star-forming region and the L-component does not extend into this region at all. Therefore we think that Shapley III belongs to D. Regarding the HSD-regions (high surface density regions) of Prévot et al. (1989) the situation is indecisive. Neither the N_{HI} distribution of D or L nor the observed radial velocities show any clearcut relation, because the intrinsic radial velocity dispersion of the stars in such regions is too large to decide this from the kinematics of the tracers. Therefore we have to leave this question unanswered.

We think, however, that his is different regarding the 30 Dor region at $(X, Y) = (1\overset{\circ}{.}34, 0\overset{\circ}{.}63)$. It is situated where the two lobes of the H I L-component come closest together, but there is no H I gas belonging to L at this position. In the D-component on the other hand, the H I column density is here close to its peak value. This is, we think, rather compelling evidence against 30 Dor being a member of the disk population.

Walborn (1984, 1991) lists a total of 111 known OB stars belonging to the 30 Dor region resulting in a flux $N_{\mathrm{Ly_C}} = 2.8 \times 10^{51}\,\mathrm{s^{-1}}$. Adding to this the flux of $2 \times 10^{51}\,\mathrm{s^{-1}}$ for R 136 a total flux $N_{\mathrm{Ly_C}} = 4.8 \times 10^{51}\,\mathrm{s^{-1}}$ results corresponding to an excitation measure $U = 530\,\mathrm{pc\,cm^{-3}}$. This will result in a fully ionzed Strömgren sphere of 230 - 1060 pc diameter, the actual value depending on the ambient H I gas density. Compared to a half-power diameter of $0\overset{\circ}{.}7$ for the thermal radio source 30 Dor (Klein et al. 1989) this is a fairly consistent result showing that 30 Dor cannot be a member of the D-component, but quite well belong to L. A strong source of UV radiation in the region of 30 Dor is visible too in the vacuum UV image of the LMC obtained by Smith et al. (1987). Figure 4 shows an overlay of the contour diagram for the L-component H I column density with an image of the LMC near 1500 Å. 30 Dor is part of a large region ($1\overset{\circ}{.}5 \times 1\overset{\circ}{.}0$) of strong UV radiation that links the two H I gas lobes so that they form one single large gas cloud.

The strong ionizing radiation flux of 30 Dor can even be used to obtain an estimate on its minimum distance to the disk, because this radiation will ionize the outer skin of the clouds. This will cause a decrease of the observed H I column density, and if we believe a reduction of 30% to be acceptable, a height of 250 pc above (or below) the plane of the disk will result.

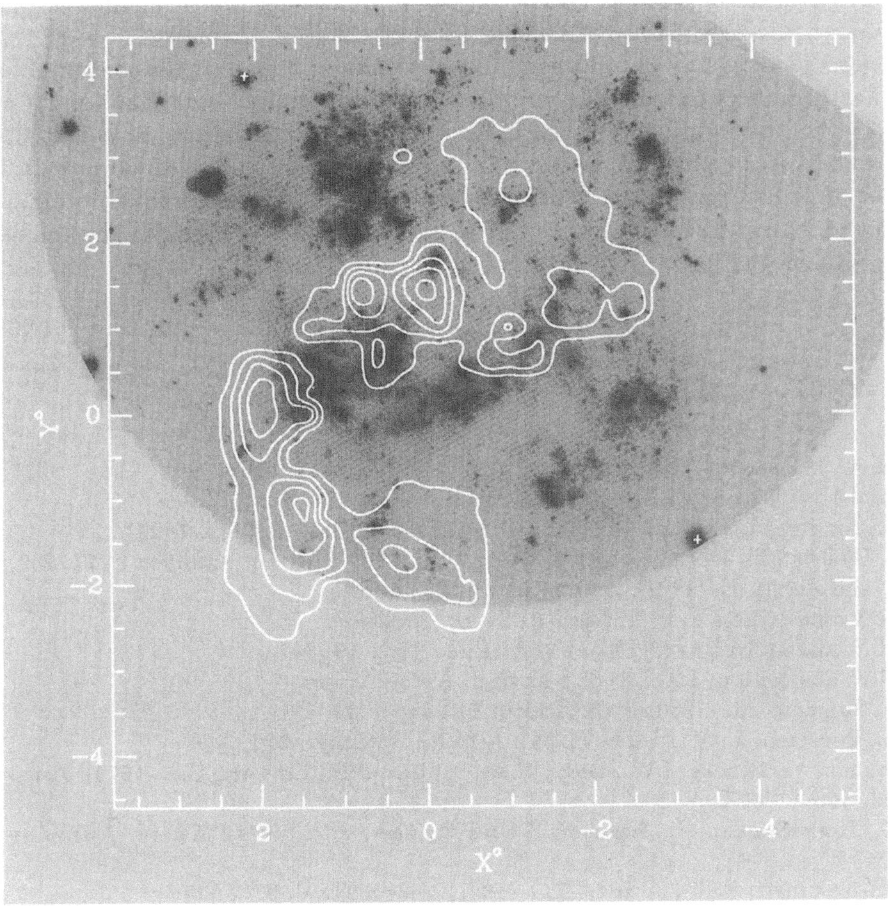

Fig. 4. Contour map of the H I column density of the L-component overlaid on the UV emission at 1500 Å(after Smith et al. 1987).

A problem separate from the question of the distance between L and D is, which of these lies in front. This has profound consequences, because due to L having a lower radial velocity than D, L in front of D means that they are currently separating from each other, this separation commenced $1 - 2 \times 10^7$ years ago. If, on the other hand, L lies behind D the two components will collide in $1 - 2 \times 10^7$ years! A possible method to decide between these two scenarios is to investigate the mutual absorption features. If 30 Dor belongs to L it should be absorbed by the LMC disk if L is behind D. And since the disk emission is strong at the position of 30 Dor a wide and strong absorption feature should be expected. But nothing of this kind is observed. In the contrary, the observed 21-cm line profile at the position of 30 Dor is well above the expected profile at all radial velocities. Indications of absorption in the spectrum of 30 Dor have been detected by Radhakrishnan et al. (1972). But these are interferometric measurements, and they refer only to a feature of small angular extent ($< 5'$ diameter).

Therefore we arrive at the following model for the H I gas distribution: there is a large, flat disk of gas in differential rotation with a rotation center at $(X, Y) = (-1\overset{\circ}{.}0, 0\overset{\circ}{.}7)$. The rotation curve is linear for $r < 0.7$ kpc and fairly symmetric, the disk has no large-scale warps. A second layer of H I gas is about 50 – 500 pc above the disk in the direction towards the Galaxy. It shows, too, signs of differential rotation.

A more extensive discussion of the new 21-cm line survey of the LMC is in press (Luks and Rohlfs 1992).

References

Dopita, M.A., Mathewson, D.S., Ford, V.L. (1985): Astrophys. J. **297**, 599

Feitzinger, J.V. (1980): Space Sci. Rev. **27**, 35

Kerr, F.J., Hindmann, J.V. and Robinson, B.J. (1954): Austral. J. Phys. **7**, 297

Klein, U., Wielebinski, R. Haynes, R.F., Malin, D.F. (1989): Astron. Astrophys. **211**, 280

Luks, Th. and Rohlfs, K. (1992): Astron. Astrophys., in press

McGee, R.X. and Milton, J.A. (1964): IAU Sypos. **20**, 289

McGee, R.X. and Milton, J.A. (1966a): Austral. J. Phys. **19**, 343

McGee, R.X. and Milton, J.A. (1966b): Austral. J. Phys. Suppl. 2

Murai, T., Fujimoto, M. (1980): Pub. Astron. Soc. Japan **32**, 581

Prévot, L., Rousseau, J., Martin, N. (1989): Astron. Astrophys. **225**, 305

Radhakrishnan, V., Brooks, J.W., Goss, W.M., Murray, J.D., Schwarz, U.J. (1972): Astrophys. J. Suppl. **24**, 1

Rohlfs, K., Kreitschmann, J., Siegman, B.C., Feitzinger, J.V. (1984): Astron. Astrophys. **137**, 343

Smith, A.M., Cornett, R.H. and Hills, R.S. (1987): Astrophys. J. **320**, 609

Walborn, N.R. (1984): IAU Sympos. **108**, 243

Walborn, N.R. (1991): IAU Sympos. **148**, 145

Radio Continuum from the Magellanic Clouds New Aspects

U. Klein [1], R.F. Haynes [2], D. Meinert [3], R. Wielebinski [3]

[1]Radioastronomisches Institut der Universität, Auf dem Hügel 71, W-5300 Bonn 1, Fed. Rep. of Germany
[2]Australia Telescope National Facility, CSIRO Radiophysics, P.O. Box 76, Epping NSW 2121, Australia
[3]Max-Planck-Institut für Radioastronomie, Auf dem Hügel 69, W-5300 Bonn 1, Fed. Rep. of Germany

Abstract: Radio continuum surveys of the Large and Small Magellanic Cloud (LMC and SMC), carried out over the last 8 years, have not only facilitated a comprehensive study of the thermal and nonthermal radio emission in these galaxies, but also provided linear polarization information about the two closest galactic neighbours. In particular, the magnetic field structure of the LMC could be disclosed. The large-scale structure of the LMC magnetic field resembles a trailing spiral. It is shown that the rotation measure and depolarization of a region $\sim$ 3 kpc in size which is located south-west of 30 Dor can be modelled in terms of a giant magnetic loop bending out of the LMC plane towards the observer. This conspicuous region shows a well-aligned magnetic field structure, and appears to be associated with anomalous HI components.

1 Introduction

Complete radio continuum maps of the Magellanic Couds have been obtained at $\lambda\lambda$21, 13, 12, 6, and 3.6 cm with the Parkes 64-m telescope (Haynes et al. 1986, 1991). These have resulted in studies of the radio spectral index, the thermal and nonthermal contributions, and in a number of comparisons with other spectral regimes (Loiseau et al. 1987; Klein et al. 1989; Meinert et al. in prep.). In particular, a thorough analysis of the well-known radio–FIR correlation could be conducted for the LMC with a spatial resolution that was hitherto not achievable (Xu et al. 1992). Studies of the sources in the Clouds (HII regions, SNR) and their spectra over a wide frequency range are another aspect of these surveys (Loiseau et al. 1987; Klein et al. 1989; Meinert et al. in prep.). Linear polarization was measured simultaneously in the surveys. While the observed degrees of polarization are extremely low in the SMC (Haynes et al. 1986; Loiseau et al. 1987), the LMC shows strong polarization in a region south-west of 30 Dor (Haynes et al. 1991), and measurable polarization elsewhere in its disk. In this paper, we shall focus on the linear radio polarization of the LMC, and discuss them in the light of its gas kinematics.

As stated above the observed linear polarization in the SMC is generally low (Haynes et al. 1986; Loiseau et al. 1987). This holds for all wavelengths (Haynes et al. 1991), implying that beam depolarization is the dominant process leading to the observed low degrees of linear polarization, which hardly exceed the 3–5% level at all wavelengths. This is in line with the large depth of the SMC, and in particular with its turbulent velocity structure (e.g. Le Coarer 1992).

2 Linear Radio Polarization in the LMC

Linear polarization was observed in the LMC at $\lambda\lambda$21, 13, 12, and 6 cm. At λ3.6 cm no linear polarization could be detected, probably due to the predominance of thermal radiation at this wavelength (Klein et al. 1989; Haynes et al. 1991).

In Fig. 1 we show the map of the linearly polarized emission at λ12 cm, smoothed to a $10'$ beam. The polarization map exhibits strong linear polarization located roughly south-west of 30 Dor. The main characteristics of these polarization features are:

— The polarization is concentrated in two elongated features of $\sim$3–4 kpc length pointing towards south-western direction, i.e. towards the direction of the SMC.
— The maximum degree of polarization in these filaments is $\sim$30% if referred to the total intensity. It is much higher if we account for the presence of large amounts of thermal bremsstrahlung.
— The contrast between the polarized features and the intervening regions is very high ($\gtrsim$4:1 in terms of polarized intensity), with steep lateral gradients showing up.
— The rough magnetic field orientation as judged from the $\boldsymbol{E}$-vectors at $\lambda\lambda$12 and 6 cm (where Faraday effects are small) points towards the SMC, too.
— The polarized filaments commence in the region of highest atomic (Rohlfs et al. 1984; Luks and Rohlfs 1992) and molecular (Cohen et al. 1988) gas density.
— This is also where optical images exhibit a sharp dust lane running roughly north-south, and starting just south of 30 Dor (see e.g. Plates 54 and 55 in Laustsen et al. 1987).
— In about the same region the radio spectral index changes from predominantly thermal (flat) to strongly nonthermal (steep) as is evident in the spectral index maps of Klein et al. (1989).
— It is also in this area where the extended X-ray emission is brightest (see Pakull 1992; Pietsch 1992).
— An intricate velocity structure is found here in the interstellar absorption lines (e.g. Molaro 1992) and in the HI line (Rohlfs 1992).

As evident in Fig. 1, there is also linear polarization elsewhere in the LMC. The omnipresence of a diffuse nonthermal component (Klein et al. 1989; Xu et al. 1992) means that magnetic fields and relativistic electrons are present throughout the LMC disk. The occurrence of linear polarization in some locations depicts regions with higher field ordering, given the relative unimportance of Faraday effects at the shorter wavelengths (except in the 30 Dor complex). The depolarized regions reflect in turn

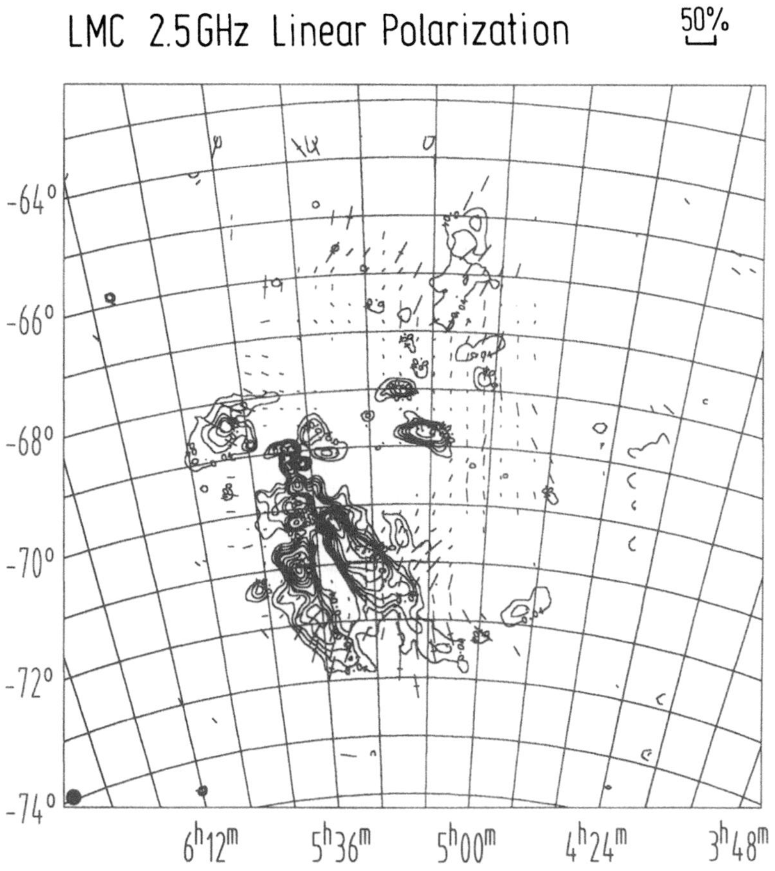

Fig. 1. Map of the linearly polarized radio emission of the LMC at 2.5 GHz, smoothed to a 10′ beam. Contours are: 40, 50, 60, ... mJy/b.a. Superimposed are the ***E***-vectors, with their lengths proportional to the fractional polarization. The scale of the latter is indicated by the bar in the upper right. The full circle in the lower left indicates the beam size.

the dominance of a turbulent magnetic field component, the turbulent scale being $\lesssim$ 140 pc (the spatial resolution of the map in Fig. 1).

The distribution of polarized emission at λ12 cm is the most sensitive one in the present investigation. While at longer wavelengths (λ21 cm) polarization is strongly reduced by Faraday dispersion in the numerous HII complexes in the LMC, the steepness of the synchrotron spectrum in the LMC ($\alpha_{\rm nth} \cong -0.8$, see Klein et al. 1989) reduces the intensity of the polarized emission at the shorter wavelengths. If smoothed to a very coarse angular resolution, low-level linear polarization is uncovered over most of the LMC disk. Assuming that Faraday rotation is unimportant at λ12 cm (with the rotation measures determined between $\lambda\lambda$21 and 12 cm this is at most 30°, see Chap. 3), we have rotated the ***E***-vectors by 90°. The magnetic field orientation obtained in this way for a 30′ resolution ($\sim$ 440 pc) is displayed in Fig. 2. The overall structure resembles a trailing spiral magnetic field, given the clockwise rotation of the

LMC. The mean degree of linear polarization at this resolution, corrected for thermal emission, is $\sim$ 5%, much less than what is typically found in bright spiral galaxies.

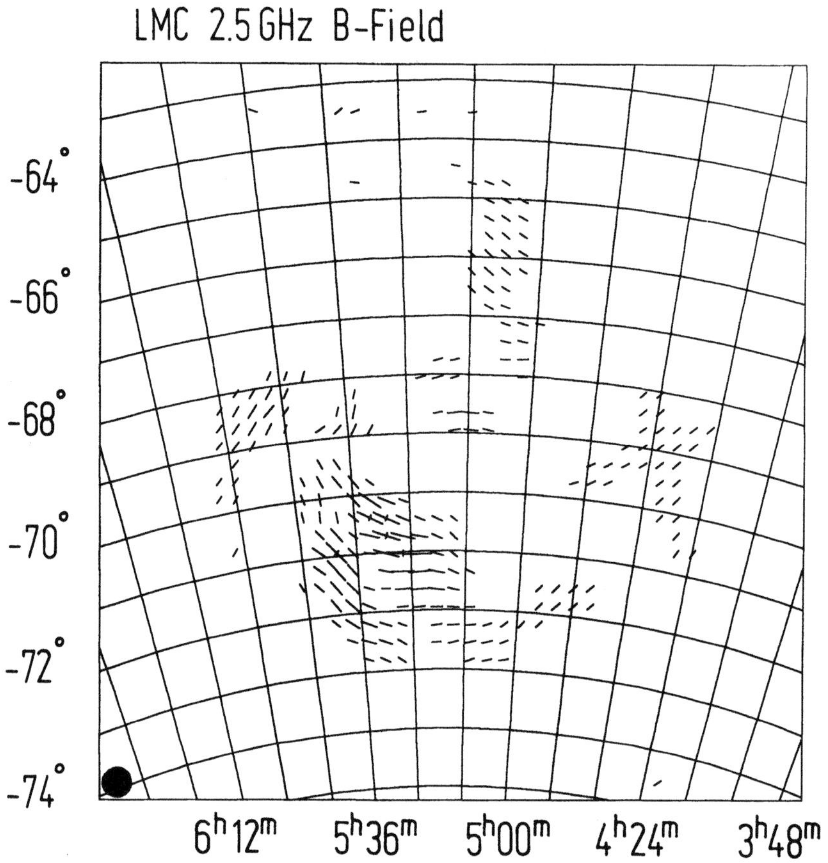

Fig. 2. Map of the large-scale structure of the magnetic field in the LMC at 30′ (440 pc) resolution. This is represented by ***B***-vectors at 2.5 GHz, which were obtained by rotating the ***E***-vectors by 90° (assuming little Faraday rotation, see text). The vectors' length is proportional to the polarized intensity.

3 Anomalies South-West of 30 Dor

We have investigated the distribution of the rotation measure (*RM*) and depolarization (*DP*) across the LMC. Here we focus on the conspicuous region containing the strongly polarized filaments south and south-west of 30 Dor, which stretch over some 3 kpc. In Fig. 3 we show the distribution of *RM* in this region, obtained between $\lambda\lambda 21$ and 12 cm (where the observed polarization is most reliable). A significant variation of the *RM* is evident, with abrupt changes in the sign immediately south of 30 Dor,

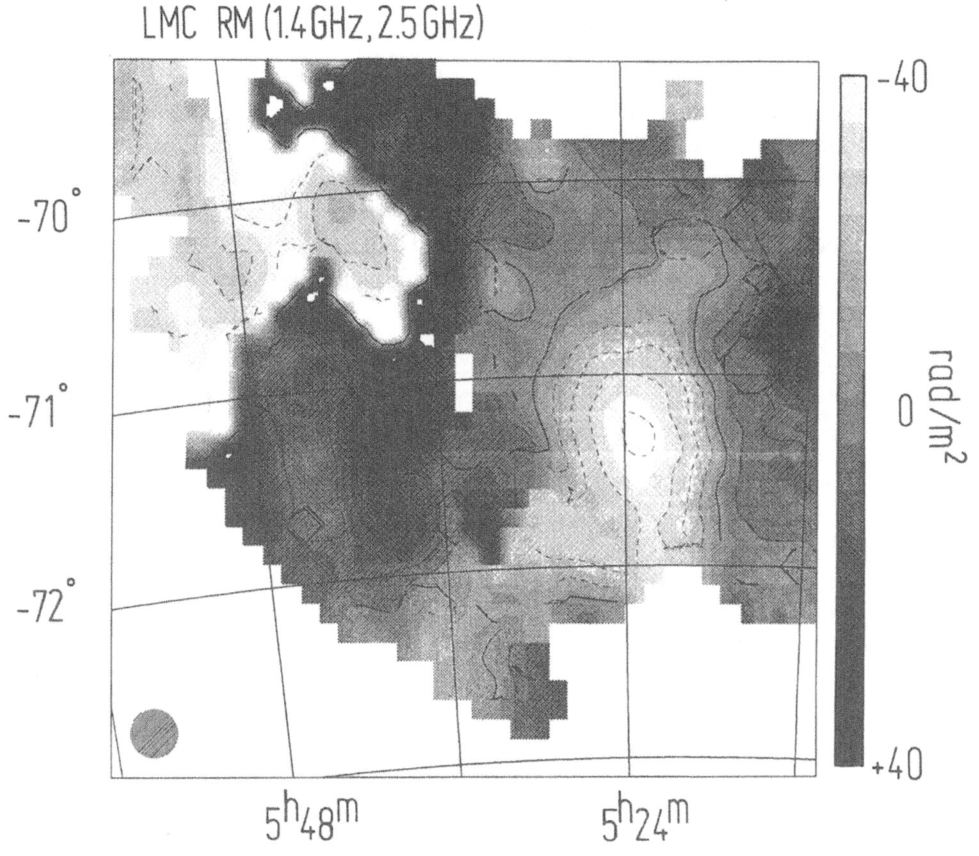

Fig. 3. Distribution of the rotation measure SW of 30 Dor, obtained between 1.4 GHz and 2.5 GHz with a resolution of 15′. The gray scale is accomopanied by contours, the values of which are: −50, −40, −30, ..., +40 rad m^{-2}. Negative contours are dashed, the zero contour is thickened.

which roughly coincide with a southern extension of thermal emission that is visible at λ3.5 cm (see Fig. 3 of Haynes et al. 1991). The outer reaches of the two polarized filaments (Fig. 1) exhibit different signs of the RM. While the termination of the western one is characterized by $RM \sim -30 \ldots -40\,\mathrm{rad\,m^{-2}}$, that of the eastern one has $RM \simeq +30 \ldots +40\,\mathrm{rad\,m^{-2}}$. Surprisingly, these locations do not coincide with any brighter HII regions. In fact, the HII region N206 A at $\alpha_{50} = 5^{\mathrm{h}}31^{\mathrm{m}}.4$, $\delta_{50} = -71^\circ.1$ does not have any obvious effect on the RM distribution in Fig. 3, neither does it affect the distribution of polarized radiation (Fig. 1). This strongly suggests that the observed variation of the RM is governed by a change in the magnetic field orientation, rather than by variations of the thermal electron density. This is further corroborated by the observed change of the sign of RM. The possible scenario thence is that the magnetic field bends out of the plane of the LMC towards the observer, such that it points out of the LMC in the eastern filament, and into the LMC in the western one. This is sketched in Fig. 4.

As outlined in Chap. 2, the whole region south and south-west of 30 Dor is special. Recently Luks and Rohlfs (1992, see also Rohlfs 1992) established the existence of a second HI component, which is detached from the LMC main disk and most likely located in front of it. It is at the very locations of strong *RM* variations (Fig. 3) where the observed velocity differences between the so-called N-component and the LMC disk are largest and have the strongest gradients. It is therefore tempting to relate the unusual magnetic field morphology with the HI gas that appears to be receding from the LMC.

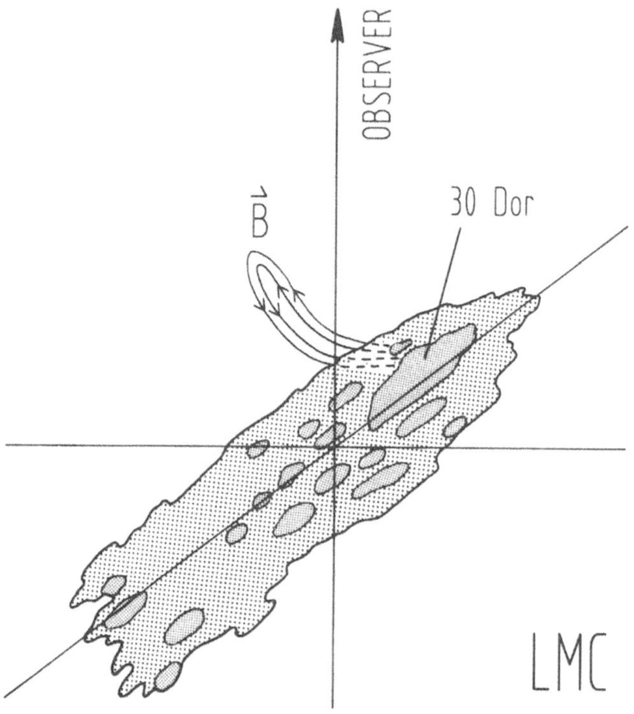

Fig. 4. Sketch of the likely magnetic field structure SW of 30 Dor. We are looking down here roughly onto the line of nodes of the LMC. The ***B***-field is bending out of the LMC south-west of 30 Dor, forming a giant loop.

At present, it is absolutely unclear what caused this peculiar magnetic field structure. In particular, the abrupt change in the polarization (polarized brightness and fractional polarization) need to be explained. A thorough study of the depolarization in this region should shed light on this question. This requires, however, the derivation of the distribution of the pure synchrotron radiation. Such maps are currently being produced (Meinert et al. in prep.). A preliminary analysis indicates that there is strong beam depolarization, i.e. a large fraction of the depolarization is caused by a random field component. The typical degrees of polarization in the two filaments are $\sim 10\%$, 30%, and 35% at $\lambda\lambda 21$, 12, and 6.3 cm, respectively, while for a uniform depolarizing slab with $RM \simeq 40\,\mathrm{rad\,m^{-2}}$ one would expect $\sim 30\%$, 68%, and 72% according to the models of Burn (1966), and supposing a maximum degree of polarization of 72% (where a spectral index of -0.7 was inferred, see Haynes et al. 1991). We are currently undertaking high-resolution observations of the linear polarization in this region of the LMC using the Australia Telescope. The only way to tackle the problem of the strong observed depolarization is via high-resolution observations.

References

Burn, B.J. (1966): Monthly Notices Roy. Astron. Soc. 133, 67

Cohen, R.S., Dame, T.M., Garay, G., Montani, J., Rubio, M., Thaddeus, P. (1988): Astrophys. J. 331, L95

Haynes, R.F., Klein, U., Wielebinski, R., Murray J.D. (1986): Astron. Astrophys. 159, 22

Haynes, R.F., Klein, U., Wayte, S.R., Wielebinski, R., Murray, J.D. et al. (1991): Astron. Astrophys. 252, 475

Klein, U., Wielebinski, R., Haynes, R.F., Malin, D.F. (1989): Astron. Astrophys. 211, 280

Laustsen, S., Madsen, C., West, R.M. (1987): *Exploring the Southern Sky* (ESO, Springer-Verlag, Berlin, Heidelberg), p. 53

Le Coarer, E. (1992): these proceedings

Loiseau, N., Klein, U., Greybe, A., Wielebinski, R., Haynes, R.F. (1987): Astron. Astrophys. 178, 62

Luks, T., Rohlfs, K. (1992): Astron. Astrophys. (in press)

Molaro, P. (1992): these proceedings

Pakull, M. (1992): these proceedings

Pietsch, W. (1992): these proceedings

Rohlfs, K. (1992): these proceedings

Rohlfs, K., Kreitschmann, J., Siegmann, B.C., Feitzinger, J.V. (1984): Astron. Astrophys. 137, 343

Xu, C., Klein, U., Meinert, D., Wielebinski, R., Haynes, R.F. (1992): Astron. Astrophys. 257, 47

The Nonthermal Radiation Component in the Large Magellanic Cloud

D. Meinert [1], U. Klein [2], C. Xu [3], R. Haynes [4], R. Wielebinski [1]

[1]Max-Planck-Institut für Radioastronomie, Auf dem Hügel 69, W-5300 Bonn 1, Fed. Rep. of Germany
[2]Radioastronomisches Institut der Universität Bonn, Auf dem Hügel 71, W-5300 Bonn 1, Fed. Rep. of Germany
[3]Max-Planck-Institut für Kernphysik, W-6900 Heidelberg, Fed. Rep. of Germany
[4]Australia Telescope National Facility, CSIRO Radiophysics, P.O. Box 76, Epping NSW 2121, Australia

Abstract: We have attempted to separate the thermal and nonthermal component of the radio emission from the Large Magellanic Cloud (LMC), using either Hα or far infrared data. On a large scale (51′ resolution $\hat{=}$ 750 pc) the differences arising from the different techniques are rather small. A significant variation of the nonthermal spectral index over the LMC is found. The main clue is a generally steeper nonthermal spectrum in the western part of LMC than in the eastern one (around the 30 Doradus region). This may be due to recent injection of cosmic ray electrons into the interstellar medium by supernovae in this active star forming region, which away from these regions age as they propagate and lose energy.

The recently obtained high-resolution maps of the Large Magellanic Cloud (LMC) in different radio frequencies (Haynes et al., 1986, 1991) offer the opportunity to separate the thermal and nonthermal component of the radiation. This poster presents the results obtained with three different methods.

The radio radiation from a galaxy is generally a superposition of thermal bremsstrahlung and nonthermal synchrotron emission. In the optically thin case the spectra of both mechanisms follow simple power-laws so that the superposition may be written as

$$S(\nu) = S_0 \left(f_0^{th} \nu^{-\alpha_{th}} + \left(1 - f_0^{th}\right) \nu^{-\alpha_{nth}} \right) \qquad (1)$$

Here $S(\nu)$ is the flux density observed at the frequency ν, S_0 is the flux density at 1 GHz, f_0^{th} is the thermal fraction at 1 GHz, α_{th} the thermal spectral index and α_{nth} the nonthermal spectral index. Basically one could apply a least-squares fit of the above form to each pixel in the radio maps. In practice this only works, however, with extremely good signal-to-noise ratios so that the method does not converge in low-brightness regions. We therefore decided to try and substitute the thermal radio emission using calibrated Hα and FIR data. The least-squares technique can still be applied to the integrated flux densities of the LMC. The results are compiled in Table 1.

Table 1. The integrated flux densities in LMC resulting from the separation. Column 1 lists the observing frequency, column 2 the integrated flux and its error. Columns 3 to 5 list the nonthermal flux derived by the different methods described in the text. Column 3 results from a least-squares fit, column 4 from the subtraction of a thermal component defined by the extinction corrected Hα flux density, and column 5 from the subtraction of a scaled FIR map.

Frequency	S_{tot} [Jy]	S_{nth}(fit) [Jy]	S_{nth}(Hα) [Jy]	S_{nth}(FIR) [Jy]
408 MHz	925±50	648	619	675
1.4 GHz	493±30	286	266	269
2.45 GHz	400±30	198	187	188
4.75 GHz	368±30	127	162	165
8.7 GHz	265±50	85	76	80
Hα	185±40			
FIR	155905±1200			
α_{nth}		0.66±0.06	0.65± 0.04	0.65±0.06

The thermal radio emission can be calculated via its proportionality to the Lyman continuum (Lyc) photon rate by using Hα or FIR data which are directly connected to the Lyc rate, too. It is then possible to find a relation between the FIR or Hα brightness and the thermal radio brightness at frequency ν of the kind $R_\nu = \frac{S_\nu}{S_\alpha}$ where S_ν is the radio brightness and S_α is the FIR or Hα brightness of the source.

The integrated nonthermal spectral indices of the LMC obtained with different methods do not differ significantly (last row of Table 1). This indicates that all methods are quite reliable in the separation, at least for the integrated flux.

Applying the methods described in the last paragraph to maps of the Large Magellanic Cloud results in maps of the nonthermal emission. Fig. 1 shows the distribution of the nonthermal radio emission resulting from the substitution of the thermal radiation component by the scaled FIR emission. The strong HII regions can still be clearly identified, indicating that not enough thermal emission has been subtracted, i.e. the scale factor R_ν is still to low or the temperature dependance of R_ν has not adequately been taken into account. The known supernova remnants that can be discerned at this resolution, and also some background sources inside the field of the LMC disk, are visible, too. The SNR are indicated in Fig. 1 by crossmarks. The overall structure of the nonthermal emission resembles a kidney-shaped distribution, with its bow towards the east. Maybe this indicates the effects of ram pressure in the east of the LMC as it moves through the galactic halo gas.

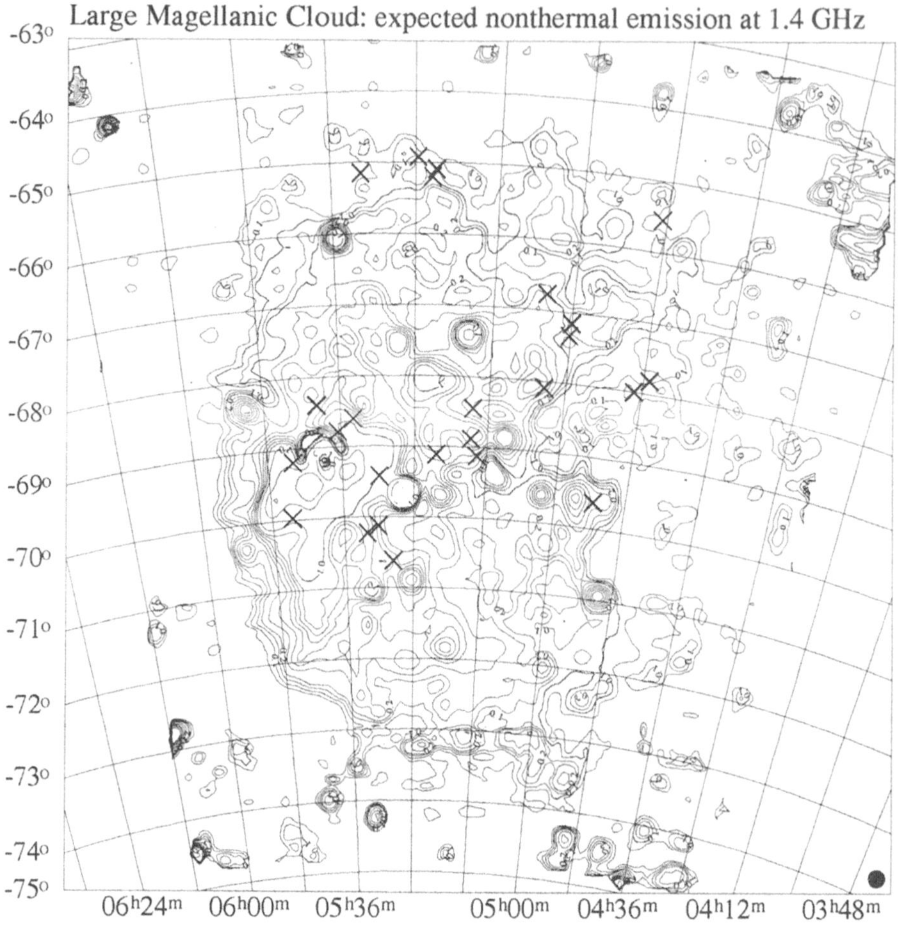

Fig. 1. LMC nonthermal emission derived by subtraction of the scaled FIR map. The contour levels are 0.05, 0.1, 0.15, 0.2, 0.3, 0.4, 0.5, 0.6, 0.7, 0.8, 1.0, 1.5, 2.0, 3.0 Jy/beam area.

References

Haynes, R. et al. (1991): Astron. Astrophys. 252, 475

Haynes, R., Klein, U., Wielebinski, R., Murray, J.D. (1986): Astron. Astrophys. 159, 22

Lonsdale, C., Helou, G., Good, J.C., Rice, W. (1985): *Catalogued Galaxies and Quasars Observed in the IRAS Survey* (Jet Propulsion Laboratory, Pasadena)

Mezger, P.G. (1978): Astron. Astrophys. 70, 565

Rohlfs, K., Kreitschmann, J., Siegmann, B.C., Feitzinger, J.V. (1984): Astron. Astrophys. 137, 343

Xu, C., Meinert, D., Wielebinski, R., Haynes, R. (1992): Astron. Astrophys. 257, 47

Interstellar Absorption-Line Observations of the Magellanic Clouds

Paolo Molaro

Osservatorio Astronomico di Trieste, Via G. B. Tiepolo 11, 34131 I, Trieste, Italy

Abstract: A brief overview of the interstellar absorption-line observations of the Magellanic clouds is given, with particular emphasis on the results obtained by the ESO Key-Program for the Magellanic Clouds. It is shown that high resolution interstellar observations towards bright MCs sources can probe the large scale distribution of the interstellar gas, revealing their 3-dimensional structure. Current views of the intermediate velocity clouds are challenged by new observations, which show that this gas is clumpy, in the foreground and of low elemental abundance. The LMC shows a complex structure and the case for a substantial amount of gas emitting in 21-cm at V$\approx$ 270 km s^{-1} behind the stellar system is receiving new support. The absorption line data accumulated so far show that the SMC is definitively a disrupted system composed of at least four separate bodies.

1. Introduction

Interstellar absorption-line observations of the Magellanic Clouds were pioneered by Feast (1960) with the first extragalactic CaII IS spectra towards bright supergiants in the LMC. But it is only in the recent years that the improvement of the ground-based instrumentation has made possible to collect interstellar CaII and NaI with a resolution of a few km s^{-1}, required to separate the fine velocity structure of the LMC gas. The most extensive surveys were made by Songaila *et al.* (1986) and by Wayte (1990). Songaila *et al.* (1986) discussed all the observations made by them in the period between 1980 and 1986, which consisted of 48 stars (30 in the LMC, and 18 in the SMC) at a resolution of $\approx$ 8-15 km s^{-1}. They also made a first effort at interpretation of the absorption lines with reference to the large-scale structure of the interstellar gas of the Magellanic Clouds. Wayte (1990) observed 24 stars in CaII (11 in the LMC and 13 in the SMC) at 5 km s^{-1} of resolution. Neither of these two surveys make a quantitative analysis of the interstellar spectra, and the discussion remains at a rather qualitative level. A number of other papers deal with only a relatively small number of lines of sight (Blades (1980), Blades and Meaburn (1980), Blades *et al.* (1980), Songaila and York (1980), Meaburn and Blades (1980), Songaila (1981), Songaila *et al.* (1981), Ferlet *et al.* 1985, Martin *et al.* 1989, Vladilo *et al.* 1989, Molaro *et al.* 1989). Recent extensive reviews of the absorption line studies have been made by Ferlet and Vidal-Madjar (1989) and de Boer (1991).

The ultraviolet domain became accessible with the IUE satellite, leading to major discoveries. It revealed the presence of highly ionized gas, of the intermediate velocity components, and allowed the measurement of the column density for a variety of elements. The data base is very limited and so far only four lines of sight have been studied in detail in the ultraviolet region. The abundance analysis, in particular for the non-refractory elements zinc and sulfur, gives abundances which are below solar by a factor 3 for the LMC, and by a factor ten for the SMC. The element-to-element abundance variation is very similar to the depletion pattern observed in our own Galaxy, suggesting that it is produced by the same kind of dust. The ultraviolet observations were reviewed by de Boer (1987, 1991) and since no new material has been brought forward in recent years, these will not be discussed here again.

SN1987A plays a special role; when it was a 4th magnitude source it provided the best optical and UV spectra we have for a line of sight towards the MCs. In addition to the strong CaII and NaI, other faint interstellar lines have been studied, as CaI, KI, [FeX], LiI (cfr. Ferlet and Vidal-Madjar (1989) for detailed references). Two recent papers have returned to these issues. Wampler *et al.* (1991) claimed that the absorption at 6379.7 Å, instead of forbidden [FeX] as proposed by Pettini *et al.* (1988), can be ascribed to the galactic DIB at 6379.3 Å, if the ratio of the DIBs 6379/6373 is anomalously high, as they observe towards some highly reddened galactic lines of sight. Direct measurements of these DIBs in the foreground galactic gas in the direction of SN1987A should be able to further test Wampler *et al.*'s claim. Baade *et al.* (1991) carried out a new search of interstellar lithium along the line of sight of SN1987A, co-adding 34 Reticon spectra taken with the CAT telescope at La Silla. Though Despite they obtain an exceptionally high signal-to-noise ratio (> 3000), the line remains undetected. The new upper limit at 3σ is 0.073 mÅ, better by a factor 2 than the previously quoted values, and the corresponding elemental abundance of lithium does not exceed $2.7\cdot 10^{-10}$, which is very close to the Pop II value and to the presently believed primordial value. Unfortunately, this upper limit is not tight, and when the cumulative errors are taken into account a Li/H as high as $\approx 10^{-9}$ is not firmly ruled out.

In the following the absorption-line observations, essentially CaII and NaI, will be considered, with some emphasis on the observations in the SN1987A field obtained in the framework of the ESO Key-programme and discussed by Molaro *et al.* (1992) and Vladilo *et al.* (1992). Two important aspects of these works, i.e. reddening and the role of the Clouds in probing the fine structure of the local gas, are omitted here.

2 Intermediate Velocity Components

The features with anomalous velocities with respect to those of the local disk arm and to the systemic velocities of the Large ($V_{hel}\approx$ 260 km s^{-1}) and Small Clouds ($V_{hel}\approx$ 150 km s^{-1}) are called intermediate velocity components (hereafter IVC). Material at $V_{lsr}\approx$ 60, 130 and 170 km s^{-1}, $\pm$ 25 km s^{-1} has been detected by Savage and de Boer (1979, 1981) by means of IUE observations of MC supergiants (velocities are given in the reference system of the original work; $V_{hel}\approx V_{lsr}+15$ for the LMC).

Later these features were detected in the optical CaII lines (Blades 1980, Songaila and York 1980), and in radio through measurements sensitive to hydrogen column densities of a few times 10^{18} cm^{-2} (Mc Gee and Newton, 1983). Towards SN1987A, components are seen at 70, 129 and 171 km s^{-1} in various ions, as FeII, MgI, MgII, MnII, CII, CIV and AlIII (Blades *et al.* 1988). The higher resolution available in the optical CaII lines made it possible to resolve these three major components into ten distinct subcomponents (Vidal-Madjar *et al.* 1987).

The nature of the IVCs is not understood and we do not know whether they are related to the 21-cm high velocity cloud phenomenon or to gravitational interaction of the Magellanic Clouds with the Galaxy. Their location is controversial. The various arguments which have been put forward to support the different interpretations are listed in Ferlet and Vidal-Madjar (1989). Savage and de Boer (1981) placed the 60 and 170 km s^{-1} components in a galactic extended halo at a distance of a few Kpcs, leaving the 170 km s^{-1} component associated to the LMC. Songaila *et al.* (1986) claimed that the IVC at 220 km s^{-1} observed in the direction of the SMC is immersed in the Small Cloud, and argued that *"the simplest assumption is that all the weak absorption along the line of sight arises in low column density material within the Magellanic Clouds".*

Two papers have recently addressed the case of the origin of IVCs by looking at their presence at 21-cm outside the face of the Clouds. de Boer *et al.* (1990), with the 30 m dish of Villa Elisa, scanned Declination -67° from about 4 to 6 hours in R.A. The $V_{lsr} \approx$ 70 km s^{-1} is found at all the positions, while material emitting at $V_{lsr} \approx$ 140 km s^{-1} is observed only in a few positions, which in same cases are more than 3 degrees beyond the boundary of the LMC. Wayte (1990) observed an extended grid ($-50 < b < -15$, $260 < l < 320$) with the Parkes radio telescope. The $V_{lsr} \approx$ 60 km s^{-1} is found everywhere, the $V_{lsr} \approx$ 130 km s^{-1} appears in correspondence with the LMC, with the Bridge, and also at about 20° to the north of the LMC; the $V_{lsr} \approx$ 170 km s^{-1} is localized in a small region centered on the LMC. Thus, these new radio observations suggest that the IVCs are **not** spatially associated with the Magellanic Clouds. However, Wayte locates the 170 km s^{-1} component *in* the LMC on the basis of observed CaII absorption lines; we will return to this point later on showing that other explanations are possible.

Abundances for the SN1987A IVCs have been derived by Blades *et al.* (1988) and used as a discriminant for their association with the MW or with the LMC. According to Blades et al (1988) the 70 km s^{-1} component has [M/H]= 0.23 $\pm$0.20 and appears composed of galactic gas with solar composition, while the 130 and 171 km s^{-1} components have -0.59$\pm$0.15 and -0.46$\pm$ 0.15, respectively, and are made up of gas processed in the LMC.

A new set of about a dozen NaI and CaII high resolution (3-7 km s^{-1}) observations concentrated in a small field of 30x30' around SN1987A have been discussed by Molaro *et al.* (1992) and Vladilo *et al.* (1992). IVCs are detected in almost all the CaII spectra, even if they are often very weak (log N(CaII)$\approx$ 11.0). A remarkable exception is that of the two close lines of sight towards Sk-69 211 and Sk-69 216, which show several, rather strong IVCs. In Fig. 1 the CaII and NaI spectra of Sk -69 211 are given as examples. A close comparison of the different lines of sight shows that the IVCs clump on scales which are smaller than the surveyed field and perhaps as small as a few arcminutes. This becomes particularly clear by comparing the lines of sight towards Sk-69 211, Sk-69 216 and SN1987A. The two Sanduleak stars are separated by about

2 arcminutes, and are about 8 arcminutes from SN1987A. Only the 64 and 170 km s^{-1} components are shared by all three lines of sight, the 90 km s^{-1} by two but several components are seen in only one line of sight, such as the 100 km s^{-1} component observed in Sk-69 216, and the 121 and 2 16 km s^{-1} components in SN1987A.

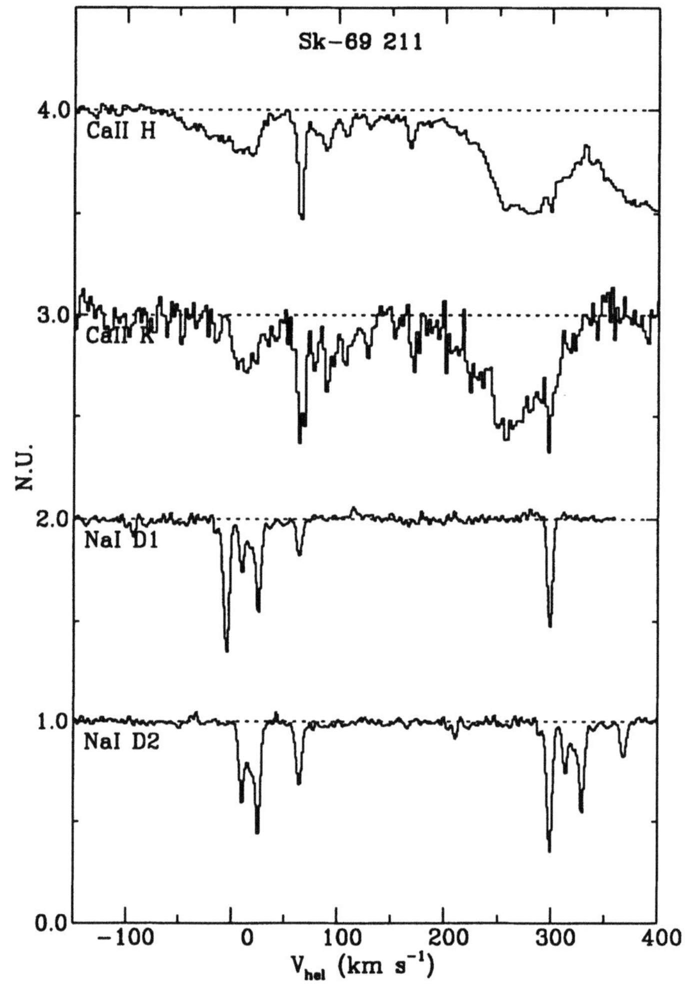

Fig. 1. CaII and NaI interstellar doublet resonance lines of Sk-69 211 in heliocentric velocities. Owing to the small separation of the NaI doublet ($\approx$ 300 km s^{-1}), a portion of the D2 line is duplicated on the D1 spectrum at negative velocities.

The observed variations do not help to solve the distance problem, since we do not know the typical linear scale (if there is one!) of the IVCs. The size of the field corresponds to about 0.45 Kpc at the LMC distance, but it is only 9 pc at a distance of 1 Kpc, and the observed variations are compatible with any distance between us and the LMC. However, when a component is shared by several lines of sight this would imply the presence of an extended body of gas at the LMC distance. An interesting case is that of the 64 km s^{-1} component, which is detected in all the stars in the southern region of our field. The measured radial velocity is exceptionally constant, the mean radial velocity of the sodium feature towards 5 lines of sight spread over 20 arcminutes in the sky being 64.4 $\pm$ 0.3 km s^{-1}. At the LMC distance the angular size implies an extension of $\approx$ 0.4 Kpc, a dimension at which it becomes very unlikely to maintain such a constant velocity, providing strong evidence that the cloud is relatively close to our own Galaxy.

An important implication of the observed clumpiness of the IVCs gas is that this makes it nearly impossible to infer safe conclusions about the IVCs location by comparing the observed features towards different lines of sight, even if they differ by only a few arcminutes. For instance, Wayte's (1990) conclusion that the IVC at $\approx$ 170 km s^{-1} belongs to the LMC is based on CaII interstellar observations of the stars SK-69 51 and Sk-69 52, which are separated by $\approx$ 12 arcminutes. Towards Sk-69 51 there are two components at $V_{hel} \approx$ 157 and 177 km s^{-1}, while only the lower velocity is seen in the other star. Wayte (1990) argued that the 177 km s^{-1} is behind the lower-velocity and the main LMC component present in both the spectra. It is evident that clumping of the gas on scales smaller than the angular separation of the two stars will invalidate Wayte's conclusion. Similar arguments can be opposed to Songaila *et al.*'s (1986) claim that the 220 km s^{-1} IVC is within the SMC. As a matter of fact, there is no unambiguous evidence that IVCs are immersed in eitherthe LMC or the SMC. On the contrary, Sk-69 195, shown in Fig. 2, provides some evidence which suggests the opposite.

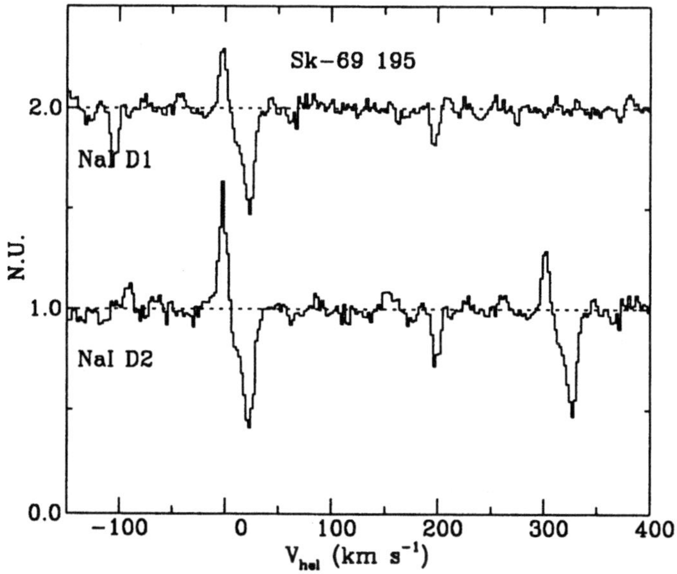

Fig. 2. NaI D1 and D2 resonance lines for Sk-69 195 in heliocentric velocities

From the absence of interstellar absorptions at the LMC velocity we know that this star is located in the foreground of the main disk component of the LMC, and therefore the IVC detected in NaI at $\approx$ 200 km s^{-1} in Sk-69 195 is, like the star, in front of and not mixed with the main LMC gas.

IVCs are only seldom detected in NaI. In Fig. 3 the CaII/NaI ratios for all the interstellar features measured in the field of SN1987A are shown. The values of the IVCs are always higher than those of the galactic or LMC gas and invariably >> 1, showing evidence for the Routly-Spitzer effect. In a few extreme cases the ratios are larger than 30, reaching the Routly-Spitzer limit regime (Hobbs 1983). This effect is understood as the result of grain destruction by sputtering in the accelerated gas. Thus the high CaII/NaI ratios strongly suggest that the IVCs are free of dust, as does the absence of any depletion pattern among the relative abundances of the various elements in the SN1987A IVCs (Blades *et al.* 1988). A remarkable exception is that of the 216 km s^{-1} component toward the SN1987a, which has a CaII/NaI value close to unity. However, this cloud appears to be special in many respects, as pointed out by Andreani and Vidal-Madjar (1988), and D'Odorico *et al.* (1991).

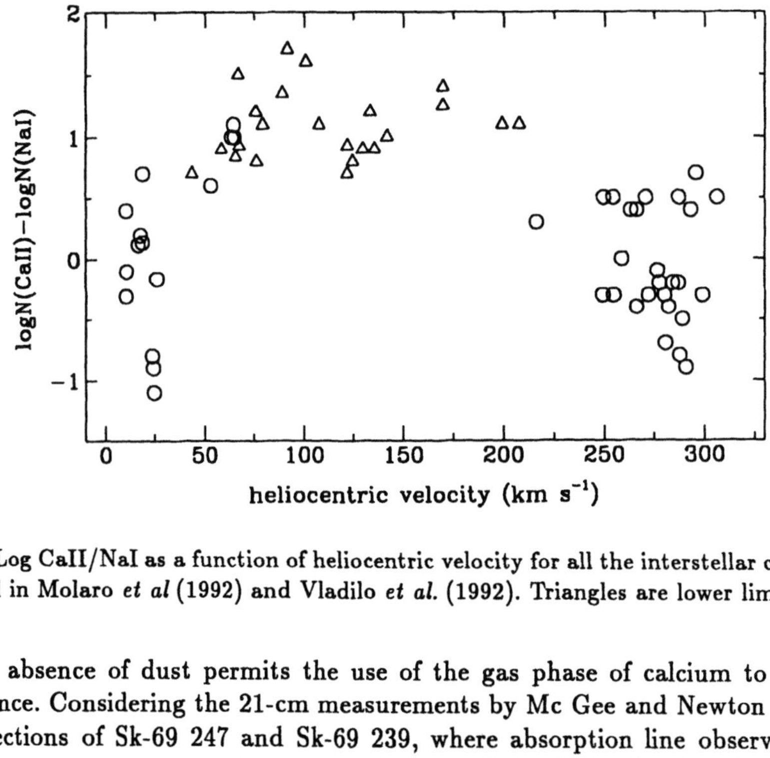

Fig. 3. Log CaII/NaI as a function of heliocentric velocity for all the interstellar components reported in Molaro *et al* (1992) and Vladilo *et al.* (1992). Triangles are lower limits.

The absence of dust permits the use of the gas phase of calcium to derive its abundance. Considering the 21-cm measurements by Mc Gee and Newton (1986) for the directions of Sk-69 247 and Sk-69 239, where absorption line observations are available, a close correspondence exists in the radial velocity for the clouds at V_{hel} of 55 , 64 and 122 km s^{-1}. Combining the column densities, we have log (CaII/HI)$\approx$ -7.7 for all the three clouds. The average galactic fraction of twice-ionized calcium is about four times that of the singly ionized species (Phillips *et al.* 1984), which is a conservative value for the IVC where the radiation field is probably diluted compared to the disk of the Galaxy. When the fraction of twice-ionized calcium is taken into account the calcium abundance is $\approx$ -1.4 dex lower than the solar value. The difference between this abundance and the values derived from SN1987A might originate from errors in the hydrogen column density. Blades *et al.* adopted an average value for HI from Mc Gee and Newton (1986), which cannot be adequate for the SN1987A direction where the IVCs appear stronger than towards other directions. Unfortunately, the particularly low abundance we found for the IVCs is not very informative of their LMC or Galactic membership.

3 The LMC

The large-scale distribution of the neutral hydrogen in the LMC cannot be derived straightforwardly. About half of the emission profiles exhibit a multi-peak morphology, which generates diverging interpretations. Mc Gee and Milton (1966) found the emission peaks at 234 ± 10 km s^{-1}, 270 ± 10 km s^{-1}, and 300 ± 10 km s^{-1} and suggested that the gas at 300 km s^{-1} and 240 km s^{-1} belongs to two spiral arms, while two smaller bodies at 273 km s^{-1} belong to two other arms on a different plane. The presence of overlapping HI sheets with sizes of a few Kpcs has been confirmed more recently by Meaburn *et al.* (1987). On the other hand, Rohlfs *et al.* (1984), considering the weighted mean radial velocity of the profiles, have been able to reconstruct a smooth rotational-like structure for the LMC (see also Rohlfs these proceedings). An important point to clarify is whether the multi-peak emission features refer to rather localized phenomena or are a global feature of the LMC. Comparison between 21-cm emission and absorption profiles might help in locating the material and in understanding if there are two or more spatially separated bodies of gas.

Songaila *et al.* (1986) found that there are many lines of sight that show absorbing components at velocities which deviate from the rotational curve derived by Rohlfs *et al.* (1984). They pointed out that the 270 km s^{-1} gas appears to be located behind the material at lower and higher velocities. Some stars (R87, R89, Sk-69 91 and Sk-68 155) located to the southwest and to the northeast of the Tarantula nebula, do not show interstellar CaII absorption at 270 km s^{-1}, even when the Mc Gee and Milton (1966) maps show 21-cm emission at this velocity. Wayte (1990) found the same result from Sk-69 51 and Sk-69 52, which are located in the far southwest of the LMC, where the 270 gas is found behind the 234 km s^{-1} material.

New evidence for the region around SN1987A is provided by Vladilo *et al.* (1992) and the cases of Sk-69 213 and Sk-69 215, where the sodium interstellar lines are compared with the closest Rohlfs (1984) profile, are shown in Fig. 3 and Fig. 4. For Sk-69 213 in correspondence with the HI maximum emission at 270 km s^{-1} there is no evidence of absorption. For Sk-69 215, a small feature is present in absorption at 270 km s^{-1}, among several other components. To account for the complexity observed in Sk-69 215 a multi-component fitting of the 21-cm profile is performed, entering the position of the detected absorptions and leaving the hydrogen column density free to vary. The deblending procedure gives the column density for each component and it is found that for the 288 km s^{-1} component, which here is the disk component, log (HI/NaI) is 8.5, but for the 270 km s^{-1} component it is ≈ 9.6, confirming that the bulk of the gas responsible for the the 21-cm emission at 270 km s^{-1} is beyond the gas seen in absorption.

Wayte (1990) claims to have found evidence that the 270 km s^{-1} layer is in front of the 290 in the northeast of 30 Dor, in contradiction with the indications of Songaila *et al.* (1986). However, a close inspection to the profiles for Sk-69 120, Sk-69 166, Sk-69 135 and Sk-69 138 shows that 21-cm emission is always present but not CaII in absorption at ≈ 270 km s^{-1}; towards Sk-68 135 both the 270 and 300 km s^{-1} components are present and so are not informative about their mutual position.

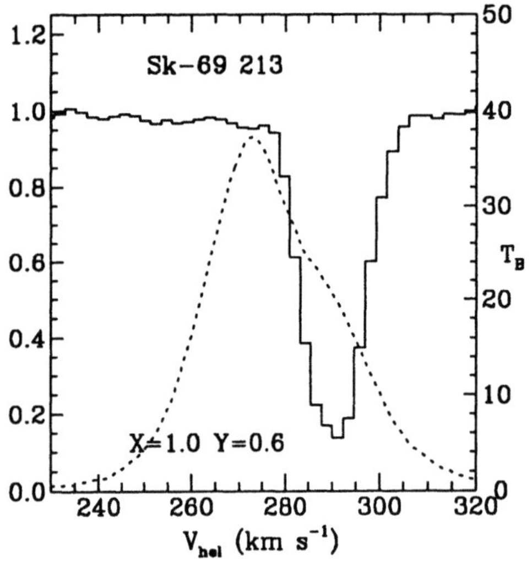

Fig. 4. NaI D1 absorption for Sk-69 213 superimposed on the 21 HI cm emission (dotted line, right hand vertical axis) at X=1.2, Y=0.6 from Rohlfs et al (1984)

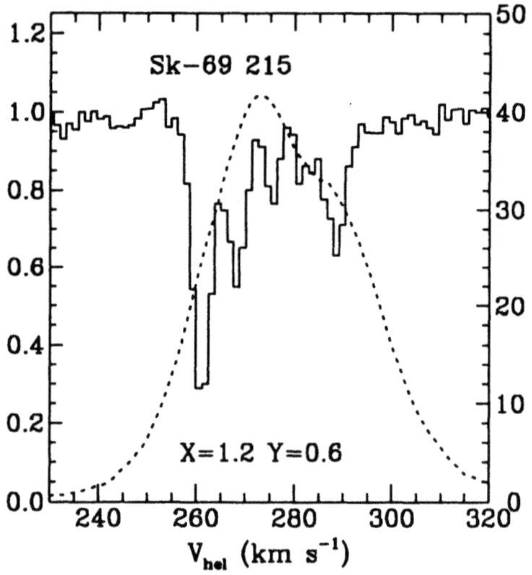

Fig. 5. NaI D1 absorption for Sk-69 215 superimposed on the 21 HI cm emission (dotted line, right hand vertical axis) at X=1.0, Y=0.6 from Rohlfs et al (1984)

Five out of 13 stars from the sample of Vladilo *et al.* (1992) show, like Sk-69 215, more than 5 resolved components spanning from 250 km s^{-1} up to 300 km s^{-1}, revealing that the gas has a more complex structure than expected on the basis of the 21-cm profiles, which in this area have two discernible components at about 270 and 290 km s^{-1}. An interesting point is to understand whether there is any ordered structure in the absorbing material. An indication in this sense comes from the clustering of the different components at V_{hel} ≈ 271, 289 and 299 km s^{-1}. The fact that these components appear to be also located in defined areas in the field and that the ratios CaII/NaI are similar for each component lends support to the existence of different macroscopic structures in the region between 30 Dor and SN1987A.

4 The SMC

In the radio survey of Mc Gee and Newton (1981) essentially all the 21-cm profiles obtained in the main body of the SMC show two, three or even four peaks. According to Martin *et al.* (1989) the velocities of these four components are V_{hel} ≈ 110 km s^{-1}, ≈ 132 km s^{-1}, ≈ 169 km s^{-1}, ≈ 190 km s^{-1} and the gas fractions are 6%, 34%, 52% and 7% respectively. The 132 and the 169 km s^{-1} features, which are the two components originally identified by Hindman (1967), share the bulk of the gas and cover the largest areas, well outside the optical border of the SMC, while the other two components appear less intense and less pervasive. The 110 km s^{-1} component is rather localized in the southwest region of the optical body of the SMC. Mathewson *et al.* (1984) suggested that the two main bodies are separate entities: the *Mini-Magellanic Cloud* (MMC, V≈ 169 km s^{-1}) and the *Small Magellanic Cloud Remnant* (SMCR, V≈ 132 km s^{-1}), which resulted from a collision between the two Clouds. Using the absorption line data from Feast *et al.* (1960) they noted that the low velocity component at 132 km s^{-1} is invariably present in the stellar spectra and deduced that the MMCR is in front of the MMC. This first suggestion is now supported by a large body of data from Songaila *et al.* (1986), Mathewson *et al.* (1988) Martin *et al.* (1989) and Wayte (1990) (see also Maurice and Silvy these proceedings). In particular Wayte (1990) added a significant number of new observations which do not leave any doubt that the component at 132 km s^{-1} is in front of the 169 km s^{-1} component. The smaller bodies at $V_{hel} \approx$ 100 km s^{-1} and at 190 km s^{-1} are seldom identified and it is not clear how they are related to the two major bodies. Some evidence is provided that the 100 km s^{-1} is the closest to us and the 190 km s^{-1} is in the rear of the SMC. Absorptions are also detected without emission counterparts at $V_{hel}<$ 100 km s^{-1} and $V_{hel}>$ 200 km s^{-1}; in particular, a feature at 220 km s^{-1} is observed in several points of the SMC. As we have already mentioned Songaila *et al.* (1986) claimed that this component is sandwiched between the two main bodies, but the observations of Wayte (1990) call into question the Songaila *et al.* (1986) classification. It might still be true that this gas at this velocity is interspersed in the Small Cloud, but its complex behaviour with respect to the other absorption layers of the SMC is much more simply explained by a clumpy foreground gas projected on the SMC.

Future observations with a greater spatial coverage could reveal velocity gradients, warps, and dimensions of the various gas structures which might provide new insights into the origin of this highly disrupted galaxy.

Acknowledgements. It is a pleasure to express my gratitude to all the members of the interstellar group of the ESO Key-Programme: G. Vladilo, S. Monai, S. d'Odorico, R. Ferlet, A. Vidal-Madjar and M. Dennefeld for their valuable collaboration, and for allowing me to present our joint work before official publication. I also warmly thank Klaas de Boer for his efficient coordination of the Key-Programme and for his keen support of this work.

References

Andreani P. and Vidal-Madjar A.:1988, Nature **333**,432.

Baade D., Cristiani S., T. Lanz, R.A. Malaney, K.C. Sahu, and Vladilo, G.: 1991, Astron. Astrophys. **251**, 253.

Blades J.C.:1980, M.N.R.A.S. **190**,33

Blades J.C. and Meaburn J.:1980, M.N.R.A.S. **190**,59

Blades J.C., Meaburn J., and Elliot K.H.:1980, M.N.R.A.S. **192**,101

Blades J.C., Wheatley J.M., Panagia N., Grewing M., Pettini M. and Wamsteker W.:1988, Ap. J. **334**,308

de Boer K.S. and Savage B.D.:1980, Ap. J. **238**,86

de Boer K.S., Jura M.A., Shull J.M., 1987b, Exploring the Universe with the IUE satellite, ASSL Vol. 129, 485.

de Boer K.S., Morras R., Bajaja E.:1990, Astron.Astrophys. **233**,523
de Boer K.S., 1991, The Magellanic Clouds and their Dynamical Interaction with the Milky Way, IAU Symp. No. 148, ed. Haynes, p. 401
D'Odorico S., Molaro P. and Vladilo G., 1991, A&A 247, L5.
Feast M.W., Thackeray A.D., Wesselink A.J.:1960, M.N.R.A.S. **121**,337
Ferlet R., Dennefeld M. and Maurice E.:1985, Astron. Astrophys. **152**,151
Ferlet R. Vidal-Madjar A.: 1989 in "Recent developments of Magellanic Clouds research" a European colloquium, de Boer, Spite, Stasinska eds.,
Hindman J. V.: 1967, Aust. J. Phys. **20**, 147.
Hobbs L.M.:1983, Ap. J. **265**,817
Martin N., Maurice E. and Lequex J.: 1989 **215**, 219.
Mathewson D.S., Ford V.L.: 1984, I.A.U. Symp. N. **108**, 125.
McGee R.X., and Newton L.M.: 1981, Proc. Astron. Soc. Aust. **4**, 189.
McGee R.X., Newton L.M. and Morton D.C.:1983, M.N.R.A.S. **205**,1191
McGee R.X., and Newton L.M.:1986, Publ. Astron. Soc. Austral. **6**,358
Meaburn J. and Blades J.C.:1980, M.N.R.A.S. **190**,403
Meaburn J., Solomon N., Laspias V. and Goudis G.:1989, Astron. Astrophys. **225**,497
Molaro P., Vladilo G., Avila G. and D'Odorico S.:1989, Ap. J. *Letters* **339**,L63
Molaro P., Monai S., Vladilo G., D' Odorico S., Ferlet R., and Vidal-Madjar, A., Dennefeld M. 1992, submitted to A&A.
Pettini M., Stathakis R., D'Odorico S., Molaro P., and Vladilo G.: 1989, Ap. J. **340**, 256.
Phillips, A.P., Pettini, M., and Gondhalekar, P.M.: 1984, M.N.R.A.S. **206**, 337.
Savage B.D. and de Boer K.S.:1979, Ap. J. **230**,L77
Savage B.D. and de Boer K.S.:1981, Ap. J. **243**,460
Songaila N. and York D.G.:1980, Ap. J. **242**,976
Songaila N.:1981, Ap. J. *Letters* **243**,L19
Songaila N., Cowie L.L. and York D.G.:1981, Ap. J. **248**,956
Songaila N., Cowie L.L., Blades J.C. and Hu E.M.:1986, Ap. J. **303**,198
Vidal-Madjar A., Andreani P., Cristiani S., Ferlet R., Lanz T. and Vladilo G.:1987, Astron. Astrophys. **177**,L17
Vladilo G., Centurion M., Molaro P. and Monai S., 1989, Recent Developments of Magellanic Clouds Research, eds. de Boer et al., p. 139
Vladilo G., Molaro P., Monai S., D'Odorico S., Ferlet R., Vidal-Madjar A. and Dennefeld M.: 1992, Astron. Astrophys. *submitted*
Wampler E. J., Chen, and G. Setti: 1991 Astron. Astrophys. **248**, 633.
Wayte S.R.:1990, Ap.J. **355**,473

Interstellar Sodium in the Magellanic Clouds at High Spectral Resolution

Eric Maurice, Jocelyne Silvy

Observatoire de Marseille, 2 Place Le Verrier, F - 13248 Marseille, France

Abstract: We have observed the interstellar sodium in front of 36 supergiants in both the Small and the Large Magellanic Cloud. The spectral resolution is of the order of 65 000 and the spectra clearly show several components due to the presence along the line of sight of four complexes of stars and gas. Comparaison is made with observations of the 21 cm HI emission line. Information on the local structure and position of these complexes is inferred from this study. Examples are given.

The structure of the Magellanic Clouds has been the object of much work and debate during the recent years. The careful study of the stellar data and of the association of the stars with the HI gas has given the indication of the existence and of the location of various complexes of HI, stars and HII regions. However, the determination of the relative location of these complexes along the line of sight remains in all cases extremely uncertain. The quasi-unique source of information is the presence and the structure of interstellar lines, and the radial velocities of their components.

We have undertaken the study the interstellar Sodium D lines in the spectra of the brighter early type supergiants of the Magellanic Clouds. The presence of the two D1 and D2 lines allows to better ascertain the structure of the line profiles and these lines are favourably situated in the region of higher efficiency of CCD detectors.

Observations were secured at ESO, La Silla in Chile with the Coudé Echelle Spectrometer (CES) equipped with a CCD at the Coudé focus of the 1.4 m Coudé Auxiliary Telescope (CAT). In these observing conditions, a signal-to-noise ratio of the order of 30 is obtained for a two hours integration of a twelfth V magnitude early type (O-B-A) supergiant.

Data reduction: In addition to the standard reduction procedure, a special care has been taken on the following points: radial velocity zero-point correction, substraction of the atmospheric absorption spectrum by the division of the spectrum by that of of a bright star, observed the same night and showing no or very little stellar and interstellar lines, correction of the "cosmic rays impacts" which are rather numerous on our two hours integrations. In a future step of our programme, we will make a more refined substraction of the atmospheric absorption spectrum using the synthetic atmospheric spectrum published by Lundström et al. (1991).

In the figure below, we show the comparison of the Sodium interstellar D lines with the 21 cm HI emission line on approximately the same lines of sight. The general agreement is excellent. In the case of the SMC, we confirm that, on the line of sight of the stars shown here, the L complex, as defined by Martin et al. (1989), is situated in front of the H complex. For other stars we have observed, this general picture is not respected; our observations reveal some local peculiar structures such as warps. These results will be presented in detail in a forthcoming paper. In the case of the LMC, the spectrum of Sk -69° 289 shows that the high velocity complex is situated in front of the low velocity one but other observations show the contrary. The aim of our work is to clarify the structure of these regions in the LMC.

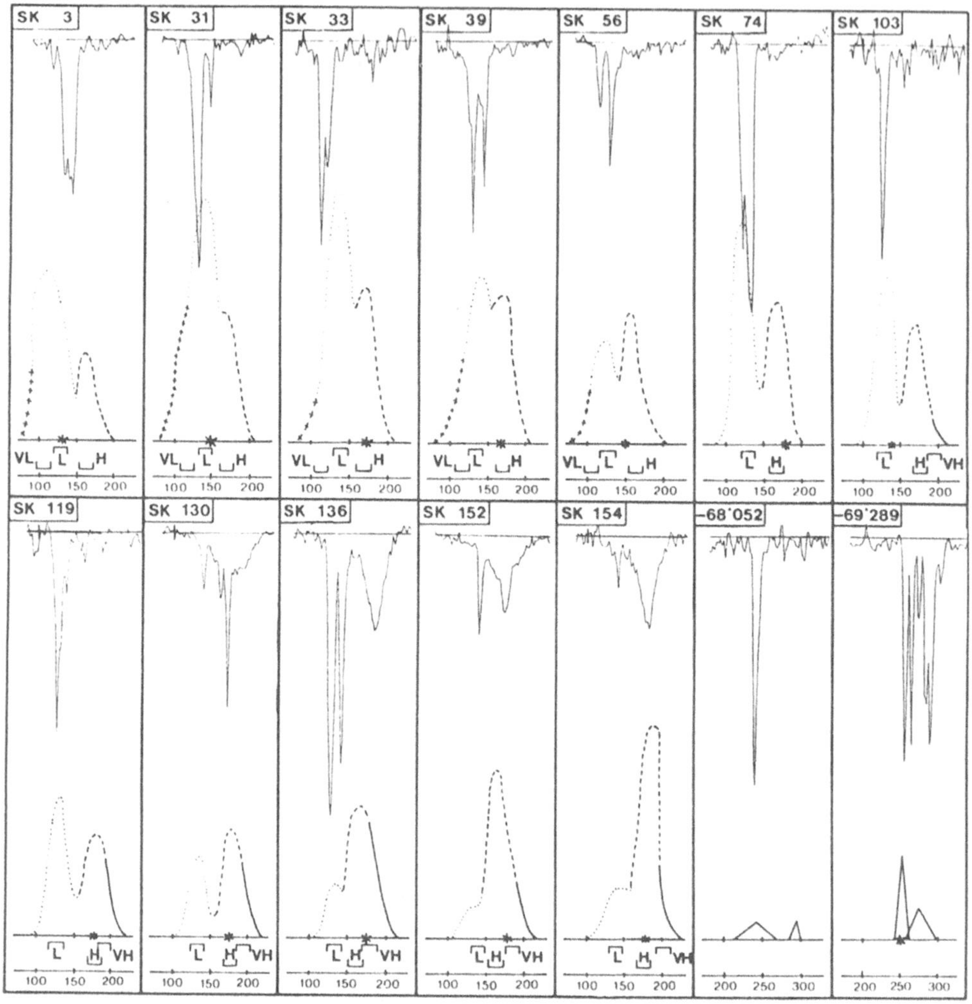

Legend to the figure: In the upper part of each diagram, the D2 line is shown except in the case of the two LMC stars (D1). It is compared to the HI emission line from Mc Gee and Newton (1981) for the SMC and from Rohlfs et al. (1984) for the LMC. The heliocentric velocity scale, in km s^{-1}, is the same for the whole figure. An asterisk indicates the stellar radial velocity from Ardeberg et al. (1972) for the LMC, and from Maurice et al. (1989) for the SMC.

In the case of the LMC, for which HI profiles were not published, these latter are schematically represented as triangles. The height of the triangle refers to the peak brightness temperature T_k and the half base of the triangle to the velocity dispersion of components in km s^{-1}.

In the case of the HI line in the SMC, crosses correspond to the very low velocity complex [VL], dotted lines to the low velocity complex [L], dashed lines to the high velocity complex [H], and continuous lines to the very high velocity complex [VH]

References

Ardeberg, A., Brunet, J.P., Maurice, E., Prévot, L.:1972, Astron. Astrophys. Suppl. Ser. **6**, 249.

Lundström, I., Ardeberg, A., Lindgren, H., Maurice, E.: 1991, Astron. Astrophys. Suppl. Ser. 91, 199.

Martin, N., Maurice, E., Lequeux, J.: 1989, Astron. Astrophys. **215**, 219.

Maurice, E., Bouchet, P., Martin, N.: 1989, Astron. Astrophys. Suppl. Ser. **78**, 445

Mc Gee, R.X., Newton, L.M.:1981, Proc. Astron. Soc. Australia, 4(3), 308.

Rohlfs, K., Kreitschmann, J., Siegman, B.C., Feitzinger, J.V.: 1984, Astron. Astrophys. 137, 343.

H II Regions
Massive Stars Affecting the ISM

Michael R. Rosa [1] *

[1]The Space Telescope European Coordinating Facility, European Southern Observatory, D-8046 Garching, Fed. Rep. of Germany
*Affiliated to the Astrophysics Division, Space Science Department, European Space Agency

Abstract: The advancement of instrumentation in sensitivity and spectral coverage has opened a new perspective on the ionized medium in galactic contexts. Studies in the Magellanic Clouds play a central role because of the external perspective. Classical H II regions are shown to be but the peak of the iceberg of an ionized ISM component to be present almost everywhere. The main bodies of H II regions themselves break up into subcomponents that are ionized, shaped and eventually dispersed by the energetic interactions of massive stars and their evolutionary products with the environment: UV radiation, fast winds and shocks from SN explosions. Spectroscopic evidence for shock ionized volumes of gas within the 30 Dor nebula is presented and discussed in the context of model nebulae.

1 Introduction

New aspects of old subjects – are H II regions in the Magellanic Clouds special ? Clearly, any object of astrophysical interest located in our neighbouring galaxies is attractive for at least two reasons. We may apply our theoretical tools and our ideas about H II regions calibrated elsewhere in order to learn more about the Magellanic Clouds, namely elemental abundances in the interstellar medium (ISM), star formation rates and star formation histories. Conversely we may benefit from having an external perspective onto two neighbouring galaxies in order to investigate the complex interrelationship between massive stars and the surrounding ISM. Added value comes from the fact that, quite opposite to the galactic case,

a. distances are rather well known and large enough to not be confused by line of sight effects in inhomogeneous objects (like in eg. the Orion nebula);
b. objects of various quality are roughly at the same distance, particularly true for the LMC with its comperatively small linear depth (Caldwell and Laney 1991);
c. galactic foreground as well as local extinction are minimal and can be estimated using non–nebular data as well (eg. stellar photometry, H I and interstellar absorption line column densities);

d. and, as a corollary, little or no confusion with foreground and background objects along the same line of sight at differing wavelengths.

Only two years have passed since IAU Symposium 148 on the Magellanic Clouds. In the following I therefore will only briefly review the more recent activities, including references that permit follow up reading in more depth. The main part of the present contribution will be devoted to the interaction of massive stars in the 30 Doradus region with the ambient ISM.

2 Star formation aspects and chemical abundances

Recognized and studied by astrophysical means early on H II regions are the classical ISM case. They are direct tracers of gas and hot stars in the two galaxies closer by than any other extragalactic system. A census of H II regions and estimation of the total Balmer line luminosity may be used to infer star formation rates. The H II region luminosity function can similarly be compared with present day stellar mass functions. These aspects have been discussed for example by Kennicutt (1991) and by Wilcots and Hodge (1991).

Physical analysis of H II regions in the Magellanic Clouds provides important insights into the chemical evolution of low mass galaxies at low absolute metal abundance. Russel and Dopita (1990) have analyzed several H II regions in each Cloud in a homogeneous way using elaborate photoionization models, and compared their data with earlier studies. The variance of O/H in the LMC seems to be significantly higher than the one in the SMC. Determinations of elemental abundances in stellar atmospheres and in gaseous nebulae are now in reasonably good agreement (Pagel 1993, this volume). C from the UV lines in H II regions might still be a problematic case as long as the scaling to O or H from ground based observations is the major uncertainty (see also Dufour 1990).

3 He III regions

Classical H II regions ionized by single stars or clusters of stars with $T_{eff} \leq 55\,000$ K do not exhibit the recombination spectra of He^+. Ionization of He^+ would require appreciable flux at 54 eV not present in such ionizing continua. However, over the recent years a number of gaseous nebulae have been found especially in the LMC and the SMC showing conspicuous or even strong He^+ 4686 Å emission. Table 1 summarizes all objects known to date together with the properties of the ionizing sources.

It is noteworthy that only one case is known in the Galaxy as compared to four objects in the LMC, one in the SMC and another one in the irregular galaxy IC 1613. With the exception of N 159 F, a nebula associated with the powerful X-ray source LMC-X 1, all remaining He III nebulae seem to be ionized by evolved very massive stars. The sheer presence of an appreciable He^{+2} zone indicates extreme UV continua with $T_{eff} \geq 60\,000$ K (Garnett et al. 1991b). It probably requires the bare core of a WO

Table 1.. He III regions and their most probable ionizing sources

System	Object	Ionizing Source	Reference
LMC	N 44 C	O 4–6	Stasinska et al. 1986
LMC	N 159 F	LMC X–1	Pakull and Angebault 1986
LMC	N 79	WN 1	Pakull 1991
LMC	N 206	WN , O6	Niemela et al. 1991
SMC	N 76	WN 1 , O6f	Garnett et al. 1991a
IC 1613	S 3	WO	Davidson and Kinman 1986
MWG	G 2.4-1.4	WO	Dopita et al 1990

star at solar metallicity to produce the same EUV color shown by less evolved stages of massive stars (WN 1, O 4) at lower metallicity. Spatially resolved spectroscopy of these LMC and SMC nebulae combined with photoionization models will give valuable insigths into the physics of stellar atmospheres and stellar evolution at low initial metallicity. Following up the survey of ring nebulae around WR stars in both Clouds (Chu et al., this volume) will very likely reveal additional nebulae with He III regions.

4 The multi–phase ISM near massive stars

Observational evidence increasingly demonstrates that the actual interstellar medium is far from a simple concept of distinct regions, zones or clouds containing and separating the volumes of different phases. Studies of the ionized ISM in the Magellanic Clouds play a central role as outlined in the introduction. Most of what we knew already on the large scale morphology of the ionized ISM in the Clouds had been based on work with interference filters and photographic plate material. The complex nature of the H^+ distribution, in particular in the larger 30 Doradus region with its 2 kpc super–shells, is well established since years (eg. Meaburn 1979).

Employing modern 2-D detectors and Fabry–Perot scanning techniques has, however, greatly increased the detection threshold and dynamical range. In addition it opened a window to obtain surface brightness maps with velocity information. Kinematical data are a prerequisit to link morphological features such as shells and bubbles with the hydro–dynamics of ionization fronts, winds and SN explosions. The deep ESO/Marseille Hα survey (cf. LeCroar, this volume) provides an invaluable data base for such studies, last but not least because of its complete and homogeneous coverage of each Cloud.

On large scales we see in both Clouds classical, high surface brightness H II regions embedded in an ocean of diffuse and filamentary H II gas, the tenuous medium being present throughout the body of both galaxies (see also Wilcots and Hodge 1991). Presumably it is a counterpart of the warm, ionized disk/halo gas in our own galaxy

(Reynolds 1991). The latter accounts for about 25 % of the atomic hydrogen in the Galaxy and seems to be ionized by Lyman continuum quanta leaking from high surface brightness H II regions. The large mean free paths of these ionizing photons, several hundred pc in the Galaxy and in the Magellanic Clouds, can only be understood if the ISM is highly structured at all spatial frequencies and if neutral blobs and ionized regimes coexist. A galactic example of how the ISM might be structured on small scales, albeit at much higher average density, is the intricate network of dense ionized filaments and partially ionized globules bathing in a more diffuse component, best seen in HST WFPC images of the Orion nebula (Hester et al. 1991).

Improved sensitivity, dynamical range and spatial resolution also reveals the complex nature of what was catalogued and believed to be "classical" H II regions. Objects such as N 120 in the LMC (Laval et al. 1992) or N 66 in the SMC (Ye, Turtle and Kennicutt 1991) turn out to be groups of gaseous nebulae of different types such as SN remnants, stellar wind blown shells, and blister H II regions at the edge of the parent molecular clouds – very much like the greater Orion region encompassing Barnards loop (cf. Kennicutt 1984). The ionized interstellar environment of single stars as well as of groups of stars in both, the Galaxy and in the Magellanic Clouds, is thus to be looked at as more than just the numerical Stroemgren volume.

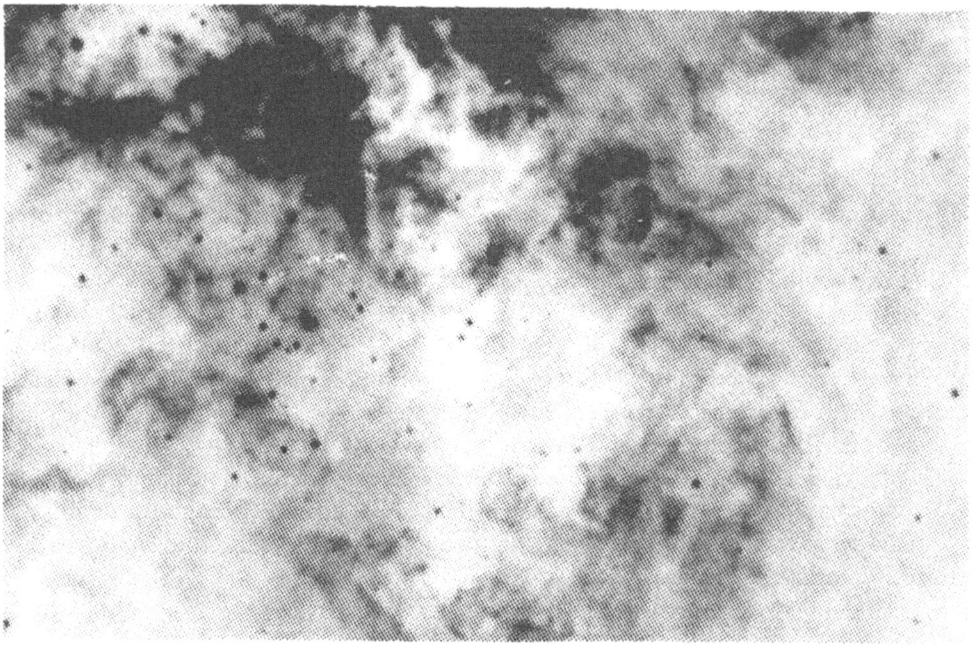

Fig. 1. A 50 pc diameter bubble south of the core of 30 Dor. Note the scalloped structure at interface between the mostly photoionized exterior and the hot, rarefied interior.

5 The 30 Dor starburst region

The 30 Dor nebula in the LMC is a prototype of the class of giant extragalactic H II regions (GEHR), gaseous nebulae ionized by large complexes of massive stars common in more distant spiral and irregular galaxies (Rosa and D'Odorico 1986). The stellar component of the 30 Dor starburst region includes numerous examples of the short lived late evolutionary stages of the most massive stars known (eg. Walborn 1991). Even a scanty look on the morphology of the brighter portions of the H II region foretells the interaction of radiation, winds, and explosions of these stars with the ISM.

Large ionized shells of slowly expanding gas on scales of 300 pc to kpc envelop rarefied bubbles of 50 – 100 pc in the periphery of the nebula proper (Meaburn 1984). Figure 1 shows an H_β image (EFOSC, 3.6m tel., ESO, La Silla) of such a bubble located about 50 pc south of the core of 30 Dor. The high surface brightness parts of the nebula are again composed out of filaments and loops connecting Orion sized (≈ 5 arcsec) patches. Emission line velocity mapping, as pioneered by Smith and Weedman (1972), shows the high surface brightness gas to be pushed around at velocities of 20 – 50 km s^{-1}. At fainter light levels the tenuous gas in the bubbles shows velocities reaching 300 km s_1 (Chu and Kennicutt 1991). X–ray images have demonstrated that hot gas at 5×10^6 K coexists with the H^+ gas at 10^4 K throughout the nebula. X–ray peaks are observed in the rarefied bubbles presumably swept by stellar winds prior to SN explosions (eg. Chu and MacLow 1990) As seen in Fig. 1 corrugated shock fronts separate the rarefied bubble interiors from the inhomogeneous H II region exterior. Energy and cooling times scales (Wang and Helfand 1991).

6 The ionization of the 30 Dor nebula

Clearly, the ISM near massive stars is far from the textbook case of a static Stroemgren sphere in isolation. Clusters of massive stars in GEHRs seem to keep the super–large volumes of gas ionized by all possible interactions with the ISM they are capable of during their entire life, namely EUV photons, fast winds during Of and WR stages, and finally shocked gas from SN explosions. If so we ought to see the contribution of shock ionization as contamination of the primarily photoionized spectra of GEHRs, as has been suggested by Peimbert, Sarmiento and Fierro (1991).

Spatially resolved spectrophotometry of 70 positions in the 30 Dor nebula is now available for a homogeneous analysis. The results on electron temperatures and chemical abundances are discussed by Kinkel and Rosa (this volume). This data base covers the face of the nebula in high and low surface brightness areas, reaching out to a distance of 300 arcsec or 75 parsec from R 136, the dense core of the ionizing cluster. Is the ionization structure of the 30 Dor nebula in accord with pure photoionization models ?

In Figure 2 the observed ionic fractions of sulphur and oxygen are shown (open and filled circles and stars), overlayed with model nebulae using Kurucz atmospheres of T_{eff} 37, 40, 45, and 50 kK and a blackbody of 100 kK as ionizing sources. The models are those of Mathis and Rosa (1991), each line connecting the theoretical

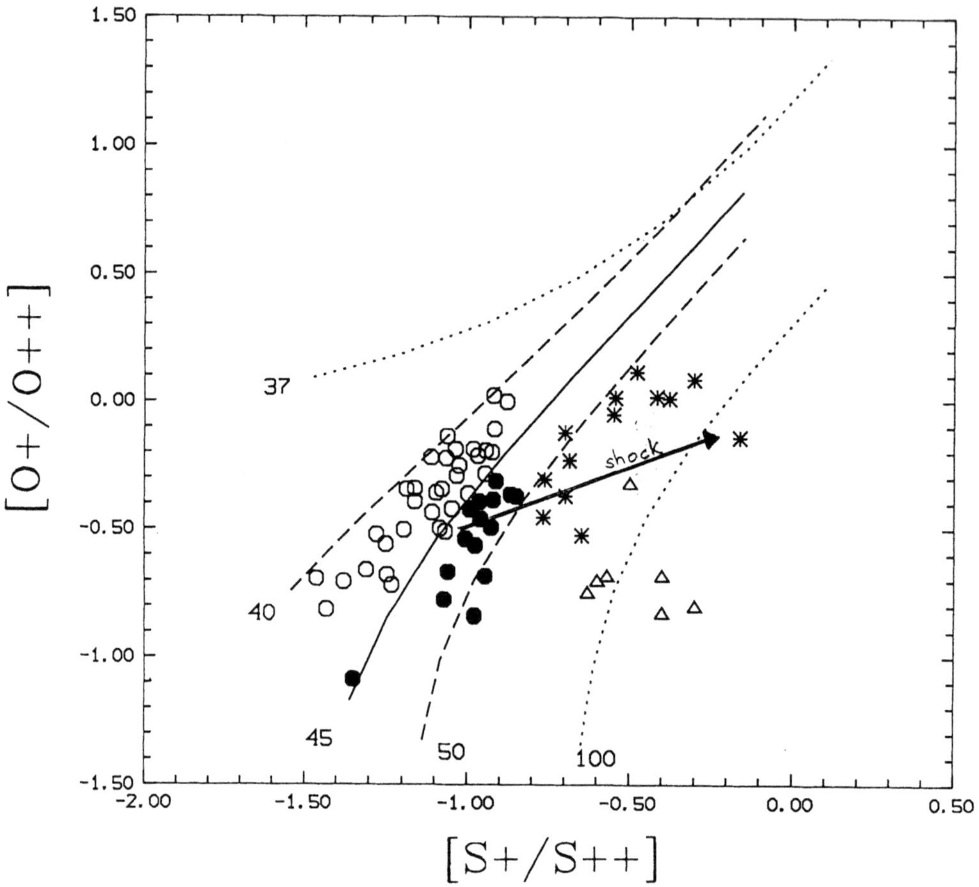

Fig. 2. The sulphur – oxygen ionization diagram of 30 Dor. Spatially resolved data for 70 positions are depicted as open circles, filled circles and star symbols as described in the text. Triangles represent integrated spectra of H II galaxies from Garnett (1989). The loci of model nebulae with differing ionizing sources are overlayed as lines. The effect of seeing a shock ionized volume of gas along the same line of sight as an average photoionized volume is shown by the arrow.

location of purely photoionized nebulae with the same metallicity and ionizing source. The ionization parameter ($\propto L_{\mathrm{Lyman}}$ / N_e^2) decreases from the lower left to the upper right along a given model. A volume element ionized primarily by the stellar radiation field, i.e."close" to the ionizing star(s), is expected to match with the lower left part of a given model nebula track. Similarly, volumes ionized primarily by the diffuse (H and He recombination) radiation field should be found at lower oxygen and sulphur ionization ratios but still along the same model track.

As can be seen, the great majority of the data is consistent with models for ionizing stars in the range 40 kK $\leq T_{\text{eff}} \leq$ 50 kK, as long as the ioniziation parameter is high, i.e. the ionization equilibrium being dominated by direct stellar radiation. Filled circles denote positions likely to be dominated by radiation from hotter than average stars, open circles those that seem to see radiation primarily from the cooler members of the cluster.

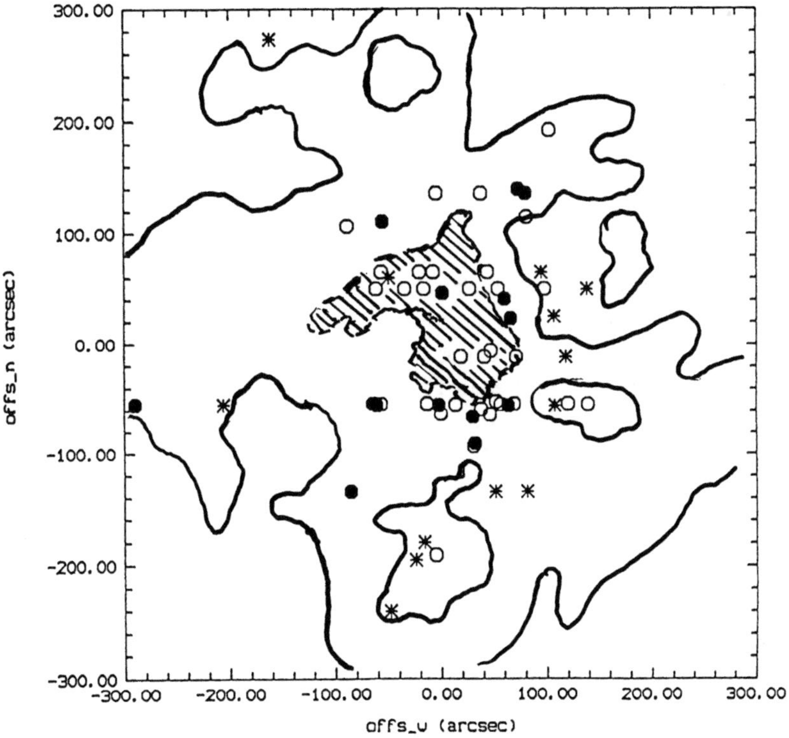

Fig. 3. The location of spectrophotometric positions in 30 Dor

7 Shock ionized gas in 30 Dor

Those positions marked with star symbols can not be dominated by ionizing sources much hotter than average. If this were the case the data would show a much higher ionization parameter. Instead, these positions are likely to be contamined by shock ionized volumes along the line of sight. Following Peimert, Sarmiento and Fierro (1991), a 140 km s^{-1} shock model from Hartigan, Raymond and Hartman (1987) has been added to an average model on the T_{eff} 45 kK track, Hβ being produced in equal proportions by the photoionized and the shock ionized volumina. The effect in the ionization diagram is shown by the arrow. Most of the suspect positions can in fact be reached from the 45 kK photoionized track by even less contamination.

The location of the suspect areas in the nebula provides additional support for the hypothesis of shock contamination. Figure 3 shows the 70 spectrophotometric positions as offsets in arcseconds from R 136, same coding as in Fig. 2. The 2nd lowest isophote from Faulkner's (1968) map of 30 Dor is drawn in this sketch, the area of the 3 highest isophotes is hatched. All but one of the suspect positions are located in the domain occupied by large bubbles, which incidentaly are also bright in X-ray emission (compare with Fig. 3 and 4 of Wang and Helfand 1991).

8 Concluding remarks on modelling of H II regions

It goes beyond the scope of this contribution to thorougly discuss other possible explanations for the mismatch of some positions in 30 Dor with photoionization models. However, it is noteworthy that the integrated spectra of many GEHRs show a similar behaviour, i.e. S^+ / S^{+2} larger than predicted by photoionization models with hot atmospheres or even infinitely hot black bodies. Garnett (1989) mentions modification of atomic parameters (here the S^+ photoionization cross section, or the dielectronic recombination rate of) in order to bring models into accord with such observations, some of which are shown as triangles in Fig. 2. The data of 30 Dor immediately show that the modification of atomic parameters determining the ionization of S do not suffice, since the only result would be a nearly horizontal shift of the entire model tracks with respect to the data. Another option is the inclusion of an inhomogeneous density structure into photoionization codes. Shadow regions behind higher than average density blobs and cloudlets are expected to mimic a lower S ionization at a given O ionization. This latter approach is certainly worth testing together with a self-consistent inclusion of shock heating. Shocks and density inhomogeneities are abundant in the outskirts of real nebulae, as the observations of H II regions, particularly those in the Magellanic Clouds, have shown.

References

Caldwell, Laney (1991): in *The Magellanic Clouds*, Proc. IAU Symp No 148, ed. by R. Haynes and D. Milne, (Kluwer, Dordrecht), p. 249

Chu, Y.-H., MacLow, M.-M. (1990): ApJ 365, 510

Davidson, K., Kinman, T.D. (1982): ApJ PASP 94, 634

Dopita, M.A., Lozinskaya, T.A., McGregor, P.J., Rawlings, S.J. (1990): ApJ 351, 563

Dufour, R.J. (1990): in *Evolution in Astrophysics, IUE astronomy in the era of new space missions*, ed. by E.J. Rolfe ESA-SP-310, p. 117

Faulkner, D.J. (1968): MNRAS 135, 401

Garnett, D.R. (1989): Apj 345, 282

Garnett, D.R., Kennicutt, R.C., Chu, Y.-H., Skillman, E.D. (1991a): ApJ 373, 458

Garnett, D.R., Kennicutt, R.C., Chu, Y.-H. (1991b): PASP 103, 850

Hartigan, P., Raymond, J., Hartmann, L. (1987): ApJ 316, 323

Hester, J.J., Gilmozzi, R., O'Dell, C.R. et al. (1991): ApJL 369, L75

Kennicutt, R.C. (1984): ApJ 287, 116
Kennicutt, R.C. (1991): in *The Magellanic Clouds*, Proc. IAU Symp No 148, ed. by R. Haynes and D. Milne, (Kluwer, Dordrecht), p. 139
Laval, A., Rosado, M., Boulesteix, J. et al. (1992): AA 253, 213
Lozinskaya, T.A. (1991): "*Supernovae and Stellar Wind in the Interstellar Medium*", AIP, New York
Mathis, J.S., Rosa, M.R. (1991): AA 245, 625
Meaburn, J. (1979): AA 75, 127
Meaburn, J. (1984): MNRAS 211, 521
Niemela, V.S., Heathcote, S., Weller, W. (1991): in *Wolf–Rayet Stars and Their Interrelations with Other Massive Stars in Galaxies*, Proc. IAU Symp No 143, ed. by K. van der Hucht and B. Hidayat, (Kluwer, Dordrecht), p. 425
Pakull, M.W. (1991)in *Wolf–Rayet Stars and Their Interrelations with Other Massive Stars in Galaxies*, Proc. IAU Symp No 143, ed. by K. van der Hucht and B. Hidayat, (Kluwer, Dordrecht), p. 391
Pakull, M.W., Angebault, L.P. (1986): Nature 322, 511
Peimbert, M., Sarmiento, A., Fierro, J. (1991): PASP 103, 815
Reynolds, R.J. (1991): in *The Interstellar Disk–Halo Connection in Galaxies*, Proc. IAU Symp No 144, ed. by J.B.G.M. Bloemen, (Kluwer, Dordrecht), p. 67
Rosa, M.R., D'Odorico, S. (1986): in *Luminous Stars and Associations in Galaxies*, Proc IAU Symp No 116, ed. by C.W.de Loore, A. Willis and P.G. Laskarides, (Reidel, Dordrecht), p. 355
Russel, S.C., Dopita, M.A. (1990): ApJS 74, 49
Smith, M.G., Weedman, D.W. (1971): ApJ 169, 271
Stasinska, G., Testor, G., Heydari–Malayeri, M. (1986): AA 170, L4
Walborn, N. (1991): in *The Magellanic Clouds*, Proc. IAU Symp No 148, ed. by R. Haynes and D. Milne, (Kluwer, Dordrecht), p. 145
Wang, Q., Helfand, D.J. (1991): ApJ 370, 541
Wilcots, E.M., Hodge, P. (1991): in *The Magellanic Clouds*, Proc. IAU Symp No 148, ed. by R. Haynes and D. Milne, (Kluwer, Dordrecht), p. 226
Ye, T., Turtle, A.J., Kennicutt R.C. (1991): MNRAS 249, 722

Chemical Homogeneity of the 30 Dor Nebula

Ulrich Kinkel [1,2] and Michael R. Rosa [3,4]

[1]Landessternwarte Heidelberg
[2]European Southern Observatory
[3]The Space Telescope European Coordinating Facility
[4]Affiliated to the Astrophysics division, Space Science Dept., ESA

1 Introduction.

The 30 Dor nebula in the LMC is of constant interest in gaseous nebula research because its proximity gives it a prototype role for "giant extragalactic HII regions". The spatial resolution available allows a test of nebular models and analysis schemes based on them and raises the hope of detecting local abundance variations which may be caused by individual objects in the exciting cluster. We address the question of chemical inhomogeneities in the 30 Dor complex as well as some problems of abundance measurements from HII regions.

2 Data, Electron Temperatures and Mean Abundances.

We have assembled optical spectrophotometric data pertaining to 70 positions in the nebula. Of these, 32 were new observations and 38 were taken from the literature as follows: Rosa & Mathis 1987 (10), Mathis, Chu, Peterson 1985 (22), Shaver et al. 1983 (1), Peimbert & Torres-Peimbert 1974 (4) and Boeshaar et al. 1980 (1). Most observations cover roughly 3600Å-7000Å, whereas a number of spectra—all of them new observations—extend to 9700Å, measuring the [SIII]9069, 9532Å nebular lines which are for the first time used in abundance analyses of 30 Dor. The data have been reanalyzed in a uniform way, using a single reddening law (Savage & Mathis 1979). A unique set of atomic parameters together with a plasma diagnostics code based on the five-level program by DeRobertis et al. 1987 was used to compute abundances of ionic species. For metals other than O, the ionization correction factors of Mathis & Rosa 1991 were used. The distribution of electron temperatures as determined from forbidden line ratios has a mean of 10510K with a standard deviation of 750K. In addition there are a few positions with a strong deviation from the mean value. However, due to the great variety in the properties of the data used, individual measurements of T_e-sensitive lines appear to suffer from varying kinds of systematic errors as is suggested by the comparison of spatially coincident observations from different sources. We therefore adopt the mean of 10510K for use in all abundance determinations. This gives the following abundances:

$$12 + \log(He/H) = 10.94, \ \sigma = 0.1$$
$$12 + \log(O/H) = 8.29, \ \sigma = 0.05$$
$$\log(N/O) = -1.48, \ \sigma = 0.2$$
$$\log(S/O) = -1.62, \ \sigma = 0.2$$
$$\log(Ne/O) = -0.78, \ \sigma = 0.1$$
$$\log(Ar/O) = -2.25, \ \sigma = 0.2$$
$$\log(Cl/O) = -3.35, \ \sigma = 0.2$$

(σ denotes the scatter among individual positions. The small scatter in $\log(O/H)$ does not represent the true error, but is a consequence of the constant temperature assumption and the 'thermostatic' action of oxygen.)

3 Abundance ratios of heavy elements with respect to oxygen.

Whereas the investigation of abundance ratios X/H depends critically on electron temperature, this is much less the case for the ratios X/O. Since individual enrichment sources are not expected to produce all heavy elements in equal proportion, inhomogeneities might also show up in these ratios. Abundances of N, S, Ne, Ar and Cl rely on correction factors for unseen stages of ionization (ICF) which is often very large. These factors depend on a combination of the degree of ionization with ionizing radiation properties and may be parametrized in terms of $f(O) = O^{+}/O^{2+}$ and $f(S) = S^{+}/S^{2+}$(Mathis,Rosa 1991). As illustrated by Fig.1 the $f(O)$,$f(S)$-dependence of the abundances is found to be weak or absent, indicating the quality of the ICF used. But the scatter about the mean is also at least mainly caused by observational error, as suggested by the following. In most data sets of our collection S^{2+} is taken from the difficult [SIII]6312Å line because of lacking IR coverage. This leads to a considerable uncertainty in $f(S)$, which is transported to the ICF. Access to the [SIII]NIR lines in fact leads to a significant reduction of the scatter in X/O abundance ratios.

4 Conclusion.

The position 4B of Rosa & Mathis (1987) remains the only evidence of chemical abundance variations within 30 Dor. This result could not be reproduced by Lortet & Testor (1991); no other comparable position has been found in our data. The scatter of the data set appears to be dominated by uncertainties in diagnostic parameters as $f(O)$ and $f(S)$. This demonstrates the importance of the measurement of the [SIII] nebular lines in the NIR not only for better sulfur abundances—as discussed by Dennefeld & Stasińska 1983—but also for the improvement of the ionization correction for other elements.

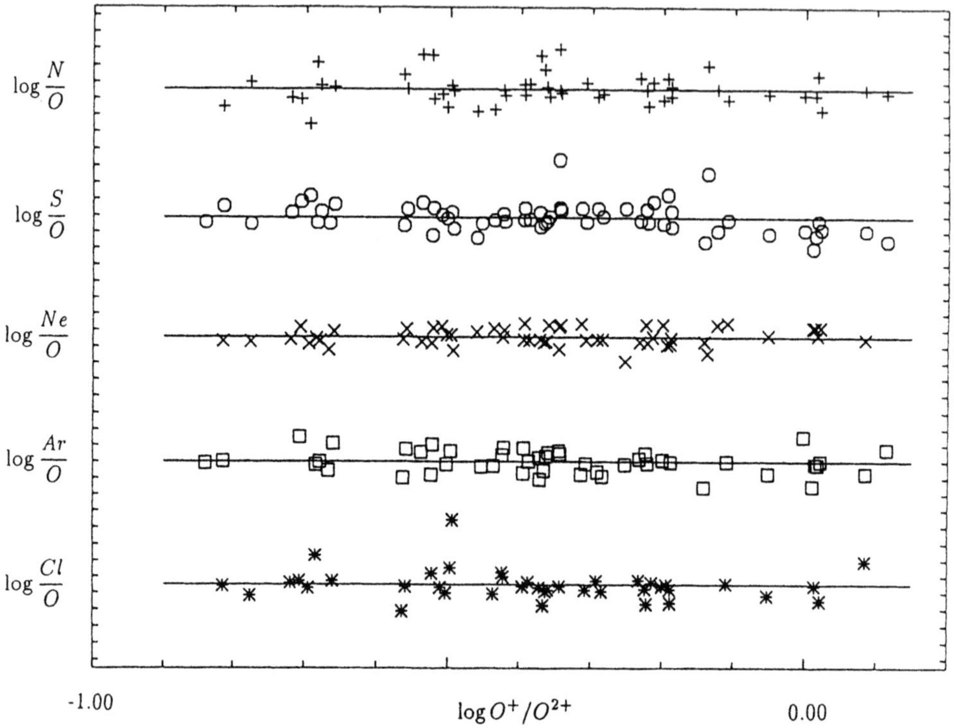

Fig. 1. Abundance ratios $\log(X/O)$ vs. degree of ionization. Horizontal lines indicate mean values as given in the text. Y-axis tickmark spacing is 0.2dex.

References

Boeshaar, G.O. et al. (1980): Ap&Space Sci. **68**, 335.
Dennefeld, M., Stasińska, G. (1983): A&A **118**, 234.
DeRobertis, M.M., Dufour, R.J., Hunt, R.W. (1987): J. Roy. Soc. Can. **81**, 195.
Lortet, M.-C., Testor, G. (1991): A&A *Suppl. Ser.* **89**, 185.
Mathis, J.S., Chu, Y.-H., Peterson, D.E. (1985): ApJ **292**, 155.
Mathis, J.S., Rosa, M. (1991): A&A **245**, 625.
Peimbert, M., Torres-Peimbert, S. (1974): ApJ **193**, 327.
Rosa, M., Mathis, J.S. (1987): ApJ **317**, 163.
Savage, B.D., Mathis, J.D. (1979): Ann.Rev.Astr.Ap **17**, 73.
Shaver, P.A. et al. (1983): M.N.R.A.S. **204**, 53.

UIT Ultraviolet Imaging of 30 Doradus

P. Hintzen [1,2], K.-P. Cheng [2], A. Michalitsianos [2], R. Bohlin [2], R. O'Connell [4], R. Cornett [5], M. Roberts [6], A. Smith [2], E. Smith [2], and T. Stecher [2]

[1]Dept. of Physics and Astronomy, UNLV, Las Vegas, NV 89154, U.S.A.
[2]NASA/GSFC/LASP, Greenbelt, MD 20771, U.S.A.
[3]Space Telescope Science Institute, Baltimore, MD 21218, U.S.A.
[4]Astronomy Dept., University of Virginia, Charlottesville, VA 22903, U.S.A.
[5]Hughes STX Corporation, 4400 Forbes Blvd., Lanham, MD 20706, U.S.A.
[6]National Radio Astronomy Observatory, Charlottesville, VA 22903, U.S.A.

Abstract: During the Astro-1 mission, near- and far-UV images of the 30 Doradus region were obtained using the Ultraviolet Imaging Telescope (UIT). These wide-field, 40' in diameter, high spatial resolution, 2-3", UIT UV images reveal a rich field of luminous UV-bright stars, clusters, and associations. There are 181 stars brighter than m_{2558A}=16.5 and 197 stars brighter than m_{1615A}=16.4 within 3' diameter of the 30 Doradus central cluster. We have derived UV fluxes emitted from the 30 Doradus central cluster and from its UV bright core, R136. The region within 5" of R136 produces ~14% of the far-UV flux (λ=1892 Å) and ~16% of the near-UV flux (λ=2558 Å) emitted from the 3' diameter central cluster. The derived UV luminosity of R136 at 1892 Å is only 7.8 times that of the nearby O6-7 Iaf star, R139, and the m_{1892}-m_V colors of R136 are similar to other O or Wolf-Rayet stars in the same region. These UIT data, combined with other published observations at longer wavelengths, indicate that there is no observational evidence for a supermassive star in R136.

The 30 Doradus central cluster, NGC 2070, is characterized by numerous O stars, B supergiants, and Wolf-Rayet stars (Melnick 1985, A&A, **153**, 235). Many faint discrete sources are also seen in the UIT images of this field. We find that R136, the brightest component of NGC 2070 at visible wavelengths, is also the brightest component in the UV (see Figure 1, where R136 is indicated by the smaller circle). Moffat et al. (1985, Ap.J., **295**, 109) have derived an approximate luminosity ratio, L_{UV}(R136, 10" diameter)/L_{UV}(30 Doradus cluster, 2' diameter)$\approx$17%, from International Ultraviolet Explorer (IUE) slit spectroscopy. However, the diameter of the UV-emitting region associated with the 30 Doradus cluster is about 3' which is significantly larger than that inferred from the mapping performed with the IUE large aperture (10"x20"). Therefore, the UIT 40' diameter images provide greatly improved UV fluxes for both R136 and for the 30 Doradus central cluster as a whole. We find that R136 (10" in diameter) contributes ~14% of the total, integrated UV-flux emitted by the 3' diameter central cluster in 30 Doradus at 1892 Å and ~16% at 2558 Å .

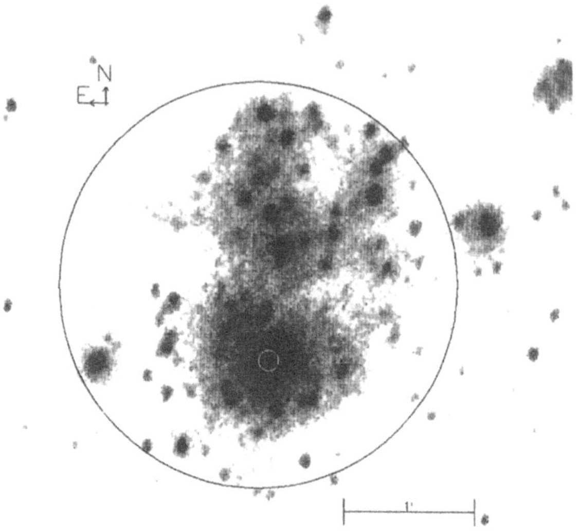

Fig. 1. The central field of the UIT Far-UV 30 Doradus image (λ_o=1615Å, FWHM=225Å, exposure time =17.4 sec). The large circle has a diameter of 3' and the small circle centered on the UV bright core R136 has a radius of 5". The UIT optics and detectors are described in detail by Stecher et al. (1992, Ap.J. Letters, in press).

R136 is a complex multiple system which has been resolved into 28 components within a radius of 4.6" (Walborn 1986, IAU Symp., **116**, p. 185). Using speckle imaging, Weigelt and Baier (1985, A&A, **150**, L18) demonstrated that R136a, the visually brightest component of R136, is a compact cluster of 8 stars (R136a1-a8) roughly 0.7" in diameter. This result has been recently confirmed by the HST-Faint Object Camera (Weigelt et al. 1991, Ap.J. Letters, **378**, L21), but the physical nature of the brightest component, R136a1, still remains uncertain. Early studies of R136a yielded mass estimates of 500-2500 $M_\odot$ (Schmidt-Kaler 1980, Transactions IAU XVIIB, p. 208; Cassinelli et al. 1981, Science, **212**, 1497). After speckle interferometry resolved R136a, Walborn (1986, IAU Symp., **116**, p. 185) calculated an upper mass limit of 250 $M_\odot$ for R136a1, the brightest sub-component of R136a. This is an order of magnitude smaller than the previous estimates for R136a but still considerably larger than the conventional theoretical upper limit on stellar masses, 60-120 $M_\odot$ (Conti and Burnichon, 1975, A&A, **38**, 467). A supermassive star with mass $\sim$250 $M_\odot$ can be expected to be much more luminous than "normal" stars with M=60-120 $M_\odot$ (Maeder 1980, A&A, **92**, 101). Also, Maeder concludes that such a supermassive star would evolve "directly toward the left in the HR diagram" as it left the main sequence and therefore might have abnormally blue colors, though an optically thick stellar wind would also have a major effect on colors. However, the derived UV luminosity of R136 at 1892Å is only 7.8 times that of the nearby O6-7 Iaf star R139, and the UV colors of R136 are similar to normal O or Wolf-Rayet WN stars. Consequently, the bolometric luminosity and mass of the visually brightest component in R136, R136a1, cannot be substantially greater than those for other luminous O stars in the 30 Doradus complex. There is therefore no observational evidence for a supermassive star in R136.

The Kinematics of red Supergiants in 30 Dor and Shapley II: What for ?

N. Martin [1], M.C. Lortet [2]

[1] Observatoire de Marseille, 2 Place Le Verrier, F-13248 Marseille, France
[2] DAEC, Observatoire de Paris, Section de Meudon, F 92195 Meudon Cedex

Abstract: We studied stellar and interstellar data of two regions of the Large Magellanic Cloud, 30 Dor and Shapley II area. We observed the splitting of the gas into a number of separate layers. Stars are generally associated with the low-velocity component, which is behind the high-velocity component.

The red supergiants (RSG) are tracers of star formation, 7 to 50 Myr ago. Indeed in the LMC, most of them are collected in geographical groups remarkable for their low radial velocity dispersion (Prévot et al. 1989): these groups reveal the places and volumes where star formation proceeded. Here we focus on the 30 Dor and Shapley II area. Additional CORAVEL velocity measurements (Lortet et al. 1992), by adding new members, have allowed a better knowledge of the corresponding groups. On the one hand, compared to the tracers of neutral gas (HI 21 cm emission, Rohlfs et al. 1984, Meaburn et al. 1987; NaI interstellar lines), an ionized gas (CaII) the rich group Shapley II and the smaller one 30 Dor 1 have been found to have both exhausted the high velocity HI gas, and to be connected by several intermediate stars. On the other hand, 30 Dor 1, related to high velocity HI gas, and the region just to the south, 30 Dor 2, related to the low velocity HI gas, originated from two different bodies of gas. The RSG studied in the core of the 30 Dor area share the same velocity as the bulk of the ionized gas: they reveal that continuous stellar formation has occured in the same cloud, a cloud related to the low-velocity HI gas and extending to the South-West.

On contrast, little star formation has occured in the high-velocity gas east of roughly 05h37m, though the total NH column density is large. The reason may be that the gas is splitted into a number of separate layers: this is indeed suggested by the discrete components found on the cut Y = 0.2, for HI by Meaburn et al. at position 44, and for interstellar NaI by Vladilo et al., Maurice & Silvy and Ferlet et al. for the stars investigated (Fig. 1). Sk-69 289 (stellar RV = 248 km s^{-1}) is an outstanding example showing that all the gas of velocity larger than 260 km s^{-1} is in front of the low-velocity component HI at RV about 250 km s^{-1}.

We would emphasize the interest of yellow stars (e.g. Sk-69 219 = G 210) for whom accurate value of the star velocity can be obtained by CORAVEL as well as interstellar lines can be studied: they may contribute safely to the question of the gas distribution along the line of sight.

Finally we would draw attention on the discovery of a new genuine S Dor star by the absence of correlation peak when observed by CORAVEL in 1990 and 1991, namely a star which was known as a F5-F8 Ib supergiant and is now a B supergiant (Parker J.Wm, Clayton G.C., Conti P.S., Winge C., 1992, in preparation for ApJ Letters).

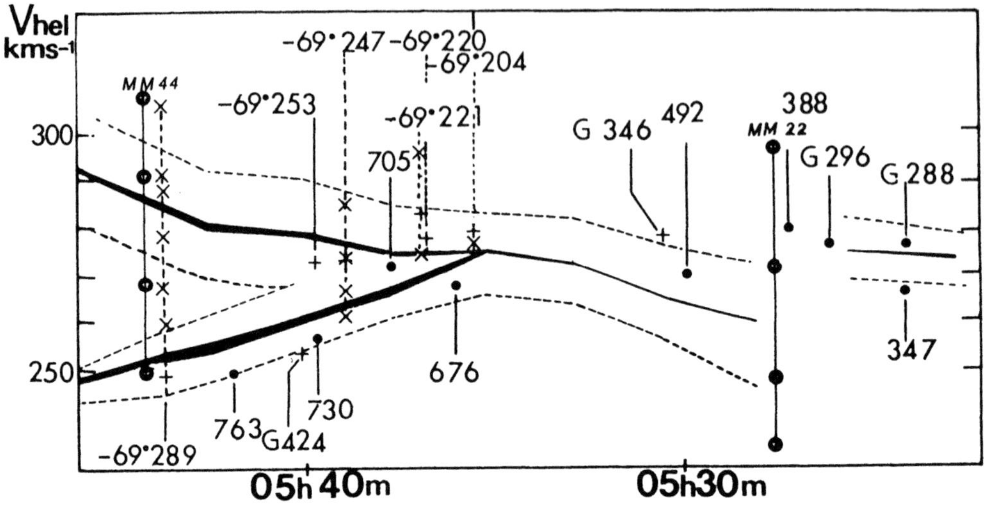

Fig. 1. East-West positional velocity array at Y = 0.2 (Declination about -69 31). The stars without acronymes are RM1, (Rebeirot et al. 1985), G stars are from Ardeberg et al. 1972.The velocities are heliocentric radial velocities. The data for neutral hydrogen are basically from Rohlfs et al. 1984: RV+σ/2, and RV-σ/2 are indicated by dashed lines, the brightness by the thickness of the black areas. Red supergiants with measured CORAVEL velocities (high accuracy radial velocity) are shown (filled circles) as well as a few OB stars, for which stellar (+ sign) and interstellar (x signs) radial velocities come mostly from Vladilo et al. 1992, Maurice and Silvy (private communication), Ferlet et al. 1985. The NaI and CaII interstellar lines found for Sk-69 220, Sk-69 247 and (even more) Sk-69 289 confirm the existence of several gas components of velocities higher than 268 km s^{-1}, as observed for neutral HI by Meaburn et al. 1987 at their position MM 44.

References

Ardeberg, A., Brunet, J.P., Maurice, E., Prévot, L.: 1972, Astron. Astrophys. Suppl. Ser. **6**, 249.

Brunet, J.P., Prévot, L., Maurice, E., Muratorio, G.: 1973, Astron. Astrophys. Suppl. Ser. **9**, 447.

Ferlet, R., Dennefeld, M., Maurice, E.: 1985, Astron. Astrophys. **152**, 151.

Lortet, M.C., Martin, N., Lindgren, H.: 1992, to be submitted.

Lortet, M.C., Testor, G., 1987, in Starbursts and Galaxy Evolution, eds. T.X. Thuan, T. Montmerle, J. Tran Thamh Van, Frontières, Gif-sur-Yvette, p. 79.

Meaburn, J., Marston, A.P., McGee, R.X., Newton, L.M.: 1987, Monthly Notices Roy. Astron. Soc. **225**, 591. (MM)

Mendoza, E.E., Gomez, T.: 1973, PASP, **85**, 439. (MG)

Prévot, L., Rousseau, J., Martin, N.: 1989, Astron. Astrophys. **225**, 303.

Rebeirot, E., Martin, N., Mianes, P., Prévot, L., Robin, A., Rousseau, J., Peyrin, Y.: 1983, Astron. Astrophys. Suppl. Ser. **51**, 277. (RM1-RM2)

Rohlfs, K., Kreitschmann, J., Siegman, B.C., Feitzinger, J.V.: 1984, Astron. Astrophys. **137**, 343.

Sanduleak, N.: 1970, Contr. Cerro-Tololo Inter-Am. Obs. No. 89.

Vladilo, G., Molaro, P., Monai, S., D'odorico, S., Dennefeld, M., Ferlet, R., Vidal-Madjar, A.: 1992, to be submitted.

Wayte, S.R.: 1990, Astron. J. **355**, 473.

Investigations of the Interstellar Matter in and around supergiant shell LMC4

H. Domgoergen [1], D.J. Bomans [1], K.S. de Boer [1], J. Koornneef [2]

[1]Sternwarte University of Bonn, Auf dem Hügel 71, D-5300 Bonn 1,
[2]ESA, Space Telescope Science Institute, 3700 San Martin Drive, Baltimore, MD 21218, USA

Abstract: In order to investigate dynamics, three-dimensional structure and physical conditions of the interstellar gas in the giant superbubble LMC4 we have analysed IUE high dispersion spectra of seven hot stars in this part of the LMC in comparison with 21cm emission data.

Introduction

Deep Hα photographs show that there exists at least five giant, ringlike structures of HII regions and filaments in the LMC (Goudis & Meaburn 1978). LMC4, located in the NE of the LMC, is with a diameter of 1.2 kpc the biggest so called supergiant shell. A 21cm survey of that region makes visible that the neutral gas is concentrated in a ring to the outside of the HII regions whereas the inner part of the LMC4 shows a deep hole in HI. This hole coincides with Shapley Constellation III, one of the largest concentrations of young stars in the LMC.

The mechanism responsible for the creation of such a big structure is still unclear. SN explosions and stellar of the young stars in the centre of the supergiant shell are not sufficient to create such a big structure. The inclusion of large-scale self-propagating star formation is needed. Indeed, there are signs for this effect in LMC4 (Dopita et al. 1985), but the pattern appears to be complicated (Reid et al. 1987, Vallenari, Bomans, de Boer 1992).

Data Analysis

We have analysed the velocity structure of the neutral gas using the low ionized elements SiII and SII seen in absorption in our IUE high dispersion spectra. A comparison of the profiles seen in different absorption lines in the spectrum of one star makes it possible to derive the velocity pattern of the clouds along the line of sight. A further comparison of the so detected velocity structure with that seen in 21cm close to the star allows to conclude which components of the gas lie in front or behind the star.

In order to study the hot gas component of LMC4 we have analysed the absorption lines of NV, CIV and SiIV seen in the spectral range of the IUE SWP camera. We compare the detected absorption profiles with these seen in the neutral gas with the aim to separate HII region gas from that not directly surrounding the star.

Conclusions

* In most cases three absorption components of low ions are detected. The lower velocity components often have no corresponding peak in 21cm, which we interpret as an optical depth effect.
* Our sample of spectra shows a complicated pattern of the velocity components. A high velocity component at 220 km s^{-1} is detected along several lines of sight.
* The naive picture: "the higher the velocity of a cloud the farther away it is" turns out to be wrong. The comparison of emission and absorption profiles shows that there exists often "infalling" gas at 260 km s^{-1}.aproching LMC4 from the back.
* The 21cm emission profiles near stars located in HII regions show clearly one very strong main component at 280 km s^{-1}. This component coincides with the strongest SiII absorption component. We therefore conclude that the young stars are located inside of this gas.
* The existence of hot gas inside of LMC4 remains unclear as the spectra often are to noisy to unambigously detect an absorption component as faint as expected. Only the spectrum of Sk-67 166 shows clear absorption of CIV, which can be attributed to hot gas from the interior of LMC4.

References

Dopita, M.A., Mathewson, D.S., Ford, V.L., 1985, ApJ, 297, 599
Goudis, C., Meaburn, J., 1978, A&A, 68, 189
Reid, N., Mould, J., Thompson, I., 1987, ApJ, 323, 433
Vallenari, A., Bomans, D.J., de Boer, K.S., 1992, A&A, submitted

A SURVEY OF THE ENVIRONMENT OF WOLF-RAYET STARS IN THE MAGELLANIC CLOUDS

M. A. Dopita, J. F. Bell
Mt. Stromlo and Siding Springs Observatories, Institute of Advanced Studies,
The Australian National University,
Private Bag, Weston Creek P.O., ACT 2611, Australia

T.A. Lozinskaya
Sternberg State Astronomical Institute,

You-Hua Chu
University of Illinois

Abstract

We have made an complete Hα survey of the environment of the Wolf-Rayet (W-R) stars known in the both of Magellanic Clouds in the search for associated nebulosity. We have discovered a number of new ring nebulae, bring the total known to 16% of the number of W-R stars. There appears to be no clear association of ring nebulae with W-R subtype other than a relative scarcity of ring nebulae associated with later-type WNstars (WN 4.5-WN10).

Observations

During the 1990-91 and 1991-2 observing seasons, have made an imaging survey in Hα and in [O III] λ5007Å of the interstellar environment of the Wolf-Rayet (W-R) stars listed in the Breysacher (1981) and Azzopardi and Breysacher (1979) calalogs. The telescope used was the the ANU 2.3m Advanced Technology Telescope. The interference filters used had 16Å FWHM, allowing the hydrogen line to be clearly separated from the nearby [N II] lines. In the 1990-91 season we used the Double-Beam Spectrograph in imaging mode, with its Photon Counting Array. This gave a field of 5.1 arc min, at a scale of 0.65 arc sec. pixel $^{-1}$. For the 1991-92 season we were able to use the new focal reducing imager with a Tekronix 1024x1024 CCD as detector. This gave a field of 6.7 arc min. at a scale of 0.60 arc sec. pixel $^{-1}$. The 90mm interference filters were placed directly in the f/18 Nasmyth beam of the telescope, and in the case of the Hα, the filter was tilited 7 degrees to tune it to the redshift of the Magellanic Clouds.

The data were all reduced using the IRAF package, and displayed using the Adobe Photoshop® software. Since the signal to noise of the images was much better with the CCD imager, we repeated observations of all objects found to have associated nebulosity in the first season. For a majority of the ring nebulae, we also obtained images in the [O III] λ5007Å line (40Å FWHM) which allowed us to form colour maps of the variations in excitation conditions across the objects.

Results

The compilation of Chu and Lasker (1980) was the most complete compilation of ring nebulae until the present work. We have discovered ring nebulae associated with AB #5 in the SMC, and Br #2, 13, 18, 25, 31, 34, 40, and 56 in the LMC. Of these, AB #5 (WN 3), Br #13 (WN 8), Br # 25 (WN 3), Br # 31 (WC5 +O9) and Br # 34 (WN3 + B3I) show clear evidence for stellar ejecta associated with the W - R star. The nebulae associated with Br #2 and 25 were identified spectroscopically by Pakull (1991). Of these Br #25 is particularly magnificent. It lies in a complex region of recent star formation (N 44), at a point of strong density gradient in the ISM. It shows evidence for ejection of a regular series of shells apparently associated with discrete mass-loss events. In addition, there is a radial plume of material in the direction of maximum density gradient.

The incidence of ring nebulae is 16% of the total number of W-R stars. In the SMC, the number of W-R stars known is too few to be able to draw any statistically significant conclusions. However, in the LMC it is possible to derive a breakdown between the various W-R sub-classes (Table 1).

Table 1: The Incidence of Ring Nebulae amongst the W-R subclasses in the LMC.

Spectral Class	Number in Br Survey	Number of Ring Neb.	Percentage With Ring Neb.
WN 3	28	4	14
WN 4	23	5	31
WN 4.5-10	22	2	10
WC 4-5	17	3	18
All Classes	100	16	16

The relative scarcity of ring nebulae associated with late-type WN stars is probably statistically significant. These stars are found to be preferentially associated with the youngest stellar clusters (Schild, Lortet and Testor; 1991), and so this may be explained by the fact that the high-energy wind phase of these objects has not continued long enough to have yet inflated a ring nebula around these objects.

References

Azzopardi, M. and Breysacher, J. 1979, *Ast. Ap.*, 75, 120.

Breysacher, J. 1981 *Ast. Ap. Suppl. Ser.* , 43, 203.

Chu, Yu-Hua and Lasker, B.M 1980 *Pub. Ast. Soc Pac.* , 92, 730.

Pakull, M.W. 1991 in *Wolf Rayet Stars and Interrelations with other Massive Stars in Galaxies*, eds K.A. van der Hucht and B. Hidayat, Kluwer:Dordrecht, p479.

Schild, H., Lortet, M.C., and Testor, G. 1991 in *Wolf Rayet Stars and Interrelations with other Massive Stars in Galaxies*, eds K.A. van der Hucht and B. Hidayat, Kluwer:Dordrecht, p479.

Emission Line Stars in LMC Dark Clouds[1]

R. Kneer [1], J. Krautter [1], D. Baade [2]

[1]Landessternwarte Heidelberg–Königstuhl,
[2]European Southern Observatory, Garching

Introduction

Here we report on spectroscopic and photometric observations of emission line stars in Hodge dark clouds in the LMC. The stars were been found on the basis of a grism survey carried out with the ESO 3.6 m telescope on La Silla and the attached EFOSC instrument. The objective was to find in the LMC pre–main–sequence stars of intermediate mass (Herbig Ae/Be stars, $3M_\odot \leq M \leq 5M_\odot$) and possibly of low mass (T Tauri stars, $M < 3M_\odot$). Four Hodge dark clouds were studied: H 44, H 52, H 53 and H 65, which have been chosen for their high density and whose position is near 30 Dor. 20 candidates were found, most of them in H 44 and H 65. For verification and detailed studies subsequent spectroscopic and photometric observations have been made.

Observation

In order to achieve a spectral classification in the MK–system and for a more detailed study of the emission line spectrum, medium resolution spectroscopy was carried out in two wavelength ranges. In the blue range 4200 Å – 5050 Å the dispersion was 59 Å/mm resulting in a spectral resolution of 2.4 Å, in the red range 5800 Å – 7000 Å the dispersion was 78 Å/mm and the resolution 3.0 Å, respectively. These observations were obtained with the Optopus multi–object–spectrograph which was attached to the ESO 3.6 m telescope. As detector a RCA CCD was used.

A problem was caused by the adjacent 30 Dor–H II region, whose emission shows highly variable intensity and strongly contaminates the star emission. Since no long slit spectroscopy is possible with Optopus, no sky subtraction was performed.

CCD–photometry in the Kron–Cousins B, V, R, and I bands was carried out with the Danish 1.5 m telescope on La Silla and the CTIO 0.9 m telescope. Infrared photometry in the JHK(L) bands was obtained with the ESO/MPI 2.2 m telescope and an InSb–detector. From the photometric data the spectral energy distribution (SED) of the stars could be obtained.

Results

A total of 15 stars could be verified as (hitherto unknown) emission line objects. Since no reasonable subtraction of the background emission could be performed (see above), the criterion for an emission line nature was that the FWHM of the emission line is significantly broader than the nebular emission from the H II region.

---

[1] based on observations collected at ESO, La Silla (Chile)

On the basis of their absorption spectrum eight stars could be unambiguously classified as B–type stars, for three others the classification is somewhat doubtful. But their spectral energy distribution strongly indicate that they are B stars too. An IR-excess is visible in the SED of most of them. Most probably these objects are comparable with 'normal' galactic Be stars, however, for several of them a PMS nature similar to Herbig Ae/Be stars cannot be excluded.

One star (H 44 – 33) in the sample shows several properties which are commonly known as indicators for pre–main–sequence objects. The red spectrum and the SED are shown in the figures. The following properties are remarkable:

- very strong emission in Hα, $EW_\alpha \approx -740$ Å
- many permitted and forbidden emission lines of single ionized metals (mainly Fe II, [Fe II])
- no absorption lines → no spectral classification was possible, however the energy distribution in the optical spectral range indicates an early spectral type
- high IR–excess
- variable intensity, $m_V \approx 18.2\,mag$
- luminosity $L \approx 1600\,L_\odot$

The spectral energy distribution is very similar to those of galactic Herbig Ae/Be stars. The large variety of emission lines and the presence of forbidden lines indicate the presence of extended circumstellar matter. On the basis of the above properties and due to its location in a dark cloud it seems safe to conclude that the star H 44 – 33 is in fact a PMS star of intermediate mass. This constitutes the first detection of such an object outside of our galaxy.

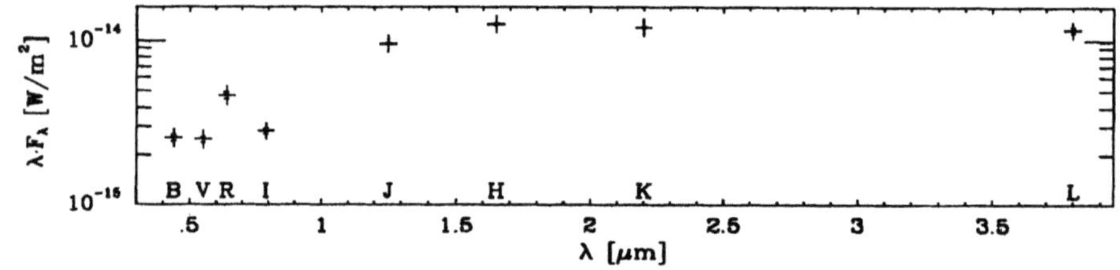

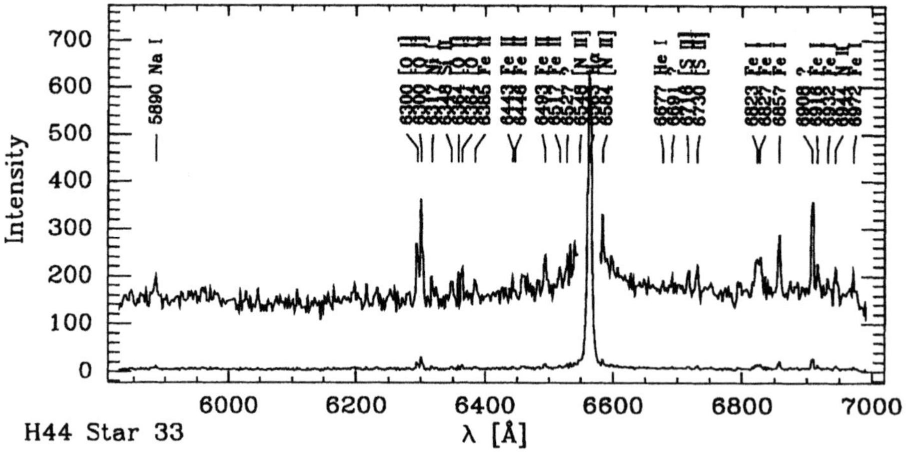

A High Resolution and High Sensitivity Survey of the SMC at 843 MHz

Taisheng Ye, A. J. Turtle

School of Physics, University of Sydney, N.S.W. 2006, Australia

Abstract: The Molonglo Observatory Synthesis Telescope has completed a survey of the Small Magellanic Cloud at a frequency of 843 MHz with a synthesised beamwidth of about 44 arcsec and a sensitivity of 2 mJy. A mosaiced map covering 28 square degrees has been produced. Catalogues of 70 H II regions, 15 supernova remnants and 1100 background sources have been compiled.

A survey of the Small Magellanic Cloud (SMC) using the Molonglo Observatory Synthesis Telescope (MOST) at 843 MHz with the angular resolution of 43 $\times$ 45 arcsec has been completed. A brief description of the observations has been given by Turtle & Amy (1991). The survey area covers 00^{H} 22^{M} to 01^{H} 40^{M} in Right Ascension and $-70°$ $20'$ to $-75°$ $05'$ in Declination (J2000).

Procedures have been developed to remove various artifacts and solar interference, reduce pointing and flux density uncertainties and improve the sensitivity by mosaicing. The final map of 28 square degrees has a sensitivity of around 2 mJy per beam area. The central part of the map is presented in Fig. 1. This radio map of the SMC has the highest resolution (corresponding to 13 pc at a distance of 60 kpc) and the highest sensitivity obtained to date. We have nearly completed a comprehensive analysis of the map. A total of 70 H II regions have been identified, 15 supernova remnants (SNRs) have been confirmed or discovered, 4 other remnants need to be confirmed and 1100 presumed background sources stronger than 5 mJy have been catalogued. The detailed results will be published in a series of papers.

A comparison of the MOST map and smoothed Hα data reveals no optically obscured extended H II regions. The detailed structure of nebulae shown in the radio and optical maps is nearly the same if the maps have the same resolution and comparable sensitivity.

We have developed a new and sensitive method of locating and separating SNRs in giant H II regions (Ye, Turtle & Kennicutt 1991) and have discovered three SNRs embedded in the N19 and N66 regions – the first and second brightest H II regions in the SMC.

There is no significant difference in the number of background sources in various ranges of flux density compared to regions away from the Clouds at high galactic latitudes.

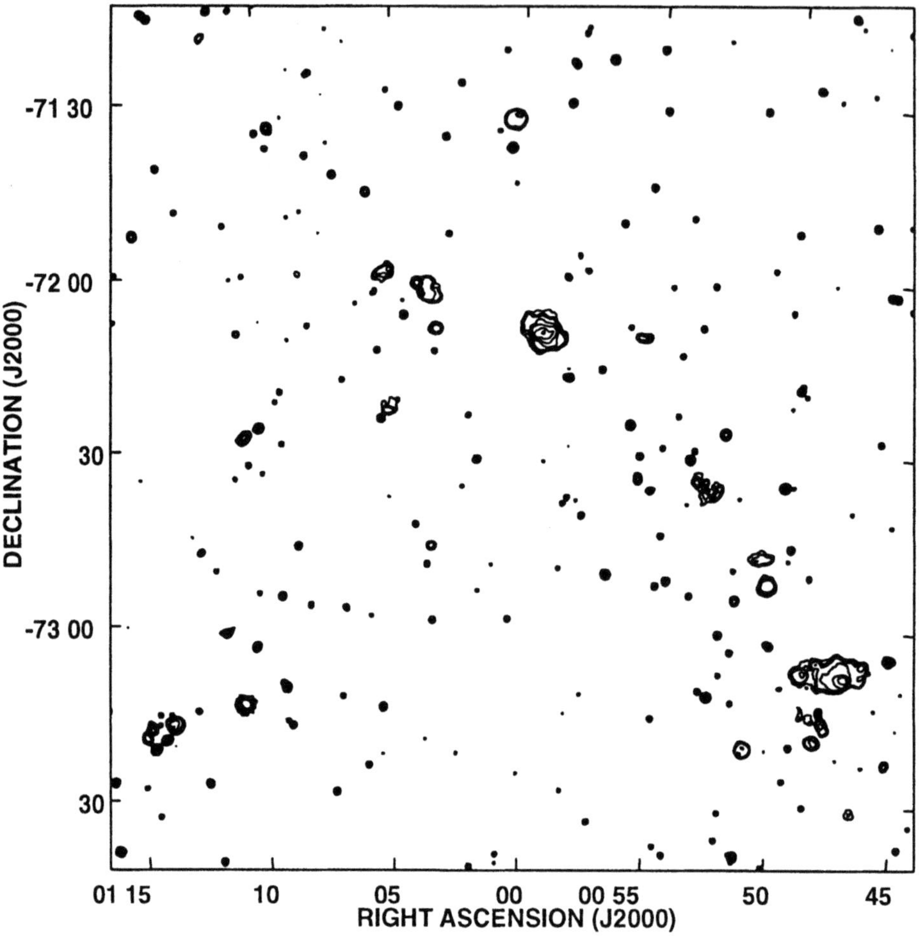

Fig. 1. The mosaiced MOST map of the main body of the SMC at 843 MHz. The lowest contour level plotted is 5 mJy/beam

References

Turtle, A.J., Amy, S.W. (1991): " The Magellanic Clouds", in IAU Symp. 148, ed. by R. Haynes and D. Milne, pp. 115–117

Ye, Taisheng, Turtle, A.J., Kennicutt, R.C. Jr. (1991): Mon. Not. Roy. astro. Soc. **249** 722

Dust and Gas Masses from mm continuum Observations of LMC and SMC

G.Dall'Oglio, P.Andreani, P.Merluzzi, L.Piccirillo, L.Pizzo, L.Rossi

Dipartimento di Fisica, Università di Roma "La Sapienza", Roma, Italy

This poster paper presents values of the dust and gas masses as found from millimetric continuum measurements of the Large and Small Magellanic Clouds. Observations were taken at the Italian Base in Antarctica with a 1m flux collector and a 2.6m telescope. Details of the experiments together with the procedures of data analysis can be found elsewhere (ref. [1], [2], [3]).
Signals at 1 *mm* and 2 *mm* of the LMC and SMC, taken with the 1m instrument, turn out to be highly (spatially) correlated with the FIR emission as a comparison with IRAS 100 μm map suggests. Because of the large beam ($\sim 1°$ FWHM), the entire emission from these galaxies has been collected and these data can be used to infer the overall properties of cold and warm dusts within the interstellar medium of these sources. Once a dust model has been adopted the amount of gas mass can be found. The FIR/mm spectra of the Magellanic Clouds show that a conspicuous *cold component* seems to be present in the interstellar medium of these galaxies. Flux density measurements at 60, 100 micron and 1 and 2 *mm* have been used to estimate the dust temperature and therefore the total amount of dust present in the interstellar medium can be evaluated from the relation:

$$F_\lambda \simeq \frac{2kc \sum_i \chi_i(\lambda) < T_i > M_d}{\lambda^4 D^2} \tag{1}$$

where $\chi_i(\lambda)$ are the grain opacities due to component i; $< T_i >$ is the average temperature (at these wavelengths it is possible to approximate the temperature distribution of the grains with a single temperature) and M_d is the dust mass.

We estimated the amount of galaxy dust by using two different models of interstellar dust ([4], [5]). These models provide values for the grain opacities at $\lambda = 1$ *mm* using different composition and dimensions of dust grains. Table I lists the dust and gas masses found by considering only 1 *mm* flux densities for the Large and Small Magellanic Clouds. Values inferred from different observations (CO line emission, HI emission) are also reported for comparison.

As it is clear from the results in table I that the gas masses inferred from this procedure are inconsistent with those found from spectroscopic observations. The biggest

Table I. Dust and Gas Masses (in $M_\odot$)

	LMC	SMC	notes
$M_{dynamics}$	6.1 10^9	1.8 10^9	
M_{H_2} (CO line)	1.4 10^8	3.0 10^7	
M_{HI} (21 *cm*)	7.0 10^8	6.5 10^8	
$M_{warm\ dust}$ (IRAS)	(7.5±2.5) 10^4	(4.1±1.1) 10^3	M-W *model* (ref[4])
$M_{warm\ dust}$ (IRAS)	(1.0±0.3) 10^5	(5.4±1.5) 10^4	D-L *model* (ref[5])
$M_{cold\ dust}$ (*mm*)	(9.1±2.8) 10^5	(8.0±2.8) 10^5	M-W *model* (ref[4])
$M_{cold\ dust}$ (*mm*)	(1.0±0.3) 10^7	(9.0±3.0) 10^6	D-L *model* (ref[5])
M_{gas}	(3.3±1.1) 10^8	(3.0±1.0) 10^8	M-W *model* (ref[4])
M_{gas}	(1.7±0.5) 10^9	(1.5±0.5) 10^9	D-L *model* (ref[5])

uncertainty in eq.(1) lies in the estimation of grain opacities at long wavelengths, $\chi(\lambda)$, which are poorly constrained by *submm* and *mm* observations. Different dust models predict grain opacities differing by a factor of seven at 1 *mm* and this fact requires a substantial revision of our knowledge of dust emission at these wavelengths. Our group is also making an effort towards a better understanding of cosmic dust properties: data on cosmic analogs made of hydrogenate amorphous carbon containing randomly oriented graphitic islands show that once their presence has been considered in the interstellar medium of the Galaxy a better match with the observed long wavelength galaxy spectrum results ([6]). Therefore, they could represent good candidates as interstellar grains and it becomes worth pursuing the investigation of their properties.

References

1. Andreani P., Ceccarelli C., Dall'Oglio G., Martinis L., Piccirillo L., Pizzo L., Rossi L., Venturino C., Astrophysical Journal, **348**, 467 (1990)
2. Andreani P., Dall'Oglio G., Martinis L., Piccirillo L., Pizzo L., Rossi L., Venturino C., Astrophysical Journal, **375**, 148, (1991)
3. Dall'Oglio G., Ade P.A.R, Andreani P., Calisse P., Cappai M., Habel R., Iacoangeli A., Martinis L., Merluzzi P., Piccirillo L., Pizzo L., Polcaro V.F., Rossi L., Experimental Astronomy, submitted (1992)
4. Mathis J.S. and Whiffen G. Astroph. J., 341, 808 (1989)
5. Draine B.T. Lee H.M., Astroph.J., 285, 89, (1984)
6. Merluzzi P., Bussoletti E., Dall'Oglio G., Piccirillo L., Astron. & Astroph., in press (1992)

Far Infrared Luminosities of GMCs in the Large Magellanic Cloud

D. A. Caldwell and M. L. Kutner[1]

[1] Physics Department, Rensselaer Polytechnic Institute, Troy, NY 12180-3590, USA.

Summary. High Resolution IRAS and CO ($J = 1 \rightarrow 0$) data in the LMC are analyzed to determine the total infrared and CO luminosities for individual molecular clouds. The star formation activity is also found for these clouds.

1. Introduction

The Magellanic Clouds currently provide a unique opportunity to study individual molecular clouds outside of the Milky Way under very different environmental conditions. Numerous studies of molecular clouds have been carried out in the inner Galaxy (see e.g. Cohen et al.). Kutner & Mead (1981) observed molecular clouds outside the solar circle and found far-infrared (FIR) emission in several of the clouds observed in CO (Mead, et al. 1987). Mead, Kutner & Evans (1990) calculated the star formation per unit mass or 'star formation activity,' defined as the ratio of FIR luminosity to CO luminosity, for several clouds in the outer Galaxy. Our study is very similar to this outer Galaxy study, involving comparison of FIR and CO luminosities of individual giant molecular clouds (GMC).

2. Observations

The CO observations are from a study of the Magellanic Clouds which has been designated a Key Program for the SEST (15m). The goal of the program is to produce fully sampled (half-beamwidth) maps of a large number of individual GMCs. The SEST provides a resolution of 40″ in the CO ($J = 1 \rightarrow 0$) line; this corresponds to a linear resolution of 10 pc at the LMC. This resolution allows us to study individual GMCs within cloud complexes in the LMC.

The FIR data are obtained from the IRAS High Resolution data. These data are the original IRAS survey scans reprocessed using the Maximum Correlation Method algorithm of Aumann, Fowler & Melnyk (1990). The algorithm relies on over-sampling in the IRAS survey, thus the final resolution will depend on the coverage of the given region. For our images the resolution is $\sim 45''$ to $1'$ at $60\,\mu$m and $\sim 1'15''$ at $100\,\mu$m, corresponding to a linear resolution of 12 to 18 pc at the LMC. Using these High Resolution IRAS data we have FIR maps of comparable resolution to our CO maps, and are able to discern individual clouds in FIR.

3. Results

We have produced FIR maps for 16 molecular clouds in the LMC for which SEST CO data were available. This includes 5 clouds in the complex south of 30 Doradus, 10 clouds in the N11 region, and the N159 region. Analysis of FIR and CO maps for these regions indicates that FIR emission does correspond quite well spatially to CO peaks and that we can compare FIR emission and CO emission for individual molecular clouds.

FIR and CO luminosities for the individual molecular clouds for which there was good CO–FIR correspondence have been measured. The FIR luminosity was calculated using the color corrected $60\,\mu$m and $100\,\mu$m fluxes. Color corrections and total L_{FIR} extrapolation are as given in the IRAS Explanatory Supplement.

4. Discussion and Future Investigations

In the clouds we studied, we found L_{FIR} to be comparable to that for clouds in the Milky Way. L_{CO} for these clouds is also comparable to that found in outer Galaxy clouds (Mead, Kutner & Evans 1990), and the less CO-luminous inner Galaxy clouds (Scoville & Good 1989; Mooney & Solomon 1988; Myers et al. 1986). The clouds in the LMC show the same wide range of values of star formation activity, virtually independent of CO luminosity, as is seen in Milky Way molecular clouds. We also see that the star formation activity in LMC clouds is comparable to that of Milky Way clouds.

We plan to expand our survey of the LMC; also, using available CO data, we will extend the study to the SMC. In order to improve our FIR data we plan to use Kuiper Airborne Observatory (KAO) observations. The KAO resolution is slightly better than the IRAS High Resolution data: ~ 45″ at 160 μm; comparable to SEST. A FIR KAO map for N159 is shown along with CO and radio continuum maps in Fig. 1. KAO observations can also be used to verify the IRAS photometry and give a more accurate measure of L_{FIR} by including longer wavelengths. We received a partial flight in March 1991, and mapped one cloud at 160 μm. Additional flights have been proposed for 1993.

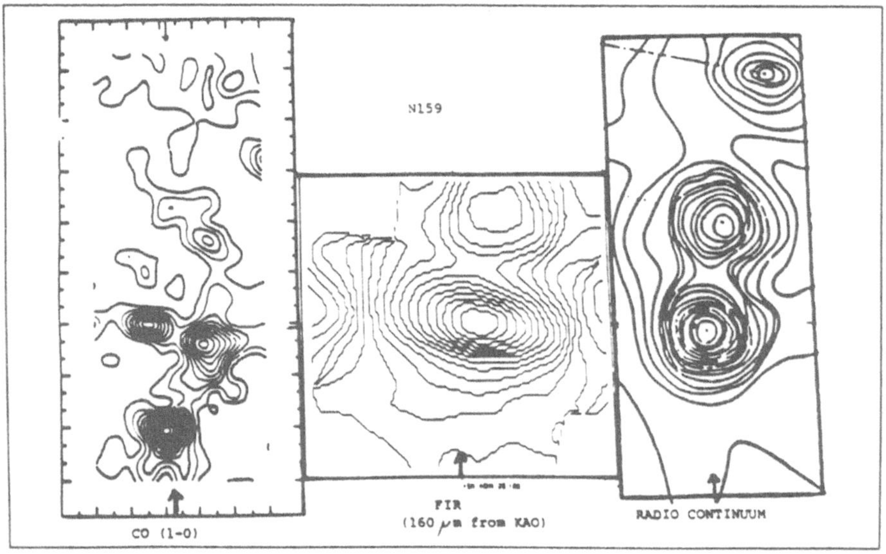

Fig. 1. 160 μm emission in the vicinity of N159. Also shown on the same angular scale are the CO ($J = 1 \rightarrow 0$) SEST map, and the radio continuum. The arrow at bottom center marks the same RA in each map. The north–south extent of the FIR map is 18′.

Acknowledgements. This research is supported by NASA grants: NAG2-677, and NAG5-1631. Special thanks to Sean Carey, Jean Chiar, and Colleen Stevens for their assistance.

References

Aumann, H.H., Fowler, J.W., Melnyk, M. (1990) **Astron. J., 99**, 1674.
Cohen, R.S., Cong, H., Dame, T.M., Thaddeus, P. (1980) **Ap. J.***(Letters)***293**, L53.
IRAS Catalogs and Atlases: Explanatory Supplement 1988, ed. C.A. Beichman, G. Neugebauer, H.J. Habing, P.E. Clegg, T.J. Chester (Washington DC: Government Printing Office).
Kutner, M.L., Mead, K.M. (1981) **Ap. J.***(Letters)***249**, L15.
Mead, K.M., Kutner, M.L., Evans, N.J., Harvey, P.M., Wilking, B.A. (1987) **Ap. J., 312**, 321.
Mead, K.M., Kutner, M.L., Evans, N.J. (1990) **Ap. J., 354**, 492.
Mooney, T.J., Solomon, P.M. (1988) **Ap. J., 339**, 149.
Myers, P.C., Dame, T.M., Thaddeus, P., Cohen, R.S., Silverberg, R.F., Dwek, E., Hauser, M.G., (1986) **Ap. J., 301**, 398.
Scoville, N.Z., Good, J.C. (1989) **Ap. J, 339**, 149.

Modelling the SMC: atomic to molecular transition and emissivities

J. Le Bourlot, G. Pineau des Forêts, E. Roueff
Observatoire de Paris, DAEC, Place Janssen, F 92195 Meudon, France

Abstract: We present few models of the atomic to molecular transition region in the interstellar medium imbedded in the SMC environment. Photodissociation in the lines of H_2, CO and its isotopes is treated with the relevant recent molecular data. Emissivities in the fine structure lines of atomic and ionic species are also presented and compared with those obtained considering a standard Galactic interstellar cloud.

1. Presentation of the model

The molecular content of magellanic clouds is of current interest and is one aim of the Key Program of ESO performed on the SEST millimeter radio telescope (Israel et al 1992). We apply to the physical conditions relevant in the SMC a model of the atomic to molecular transition originally devoted to Galactic interstellar clouds (Le Bourlot et al. 1992a). The main features of this model are the realistic treatment of the photodissociation of H_2, CO and its isotopes coupled to the chemical processes and to the thermal balance for a semi-infinite plan parallel cloud. 75 molecular species including H, He, C, ^{13}C, O, ^{18}O, and a representative metal are considered in a chemical network of 805 reactions. The details of the model are described in Abgrall et al. (1992) and in Le Bourlot et al. (1992a). We compare the different behaviours of a standard Galactic interstellar cloud of moderate extinction and density (ISM) with a typical magellanic interstellar cloud (SMC). Table 1 gives the typical parameters describing these two environments.

Table 1

	SMC	ISM	Comments
χ	10	1	Radiation field intensity scale
O / H	$1.1\ 10^{-4}$	$8.5\ 10^{-5}$	Oxygen abundance
C / H	$1.4\ 10^{-5}$	$3.6\ 10^{-5}$	Carbon abundance
M / H	$6.4\ 10^{-6}$	$1.5\ 10^{-8}$	metal abundance
R	2.7	3.1	A_V / E (B - V)
N_H / A_V (cm^{-2})	$4.5\ 10^{22}$	$5.8\ 10^{21}$	extinction law
grains	10^{-3}	10^{-2}	grain to gas mass ratio
C in PAH	0	10%	no PAH in SMC
n_H (cm^{-3})	10^4	10^4	isochoric stationary model

2. Results

We restrict our discussion to two main topics: the predicted intensities of the fine structure lines and the CO to H_2 ratio. A more complete discussion of the various parameters will be presented in a forthcoming paper (Le Bourlot et al 1992b)
Table 2 gives the intensities of the fine structure lines for the two cases considered here:

	λ (μm)	SMC (erg cm^{-2} s^{-1} sterad^{-1})	ISM (erg cm^{-2} s^{-1} sterad^{-1})
O (1 -> 2)	63.2	5.3 10^{-7}	1.2 10^{-8}
O (0 -> 1)	145.5	7.5 10^{-9}	1.5 10^{-10}
C (1 -> 0)	609.1	1.6 10^{-6}	2.0 10^{-7}
C (2 -> 1)	370.4	8.6 10^{-7}	2.4 10^{-7}
C^+ (3/2 -> 1/2)	157.7	1.4 10^{-5}	1.4 10^{-6}

Table 2

The C^+ fine structure line which is predicted to be quite intense in the SMC's has been observed by the Kuiper Airborne Observatory (Israel at this meeting) and a map in the lines of C and C^+ should help to understand the structure of the C^+ to CO transition.
The CO / H_2 ratio is often taken as an universal parameter which is used to determine the molecular mass content of a galaxy. We plot in Fig. 1 the predicted CO / H_2 column density ratio as a function of N(H_2) which is seen to vary much more considerably over a short range in the SMC than in the MIS case. This behaviour occurs mainly from the differences in the radiation fields, the extinction law and the depletions.

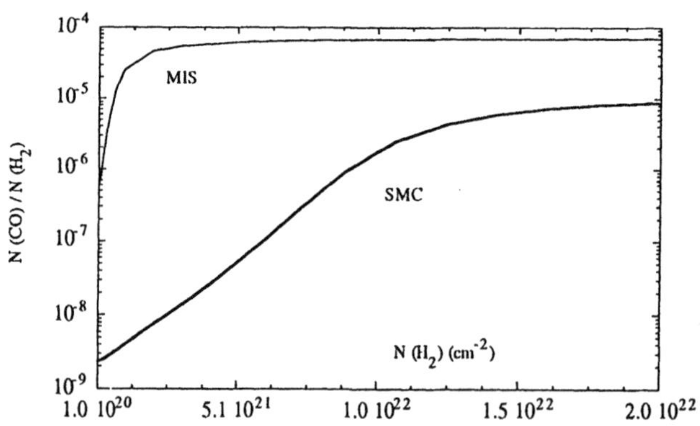

References

Abgrall H., Le Bourlot J., Pineau des Forêts G., Roueff E., Flower D.R., Heck L., *A&A* **253**, 525 (1992)
Israel F.P., Johansson L.E.B., Lequeux J., Booth R.S., Nyman L.A., Crane P., Rubio M., de Graauw Th., Kutner M.L., Gredel, R., Boulanger F., Garay G., Westerlund B., *A&A*, in press. (1992)
Le Bourlot J., Pineau des Forêts, G., Roueff E., Flower D.R., *A&A*, in press. (1992a)
Le Bourlot J. et al. to be published (1992b)

A Survey of Planetary Nebulae in the LMC

D.H.Morgan

Royal Observatory, Blackford Hill, Edinburgh EH9 3HJ, UK

Abstract: This paper describes the overall surface distribution of planetary nebulae in the LMC based on a uniform survey of objective prism plates taken with the UK 1.2m Schmidt Telescope.

1 The Survey

Morgan and Good (1992 - Paper 1) recently published a list of 86 new planetary nebula (PN) candidates located in the central $10° \times 6°$ region of the Large Magellanic Cloud (LMC) and which had been identified on two deep objective prism plates taken with the UK 1.2m Schmidt Telescope (UKST). The outer regions of the LMC have now been covered by a set of similar objective prism plates taken with the UKST, thereby allowing the completion of the UKST survey for PN in the LMC. The plates have been searched by systematic visual scanning and various checks have permitted the elimination of faint M and C stars, compact emission-line galaxies and variables such as novae. The new plates have revealed about 30 new candidate PN (Morgan 1993 - in preparation - Paper 2). Although a few further checks will be made before publication, the final list is unlikely to be significantly different from the present version.

The total number of planetary nebulae identified in the UKST survey is about 240 and consists of the new objects listed in Papers 1 and 2 plus most of the objects listed in earlier catalogues.

Most UKST plates include significant areas of overlap with plates of neighbouring fields. These overlap regions have been scanned on each plate and plate-to-plate comparisons have been made in order to determine the level of completeness of the survey. First, when the images were compared visually, it was found that most of the plates were of similar depth and resolution, with three marginally poorer. Secondly, the level of completeness of the scanning was checked by comparing the numbers of PN found in scans of areas common to two or more plates. It turned out that $\sim$20% had been found on one plate rather than two. Approximately 45% of the total survey area is covered by single plates only, though this is mostly at a low surface density and applies to $\sim$ 10% of the total number of PN.

2 Surface distribution

With a reasonably uniform survey now complete (apart from some final checking), it is possible to look at the surface distribution of PN across the LMC. There are three distinct levels of PN surface density. First, the density is highest in the Bar, despite the greater difficulty of identifying objects there. Secondly, there is a region of moderate surface density which approximately coincides with the inner elliptical cluster system defined by Kontizas et al. (1990); thirdly, there is a low density outer region extending to $\sim 7.5°$ from the LMC centre. However, the outer elliptical structure seen so clearly in the cluster distribution is not evident in the PN, for the outer PN extend farther to the east and west of the clusters, but not to the north and south, with the result that the outer system is circular in appearance. The centre of the outer envelope is close to that of the outer cluster ellipse and nearly 1° to both east and north of the centre of the PN system as a whole. The latter is roughly 0.6° north of the optical centre of the LMC (de Vaucouleurs and Freeman 1972). On the other hand, the PN distribution is markedly different from that of the CH stars which shows high density levels in the outer eastern reagions (Hartwick and Cowley 1988).

It is possible to distinguish between bright and faint PN according to the diameters of the [OIII]λ5007 lines on the objective prism plates. It transpires that there are some differences between the distributions of the bright and faint PN. Overall, the bright images constitute $\sim 40\%$ of the PN population, whereas eastward of 6^{hr}, $\sim$55% are bright, and north of -66°, only $\sim 15\%$ are bright. Both these zones are in the overall low density region.

Acknowledgements The author is grateful to the staff of the UK Schmidt Telescope Unit for taking and providing the excellent photographic plates and films.

References

Hartwick F.D.A., Cowley A.P. 1988, Astrophys. J., 334, 135.

Kontizas M., Morgan D.H., Hatzidimitriou D., Kontizas E., Astron. Astrophys. Suppl., 1990. 84, 527.

Morgan D.H., Good A.R., 1992. Astr. Astrophys. Suppl., 92, 571.

de Vaucouleurs G., Freeman K.C., 1972, Vistas Astron., 14, 163.

New CO and Hα observations of Magellanic-type irregular galaxies

Ralf-Jürgen Dettmar [1], Renate Becker [2], and Martin Shaw [3]

[1]Radioastronomisches Institut der Universität Bonn, Auf dem Hügel 71, D-5300 Bonn 1, FRG
[2]IRAM, Domaine Universitaire de Grenoble, F-38406 St. Martin d'Hères Cedex, France
[3]University of Manchester, Dept. of Astronomy, Manchester M13 9PL, U.K.

1 Introduction

Irregular galaxies (Irr's) are known to exhibit a considerable deficiency in CO line emission when compared with spiral galaxies of similar star-formation activity (e.g. Verter 1987). While the disks of most late type-spiral galaxies appear to share the molecular cloud characteristics of the Galaxy, irregulars are believed to contain either relatively little molecular material or to contain individual molecular clouds which differ in size, density or temperature.

Within the class of actively star-forming galaxies they comprise the most abundant species. In Irr's the formation of molecular clouds and stars is not organized by density waves, they rotate slowly and they are of low metallicities. Hence, they provide an star-forming environment with even more extreme conditions than the outer Galaxy.

An important difference to spiral galaxies seems to be that molecular cloud complexes in Irr's can be extremely large. This is particularly true in the presence of an off-centered bar, i.e. in Magellanic-type Irr's (IBm's). In these objects large amounts of gas can be concentrated at one or both ends of the bar where also the most active regions of star formation are found (Cohen et al. 1988, Dettmar and Heithausen 1990, Wilson and Reid 1991).

2 Observations

In order to study the star-forming regions and their relation to molecular cloud complexes in IBm and Im galaxies we used the ESO 1.5 m, 2.2 m, and 3.5 m NTT telescopes at La Silla to obtain Hα images of several southern objects.

In addition, the SEST telescope was used to detect and map some of these objects in the J=1→0 CO line. The weak CO emission of most Irr's made it necessary to integrate for 2–8 hours per position. A rms-noise of 8–25 mK/channel could be reached with a resolution of $1.8\,\mathrm{km\,s^{-1}}$.

3 Results

Here we describe some preliminary results for IC 4662, DDO 70 (Sex B), and IC 5052.

A deep Hα image of the Im galaxy IC 4662 was obtained with EMMI at the ESO NTT using a 2048×2048 FA CCD. The Hα distribution is dominated by a large superassociation comparable to 30 Dor in the LMC (Heydari-Malayeri 1990). The Hα image reveals the presence of spectacular *froth* structures in the diffuse ionized interstellar medium which are typical for Irr's (Hunter and Gallagher 1990). In IC 4662 individual froth filaments can be traced as far as 600 pc (assuming a distance of 2 Mpc) from the ionizing O stars. Only upper limits for the CO luminosity of IC 4662 were reported before (Heydari-Malayeri 1990). By adding up three different positions observed with SEST a rms-noise of 7 mK/channel could be reached and we obtained a CO spectrum with a velocity integrated main beam brightness temperature of $I_{CO} = 0.24\,\mathrm{K\,km\,sec^{-1}}$.

The Danish 1.5 m telescope was used to obtain Hα imaging of DDO 70 (Sex B). The total Hα luminosity is determined to $6.2 \times 10^{37}\,\mathrm{erg\,sec^{-1}}$ (Becker 1990). This can be converted into a star formation rate of $4.4 \times 10^{-4} M_{\odot} \mathrm{yr}^{-1}$ following Hunter and Gallagher (1986). A CO line with $I_{CO} = 0.14\,\mathrm{K\,km\,sec^{-1}}$ was detected on a 4σ level after an integration time of 260 minutes with the SEST. This is the second detection of CO emission in a dwarf irregular of Magellanic-type beyond the Small Magelanic Cloud after Wilson (1992) recently detected CO in NGC 6822.

IC 5052 is classified as type Sdm and icluded in our sample for comparison. The radio continuum emission studied by Harnett and Reynolds (1991) demonstrates the presence of ongoing star-formation. The Hα distribution obtained with the ESO/MPI 2.2 m telescope also shows a number of HII regions, some of them are clearly residing above the plane of this almost edge-on galaxy. The galaxy was mapped in the CO line. Only a weak CO emission line was detected in one of five positions. However, the measured velocity of this line is not consistent with published radial velocities from 21 cm HI and optical emission line observations.

References

Becker, R. 1990, PhD thesis, Univ. of Bonn

Cohen, R. S. et al. 1988, ApJ **331**, L95

Dettmar, R.-J., Heithausen, A. 1990, ApJ **344**, L61

Harnett, J. I., Reynolds, J. E. 1991 A&AS **88**, 73

Hunter, D. A., Gallagher, J. S. 1986 PASP **98**, 5

Hunter, D. A., Gallagher, J. S. 1990 ApJ **362**, 480

Heydari-Malayeri et al. 1990 A&A **234**, 99

Verter, F. 1987, ApJS **65**, 555

Wilson, C. D. 1992, ApJ **392**, 144

Wilson, C. D., Reid, I. N. 1991, ApJ **366**, L11

Novae, Supernovae, Supernova Remnants

Roy S. Booth, (Jan Koornneef), Monica Rubio

Willem Wamsteker, Thijs de Graauw

Ultraviolet Studies of Novae in the Large Magellanic Cloud

Sumner Starrfield [1,6], Peter H. Hauschildt [1], Steven N. Shore [2,6], George Sonneborn [3,6], R. Gonzales-Riestra [4], and Warren M. Sparks [5,6]

[1]Department of Physics and Astronomy, Arizona State University, Tempe, AZ 85287-1504,
[2]Goddard High Resolution Spectrograph Science Team- Computer Sciences Corporation, Code 681, Goddard Space Flight Center
[3]Laboratory for Astronomy and Solar Physics, Code 681, Goddard Space Flight Center
[4]IUE Observatory, European Space Agency, Madrid
[5]Applied Theoretical Physics Division, Los Alamos National Laboratory, Los Alamos, New Mexico
[6]Guest Observer, *International Ultraviolet Explorer* Observatory

Abstract

We report on the ultraviolet observations of five novae in the LMC. They are LMC 1988 #1, which was a CO nova similar to OS And; LMC 1988 #2, which was an ONeMg nova similar to Nova Her 1991; LMC 1990 #1, which was a very fast ONeMg nova similar to V693 CrA 1981; LMC 1990 #2, which was a recurrent nova similar to U Sco; and LMC 1991 which reached V$\sim$9.

1 Introduction

Before the discovery of Nova LMC 1988 #1, the studies of novae in external galaxies had, with a few exceptions, been limited to determinations of their positions within the galaxy and their light curves. The primary exceptions to this statement were the spectroscopic investigations of novae in M31 by Ciardullo, Ford, and Jacoby (1983) and Cowley and Starrfield (1987). However, because the novae in M31 only reach V$\sim$16 at maximum, it is very difficult to follow them through their evolution to the nebular stage when it becomes possible to obtain the abundances in the ejecta by standard techniques.

However, because the LMC is much closer, novae can reach to V~11, or brighter, and it become possible to follow their optical spectral evolution until very late stages in their outbursts. In addition, at V~11, they are bright enough to obtain high quality spectra with the International Ultraviolet Explorer (hereafter **IUE**) satellite in a reasonable amount of exposure time. The recent surveys of the LMC for supernovae and novae have resulted in the discovery of five novae since 1988 and we have followed their evolution with the **IUE** satellite. As a direct result of studies of the LMC novae, we have: (1) found that, at maximum light, the outbursts of the fastest classical and recurrent novae exceed $L_{\rm Edd}$, for a 1.0M$_\odot$ white dwarf, (2) found and analyzed analogous outbursts to galactic novae, (3) established accurate distances to galactic novae, (4) found evidence at late times for a high temperature ($T > 2 \times 10^5$ K) source within the system, and (5) confirmed that there are two classes of recurrent novae. One class, the U Sco class, consists of a high mass white dwarf with a compact but evolved secondary, while the second class, the RS Oph class, consists of a white dwarf with a giant secondary.

At the time of writing of this review, we have obtained **IUE** UV data on five LMC novae and each of these objects has exhibited a unique outburst. In chronological order, they are: (1) LMC 1988 #1, which was a moderate speed CO nova that formed dust and whose outburst behavior resembled that of OS And, (2) LMC 1988 #2, which was a fast ONeMg nova whose outburst resembled that of Nova Her 1991, (3) LMC 1990 #1, which was a very fast ONeMg nova with similar spectral evolution to V693 CrA 1981, (4) LMC 1990 #2, which was a very fast recurrent nova with an outburst that was very similar to that of U Sco and V394 CrA, and (5) LMC 1991, which was an extremely fast CO (probably) nova that reached V~9 and, therefore, at maximum its luminosity must have exceeded $L_{\rm Edd}$ for a 1.4M$_\odot$ white dwarf.

2 Results for Individual Novae

As we have emphasized in a number of publications, each galactic nova that has been well-studied in the UV has suffered a unique outburst. This statement is also true for the LMC novae. We provide supporting evidence for this statement by discussing the observations of all five of the LMC novae. All of the observations reported here have been obtained with the **IUE** satellite in low resolution mode (R≈300) between 1150Å-3400Å. The data have been processed using standard software available at the Regional Data Reduction Facility (RDAF) at Goddard Space Flight Center. The journal of observations will be given elsewhere (Shore et al. 1992, in preparation). Here, we present only a summary of the basic results for each nova.

2.1 Nova LMC 1988 #1

This nova was discovered by Garradd on March 21.484, 1988 (IAUC 4568). Our first observation with **IUE** was obtained on March 25, 1988. Figure 1a shows a merged SWP plus LWP spectrum obtained 5 days later. Note that the continuum has a very steep slope to the red and appears to be broken by numerous emission lines. As we have now shown (Hauschildt et al. 1992a,b), this impression is wrong. Most of the "emission" features which appear in the spectrum are regions of transparency which fall between overlapping iron lines. We refer to the appearance of this characteristic spectrum as the "iron curtain." We have only observed it in those novae that eject sufficient material so that the expanding envelope remains optically thick through maximum light in the UV. In contrast, there are other classes of novae for which far less material is ejected and the expanding gas has become optically thin, in the lines, at the time that maximum light is reached in the UV.

Because the energy distribution resembles that of a stellar atmosphere, we have now begun a series of stellar atmosphere studies of these novae (Hauschildt et al. 1992b) and have already found that the theoretical models reproduce the observations quite well if non-LTE effects as well as metal line blanketing are included in the model construction and synthetic spectra. We are just at the beginnings of this study but expect to be able to provide detailed comparisons of the ejected abundances between galactic and LMC novae (Hauschildt et al. 1992b).

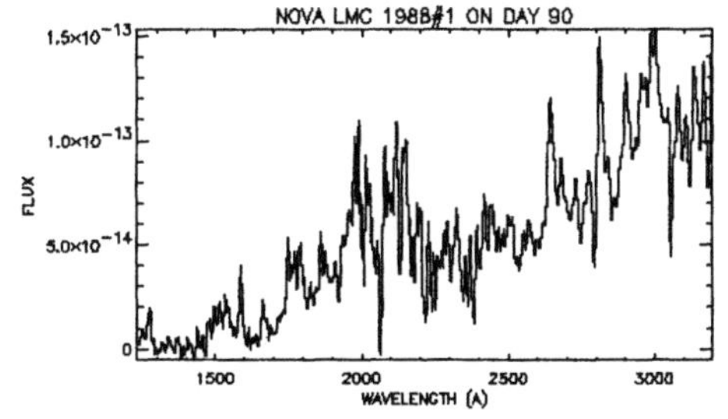

Figure 1a. The merged spectrum of LMC 1988 #1 on Day 90. Note the slope of the continuum to the red. The expanding material is still optically thick.

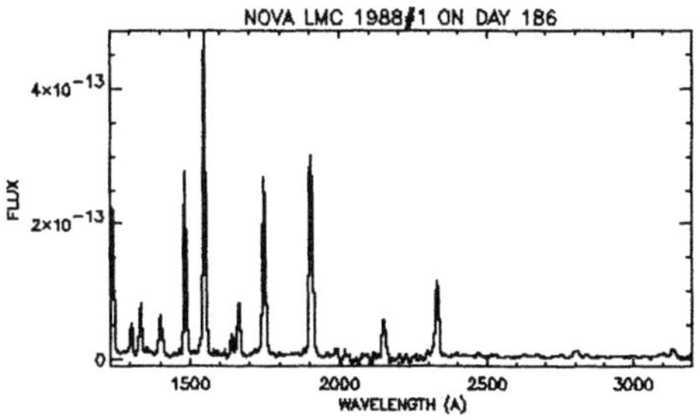

Figure 1b. The merged spectrum of LMC 1988 #1 on day 1988/186. The strongest line is CIV followed by CIII].

Nevertheless, there are a few emission lines present in this spectrum (Figure 1a). Mg II 2800Å is strong and exhibits a P-Cygni profile. In addition, the broad feature around 2600Å is, most likely, Fe II multiplet 1. But, as is shown in our companion poster paper (Hauschildt et al. 1992b), the pronounced features in the SWP portion of the spectrum are formed by the iron curtain.

We continued to obtain ultraviolet data over the next few months until it became too faint for observations with the spacecraft. A light curve showing the integrated UV flux is given in Figure 1c which also gives the light curves for three other LMC novae. Each data point is obtained by summing the SWP plus LWP spectrum with the geocoronal Lyα line removed. In all cases, the spectra were obtained on the same day. Saturated pixels are excluded from the sum so, for some of the overexposed spectra, we report only lower limits. Reddening corrections were applied using the field (non-30 Doradus) LMC extinction law (Fitzpatrick 1986). We assume $E(B-V) = 0.^m15$ for all novae in order to simplify the analysis. An earlier form of the light curve for LMC 1988 #1 (plus other novae), with no correction for reddening, can be found in Austin et al. (1990). Note that in their publication the light curve is plotted in magnitude units.

The most interesting feature of the light curve is that it exhibits an initial decline followed by a steep recovery and then the decline continues. The beginning of the

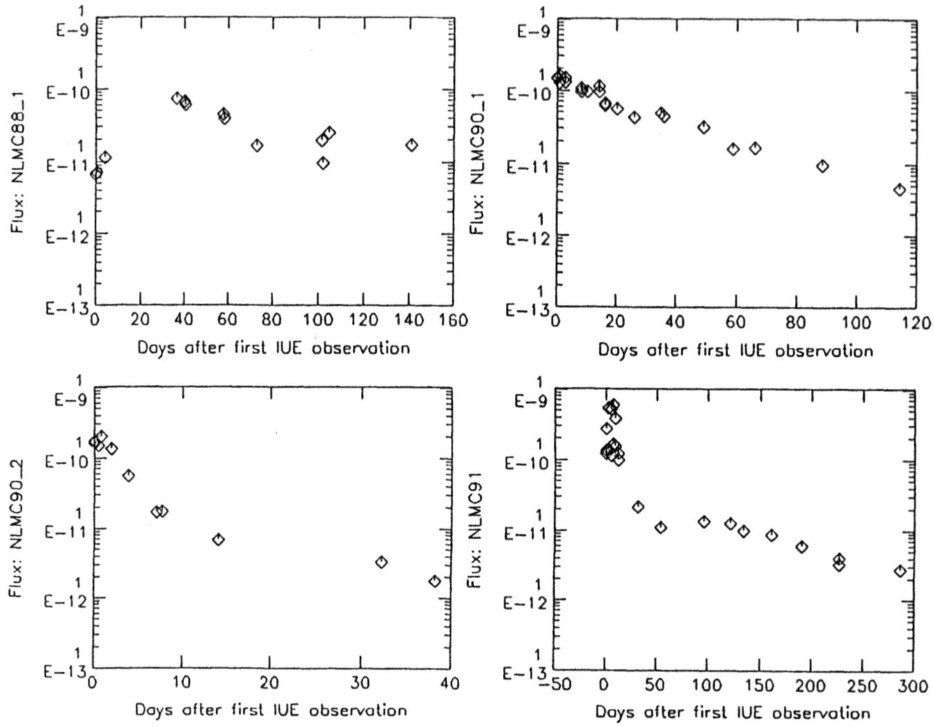

Fig. 1c. This figure shows the UV light curves of four novae. Each nova is identified in the region following "Flux:". A description of how each data point is obtained can be found in the text.

initial decline occurs at the time that a dust shell formed in the ejecta. With the effects of dust shell formation removed, the light curve evolves in a very similar fashion to Galactic CO novae (Austin et al. 1990).

Three months later, we obtained the merged spectrum shown in Figure 1b. We now observe a spectrum characteristic of an optically thin gas with a broad range in ionization. The easily identifiable emission lines are: N V 1240Å, O I 1304Å, C II 1335Å, O IV] 1402Å (probably), N IV] 1486Å, C IV 1550Å, He II 1640Å, [O III] 1660Å,1666Å, N III] 1750Å, C III] 1909Å, N II] 2140Å, C II] 2326Å, Mg II 2800Å (very weak), and O III 3133Å (also very weak). Note that, except for He II and Mg II, all of the emission lines are from various ionization stages of the CNO elements. Although the analysis is not yet finished, it is clear that this nova is a member of the CO composition class and will exhibit overabundances of the CNO nuclei. None of the lines which indicate an ONeMg novae are present (we will list some of the ONeMg indicators when we discuss LMC 1990 #1). Second, we note the very large number of ionization stages present in this spectrum. This is very characteristic of novae spectra and its cause was found by Hauschildt et al. (1992a). They discovered that the non-LTE, spherical, expanding atmospheres of novae are very extended and one is simultaneously sampling a very broad range of electron temperatures and densities. This result must be included in any analyses of novae ejecta.

We obtained more spectra after those shown in Figure 1b and they indicate that the density of the expanding material has continued to drop and the electron temperature to increase with time. The electron temperature increases because of the heating produced by the luminous underlying source which is still radiating at or near the Eddington limit (Starrfield et al 1990a,b; Starrfield 1992).

2.2 Nova LMC 1988 #2

The second nova to be found in outburst in the LMC during 1988 was discovered on October 12.5, 1988 (U.T.) at V~11.3 (IAUC 4663). This nova was at V~14.5 on October 7.5, 1988 (IAUC 4663) so its rise to maximum was not particularly rapid. UV spectra were first obtained on October 14.6, 1988 (IAUC 4669) and they showed an expanding, optically thick, pseudo-photosphere with, possibly, a few emission lines (Figure 2a). However, the spectrum was much flatter than that of LMC 1988 #1 which implies that the material was hotter at the time that the spectra were obtained. It also appears that some of the features seen in the early spectra are emission lines and not windows in the iron curtain. The feature at 1860Å must be Al III and there is an unidentified feature at about 1960Å that was also seen in SN 1987A.

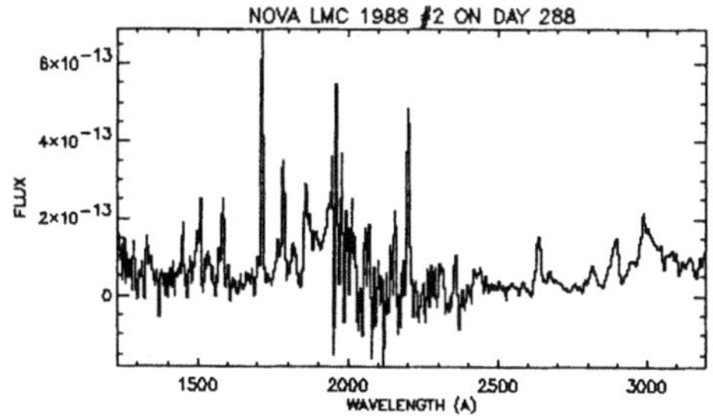

Figure 2a. The merged spectrum of LMC 1988 #2 on Day 1988/288. Note how flat the continuum is as compared to LMC 88 #1 - it must be much hotter.

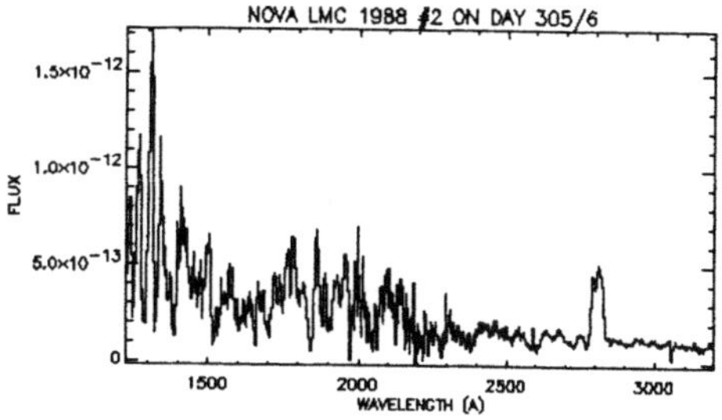

Figure 2b. The merged spectrum of LMC 1988 #2 on day 1988/305. Note the great strength of Mg II 2800. The neon lines do not appear until later.

By October 21.8 (Figure 2b), this nova showed the typical lines emitted by an optically thin gas (Cassatella and Gonzalez-Riestra 1990). Because it has not been shown before, we present the SWP spectral evolution of this nova in Figure 2c. As the density drops, we begin to see forbidden lines and the spectrum obtained on 29 December 1988 shows a strong line at 1602Å which we identify as [Ne IV]. The next spectrum, which was obtained about 5 months later, does not show [Ne IV] at 1602Å but does show [Ne V] at 1575Å. If we use as a finding list those lines observed in the UV spectra of V693 CrA (Williams et al. 1985), we find: N V 1240Å, a line at about 1310Å which we identify as Si II (a weak feature at about 1814Å may also be Si II), an unidentified line at ~1350Å, O V 1371Å, Si IV 1397Å, O IV] + Si IV]

1400Å, N IV] 1487Å, Si II 1531Å, C IV 1550Å, [Ne V] 1575Å, [Ne IV] 1602Å, He II 1640Å, [O III] 1660Å,1666Å, N IV 1719, N III] 1750Å, [Ne III] 1815Å, Al III 1857Å, and C III] 1909Å. There are other weak features that are unidentified. These emission lines are characteristic of material ejected during the outburst of an ONeMg nova. Therefore, these spectra, plus the LWP spectra that we do not show, identify this nova as an ONeMg nova intermediate in UV speed class between Nova Her 1991 and V693 CrA 1981. The optical spectra obtained by Sekiguchi et al. (1989) also demonstrate that this was an ONeMg nova. They show the typical [Ne III] and [Ne V] emission lines in the optical ultraviolet.

2.3 Nova LMC 1990 #1

This nova was discovered in the Large Magellanic Cloud by Garradd (McNaught 1990) on 16.47 January 1990 at V~11.5 and **IUE** observations began on 17.95 January, within hours of the announcement of the discovery. The initial ultraviolet spectra showed a hot continuum and strong emission lines (N V, Si IV, C IV, and Al III) with P-Cygni profiles (Shore, Sonneborn, and Starrfield 1990a, 1990b). This spectrum is shown in Figure 3a and it was completely unlike any of the spectra that have been shown in this paper. The P-Cygni lines had flat-bottomed absorption troughs extending to more than -6000 $\mathrm{km\,s^{-1}}$. This remarkable spectrum and large expansion velocities are very reminiscent of the first spectrum obtained for V693 CrA (Nova CrA 1981) an ONeMg nova (Williams et al. 1985). It seemed clear to us that we were observing a fast ONeMg nova in the LMC. Confirming evidence, that Nova LMC 1990 #1 was an ONeMg nova, came from Dopita and Rawlings (1990), who reported that [Ne III] 3868Å appeared on 22 January and [Ne V] 3426Å on 29 January. Dopita and Rawlings (1990) also reported that by 13 February 1990 [Ne V] was the strongest emission line in the visual spectrum. Nova LMC 1990 #1 was, beyond doubt, another ONeMg nova.

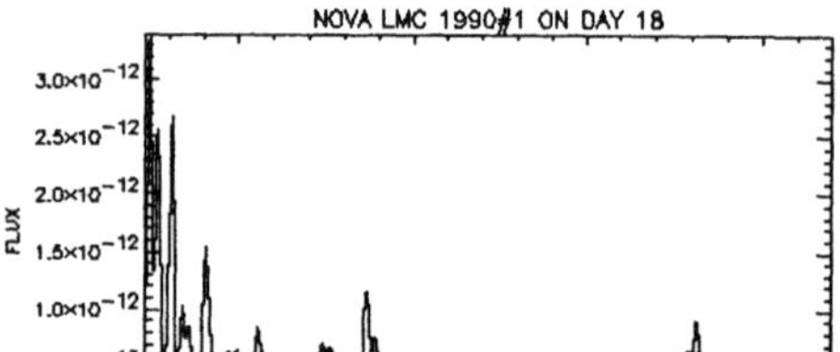

Figure 3a. The merged spectrum of LMC 1990 #1 on day 1990/18. Note the strength of MgII, NV, and CIV and the P Cygni profiles associated with these lines.

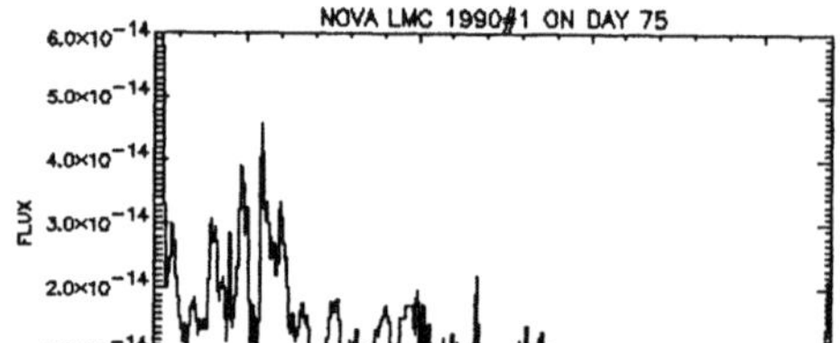

Figure 3b. The merged spectrum of LMC 1990 #1 on day 1990/75. Note that Mg II has disappeared but NIV and NIII] are still present.

In a recent abundance analysis of this nova, Dopita et al. (1992) have found that most of the intermediate mass nuclei are enhanced which is conclusive evidence that this was an explosion that occurred on an ONeMg white dwarf (Starrfield, Truran, and Sparks 1986; Weiss and Truran 1990; Nofar, Shaviv, and Starrfield 1991). Therefore,

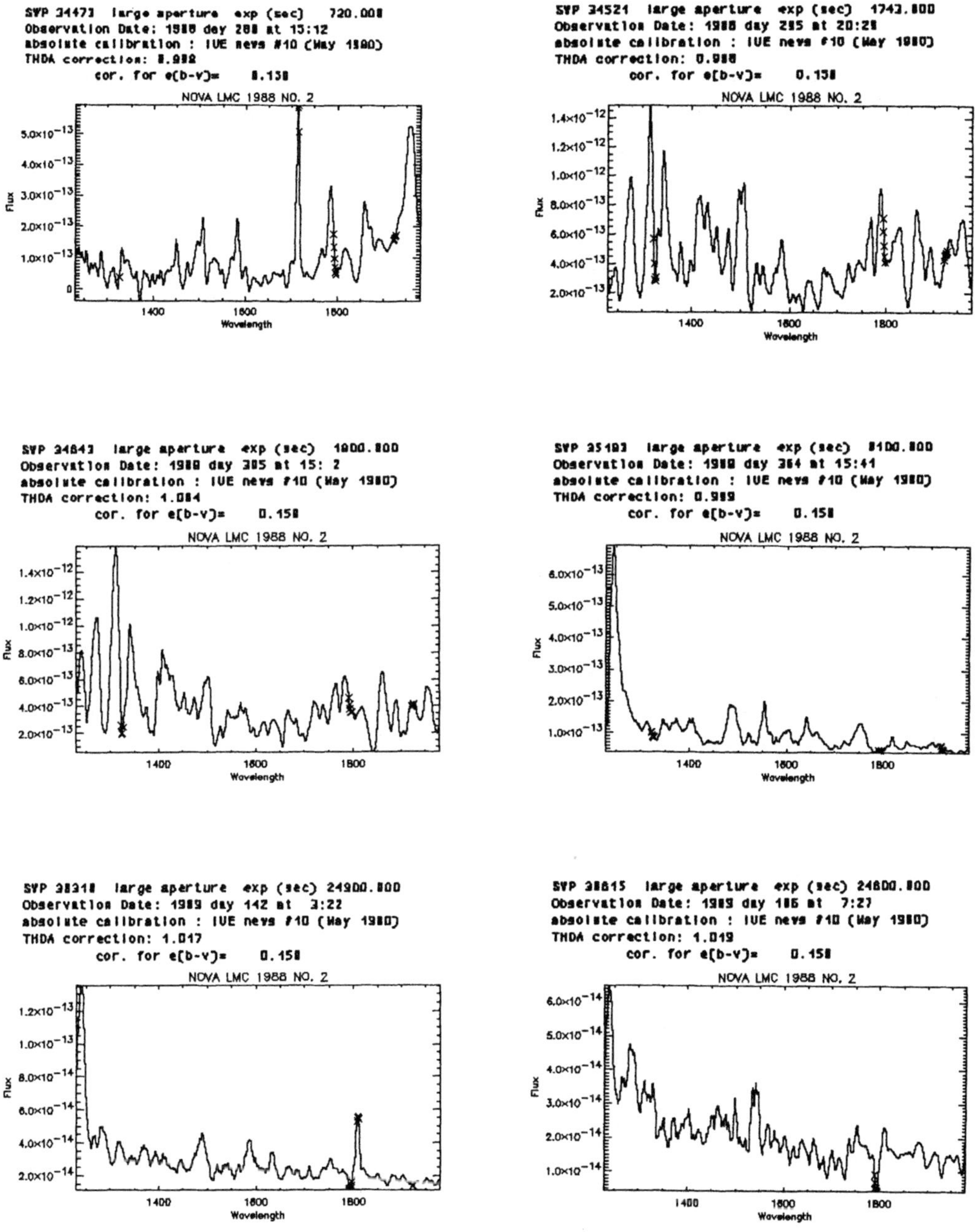

Fig. 2c. The SWP spectral evolution of LMC 1988 #2. The bottom two spectra in this figure show [Ne IV] 1602Å and [Ne V] 1575Å, respectively.

two out of the five LMC novae, studied in the ultraviolet, have been found to be ONeMg novae.

The spectral evolution of LMC 1990 #1 was presented by Sonneborn, Shore, and Starrfield (1990). A more detailed analysis is in progress (Sonneborn, Shore, and Starrfield 1992; in preparation). In contrast to the spectral evolution of the two novae studied in 1988, the SWP continuum was dominated by P-Cygni absorption lines on all of the resonance transitions. At no time in the evolution of this nova did we observe the characteristic iron-curtain features observed in LMC 1988 #2. Also in contrast to the novae studied in 1988, the delay from optical maximum to UV maximum was approximately 2 days which is much shorter than found for other novae. The maximum integrated UV luminosity of this nova reached $7 \times 10^4 L_\odot$ and it was this bright for approximately three days. A light curve for this nova can also be found in Figure 1c.

Another interesting feature of the spectral evolution of this nova is that there were remarkable variations in the appearance of the spectrum on a time scale of days. The variations were most pronounced in the Si IV 1400Å and C IV 1550Å line profiles. Between 23 and 25 January the optical depth in the high-velocity portion ($v < -4,000$ km s^{-1}) of the C IV P-Cygni absorption increased by about a factor of five. At the same time, the Si IV absorption trough disappeared, the Al III absorption trough weakened, and the total C IV, N V, and Mg II emission increased by $\sim 50\%$. In addition, the total UV flux increased by a similar amount. These latter changes are evident in the integrated and line fluxes shown in Sonneborn, Shore, and Starrfield (1990). By 26 January the fluxes and line profiles had returned to their 23 January state. This may have been a secondary ejection event.

We followed the spectral evolution of this nova for about two months and one of the final combined spectra is shown in Figure 3b. Mg II has nearly disappeared, presumably since the material has become much hotter, and there is emission around 1600Å that suggests (although the spectrum is quite noisy) that we are seeing blended [Ne V] 1575Å and [Ne IV] 1602Å.

2.4 Nova LMC 1990 #2

The optical outburst of Nova LMC 1990 # 2 was discovered by Liller (1990) as the result of photographic nova patrol observations. The position of the nova was close to that of a nova that was reported in the LMC in 1968 (Sievers 1970) and which reached a maximum visual magnitude of about 10.7 mag. Shore et al. (1991) remeasured the original plates, plus more modern surveys of that region of the LMC, and the two novae were found to coincide in position to within $0.^s5$ in RA and 6 arcsec in declination. Thus, LMC 1990 #2 represents the first extragalactic recurrent nova (RN).

The spectral development of this nova was as unusual as that of Nova LMC 1990 #1. It neither developed P-Cygni profiles on the strong emission lines nor did it display an optically thick iron curtain. Shore et al. (1991) proposed that the explanation of this behavior was that it had ejected a very low mass shell. Optical spectra can be found in Shore et al. (1991) and Williams et al. (1991). While an extensive analysis of this nova has already appeared (Shore et al. 1991), we remark here on some additional features. The nova was clearly super-Eddington (for a $1.4 M_\odot$ star with solar abundances) at maximum UV light. High ionization emission was discernable from the first spectrum,

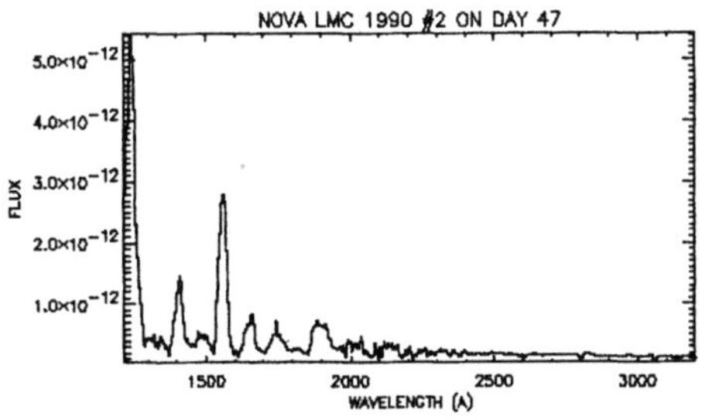

Figure 4a. The merged spectrum of LMC 1990 #2 on day 1990/47. All of the nitrogen lines are present and strong. This spectrum closely resembles the first spectrum of U Sco.

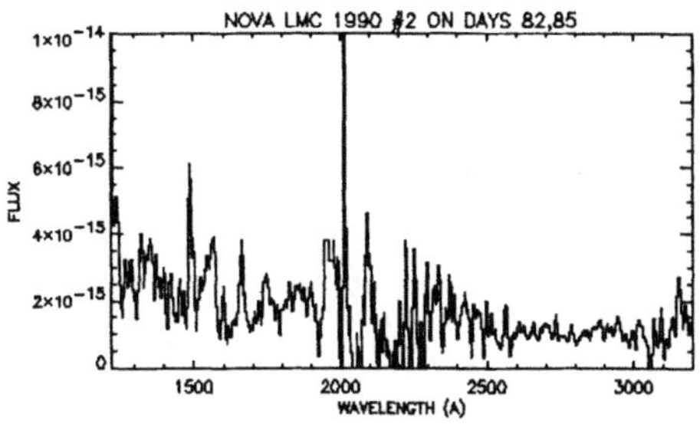

Figure 4b. A late merged spectrum of LMC 1990 #2 on day 1990/82 (SWP) and 1990/85 (LWP). It is fairly noisy but NV is still present.

which is shown in Figure 4a. Strangely enough, He II and N V were present and strong in the first observation. This nova was also especially noteworthy for the amount of flux visible shortward of Lyα, although, with only low resolution spectra, we cannot separate out the contributions from emission from Si III 1206Å, Lyα 1216Å, and N V 1240Å. The velocities observed in this nova were among the highest we have ever recorded, for novae in the UV, and may have reached close to $10^4\,\mathrm{km\,s^{-1}}$. The line profiles were unusual, both in the UV and optical. There was a narrow component of the profiles that developed within the first two weeks of the outburst, and the likely explanation is that the shell was not ejected spherically. Because of its very rapid decline, we were only able to follow it for a few weeks and Figure 4b shows one of the last sets of spectra that we were able to obtain. It is very noisy but N V is still clearly present as is N III] 1750Å and [O III] 1660Å, 1666Å. The sharp feature at 1500Å is a radiation "hit".

The spectral development of LMC 1990 #2 was very similar to that of U Sco 1979 and V394 CrA, with the single exception of the early P-Cygni profiles that were seen in U Sco (Williams et al. 1981) and not in this nova. The He/H ratio was quite similar with both novae being enhanced (by about a factor of 3) in helium. The N/C ratio was also large, and the absolute abundance of nitrogen was at least an order of magnitude above solar.

2.5 Nova LMC 1991

This nova was discovered by Liller (1991; IAUC 5244) on 16 April 1991 at V $\sim$ 12. The first IUE spectrum was obtained on 20 April 1991 and showed a flat but optically thick continuum (Figure 5a). Mg II 2800Åwas present and showed a deep P-Cygni profile. The early **IUE** spectra also showed numerous lines from singly ionized elements such as iron and silicon in addition to magnesium. We used the Fine Error Sensor (FES) to obtain the optical light curve of this nova and found that optical maximum occurred on about 24 April 1991 at V$\sim$9.0. At a distance modulus of $\sim$18.5, this corresponds to a maximum luminosity which exceeds the Eddington luminosity for a $1.4M_\odot$ white dwarf. If one includes the energy radiated in the UV along with that in the optical, and assumes a distance of 55 kpc to the LMC, then the luminosity of the nova at maximum exceeded $\sim 2 \times 10^5 L_\odot$ (Shore, Starrfield, and Sonneborn 1991). This luminosity makes LMC 1991 one of the brightest novae ever studied.

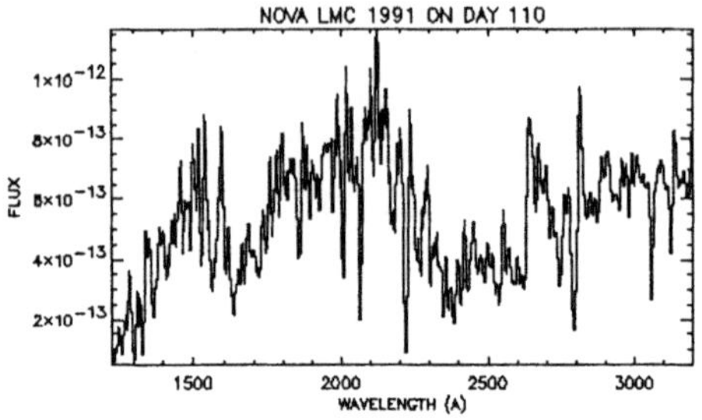

Figure 5a. A merged spectrum of LMC 1991 obtained near maximum light on day 1991/110. Note the deep P Cygni profile on MgII and AlIII. The iron curtain was very strong.

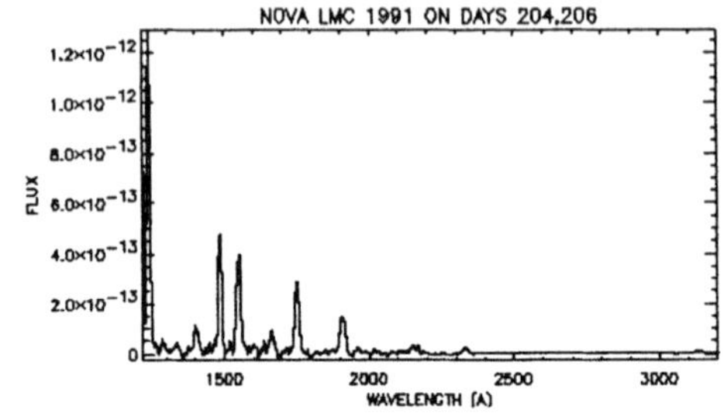

Figure 5b. A very late spectrum of LMC 1991 showing the strength of the nitrogen and carbon lines. MgII is completely absent.

We continued to study it with the **IUE** satellite and found that UV maximum occurred on 30 April 1991, or about 6 days after optical maximum. This shows, that even for the fastest and brightest novae, that optical maximum precedes UV maximum. Based on the delay between optical and UV maximum, we can state that a very massive shell was ejected in the outburst. The velocities measured from the FWHM of the UV emission lines were reasonably high, about $3300\,\mathrm{km\,s^{-1}}$, and similar measurements of the optical Balmer lines gave velocities of $2500\,\mathrm{km\,s^{-1}}$. We note that the ejection velocities measured for the fastest recurrent and ONeMg novae are higher than these values. However, they probably eject smaller mass envelopes so that the total kinetic energy of the ejecta is either smaller or comparable. For nova LMC 1991, it appears that the kinetic energy was comparable to the energy radiated during the early part of the outburst.

By 2 September 1991, the UV luminosity had fallen to $\sim 2 \times 10^3 L_\odot$ and the spectrum showed strong UV emission lines of N IV, N V, C IV, N III], and C III] (see Figure 5b). There is no indication, whatsoever, of the emission lines seen in ONeMg novae so that we classify this outburst as having occurred on a CO white dwarf. The analysis of the outburst of LMC 1991 indicate that the abundances in the ejecta are probably those of a "relatively" ordinary CNO nova. However, there was clearly a very hot photoionizing source inside the ejected shell of gas just as was found for LMC 1990 #2. It also seems likely, although the analysis has not been finished, that nitrogen was significantly enhanced in this nova.

The UV light curve for this nova is also shown in Figure 1c and the peak flux reached to nearly 10^{-9} erg cm^{-2} s^{-1}. At the distance of the LMC, this corresponds to a luminosity exceeding $10^5 L_\odot$.

4 Implications for the Luminosities and Distances to Classical and Recurrent Novae

We begin this section by again stressing the importance of discovering and studying novae in the LMC. Not only are they all at the same distance but that distance is well known from a variety of methods. In addition, their reddening is small. Because they have exhibited a variety of outburst characteristics, they allow us to compare them to Galactic novae with similar behavior which improves our understanding of Galactic novae. We will demonstrate this statement by brief discussions of two of the outbursts, that of LMC 1990 #1 and LMC 1990 #2.

4.1 The Outbursts of LMC 1990 #1 and Nova Her 1991

Both LMC 1990 #1 and Nova Hercules 1991 (hereafter: Her91) were fast ONeMg novae. Her91 was discovered on 1991 March 24 at $V \sim 5$ (Sugano and Alcock 1991). It exhibited one of the fastest declines on record. Early optical spectra showed very strong hydrogen lines in emission, with broad P-Cygni profiles extending to a velocity of about $7000\,\mathrm{km\,s^{-1}}$ (Della Valle and Turatto 1991). Ultraviolet spectra of Her91 obtained with **IUE** on 1991 March 25 revealed an optically thick shell spectrum sharply attenuated shortward of 1700Å(Sonneborn, Shore, and Starrfield 1991), in contrast to LMC 1990 #1 and V693 CrA 1981 which were already optically thin at maximum. Mg II 2800Å was in emission, with a FWZI of $7300\,\mathrm{km\,s^{-1}}$. The line spectrum of the ejecta did not become optically thin until almost a week later. Her91 was also detected in the infrared and displayed optically thin dust within a few days of maximum (Woodward et al. 1992). No other fast nova has formed dust this early in the outburst. Once Her91 entered the optically thin phase of evolution, its spectral evolution and rate of decline closely resembled that of LMC 1990 #1.

We were able to use the similarity in the evolution of these two novae to obtain a distance to Her91 (Starrfield et al. 1992). Our method depends on the fact that absolutely calibrated fluxes are provided by the **IUE** Observatory and relies on the ratio of the UV flux above and below 2000Å (the LWP/SWP flux ratio). We assumed that, at late times in both outbursts, the differences in this ratio from nova to nova were caused only by differences in the reddening. Note that this ratio is *independent* of distance. We also assumed that the dereddened values should be the same for both novae since the spectroscopic development and decline rates were very similar (Starrfield et al. 1992). Fortunately, the reddening to the LMC is both small and well known (Fitzpatrick 1986). The ratio of the LWP to SWP integrated fluxes for LMC 1990 #1, corrected for reddening with $E(B-V)=0.15$ and the Fitzpatrick (1986) extinction law, reached an asymptotic value of 0.19. The LWP/SWP flux ratio for the Her91 spectra also approached an asymptotic value. Comparing the asymptotic value of the ratio for Her91 with that of LMC 1990 #1, allowed us to rule out $E(B-V) < 0.5$ (the LWP/SWP ratio for Her91 would exceed 0.25) and $E(B-V) > 0.7$ (the LWP/SWP ratio for Her91 would be smaller than 0.14). Agreement in the LWP to SWP flux ratio between Her91 and LMC 1990 #1 was obtained for $E(B-V) = 0.6 \pm 0.1$ mag. Using this value plus the bolometric fluxes for the two novae, we derived a distance of 3.4 ± 1.6 kpc to Her91. This result is in good agreement with the determination of Woodward et al. (1992) and shows that we can use the LMC novae to obtain distances to novae with similar spectral evolution in the galaxy. This is an extremely important result since distances to Galactic novae are, in general, poorly known.

4.2 The Outbursts of LMC 1990 #2 and U Sco

Although the first spectra obtained for Nova LMC 1990 #2 resembled those of Nova LMC 1990 #1, as time progressed it became clear that they were two very different types of novae. The reason for the initial similarity was that Nova LMC 1990 #2, as a RN, did not eject very much material and that material was optically thin in the lines at optical maximum. The UV spectral and temporal development of Nova LMC 1990 #2 was very similar to that of the galactic recurrent nova U Sco during its 1979 outburst; the notable exception was the lack of P-Cygni absorption, at maximum, on the strong resonance lines of Nova LMC 1990 #2 (see Shore et al. 1991). The light curve development of Nova LMC 1990 #2 was so close to that of V394 CrA during its 1985 outburst that we were able to use the light curve of V394 CrA to successfully predict **IUE** exposure times during the decline of Nova LMC 1990 #2.

Shore et al. (1991) also found that the temperature of the underlying hot white dwarf was well determined by the photoionization analysis of the **IUE** spectra. NV 1240Å is an excellent indicator of the amount of flux emitted in the Lyman continuum. Shore et al. (1991) reported that decreasing the T_{eff} of the ionizing source from 2×10^5K to 10^5K decreased the predicted N V line strength by more than an order of magnitude. Consequently, for temperatures below 2×10^5K, no reasonable match could be obtained with either the observed N V line strength, or for the N V/N IV ratio, for any consistent set of abundances. Their abundance analysis also found that both helium and nitrogen were strongly enhanced over a solar mixture of elements.

Once it was realized that LMC 1990 #2 was a RN, Shore et al. (1991) utilized archival material to demonstrate that the secondary of this nova was too faint to be a giant. This is an important result because the theoretical studies of RN show that the short recurrence times require that the secondary be evolved (Starrfield, Sparks, and Truran 1985; Starrfield, Sparks, and Shaviv 1988). This is certainly true for RN such as T CrB, V745 Sco, and RS Oph in which the secondary is a giant but harder to confirm for RN such as U Sco, V394 CrA, and LMC 1990 #2 which, in the first two cases, have confirmed short orbital periods (Shaefer 1990). However, optical spectroscopic studies of U Sco and V394 CrA show that the material that is being transferred is significantly enriched in helium implying that the secondary in these systems is evolved even if it is small in size. Since the spectroscopic characteristics and evolution of LMC 1990 #2 were very similar to U Sco and V394 CrA, and since it does not have a giant secondary, it must also have an evolved but compact secondary. Therefore, studies of galactic novae provided information about this extragalactic nova that we could not have possibly learned from any ground based or satellite studies.

5 Summary

Over the past four years, we have used the **IUE** satellite to study the ultraviolet spectral and photometric evolution of five novae in the LMC. Each of these five novae has suffered a unique outburst. LMC 1988 #1 was a moderate speed class CO nova that showed evidence for dust formation in its UV light curve. LMC 1988 #2 was a fast ONeMg nova that exhibited an optically thick shell at maximum similar to Nova Her 1991 or Nova Pup 1991. LMC 1990 #1 was a very fast ONeMg nova and LMC 1990 #2 was a recurrent nova of the U Sco Class (evolved but compact secondary). Finally, LMC 1991 was an extremely fast and bright CO nova whose peak luminosity at maximum greatly exceeded the Eddington luminosity for a $1.4 M_\odot$ white dwarf. Of equal importance is that these outbursts have their analogs in galactic novae. By combining the results of studies of galactic and the analogous extragalactic novae, we have been able to greatly improve our distance estimates to the galactic novae. For example, Nova Her 1991 is the second closest ONeMg nova (Nova Cyg 1992 is the closest) at a distance of ~3 kpc. U Sco is at a much greater distance and must be located in the galactic halo.

This work was supported in part by NSF and NASA grants to Arizona State University and by the DOE.

References

Austin, S., Starrfield, S., Saizar, P., Shore, S. N., and Sonneborn, G. 1990 in *Evolution in Astrophysics* (ed. E. Rolfs) (ESA SP-381), p. 367.

Cassatella, A., and Gonzalez-Riestra, R. 1990, in *Physics of Classical Novae*, ed. A. Cassatella and R. Viotti, (Heidelberg: Springer-Verlag), p. 115.

Ciardullo, R., Ford, H., and Jacoby G. 1983, ApJ, **272**, 92.

Cowley, A. P., and Starrfield, S. 1987, Pub. ASP, **99**, 854.

Della Valle, M. and Turatto, M. 1991, IAU Circ., No. 5223.

Dopita, M. A., Meatheringham, S. J., Sutherland, R., Williams, R. E., Starrfield, S., Sonneborn, G., and Shore, S. N. 1992, these proceedings.

Dopita, M.A., and Rawlings, S.J., 1990, IAU Circ. No. 4964.

Fitzpatrick, E. 1986, AJ, **92**, 1068.

Hauschildt, P. H., Wehrse, R., Starrfield, S., and Shaviv, G. 1992a, ApJ, *in press*.

Hauschildt, P. H., Starrfield, S., Wehrse, R., and Shaviv, G. 1992b, these proceedings.

Liller, W. 1990, IAU Circ. No. 4964.

Liller, W. 1991, IAU Circ. No. 5244.

McNaught, R. 1990, IAU Circ. No. 4946.

Nofar, I., Shaviv, G., and Starrfield, S. 1991, ApJ, **369**, 440.

Sekiguchi, K., Kilkenny, D., Winkler, H., and Doyle, J. G. 1989, MNRAS, **241**, 827.

Shaefer, B. E. 1990, ApJ Letters, **355**, L39.

Shore, S. N., Sonneborn, G., Starrfield, S., Hamuy, M., Williams, R. E., Cassatella, A., and Drechsel, H. 1991, ApJ, **370**, 193.

Shore, S. N., Starrfield, S., and Sonneborn, G. 1991, IAU Circ. 5257.

Shore, S.N., Sonneborn, G., and Starrfield, S.G., 1990a, IAU Circ. 4947, 4949.

Shore, S. N., Sonneborn, G., and Starrfield, S. 1990, in *Evolution in Astrophysics* (ed. E. Rolfs) (ESA SP-381), p. 447.

Sievers, J. 1970, Info. Bull. Var. Stars No. 448.
Sonneborn, G., Shore, S. N., and Starrfield, S. 1990, in Evolution in Astrophysics: IUE Astronomy in the Era of New Space Missions, ed. E. Rolfe (ESA SP-310; Noordwijk), p. 439.
Sonneborn, G., Shore, S. N., and Starrfield, S. 1991, IAU Circ., No. 5226.
Starrfield, S., Truran, M., Politano, M., Sparks, W. M., Nofar, I., and Shaviv, G. 1992, in Proceedings of a Workshop to Honor W. A. Fowler on his 80th Birthday, ed. D. Schramm and S. Woosley, in press.
Starrfield, S., Truran, J. W., Sparks, W. M., and Krautter, J. 1991, in *Extreme Ultraviolet Astronomy* ed. R. F. Malina and S. Bowyer (Pergamon: New York), p. 168.
Starrfield, S., Truran, J. W., Sparks, W. M., Krautter, J., and MacDonald, J. 1990, in *Physics of Classical Novae*, ed. A. Cassatella and R. Viotti (Heidelberg: Springer-verlag), p. 306.
Starrfield, S., Sparks, W. M., and Truran, J. W., 1986, ApJ Lett., **303**, L5.
Starrfield, S. 1992, in *Binary Stars*, ed. J. Sahade and Y. Kondo, (Kluwer), in press, 1992.
Starrfield, S., Sparks, W. M., and Shaviv, G. 1988, ApJ Lett., **325**, L35.
Starrield, S., Sparks, W. M., and Truran, J., 1985, ApJ, **291**, 136.
Sugano, M., and Alcock, G. 1991, IAU Circ., No. 5222.
Weiss, A. and Truran J. W. 1990, A & A, **238**, 178.
Williams, R. E., Sparks, W. M., Gallagher, J. S., Ney, E. P., Starrfield, S., and Truran, J. W. 1981, *Ap. J.*, **251**, 221.
Williams, R., Ney, E., Sparks, W., Starrfield, S., Truran, J. 1985, MNRAS, **212**, 753.
Woodward, C. E., Gehrz, R. D., Jones, T. J., Lawrence, G. F. 1992, ApJ, **384**, L41.

PHOTOIONISATION MODELLING OF NOVA LMC 1990 #1

M. A. Dopita , S. J. Meatheringham , R.Sutherland
Mt. Stromlo and Siding Spring Observatory,
Private Bag, Weston Creek P.O., ACT 2611, Australia.

R.E. Williams
Cerro Tololo Interamerican Observatory

S. Starrfield
Arizona State University

G. Sonneborn and S. Shore
NASA Goddard Space Flight Center

Abstract

Nova LMC 1990A was a very fast Ne-O-Mg nova, for which a particularly dense coverage of spectral observation in both the UV and optical was obtained. The data for the nebular phase were subjected to an analysis by the photoionisation modelling code MAPPINGS 2. The following parameters were obtained: $L_{max} = 8x10^4 L_\odot$, $T_{eff} = 2x10^5$K and the mass of ejecta = $5.5x10^{-5} M_\odot$. The abundance ratios in the ejecta were similar to those obtained by Williams *et al.* (1985) in the case of V693 CrA 1981. The N/O ratio and the overabundance of Al is consistent with outburst on a ONeMg white dwarf of mass ~$1.2 M_\odot$, but the super-Eddington luminosity, and amount of mass ejected presents some problems to theory.

Observations

Nova LMC 1990A was amongst the fastest of Novae, with a t_3 = 8 days, and it reached a peak magnitude of about V=11.2. A very dense coverage of spectra ($\Delta\lambda$ = 4Å) were obtained in the optical and the UV after the first week, using IUE, the double beam spectrograph on the 2.3 m Advanced Technology Telescope operated by the ANU at Siding Spring, and on the Cassegrain Spectrograph of the 1.8 m at Mt. Stromlo ($\Delta\lambda$ = 8Å). In addition, three spectra were obtained at Cerro Tololo, including a vital one showing the emergence of [Ne V] in the early nebular phase on Feb 6, 1990 (day ~ 24).

The earliest optical spectra obtained, about a week after outburst, already show a pure emission-line structure, with no trace of P-Cygni profiles. In the UV, P-Cygni profiles in the resonance lines disappeared by about day 24. The emergence of very strong lines of [Ne III] and [Ne V] in the nebular phase place Nova LMC 90#1 in the class of Ne-O-Mg novae (Perrine 1903, Ferland and Shields 1978, Williams *et al.* 1985, Dopita and Rawlings 1990). The reddening to this nova, E(B-V) = 0.19±0.05 appears to be typical of the LMC. It has been determined by three independent techniques: from the ratio of the optical to the UV lines of He II,the Balmer Decrement in the nebular phase, and from a fit of the continuum between 1200 and 9000Å to a theoretical nebular continuum on day 24.

Nebular Modelling

The data obtained for the nebular phase were subjected to an analysis using our photoionisation modelling code MAPPINGS 2. We find that when the nebula is optically thick to the ionising radiation, the flux in Hα is directly proportional to the stellar luminosity, at a given temperature. The sensitivity of the flux conversion factor is very little dependent upon stellar temperature, and can in any event be calibrated from detailed photoionisation models. Extrapolating the Hα flux measured from the Cerro Tololo spectra to the visual maximum, we estimate a peak luminosity $L_{max} = 3 \times 10^{38}$ erg.s $^{-1}$ ($8 \times 10^4 L_\odot$), which agrees well with the estimate derived from the integrated 1200-2000Å flux; $L_{max} > 2.8 \times 10^{38}$ erg.s $^{-1}$ .The degree of excitation in the nebular phase indicates an effective temperature $T_{eff} = 2 \times 10^5$ K.

In a similar way to the Hα luminosity, the UV luminosity is found to give a good estimate of the total luminosity of the nova, provided that the nova shell is optically thick to the ionising radiation. When it becomes optically thin, the shell luminosity declines as t^{-3}. We can therefore regard the UV light curve in the nebular phase as being composed of a sum of segments of constant luminosity followed by a t^{-3}.decline, corresponding to each part of the shell. The column density at the point when a portion of the shell becomes optically thin is simply proportional to the radius of the shell at this point. Therefore, a fit to the UV light curve can provide an accurate estimate of the mass of the ejected shell (M_{ej}). We have applied this novel technique to Nova LMC 90#1 to determine $M_{ej} = 5.5 \times 10^{-5} M_\odot$. This model showed that some portions of the shell became optically thin after only ~8 days, whereas others were optically thick at day ~57. The difference in density between different portions of the shell implied by these figures is very large, about 50. If this corresponds to the difference between the density of the material ejected in the polar direction and the equatorial direction, then the asymmetry in the ejecta is also very large.

Finally by fitting these many components to a simplified four-component model, we were able to generate theoretical spectra which simulated the observed variations in the optical spectra through to day 47, and which fitted the observed 1200-9000Å spectrum on day 24. The derived mass ratios in the ejecta were (wrt.H = 1.0): He 1.6; C 0.0016; N 0.0196; O 0.016; Ne 0.034; Mg 0.006; Al 0.0156; Si 0.01. These mass ratios are very similar to those obtained by Williams *et al.* (1985) in the case of V693 CrA 1981. The N/O ratio and the overabundance of Al is consistent computations of outburst on a ONeMg white dwarf of mass ~$1.2 M_\odot$, but the super-Eddington luminosity implied, and the amount of mass ejected presents some problems to theory (Starrfield *et al.* 1992).

References

Perrine, 1903 Lick Obs. Bull. 2, 32, 130.
Dopita, M.A., and Rawlings, S.J., 1990, IAU Circ #4964.
Ferland, G.J., and Shields, G.A. 1978, Ap. J., 226, 172.
Williams, R.E., Ney, E.P., Sparks, W.M., Starrfield, S.G., Wyckoff, S. and Truran, J.W. 1985, MNRAS, 212, 753.
Starrfield, S., Truran, J.W., Politano, M., Sparks, W.M., Nofar, I, and Shaviv 1992, (in press)

Non-LTE, Expanding Model Atmosphere Studies of LMC Novae

P.H. Hauschildt [1], S. Starrfield [1], R. Wehrse [2] and G. Shaviv [3]

[1]Dept. of Physics & Astronomy, ASU, Tempe, AZ 85287-1504, USA
[2]Inst. für Theoretische Astrophysik, INF 561, W-6900 Heidelberg, FRG
[3]TECHNION, IL 32000 Haifa, Israel

Abstract

We present non-LTE model atmosphere calculations for the LMC nova 1988 #1 during the optically thick phase shortly after the outburst. The models are able to reproduce the characteristic features of the observed spectra and will be used to determine the atmospheric parameters and chemical abundances of LMC novae.

1 Introduction

The atmospheres of novae during the early stages can be treated as spherical and expanding, but stationary, configurations (cf. Bath & Shaviv 1976). To model the atmospheres, we have included (a) non-LTE effects, (b) line-blanketing by UV metal lines, in particular Fe II, (c) proper treatment of the radiative transfer for lines and continua, and (d) a self-consistent solution of the radiative energy equation (e.g., radiative equilibrium in the comoving frame).

Our models show the following characteristic features: (1) a very large extension of the atmosphere, typically the relative radial extension of the line and continuum forming layers is about 100, (2) large departures from LTE, (3) very large temperature gradients throughout the envelope, the electron temperatures for a model with $T_{\rm eff} = 20,000\,$K range from $5,000\,$K to $150,000\,$K, (4) therefore, multiple ionization stages are simultaneously present in the atmosphere, (5) the UV spectra are dominated by line-blanketing, in particular by the Fe II-curtain and, (6) the observed "emission lines" are merely "holes in the iron-curtain".

2 Numerical Methods

- Approximate Λ-operator iteration (ALI) method for the solution of the spherically symmetric, special relativistic equation of radiative transfer (SSRTE) for continua, LTE, and non-LTE lines (see Hauschildt 1992a),
- ALI method for the solution of the special relativistic, multi-level, non-LTE continuum and line transfer problem using a direct extension of the method of Rybicki & Hummer (1991) including overlapping lines and active continua (Hauschildt, in preparation),
- a hybrid method (ALI combined with partial linearization) for the self-consistent solution of the radiative equilibrium equation in the comoving frame (Hauschildt 1992b).

3 Conclusions

We have presented our initial results for non-LTE, expanding stellar atmosphere calculations for nova LMC 1988 #1. The earliest spectra of this nova show much higher effective temperatures than we found for the galactic nova PW Vul (Hauschildt et al. 1992). Our synthetic spectra reproduce the observed IUE spectra of LMC 1988 #1 well if non-LTE effects as well as metal line blanketing are included in the model construction and the synthetic spectra. LMC 1988 #1 shows strong evidence for enhanced CNO and smaller Fe abundances as compared to the standard LMC metal abundance of 1/3 the solar value. A more detailed analysis of the early spectra of several LMC novae is currently in preparation.

An important result of our initial modeling is the fact that non-LTE effects and coupling between different species is of extreme importance in nova atmospheres. Therefore, a self-consistent treatment of overlapping lines and continua and the best available model atoms are essential in the modeling of the early spectra of novae.

References

Bath, G. T. & Shaviv, G. 1976, MNRAS **197**, 305.
Hauschildt, P. H. 1992a, JQSRT **47**, 433
Hauschildt, P. H. 1992b, ApJ (in press).
Hauschildt, P. H., Wehrse, R., Starrfield, S. & Shaviv, G. 1992, ApJ **393**, 307.
Rybicki, G. B. & Hummer, D. G. 1991, A&A **245**, 171.

SN1987A: A View from Space

Willem Wamsteker

ESA IUE Observatory

1 Introduction

The appearance of SN1987A in February 1987, discovered independently by Shelton and Jones [1], has presented a major advance in the observational material relevant to the nature of Type II Supernovae. With its maximum brightness at $M_v = 3.0$ and the known distance to the Large Magellanic Cloud, it supplied a unique observational opportunity which has been amply used by astronomers in all branches of Astronomy. This is well illustrated by the many papers and conferences related to SN1987A between the date of explosion in February 1987 and now. The main advantage of the nearness and subsequent apparent brightness of SN1987A, even though its intrinsic luminosity was unexpectedly low at $M = -15.0$, has been the fact that one has been able to study the evolution of a Supernova from the moment of core collapse –through the neutrino detection [2]– all the way through the current time (already more than 1900 days).

2 The Space Observatories

The availability of various operational spacecraft at that time has allowed an unprecedented observational study of SN1987A. It was not only fortunate that at the time of explosion, spacecraft like the Solar Maximum Mission (SMM), the X-Ray satellite GINGA and the International Ultraviolet Explorer (IUE) were in orbit and fully operational, but also the continued observability of SN1987A has allowed to launch many special purpose missions, balloons etc. for additional observations in the Gamma and X-Ray domain. Also the launch and operations of the Hubble Space Telescope (HST), the Roentgen Astronomical Satellite (ROSAT), the Roentgen Observatory on the MIR module KVANT and later the Compton Gamma Ray Observatory (CGRO) have made important contributions to the space data obtained on SN1987A. This made it possible that between 1987 and 1992 a large variety of space instruments, each with its own unique capabilities observed SN1987A and some spacecraft still continue to do so. Unforeseen discoveries were made as well as an impressive confirmation of the general theoretical concepts developed over the last 50 years on the nature of Type II supernovae. The large effort expanded in these observations was not only important for the understanding of the SN phenomenon itself, but, being the final stage of the stellar evolution of massive stars, it has also supplied new insights in the preceding stages of the evolutionary life of Type II Supernova progenitor stars. Of course the fact that we had –for the first time– information about the progenitor star [3] and the first detection of the neutrino's associated with the core collapse in the SN1987A progenitor were extremely important. The enormous amount of observations obtained at many ground observatories contributed also in an important way to our understanding. These will be discussed elsewhere in this volume by Danziger. A major bonus from the observations with these many space observatories has been that the variety of instruments available was sufficiently large to study the full bolometric behavior, and the different energy regimes could be followed in great detail throughout the increase and the decrease in brightness of the Supernova at various stages. The rapid and at times dramatic variations of the bolometric correction have been measured accurately and with sufficient time resolution to derive for the first time a reliable bolometric lightcurve. It will be obvious that it is impossible to discuss here in detail all the results obtained. We will therefore limit ourselves to an overview of all Space Facilities which have contributed (Table 1) and summarize some especially important new results obtained with space instrumentation. A discussion of the details of the early (< 800 days) bolometric lightcurve and its implications can be found in [4], while the later stages (from 800 to 1500 days) are discussed in [5] as well as elsewhere in this volume by Danziger.

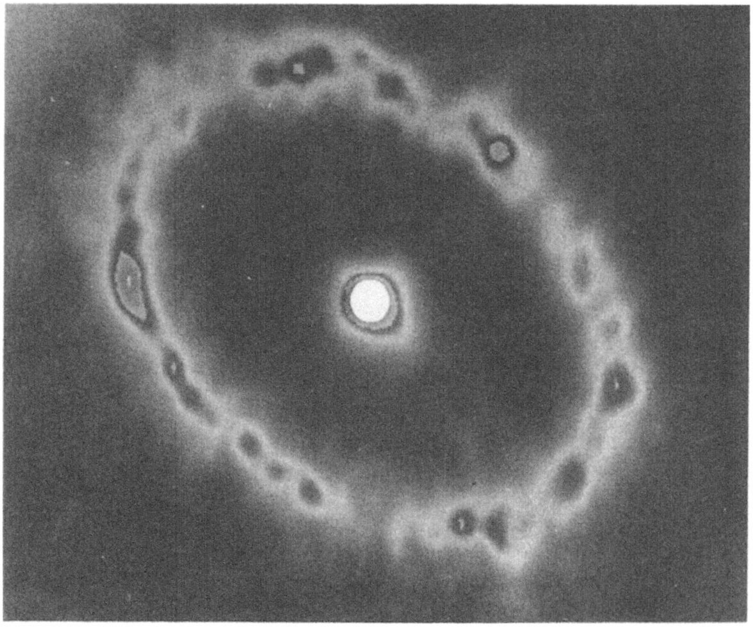

Fig. 1. Image of SN 1987A taken with the Hubble Space Telescope FOC on August 24, 1990. This image taken in the 5007 line of [OIII] shows beautifully the ring around the SN, first discovered with IUE. It is associated with the shock between the material from the different mass loss phases of the SN progenitor star Sk -69 202.

Table 1. Spaceborne instruments and Balloon experiments involved in observations of SN 1987A.

Photon energy range	Instrument	Platform	Refer.
Gamma Rays			
100 - 1300 keV	GRIS	Balloon	[6]
800 - 1400 keV	HRGRS	Balloon	[7]
100 - 1000 keV	HRGRS	Balloon	[8]
	GRAD	Balloon	[9]
700 - 7000 keV	DSGRS	Kosmos 1870	[10]
30 - 2000 keV	SIGMA	GRANAT	[11]
300 - 8500 keV	GRS	SMM	[12]
100 - 10,000 keV	OSSE	CGRO	[13]
	GRIP	Shuttle	[14]
X-Rays			
6 - 28 keV		GINGA	[15]
15 - 105 keV	HEXE		
50 - 1,500 keV	PULSAR X-1	KVANT	[16]
2 - 32 keV	TTM		
.2 - 2.1 keV		Rocket	[18]
.1 - 2 keV	XRT	ROSAT	[19]
Ultraviolet			
115 - 320 nm		IUE	[20,21,22]
260 - 350 nm		ASTRON	[23]
115 - 320 nm	FOS	HST	[ ]
115 - 350 nm	FOC	HST	[25]
147 - 172 nm	UIT	ASTRO (Shuttle)	[26]
70 - 140 nm	HUT	ASTRO (Shuttle)	[27]
Optical			
350 - 700 nm	FOC	HST	[25,28]
500 nm	FES	IUE	[20,29]

3 Observational Summary

Before discussing the results on SN1987A with the various Space Observatories, it is convenient to summarize the full extend of observational facilities which have been involved in making observations from space of SN1987A in the Gamma Rays, Hard and soft X-rays, Ultraviolet and optical domain. This summary of the instruments which have been used to observe SN1987A is given in Table 1. We have included in this list the observations made from Balloon platforms but excluded observations from airplanes, such as the IR observations made from the KAO which were considered to belong rather in the realm of ground-based observations. I summarize below, before going into some of the very exciting results in detail, in a concize form why the availability of these space instruments formed such an important complement to the more conventional instrumentation.

- Important emission lines of ^{56}Co, ^{57}Co and other products of the nuclear reactions and subsequent radioactive decay taking place after the core collapse are mostly in the gamma ray range. The observations with balloons and later with the Compton GRO allow nearly direct comparison of the energy deposition by the decay of the main radioactive nuclei ^{56}Ni,^{56}Co and ^{57}Co in the SN atmosphere, with that derived from the bolometric lightcurve matching through all stages of atmospheric expansion. The determination of ^{56}Co/^{57}Co ratio is critical for the energy balance at later stages ($>$1500 days).
- The high temperatures expected in the early stages of the atmospheric expansion can only be observed in the Ultraviolet as observable with IUE. The fast evolution of the SN1987A in the UV has allowed the unambiguous identification of the progenitor already after a few weeks [30] and the emission associated with the interaction of the UV flash with the circumstellar material has allowed a detailed study of the SN progenitor environment [31].
- Compton scattering of gamma rays into the X-ray domain supplies fairly direct information on the actual energy deposition in the atmosphere by the radioactive decay of ^{56}Co [15].
- The large temperature evolution of the SN atmosphere in a relatively short time and the enormous variations in atmospheric opacity, make it imperative to study all energy domains to establish the bolometric lightcurve of a Supernova. The bolometric lightcurve is a major piece of information needed to establish the consistency between fundamental parameters such as the mass of heavy elements produced in the SN explosion and subsequently the mass of the SN progenitor[4].
- The superior spatial resolution supplied by instruments outside the atmosphere, such as supplied by HST, allows already in early stages subarcsecond studies of the close SN environment and its atmosphere, which permits direct measurements of the size variation of the Supernova and therefore measure the expansion velocity [25] directly, as well as determine possible asymmetries in the SN explosion.
- The SN can be used as a background ultraviolet light source for the study of the Interstellar medium through the halo of our Galaxy as well as in the MC's [32].

4 Results from Space

In this section we will summarize the principle results from the Space borne observations of SN1987A. The choice by observing mode or wavelength range is somewhat artificial, while on the other hand in many cases a close physical relation exists between the results in different wavelengths domains. Therefore we will try to maintain in the following the separation by photon energy –in practice also by instrumental technique or spacecraft–, but try to maintain the relation between the different regimes and compare the results.

4.1 UV and Optical: Identification Progenitor; UV Echo's; Mass Loss ring

Although the extensive, and still continuing monitoring of SN1987A with IUE [20;21;22;29] and also that of shorter duration by ASTRON [23] and the detailed analysis of the lightcurves in the UV has been a major contribution to the understanding of the various peculiarities of SN1987A, a very important result has been the, actually secondary, result from the extremely fast decrease of the UV output. Due to the fact that already one week after the explosion, the SN had dropped below the sensitivity of the IUE Cameras especially in the shortest wavelengths (<1700Å), the spatial resolution in the IUE spectra allowed to make a detailed analysis of the spectra of the two stars which had remained visible in the ultraviolet. Careful comparison of the spectral types and comparison with the early astrometrical results [33] allowed Gilmozzi et al. [30] to establish unambiguously that the 10th magnitude star Sk -69 202 had indeed disappeared was therefore the progenitor star of SN1987A.

The early disappearance allowed for another unforeseen result associated with the UV flash of the SN1987A. Already early (arond day 80), weak emission lines were detected in the IUE spectra in the range between 1200 Å and 2000 Å [34]. These emission lines were found to be very narrow and stationary at the radial velocity of the LMC. The lines covered a wide range of ionization and were variable. The extensive study of their variation (see Figure 2) lead to the conclusion that they most likely corresponded to the emission from a shell around the SN ionized by the UV flash [21]. This shell corresponds to the shock between the slow, dense, Red Supergiant wind material and the fast low density Blue Supergiant wind, later in the progenitor evolution. The time evolution of some of the lines is shown in Figure 2. Note that the OIII] lines persist considerably longer than the lines of NIV]. This could be an indication for not pure recombination and has been suggested to indicate the posible presence of an additional source of energy maintaining the high excitation. The direct relation between the very short UV Flash,responsible for the ionization of this shell, and the possibility to derive from the lightcurves in Figure 2, the physical size of the shell on the basis of simple geometrical light-travel arguments, was used by Panagia et al. to derive an independent distance to SN1987A. [35]. In combination with an image of SN1987A taken with HST (Figure 1), clearly confirming the existence of the ring (or shell) around SN1987A. They found a distance to the Supernova of d=51.2±0.3 Kpc, corresponding to a distance to the barycenter of the LMC of 50.1 Kpc (D.M.=18.50±0.13), combining the angular measure of the shell from HST

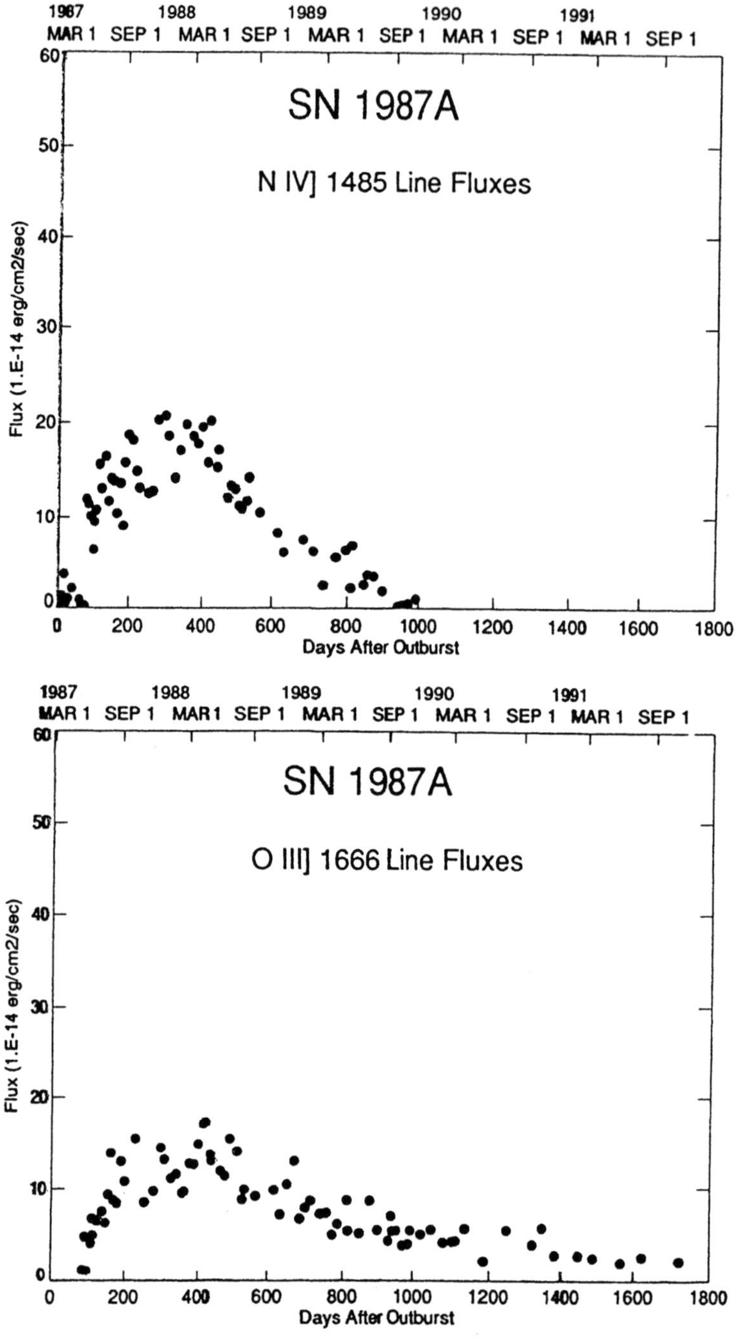

Fig. 2. Lightcurves of some of the narrow UV emission lines from the ring around SN1987A (Figure 1).The variation of the lines can be modelled with pure recombination (with two density regimes) with the ionization supplied by the UV Flash of the Supernova explosion [21]. Note however the disappearance of the NIV] lines around day 1000, while the OIII] lines remain clearly present. This could suggest an additional energy input maintaining the high excitation even at these late stages [38].

and the physical size of the shell derived from IUE. Recent observations with HST by Jacobsen et al. [25] on days 1278 and 1754 have allowed to measure directly the expansion of the SN atmosphere, which was found to have increased by 40% over this period of 476 days. The results were combined with earlier Speckle observations at days 265 and 650 [36]. The consistency of these four measurements suggests a linear expansion rate. Some evidence was found for wavelength dependence of the radius of the SN from the HST results [25], most likely illustrating the higher atmospheric opacities in the UV as compared to the longer visual wavelengths. Another unexpected result was the detection of the UV Echo, which was observed with IUE after their optical discovery [37]. The detection of the UV Echo was unanticipated from the brightness of the optical rings, but the combined effect of the brightness and short duration of the UV flash, made a detection feasible since the position of the UV Echo is displaced with respect to the optical one and can be predicted accurately. The thickness of the scattering cloud was found by Panagia and Gilmozzi to be 40 pc [37]. More recently UV Echo observations were made with UIT [26]. Although the detection was only marginal, the results are consistent with the IUE detection and suggest a scattering ratio UV/Optical of about 1, indicating that either the particles are not scattering in the Rayleigh domain, or that the scattering phase function is strongly wavelength dependent (narrower in the UV) or that the UV flash was weaker than was predicted.

4.2 Gamma Rays and X-Rays: Implications for the SN nuclear processing

Observations in the X-Ray domain of SN1987A were considerably complicated by the presence of background sources in the field of view of the GINGA satellite [15] To a lesser extend this was also valid for the observations with the HEXE imstrument which was sensitive to somewhat higher photon energies (6-28 keV for GINGA versus 15-105 keV for HEXE). This has generated an intrisic limit to the accuracy with which, especially the observations with GINGA, can be analyzed with spectral fitting of the data. The X-Ray lightcurve of SN1987A was first detected on day 131 and increased in both GINGA windows to reach its maximum in the hard window(16-28 keV) somewhat earlier than in the soft window which occurred on day 330 [15]. The spectral shape of the X-rays was somewhat anomalous but due to the uncertainty in the data no firm conclusions could be drawn from the GINGA X-Ray spectrum. The general behaviour of the lightcurve was consistent with that expected from the Thompson scattering of the gamma rays from the decay of ^{56}Co,but the emergence of the X-Ray lightcurve occurred much earlier than was predicted. The spectrum of the lightcurve appeared to soften during evolution. Most of these results are fully consistent with the results obtained from the observations with the HEXE instrument at somewhat higher photon energies, which continued to detect SN1987A nearly a year longer than GINGA. The flare seen in the GINGA results in the soft X-Ray band was not confirmed in the data obtained with HEXE, suggesting that this might have been associated with a EUV associated occurrence, although the general results obtained in the different energy domains bridging the soft X-Ray band do not exclude an observational anomaly. Although both the GINGA and the HEXE results are consistent with the general picture of ^{56}Co decay, both the early appearance and the relative slow development of the X-Ray lightcurves are rather unexpected. Both effects can be explained by the existence of much more mixing in the SN shell than was anticipated. With relatively simple Gaussian two-zone models Sunyaev obtained acceptable correspondence between the observations and a ^{56}Co decay model between days 100 and 800 [16].

Similar results were obtained by the analysis of the Gamma ray lightcurves from SMM by Leising and Share [12]. The decay in the lightcurves measured in the lines determined by power law fits to the local continua upon which lines at 847 keV, 1,238 keV, 2,599 keV and the unresolved blend at 3,250 keV of ^{56}Co were superposed. Also the results for all of these measurements were incompatible with a model with a single component at a –decreasing– uniform optical depth. More complex models involving at least two different optical depths zones represented the data well. These results confirm the extended mixing required for ^{56}Co from the X-Ray measurements. Similar conditions were suggested by the direct gamma-ray escape measurements of the emission line of ^{56}Co made by Rester [9], who found that the width of the line required significant mixing.

The spectral resolution supplied by the HEXE instruments allowed modeling of the spectrum including the effects of the decay of both ^{56}Co and ^{57}Co. This allowed to estimate the ratio ^{56}Co/^{57}Co. From these observations a value of approximately 1.5 times the solar ^{56}Fe/^{57}Fe was derived [16]. More recent results have been obtained with the OSSE instrument on the Compton Observatory on the emission lines of Cobalt in the gamma rays. Kurfess [13] reports that his recent measurements are consistent with the HEXE results and also suggest a Cobalt ratio of approximately 1.5. These results contrast strongly with ^{56}Co/^{57}Co ratio derived from the bolometric lightcurve by Suntzeff et al. [5], who found that to match the late stage ($>$ 800 days) bolometric lightcurve, using the mass of ^{56}Co derived from the bolometric lightcurve before that time of 0.07 $M_{\odot}$, an additional energy source is required. They could obtain a satisfactory match to the late lightcurve by introducing a considerably higher ^{56}Co/^{57}Co ratio of some 5 time the solar value. This discrepancy between these two independent measurements of the ^{56}Co/^{57}Co ratio is an indication of the fact the either the line modeling of the Gamma Ray lines is not valid with the optically thin models applied currently, or could suggest the presence of new enrgy sources inside the Supernova such as can be supplied by a young pulsar.

5 Conclusion

In summary we can see that the explosion of Supernova 1987A and the subsequent extensive observations with different space instruments have contributed enormously to our understanding of the physical mechanisms underlying the SN phenomenon and the late evolution stages of massive stars. It is gratifying to realize that all these observations have supplied impressive confirmation of previous theoretical concepts, but at the same time SN1987A remains a source of surprizes where the continuation of its observations can be expected to remain a source of challenges for the future.

References

1. IAU Circ. No. 4316.
2. Burrows,A.S., 1990, in Supernovae, A&A Library,Ed. A.G.Petshek, Springer Verlag, pg. 163.
3. Walborn,N.R., et al.,1989, Astron.Astrophys.,219,229.
4. Suntzeff,N.B.,Bouchet,1991,in Supernovae, ed. E.Woolsley, Springer Verlag,pg 3.
5. Suntzeff,N.B., et al.,1992,Ap.J.,384,L33.
6. Batholomy,S.,et al.,1988, IAU Circ. No. 4593.
7. Mahoney, et al., 1988,Ap.J.Letters, 334,L81.
8. Matteson,J.,et al.1991,in Supernovae, ed. .E.Woolsley, Springer Verlag,pg 283.
9. Rester,A.S., 1989, Ap.J.,342,L71.
10. Efremov,L.S., Sov.Astron.Lett.,16,337.
11. Durouchnoux,P,Matteson,J.,1991,in Supernovae, ed. S.E.Woolsley,Springer Verlag,pg 291.
12. Leising,M.D.,Share,G.H.,1990, Ap.J.,357,638.
13. Kurfess, et al. 1992, "Compton Observatory Newsletter", Vol. 2 , No.2.
14. Cook,W.R., et al.,1988,Ap.J.Letters,334,L87.
15. Tanaka,Y.,1991,in Supernovae, ed. S.E.Woosley,Springer Verlag,pg 269.
16. Sunyaev,R.A., et al., 1990, Sov. Astron Lett. 16, 171.
18. Aschenbach,B., et al., 1987, Nature,330,232,19.
20. Kirshner,R.,Gilmozzi,R.,1989, in Exploring the Universe with the IUE satellite, eds Y.Kondo, Reidel Acad. Publ. Co., pg 771.
21. Fransson,C. et al, 1989, Ap.J.,336,429.
22. Sanz de Cordova, L.S.,1992, Astron.Astrophys.,in press.
23. Lyubimkov,L.S.,1990, Sov.Astron.,34,239.
25. Jakobson,P.Machetto,F.,Panagia,N.,1992, Ap.J.,in press.
26. Crotts,A.P.S.,et al., 1992, Ap.J.Letters, Aug 10 issue.
27. Davidsen,A.,1992,priv. communication.
28. Panagia et al., 1992, Ap.J.,380,l23.
29. Wamsteker,W., et al. 1987, Astron. Astrophys., 177,l21.
30. Gilmozzi et al., 1987, Nature, 328,318.
31. Fransson, C., et al.,1989, Ap.J.,336,429.
32. Blades,J.C.,et al.,1988, Ap.J., 332, L75.
33. West,R. et al., 1987, Astron.Astrophys., 177,L1.
34. Wamsteker et al.,1978, IAU Circ., No. 4410.
35. Panagia,N.,et al.,1991,Ap.J.,380,l23.
36. Karovska,M.,Nisenson,P.,1992,Astron.Astrophys.,253,210.
37. Panagia,N.,Gilmozzi,R.,1991, in ESO/EIPC Wkshp. "SN1987A and other Supernovae", pg 381.
38. Sonneborn, G.,et al., 1992, in preparation.

SN 1987A After Five Years

I.J. Danziger and P. Bouchet

European Southern Observatory,
W–8046 Garching bei München, Germany

Abstract: We present a brief review of the most recent results concerning molecules and dust in SN 1987A. The role of dust and its impact on the bolometric light curve is also discussed. Abundance results which have accumulated over the past 5 years are reviewed in order to highlight the importance of this supernova for the chemical evolution of the LMC. Attention is directed to some outstanding problems in modelling.

Manifestations of Dust

In order to understand many aspects of the observed behaviour of SN 1987A at later phases, one must appreciate the role of dust in the expanding ejecta of the supernova. Molecules such as carbon monoxide (CO) and silicon monoxide (SiO) were formed at a very early phase (< 100 days) after outburst (Bouchet and Danziger 1992).

The ongoing studies of the CO spectrum reveal that the derived masses of this molecule have changed as the degree of sophistication of the analyses has improved. Liu *et al.* (1992) have recently reported a derived mass of CO $\sim 10^{-3}$ $M_{\odot}$, remaining approximately constant during the interval 100–400 days. More accurate treatment of optical depth and non-LTE effects have been responsible for the factor 3 increase in the derived mass compared to that first proposed by Oliva *et al.* (1987). It is our view that the only other molecule detected with certainty is SiO, for which analyses remain to be done. A claim for the detection of H_3^+ has recently been made by Miller *et al.* (1992) on the basis of fitting several emission features in the 3–4 μ region of the spectrum. Only one of the prominent features defies identification with plausible known lines other than those of the H_3^+ molecule, while some prominent features of the H_3^+ spectrum are not readily apparent in the admittedly complex supernova spectrum. For these reasons the identification, though possibly correct, is not beyond doubt.

Probably as a result of the presence of molecules, dust formed at approximately day 530 and has continued to play a dominant role in thermalizing the energy input at shorter wavelengths so that it can be radiated at IR wavelengths appropriate to the ambient temperature of the dust, currently < 150 K. Until the occurrence of SN 1987A it had been impossible, with other supernovae where IR excesses had been reported, to distinguish between dust in the envelope and IR echoes from surrounding material. In SN 1987A the spectroscopic signature of dust causing an apparent blueward shift of the peak of various broad emission lines ([O I] 6300, Mg I] 4571, [C I] 9850) proved conclusive (Danziger *et al.* 1989), together with the appreciation that the IR excess created after dust formation had to be included in the energy budget to obtain the most plausible bolometric energy (Lucy *et al* 1989). Analysis of the shape of the line profiles revealed that the dust was most probably distributed throughout the metal rich ejecta and not confined to a thin shell.

While the wavelength dependent part of the observed extinction suggested that amorphous silicate dust might be the dominant component, a wavelength independent part was most plausibly explained if the dust existed also in clumps whose optical depths could be very large. These dense thick clumps may still be preventing us from seeing to the centre of the supernova where a compact remnant (neutron star) is predicted to lie.

Reliable masses of the dust have not been determined, but it has been argued (Danziger *et al.* 1991b, Lucy *et al.* 1991) that if the sudden decrease in the luminosity of the [Si I] 1.64 μ line after day 530 is caused even partly by depletion of silicon into dust grains, then the mass of dust could be as high as ~ 0.1 $M_\odot$.

There is increasing evidence, when the somewhat fragmentary data on IR excesses and line shifts in past supernovae is reviewed afresh, that dust formation has been a feature of many supernovae. One might wonder also whether the shape associated with plateau-like light curves of some supernovae has something to do with dust formation. Consequently one might conjecture that some significant proportion of the interstellar dust in the LMC has been produced in the envelopes of all the previous generations of supernovae.

A second major change in the shape of some emission line profiles ([O I] 6300) was observed near day 760. This change resulted in an apparent broadening of the line at least on the short wavelength side. At about the same time (within a time period of approximately 10 days) there occurred a decrease in the rate at which the various emission line luminosities were decreasing. The degree of change was observed to be different for different lines such that lines of [Fe II] and [Si I] were the least affected. A simple 2-zone model of different temperatures was hypothesized to account for this behaviour (Danziger *et al.* 1991a). Whereas before the change the dominant luminosity in the lines came from the lower velocity lower temperature region, after the change when the cooler region had undergone drastic cooling (an IR catastrophe), this left the hotter region (cooling more slowly) dominating the luminosity and hence the line profiles. The fact that [Fe II] and [Si I] line luminosities were least affected is consistent with their having the highest radiative emissivity per unit mass

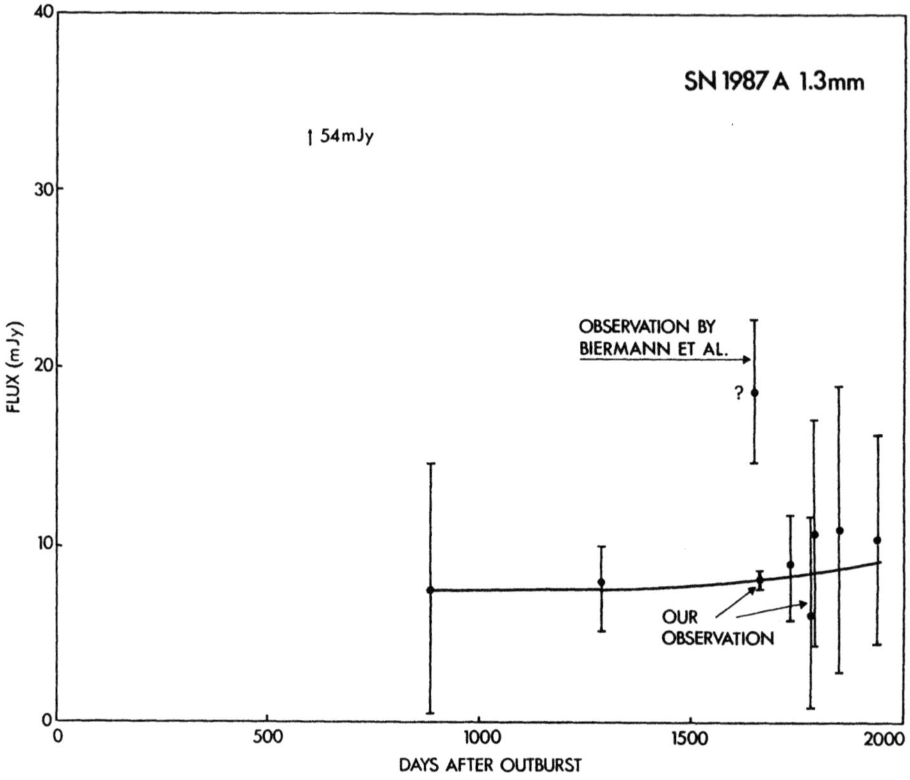

Fig. 1. This shows the temporal behaviour of the flux at 1.3 mm measured with the SEST telescope.

compared to other observed species. This change was completed on a time scale of $\sim$ 30 days.

A plot of the broad band IR spectra of SN 1987A at day 1316 is instructive for several reasons. It shows that all the accessible ground-based observations ($< 20\,\mu$) fall on the short wavelength side of a black-body fit using a temperature of 145 K. Thus the bulk of the radiation is inaccessible to observations from the ground and hence there are possibly large uncertainties in deriving the optimum fit of black body radiation with one temperature. This uncertainty is important in the determination of the bolometric luminosity at late phases. The existence of a positive detection of the supernova at 1.3 mm and an excess over a black-body fit in the L band ($\sim 3.5\,\mu$) raises the question whether there is, albeit relatively faint, a component due to free-free emission underlying the black-body emission. The origin of the 1.3 mm radiation hinges on demonstrating whether the clumped dust is an efficient emitter at this wavelength. A measurement of the spectral index of this radiation near 1.3 mm would help to elucidate this problem.

The temporal behaviour of the flux at 1.3 mm is shown in Figure 1. This plot contains all the published data (Biermann *et al.* 1990a,b, 1992) as well as new data by Bouchet and Danziger (in preparation). After day 800 there appears to be very little evolution of this flux (within unavoidably large errors). The unusually high point near day 1645 may be spuriously high because the next point some days later when the atmospheric conditions were excellent (resulting in a very small uncertainty) is significantly lower and consistent with all subsequent observations. Thus although clumpiness is undoubtedly a feature of the expanding envelope (Lucy *et al.* 1991) the observations of the 1.3 mm fluxes alone do not demonstrate this point. Therefore not only spectra at 1.3 mm but a continuing temporal sequence (with reduced errors) could help to clarify its origin.

The Bolometric Light Curve and the Radioactive Isotopes

The most recent analyses of the bolometric light curve prior to day 600 agree (to within 5 percent) that the dominant energy input was coming from the decay of an initially produced 0.07 $M_\odot$ of ^{56}Co (Suntzeff and Bouchet 1990; Bouchet *et al.* 1991; Suntzeff *et al.* 1991). After day 900 a flattening of the bolometric curve became apparent, although the degree of levelling has been a source of some debate between the CTIO and ESO observers. This arose because over restricted intervals of time, though never on a date in common, the ESO N (10 μ) and Q (20 μ) band photometry was systematically brighter than the CTIO measurements. Although this discrepancy has not yet been resolved, both observatories now agree that extending out to 1800 days there has been in the bolometric light curve an excess above the line required by decay of ^{56}Co alone. Suntzeff *et al.* (1992) and Dwek *et al.* (1992) have argued that an amount of ^{57}Co/^{56}Co $\sim$ 4–6 times the solar ratio of 57/56 stable nuclides would adequately fit the observed light curve. This is in contradiction with the facts that a value of 57/56 $\sim$ 1.5 solar was obtained by Danziger *et al.* (1991b) from a study of the temporal variation of the [Co II] 10.52 μ line, a similar value of 1.5 was obtained by Varani *et al.* (1991) from a study of the [Co II] 1.54 μ line, and an upper limit of 1.5 solar was obtained by Sunyaev *et al.* (1991) from attempts to observe the X-rays resulting from the comptonization of γ-rays from the ^{57}Co decay. It is now also in contradiction with a value of 1.5 solar resulting from the actual detection of ^{57}Co γ-rays with the Compton Gamma Ray Observatory (Kurfess *et al.* 1992) and referred to by a previous speaker.

Thus the excess in the bolometric light curve not accommodated by the now seemingly well-determined mass of ^{57}Co requires another explanation. Probably the energy output has dropped significantly below the Eddington luminosity ($\sim 10^{38}$ ergs/sec) for a 1 $M_\odot$ neutron star with an unstable accretion disk supplying matter. A pulsar either beamed away from our line-of-sight or continually obscured by the dense clumps of dust is still a possibility,

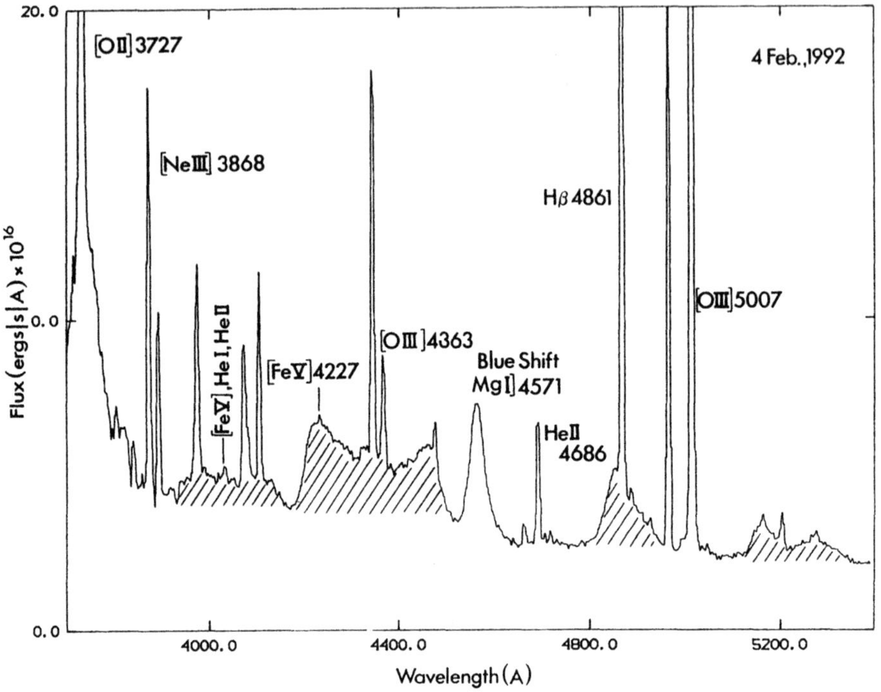

Fig. 2. A spectrum of SN 1987A taken at day 1806 at ESO, La Silla. Narrow lines from the CS gas are identified as are some broad features formed in the expanding envelope. Other broad features with hatching are the as yet unidentified features mentioned in the text.

as are other sources of radioactive decay, namely ^{44}Ti and ^{22}Na. At this stage, the amount of these radioactive species required would still be well above that predicted by theory (Woosley and Hoffmann 1991) and must be considered an unlikely possibility in view of the success of the theoretical work (Woosley 1991, Thielemann *et al.* 1991) in deriving abundances that match the observations (Danziger *et al.* 1991b).

At day 1806 (February, 1992) the spectroscopic indicators show that dust is still present in the expanding envelope of SN 1987A (see Figures 2 and 3). The peak emission of [O I] 6300 is blueshifted 495 km/sec and that of Mg I] 4571 is blueshifted 810 km/sec. Under the assumptions of the model proposed by Lucy *et al.* (1989, 1991) these blueshifts translate into optical depths of 0.8 and 1.85 respectively, at least as large as any reported in the past and still indicating a colour effect of the kind mentioned above for amorphous silicate

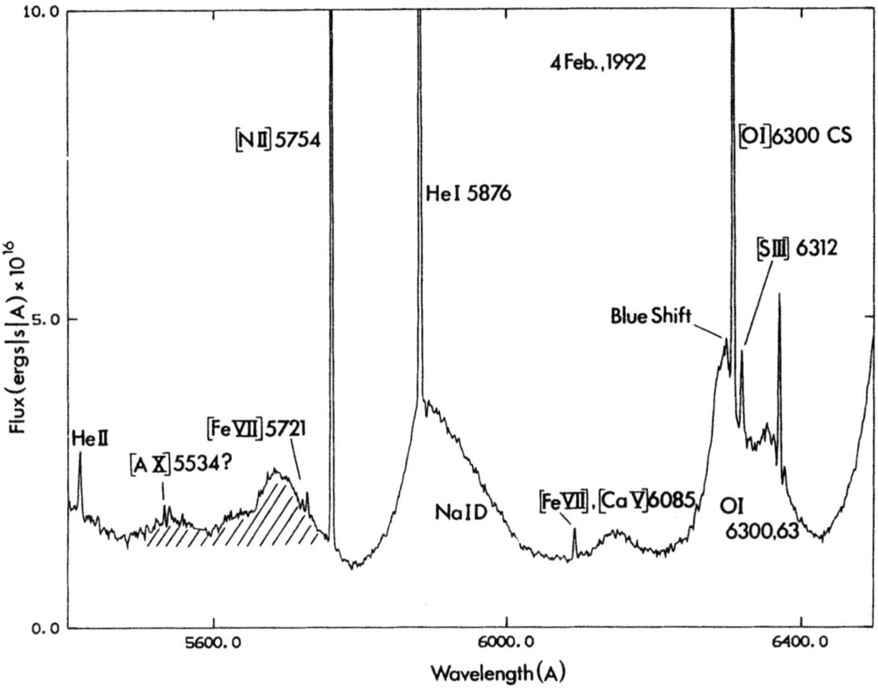

Fig. 3. Same as for figure 2.

dust. While great caution is now required in using this model because the original assumptions may be overstrained, there remains ample evidence of the presence of copious amounts of dust.

Abundance Studies

The abundance work on SN 1987A which we describe here has been published in various journals. We summarize it here because it has a relevance and interest for astronomers observing and interpreting the chemical abundances in stars and the interstellar medium of the LMC.

Narrow emission lines of helium, carbon, nitrogen and oxygen were discovered with the IUE Observatory at an early stage and analysed by Fransson *et al.* (1988). The low velocity widths of the lines (< 30 km/sec), the temporal behaviour of the luminosities, the densities and the abundances are consistent with a model of circumstellar material significantly processed by the CNO cycle and ionized by the UV flash at the time of outburst. Indeed, as the supernova has faded and the narrow emission-line spectrum has become more prominent, unambiguous detections of even coronal lines have demonstrated that this CS shell could not be photoionized by a normal hot star. Some of these lines are illustrated in Figures 2 and 3 as well as by Wang (1991). The measured abundance ratios in this material which must have been ejected prior to the explosion are as follows: N/C = 7.8, N/O = 1.6.

While modelling of the photospheric phase of the envelope of the supernova (Lucy 1987, 1988) has been successful using abundances of the metals typically expected in the atmosphere of a young hot star in the LMC, the problem of barium and possibly one or two other s-process elements deserves special attention. A prominent absorption line in the early spectrum was identified by Williams (1987) to be due to the Ba II 6142 transition. A sophisticated analysis of this line together with most other conceivable transitions by Mazzali, Lucy, and Butler (1992) using a non-LTE code reveals the following abundance pattern: Ba/Fe = 5, Sr/Fe = 2, Si/Fe = 1 relative to the cosmic ratios. Because of the shape of the line showing a lack of absorption at high negative velocities, it is concluded that the barium is confined within the inner 89 percent of the mass of the progenitor, a result in accord with models for s-process production in high mass stars by Prantzos *et al.* (1988). Because of the absolute abundances and the distinctive ratios of s-process material these results suggest a rationale for analysing stable LMC stars of mass ($\sim 20\ M_\odot$) similar to that of the progenitor of SN 1987A.

Abundance results determined for the nebular phase near day 410 have been presented and discussed by Danziger *et al.* (1991b). Good agreement for most elements has been achieved with the theoretical predictions for example of Thielemann *et al.* (1989) and Woosley (1991). Unfortunately the most abundant element produced, oxygen, suffers observationally from the largest uncertainty. Depending on the temperature used (one unique temperature for all of the oxygen is probably incorrect anyway) one can obtain masses of oxygen in the envelope in the range $0.3 - 3\ M_\odot$. A precise reliable determination of the mass of oxygen is now a challenge best addressed by models.

There are indeed other major outstanding puzzles in the interpretation of the spectra which in the nebular phase are partially exemplified in Figures 2 and 3. The broad emission line features that are hatched are lines or combinations of lines emitted by the envelope and defying a convincing even qualitative interpretation. While lines of Fe II and [Fe II] may contribute to these features, there remain so many wavelength discrepancies to convince one that other unknown transitions play a significant role. In any case more detailed modelling is called for here also. It would be interesting to test with models whether this lack of a strong correlation with lines of ionized iron results from the cooling proposed by Spyromilio and Graham (1992) whereby the bulk of the $0.07\ M_\odot$ of iron is too cold to emit at these and near-IR wavelengths.

References

Biermann, P.L., Chini, R., Greybe-Götz, A., Haslam, G., Kreysa, E., Mezger, P.G. 1990a, Astron. Astrophys., 227, L21

Biermann, P.L., Chini, R., Haslam, G., Kreysa, E., Lemke, R. 1990b, Astron. Astrophys., 236, L17

Bouchet, P., Danziger, I.J., Lucy, L.B. 1991, Astron. J. 102, 1135

Bouchet, P., Danziger, I.J. 1992, Astron. Astrophys. in press

Danziger, I.J., Gouiffes, C., Bouchet, P., Lucy, L.B. 1989, IAU Circ. No. 4746

Danziger, I.J., Bouchet, P., Gouiffes, C., Lucy, L.B. 1991a, in The ESO/EIPC Workshop: SN 1987A and other Supernovae, Marciana Marina, Isola d'Elba, 17–22 September 1990, ed. I.J. Danziger, L.B. Lucy, and W. Hillebrandt

Danziger, I.J., Lucy, L.B., Bouchet, P., Gouiffes, C. 1991b, in "Supernovae" (Proc. of Tenth Santa Cruz Summer Workshop) ed. S.E. Woosley (Springer Verlag, N.Y.), p. 69

Dwek, E., Moseley, S.H., Glaccum, W., Graham, J.R., Loewenstein, R.F., Silverberg, R.F., Smith, R.K. 1992, Astrophys. J., 389, L21

Fransson, C., Cassatella, A., Gilmozzi, R., Kirshner, R., Panagia, N., Sonneborn, G., Wamsteker, W. 1989, Astrophys. J., 336, 429

Kurfess, J.D., Johnson, W.N., Kinzer, R.L., Kroeger, R.A., Strickman, M.S., Clayton, D.D., Leising, M., Cameron, R.A., Jung, G.V., Grabelsky, D.A., Purcell, W.R., Ulmer, M.P., Grove, J.E. 1992, Bull. A.A.S. 24, 750

Liu, W., Dalgarno, A., Lepp, S. 1992, Astrophys. J., in press

Lucy, L.B. 1987, in ESO Workshop on the SN 1987A, ed. I.J. Danziger (ESO, Garching b. München), p. 417

Lucy, L.B. 1988, in "Supernova 1987A in the Large Magellanic Cloud", eds. M. Kafatos, and A. Michalitsianos (Cambridge University Press, Cambridge), p. 323

Lucy, L.B., Danziger, I.J., Gouiffes, C., and Bouchet, P. 1989, in "Structure and Dynamics of the Interstellar Medium", Lecture Notes in Physics, ed. G. Tenorio-Tagle, M. Moles, and J. Melnick (Springer, Berlin)

Lucy, L.B., Danziger, I.J., Gouiffes, C., Bouchet, P. 1991, in "Supernovae" (Proc. of Tenth Santa Cruz Summer Workshop) ed. S.E. Woosley (Springer Verlag, N.Y.), p. 82

Mazzali, P.A., Lucy, L.B., Butler, K. 1992, Astron. Astrophys., 258, 399

Miller, S., Tennyson, J., Lepp, S., Dalgarno, A. 1992, Nature, Vol. 355

Oliva, E., Moorwood, A.F.M., Danziger, I.J. 1987, The Messenger, 50, 18

Prantzos, N., Arnould, M., Casse, M. 1988, Astrophys. J., 331, L15

Spyromilio, J., Graham, J.R. 1992, Mon. Not. R. astr. Soc., 255, 671

Suntzeff, N.B., Bouchet, P. 1990, Astron. J., 99, 650

Suntzeff, N.B., Phillips, M.M., Depoy, D.L., Elias, J.H., Walker, A.R. 1991, Astron. J. 102, 1118

Suntzeff, N.B., Phillips, M.M., Elias, J.H., De Poy, D.L., Walker, A.R. 1992, Astrophys. J. Lett. 384, L33

Sunyaev, R.A., Kaniovsky, A.S., Efremov, V.V., Grebenev, S.A., Kuznetsov, A.V., Englhauser, J., Doebereiner, S., Pietsch, W., Reppin, C., Truemper, J., Kendziorra, E., Maisack, M., Mony, B., Staubert, R. 1991, in "Supernovae" (Proc. of Tenth Santa Cruz Summer Workshop) ed. S.E. Woosley (Springer Verlag, N.Y.), p. 767

Thielemann, F.-K., Hashimoto, M., Nomoto, K. 1990, Astrophys. J., 349, 222

Varani, G.-F., Meikle, W.P.S., Spyromilio, J., Allen, D.A. 1990, Mon. Not. Roy. Ast. Soc. 245, 570

Wang, L. 1991, Astron. Astrophys., 246, L69

Williams, R.E. 1987, Astrophys. J., 320, L118

Woosley, S.E. 1991, in "Supernovae" (Proc. of Tenth Santa Cruz Summer Workshop) ed. S.E. Woosley (Springer Verlag, N.Y.), p. 202

Woosley, S.E., Hoffman, R. 1991, Astrophys. J. Lett. 368, L31

The Abundances of SN1987A

Rainer Wehrse

Institut für Theoretische Astrophysik, Im Neuenheimer Feld 561, D 6900
Heidelberg, Germany

Abstract: The spectra of SN 1987A have been analyzed and modeled by various authors. However, in contrast to the other objects discussed in this volume there is not yet a comprehensive abundance analysis for this object. The reasons for this situation are manifold and depend on the expansion phase. We emphasize here the difficulties of spectral diagnostics during the coasting phase. Taking the resulting uncertainties into account we discuss some recent determinations of the composition.

1 Introduction

Values for element abundances of SN1987A are found in a large number of papers. They refer to various phases in the evolution and are determined both from observed spectra and from hydrodynamical modeling. This reflects the strong interest in these data since they contain an enormous amount of information (i) on the general abundance pattern in the LMC, (ii) on the evolution of the progenitor star Sk -69^0 202 in the environment of the LMC, (iii) on the production of heavy elements in the explosion, (iv) the mixing of processed species with unprocessed matter, etc.

Unfortunately, in spite of all these activities there is not yet a full comprehensive abundance analysis of SN1987A. This situation is caused by number of specific difficulties that mainly result from the large expansion velocities encountered and from the large geometrical extension of the object which make it difficult, if not effectively impossible to determine the basic model parameters like "the radius" and "the effective temperature" (cf. Baschek, Scholz, Wehrse, 1991, for less extreme cases). In this paper we will discuss in some detail these various aspects mainly for the coasting phase which in the case of SN1987A comprises the first few days of observations. We choose this period since the corresponding spectra are believed to represent essentially the matter of the outer layer of the parent star before the explosion and since modeling at first glance seems to be quite straightforward.

2 Abundance Determinations from SNe II during the Coasting Phase and their Uncertainties

2.1 Construction of Model Atmospheres for SNe II during the Coasting Phase: an Overview

In the construction of model atmospheres and the corresponding synthetic spectra for the coasting phase the following assumptions are usually made (later called "standard assumptions"):

(i) the object is in a stationary state;
(ii) the radiating matter is spherically distributed;
(iii) the matter is in radiative equilibrium, i.e. the same amount of energy that enters the atmosphere at the bottom is radiated away from the surface;
(iv) the expansion velocity increases linearly with the radius;
(v) the density distribution varies with a power of the radius.

The model parameters usually chosen are the radius R at a predetermined radial optical depth (corresponding to a suitable mean or to a given wavelength), the effective temperature T_{eff}, the expansion velocity v_0 at R, the microturbulence velocity v_m, the power index n for the density distribution, the element abundances, and the pressure at the outer surface (in most cases it does not have a significant influence).

The actual model construction proceeds in a way analogous to the case of compact static atmospheres: with an estimated temperature distribution the run of the pressure is determined by means of the definition of the optical depth, the equation of state and the density-radius relation given above. Subsequently, the rate equations for the species considered and the radiative transfer equation for all frequencies are solved. This gives the emitted spectrum and a check on the fulfillment of the energy and particle conservation. From the deviations, corrections for the temperature distribution and the level occupation numbers are derived and the described steps are performed again. The iteration is stopped when the errors are smaller than a prescribed limit.

The determination of the element abundances proceeds then in the same way as for static atmospheres: the model parameters are varied until a "satisfactory agreement" between the synthetic and the observed spectra are obtained. The criterion for the "satisfactory agreement" are usually quite subjective (however, in most cases both flux distributions are shown in graphical form, see Fig. 1 as an example).

2.2 Basic Uncertainties

2.2.1 Parameter Space

A very basic uncertainty results from the fact that the number of parameters needed to specify a supernova photosphere is significantly higher than for stars in hydrostatic equilibrium: whereas it is 6 for SNe (in addition to abundances) it is only 3 (effective temperature, gravity, and microturbulence) for stars with compact atmospheres e.g. like the Sun or Sirius. Even in the latter case, most abundances can only be determined with an accuracy of 0.3 dex (c.f. Sadakane, 1990, Spite, 1990). Even if all other conditions were kept constant, the additional parameter space dimensions will reduce the accuracy of element abundances by a significant amount.

2.2.2 Information Content of Spectra

The information content of supernova spectra is smaller than that of spectra from stars in hydrostatic equilibrium (with the possible exception of white dwarfs) since all lines are very much broadened by the expansion (and perhaps by the microturbulence) so that they overlap strongly and individual transitions can rarely be identified in the spectra. This implies e.g. that in the ultraviolet range the opacity is always a sum of contributions from many species without a dominant term and that the contributions of particular elements have to be determined from flux differences at many wavelength points. The corresponding matrices are ill conditioned so that the derived abundance values, i.e. the solution of the inverse problem, is largely determined by the noise in the spectra. A quantitative description of these effects (Hafner and Wehrse, 1992) will be published elsewhere.

2.2.3 Uncertainties in the Absolute Masses

For the validation of hydrodynamical supernova models it is of primary importance to obtain empirically from spectra the ejected absolute masses M_i for a number of elements i (A. Maeder, private communication). In order to calculate these values it is necessary to determine the relative abundances first. Then the amount of matter has to be calculated that passes through a shell of prescribed radius during the period of interest. Therefore all the errors of the relative abundances apply and in addition the uncertainty in the mass loss rate

$$\Delta \log \dot{M} \geq (n-3) \Delta \log r$$

(n=power law index of density) applies. Since the radius has to be found spectroscopically (radii determined otherwise lead to inconsistencies with the velocities) and this can hardly be done with an accuracy better than 0.2 dex and since furthermore $n \approx 10$) it is questionable that the corresponding $\dot{M}_i$ values found in the literature pose significant constraints on the theory.

2.3 Uncertainties Resulting from the Model Construction

2.3.1 Are the Standard Assumptions fulfilled?

The validity of the assumptions listed in Section 2.1 has to be checked by detailed hydrodynamical modeling and by comparing the relevant timescales: Since the timescales for the adjustment of excitation and ionization equilibria as well as for the propagation of light are much shorter than the hydrodynamical time scales, it seems certain that the photospheres of SNeII behave spectroscopically like stationary media. The assumption of sphericity is probably less accurately fulfilled, since linear and circular polarization have been observed from SN1987A (Schwarz and Mundt, 1987) and 2 dimensional hydrodynamical models (cf. Müller et al., 1991) show that strong mixing occurs and that therefore the expansion is anything but laminar. These models imply that the heating of the photosphere by γ rays from radioactive nuclei starts earlier than expected from 1 D models; however, during the first few days the photosphere should not yet be affected. Since, in addition, there seem to be no indications of dissipation of bulk motion energy, radiative equilibrium appears to be a safe assumption.

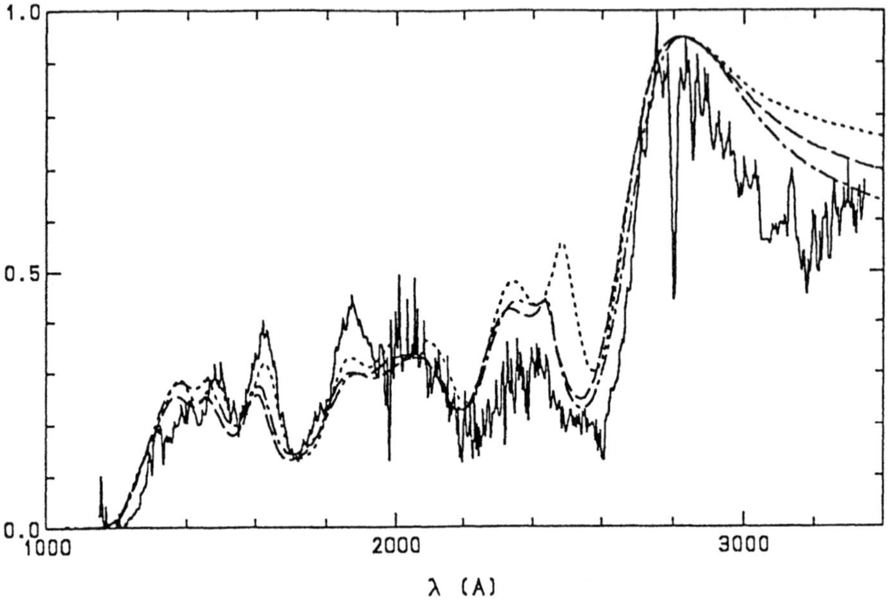

Fig. 1. Comparison of the UV spectrum of SN1987A as observed by IUE on February 25, 1987 (full curve), and synthetic spectra by Hauschildt (1991) who assumed that the abundances of the metals are reduced by a factor three compared to the solar values (except for Mg that has a relative overabundance of 1 dex), see Table 1.

In contrast, the density profile and the expansion velocities depend on the structure of the progenitor star, its chemical composition, and on the explosion energy so that only the general form of the depth dependence can be predicted by hydrodynamical calculations, the parameters have to be determined empirically.

2.3.2 Relativistic Effects in Radiative Transfer and the Independence of Parameters

The expansion velocities of the SN1987A atmosphere reach about $30,000$ km/s ≈ 0.1 speed of light. They imply errors in the mean intensities of about 50% (Mihalas, 1980) and up to a factor 10 in the abundances (Hauschildt, Best and Wehrse, 1991) if relativistic effects in the radiative transfer are neglected. Therefore, the correct treatment of the full special relativistic transfer equation and of the relativistic energy equation (that determines the depth dependence of the total flux) are of crucial importance.

Furthermore, these relativistic effects lead to the following problem: The luminosity L_{base} measured in the comoving frame at the base of the atmosphere is no longer equal to the luminosity L_{obs} as seen by a distant observer, but depends in addition on the velocity and on the radius so that various combinations of these quantities lead to the same L_{base} and therefore to very similar temperature distributions. Since the run of the temperature in turn determines essentially the emitted spectrum, it is evident

Table 1. Comparison of element abundances; for details see text.

element	logarithmic abundances				
	Sun	SN early ph.	SN nucl. synth.	SN day 410	
H	+3.44	+3.92			
He	+2.43	+2.91			
C	0.00	0.00	0.00	0.00	
N	-0.51	-0.51	-1.69		
O	+0.37	+0.37	+1.00	+1.80	(OI)
Ne	-0.47	-0.47	+0.13		
Na	-2.25	-2.25	-2.28		
Mg	-0.97	+0.03	-0.10		
Al	-2.08	-2.08	-1.22		
Si	-1.00	-1.00	-0.42	-0.39	(SiI)
S	-1.29	-1.29	-1.10		
Ca	-2.22	-2.22	-2.06	<-1.23	(CaII)
Sc	-5.47	-5.47	-10.80		
Ti	-3.62	-3.62	-3.36		
Cr	-2.87	-2.87	-2.71		
Mn	-3.02	-3.02	-3.36		
Fe	-1.05	-1.05	-0.83	-0.15	(FeI+II)
Co	-3.65	-3.65	-4.13	-1.95	(CoII)
Ni	-2.31	-2.31	-1.60	-1.92	(NiI+II)

that it is virtually impossible to determine the supernova parameters uniquely from a given spectrum. In particular, variations in the radius and in the effective temperature (as defined above) may no longer be distinguished uniquely. Details of this irritating consequence of the fast expansion will be given by Hafner and Wehrse (1992).

3 Discussion of Abundances of Selected Elements

In Table 1 we list element abundances for SN 1987A as obtained from a model fit to early UV spectra (Fig. 1., from Hauschildt, 1991), from modeling the nucleosynthesis together with the hydrodynamics (Nomoto et al., 1991) and from the analysis of spectra of day 410 (Danziger et al., 1991). For comparison the corresponding data for the sun are also given. All abundances refer to the logarithm of the number density and are normalized to carbon. The data for the early phase should represent essentially matter unprocessed by the explosion and on day 410 most of the nucleosynthesis products should be visible so that columns 4 and 5 should correspond to each other. However, it should be noted that the data of column 5 have been determined only from the ions given in brackets and that no ionization correction has been applied so that the total element abundances may be higher.

The example of Fig. 1 and additional data given by Hauschildt (1991) show that the early spectra are very well reproduced with the "classical LMC metal abundances" (1/3 solar) with the possible exception of magnesium that seems to be about 1 dex overabundant compared to the other heavy elements. However, it should be pointed out that it has not been checked in detail that all abundances *have to be* as high as the numbers indicate. On the other hand, in view of the discussion in Section 2 also a metallicity as high as that of the sun cannot be excluded. Therefore, it appears that at least at present properties of Sk -69^0 202 can hardly be inferred from the early SN1987A spectra.

Note in addition that even the abundances predicted by nucleosynthesis (with the exception of scandium and the possible exception of nitrogen) are compatible with those of the coasting phase so that possible mixing at a very early phase cannot be excluded from these data. The values obtained for day 410 are in general higher than those for the early stages and than those of the nucleosynthesis calculations. The discrepancy is in particular important for cobalt. Note however, that recent observations (GRONEWS, Aug. 1992) of γ rays from cobalt decays indicate that the mass of Co produced in the explosion is at least a factor of three lower than deduced from the late spectra and the lightcurve.

4 Conclusions

In this paper we have reviewed the determination of abundances from SN1987A with special emphasis on the coasting phase. It has been demonstrated that -due to the very high expansion velocities encountered- the information content of SN spectra for this it might not be possible to determine the basic parameters uniquely if only spectroscopic evidence is employed. The problem is magnified by the large number of free basic parameters (at least 2 more than for hydrostatic atmospheres). It is therefore evident that the accuracy of element abundances determined from spectra of the coasting phase is considerably lower than in the case of hydrostatic stars (with the possible exception of M dwarfs). Realistic error estimates can presently not be given. The situation is somewhat better for later stages, however, even here separate observations indicate large discrepancies. In view of these uncertainties it seems only safe to say that the chemical composition of the SN1987A atmosphere during the early phase was essentially solar and that later the abundances of the iron group might be increased.

Acknowledgment: I thank B. Baschek for critically reading the manuscript. I am indebted to M. Best, M. Hafner, P. H. Hauschildt, D. Kelly, and G. Shaviv for many helpful discussions. This work has in part been supported by the Deutsche Forschungs gemeinschaft (Project We 1148/1-2).

References

Baschek, B., Scholz, M., Wehrse, R. 1991, A&A 246, 374

Danziger, I.J., Lucy, L.B., Bouchet, P., Gouiffes, C. 1991, in: Supernovae, Woosley, S.E., ed., Springer Verlag, Heidelberg, p. 69

Hafner, M., Wehrse, R. 1992, in preparation

Hauschildt, P.H. 1991 Ph.D. thesis Heidelberg University

Hauschildt, P.H., Best, M., Wehrse, R. 1991, A&A 247, L21

Mihalas, D., 1980, ApJ 237, 574

Müller, E., Fryxell, B., Arnett, D., 1991, A&A 251, 505

Nomoto, K., Shigeyama, T., Kumagai, S., Yamaoka, H., 1991, in: Supernovae, Woosley, S.E., ed., Springer Verlag, Heidelberg, p. 176

Sadakane, K. 1990, in: Accuracy of Element Abundances from Stellar Abundances, Wehrse, R., ed., Springer Verlag, Heidelberg, p. 57

Schwarz, H.E., Mundt, R., 1987, A&A 177, L4

Spite, F. 1990, in: Accuracy of Element Abundances from Stellar Abundances, Wehrse, R., ed., Springer Verlag, Heidelberg, p. 69

Dust Formation in the Ejecta of SN 1987A

P. Woitke, C. Dominik, J.M. Winters, E. Sedlmayr

Institut für Astronomie und Astrophysik, Technische Univerisität Berlin
Sekr. PN 8-1, Hardenbergstraße 36, D-1000 Berlin 12

Abstract: Model calculations for dust formation in the ejecta of SN 1987A are presented. The outer layers of the ejecta are found to condense earlier, whereas the inner layers condense more effectively. The resulting mean degree of condensation is rather low ($\approx$ 15%) and the dust particles remain extremely small ($< 0.005\,\mu$m). Synthetic IR spectra are calculated and compared with the ESO infrared data set.

1. The Model

We consider dust formation in the ejecta of SN 1987A, which is assumed to be spherically symmetric. The formation and growth of dust is calculated with a system of moment equations according to the theories of Gail & Sedlmayr (1988) on the basis of chemical equilibrium and classical nucleation theory. The underlying density- and temperature-structure is taken from NLTE comoving-frame radiation transfer calculations of Höflich (1990) for day 220, and is assumed to expand homologously and to cool down with a powerlaw in time ($\varrho_M \propto t^{-3}$, $T_M \propto t^{-3(\gamma-1)}$, where t: time since the explosion, ϱ_M: mass density and T_M: gas temperature of a considered layer with enclosed mass M). γ is treated as a free parameter (constant throughout the ejecta). These assumptions have also been used by Kozasa et al. (1989).

To calculate IR-spectra we have applied the discrete Λ-operator method of Avrett & Loeser (1985) for the radiation transfer, which includes isotropical scattering. The dust absorption and scattering coefficients are calculated according to Mie-theory in the small particle limit with optical data for astronomical silicate. Only the macroscopic grains with $N > N_l$ are taken into account, where N is the number of monomers a dust grain contains and N_l is the limiting size for dust grains with the absorption properties of solid particles rather than of molecules.

2. Results

• Due to the lower gas temperatures, condensation takes place in the outer layers earlier than in the inner layers. Due to the lower densities in the outer layers M > $10\,M_\odot$, however, the growth timescale is much larger than the expansion timescale. Therefore, the formation of macroscopic dust is almost suppressed in the outer layers. The resulting mean degree of condensation for all layers is found to be quite low ($\approx$ 15%).

• Even in the inner layers the dust grains remain extremely small ($a \leq 0.005\,\mu$m, i.e. the largest grains contain only a few 10 000 atoms). Therefore, the limiting size N_l is a decisive parameter to our model.

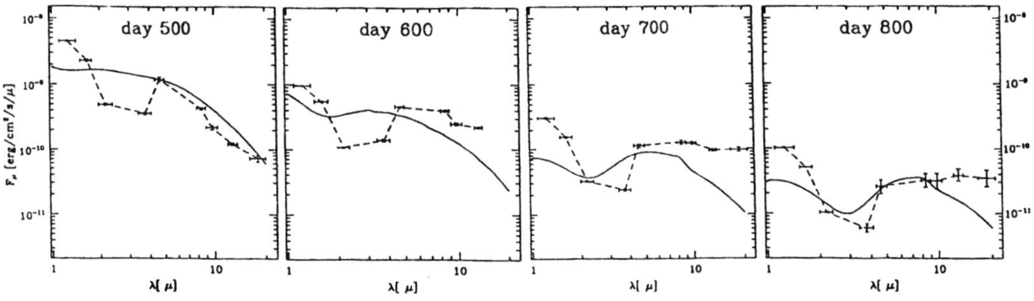

Fig. 1-4. Synthetic IR-spectra for SN 1987A (full lines) in comparison with the ESO broad-band photometric data (Bouchet et al. 1989) connected with dotted lines. Parameters: $\gamma = 1.8$ and $N_l = 1000$

• The starting-time for the development of the IR-excess as well as the level of the IR-flux can be simultaneously fitted to observations with $\gamma = (1.7, \ldots, 1.9)$.
• The model presented here can fit the mid IR-spectra but cannot satisfactory explain the coexistence of UV, optical and far IR brightness (see Figs. 1-4). For a larger choice of N_l only a negligible part of the condensed particles grow to "dust" grains, producing only minor dust emission in the far IR — for a smaller choice of N_l a total black-out in the UV, optical and near IR is the consequence.

3. Conclusions

• From the theoretical point of view, dust formation in the ejecta of SN 1987A necessarily occurs at least for the layers $M \approx (5 - 10)\,M_\odot$. Due to the short expansion timescales, the outer layers (about one half of the ejecta in mass), make no significant contribution to the injection of newly formed dust into the Interstellar Medium.
• The coexistence of UV, optical and far IR brightness observed after the onset of dust formation implies the following conclusion:
- The dust is distributed in a clumpy structure, or/and
- the origin of the UV and optical brightness is located essentially outside the dust shell. This point is strongly supported by the observed blueshift of optical emission lines: The blueshift is caused by the greater attenuation of redshifted radiation received from the far and therefore receding matter. The line is only blueshifted, if the dust is located inside the line-emitting region — a dust shell around this region would attenuate the total line-flux, but would not change the line profil.

References

Avrett, E.H.; Loeser, R.: 1984, in: *Methods in Radiative Transfer*, ed. W. Kalkhofen, Cambrigde university press
Bouchet, P.; Danziger, I.J.; Lucy, L.B.: 1989, in: *Supernovae*, ed. S.E. Woosley, 10th Santa Cruz Workshop, Springer-Verlag
Gail, H.-P., Sedlmayr, E.: 1988, *Astron. Astrophys.* **206**, 153
Höflich, P.: 1990, *MPA 563*, Max Planck Institut, Garching
Kozasa, T.; Hasegawa, H.: 1989, *Astrophys. J.* **344**, 325

Supernova Remnants in the Magellanic Clouds

Margarita Rosado [1], Etienne le Coarer [2], Yvon Georgelin [2], Annick Viale [2]

[1]Universidad Nacional Autonoma de Mexico, Instituto Astronomia. Apartado Postal 70-264, Ciudad Universitaria, CP-04510 Mexico,D.F.
[2]Observatoire de Marseille 2,pl Le Verrier F-13248 Marseille France

1 Introduction

Supernova Remnants (SNRs) in the Magellanic Clouds(MCs) have been the subject of numerous studies, both as individual SNRs and as a class of objects. Recent references to these works are sumarized in the IAU Symp. 148 on the MCs (Haynes and Milne,1991), in the IAU Colloquium 101(Roger and Landecker, 1987),and in several papers which will be reviewed below. Here, we review some of the most important conclusions reached from these works and discuss several aspects of our own current research.

2 Criteria of Identification and Catalogues of SNRs in the MCs

The most exhaustive list of SNRs in the MCs has been established by Mathewson and co-workers in a series of papers (Mathewson et al., 1983, 1984, 1985) based on the X-ray observations with the Einstein Satellite (Long et al., 1981, Seward and Mitchell, 1981, Inoue et al., 1983), optical narrow-band images and previous (Mills et al., 1982) ,archival and some additional radio observations. There are 28 (from a total of 32 listed) well identified SNRs in the Large Magellanic Cloud (LMC), and 11 in the Small Magellanic Cloud (SMC).

In our Galaxy, a SNR is usually identified by its non-thermal radio emission. Several of the SNRs identified this way also have extended soft X-ray emission (Seward, 1990) and some others have also optical counterparts, which reveal large internal motions and/or an emission characteristic of radiative shocks.

In the MCs, the use of the radio discriminators is not as easy because of the low angular resolution of radio telescopes at the southern hemisphere which prevents the detection of small diameter SNRs (i.e., unresolved SNRs could be confused with background extragalactic sources such as radio galaxies). The Australia Telescope Compact Array, which has a 2" angular resolution, will surely improve this situation and the work on SNRs in the MCs will be stimulated.

In order to avoid misidentifications, a series of criteria in several spectral ranges have been defined to identify SNRs in the MCs.According to Mathewson and collaborators,the candidate must satisfy at least 2 of the following conditions:

1) To be an extended soft X-ray source appearing in the surveys of Long et al. (1981), for the LMC, or Seward and Mitchell (1981) or Inoue et al. (1983) for the SMC.
2) To be a non-thermal radio source (or at least, to have a "typical" appearence of a SNR at radio wavelenghts, which is rather arbitrary).
3) To have optical emission with the following characteristics:
   a) $[SII]/(H\alpha + [NII]) \gtrsim 0.6$ or
   b) $H\alpha$ with looped or filamentary structure

   and

   c) An Hα flux different to the one expected for the radio flux in the case of a thermal source (in the case of finding an HII region coincident with radio sources).

These criteria are rather secure and have proved to be the best ones. Nonetheless, they are biased towards radio properties and further kinematical information is probably required (for instance, the last 4 candidates, in the LMC, adopted as "secure" are indeed dubious). Furthermore, one must keep in mind that several SNRs in the MCs could not fulfill 2 of the 3 criteria given above because of the following:

- Small diameter SNRs could be missed by the Image Proportional Counter (IPC) of the Einstein Observatory (angular resolution of 1') if their X-ray emission is not very high (or if it is attenuated by absorbing matter). It is interesting to note that only one of the 11 SNRs in the SMC has a small diameter, but its emission is equivalent to Cas A SNR (one of the most intense X-ray sources in our Galaxy).

- Very large diameter SNRs could have non-thermal radio emission or X- rays below the sensitivity limits of existing detectors, or their "characteristic" emission could be masked by foreground or background emissions. Two examples illustrate this problem: 1) SNR N11L was missed with the IPC, but it was detected with the HRI. 2) Chu and Mac Low (1990) have found X-ray emission in several HII bubbles of the LMC where deep exposures with the IPC were taken. These deep exposures do not cover the whole galaxy as the IPC survey does.

- SN exploding inside bright HII regions cannot be easily identified with these draconianic criteria. Some additional criteria, which can complement the search for SNRs in these galaxies, have been also proposed and should be included in order to have a more complete sample:

- Optical studies (Rosado et al., 1981, Rosado et al., 1982, Rosado, 1986, Chu and Kennicutt, 1988) show that large velocities ($V_{exp} \gtrsim 100 kms^{-1}$) or large velocity gradients are present in SNRs in the LMC (wind-blown bubbles of the same diameter show lower expansion velocities).The presence of a SNR is traced by the complexity of its velocity profiles (showing humps, splitting, or high velocity wings) when obtained by an echelle or a scanning FP or by the presence of emission in several of the λ-

maps of a scanning FP observation (corresponding to slices of the nebula at different velocities).

- In addition, the "scaled subtraction" method uses the $H\alpha/S\nu$ ratio in a very local analysis, avoiding the problems of extinction (Ye et al., 1991). The Hα intensity is very high in giant HII regions, and the variation of the $H\alpha/S\nu$ ratio within the region due to the presence of SNRs may be small. When the Hα map is scaled and subtracted from the radio map, the difference map reveals the presence of non-thermal sources.

- Similarly, the [FeII]$(1.6435\mu)/Br\gamma$ line-ratio is a useful tool because it traces shock emission (Oliva et al., 1989)

3 SNRs in HII Regions or Inside Wind–Driven Cavities

Supernovae (SN) SNIb and SNII are thought to be originated by the explosions of massive Population I stars and these types of stars tend to be associated to rich stellar clusters thus, it seems natural to think that: a) several of the declared SNRs would be associated to HII complexes, and, their evolution have proceeded in a modified interstellar medium(ISM) due to the influence of the SN progenitor or the whole stellar cluster, and b) there could remain several SNRs to be discovered inside the HII complexes. Obviously, the criteria based on radio and X-ray emissions have large limitations in these cases, and one must use the kinematic criteria or the scaled subtraction method.

From the theoretical point of view, there are now models of SNR evolution in modified ambients including the presence of a circumstellar shell and a wind-driven cavity (Chevalier,1990, Tenorio-Tagle et al., 1990, Franco et al., 1991).

From the observational point of view, there are several pieces of evidence indicating that actual SN may be evolving under these conditions:

- Several large diameter bubbles or HII complexes have soft X-ray emission more brighter than what is expected for wind-driven bubbles (Chu and Mac Low,1990, Wang and Helfand, 1991) and, as stated before, the X-rays were detected in those bubbles where the exposure times of the IPC were longer. One possibility, for explaining this, is off-centerd SN explosions (see Rozyczka et al., 1992). The kinematics of some of these regions shows that violent motions held inside sometimes with a good spatial correlation with the X-ray emission region (Rosado et al.,1990 for N186E, Chu and Kennicutt, 1988 for N66, Rosado et al.,1981 for N70, Chu et al.,1992, for 30 Dor, Lasker, 1980 and Meaburn and Terrett, 1990, for N51D, Goudis and Meaburn, 1984 and Georgelin et al., 1983, for N44).

- Large HII complexes show large velocity motions which are similar to those found in SNRs. Although it is possible to energize these complexes with stellar winds, very large velocities are observed only in SNRs. While it is an open question which deserves more work, the fact of finding large velocities inside a nebula is likely associated with the presence of a SNR.

- The velocity profiles of SNRs in the MCs are usually contaminated with HII region emission and, in some cases, it is possible to show that the HII region is associated with the SNR (the case of the SNR N120 in the LMC, Rosado et al.,1992a, Rosado, 1989, Laval et al., 1992 the SNR N186D in the LMC, Laval et al., 1989, the

SNRs in the SMC, Rosado et al, 1992b, this work, some SNRs in the MCs, Chu and Kennicutt, 1988). The possibility that SNRs show the classical criteria would depend on the brightness of the associated HII region and on the place where the SN explodes (i.e.,inside or at the edge of the HII region).

- Some of the conclusions of the N-D and Vr-D relations found in the past (Mathewson et al., 1983, Rosado, 1986, Danziger and Leibowitz,1985) can be reconciled if we interpret the diameter of the SNR with the diameter of the wind-driven shell formed before the SN explosion.

4 The $\Sigma - D$ Relation and Other Relations

From their first list of SNRs in the LMC and in the SMC, excluding the Balmer dominated type, Mathewson et al. (1983) derived a $\Sigma - D$ relation:

$$\Sigma_{408MHz} = 1.510^{-16} D^{-2.6} Wm^{-2} Hz^{-1} sr^{-1}$$

It is important to emphasize that the 6 SNRs known at that time in the SMC did not show any important dispersion around this relation although a) they belong to a different galaxy (the SMC) and consequently the errors in distance determinations on both the LMC and the SMC are included in this fit and b) there is some controversy about the depth of the SMC which would introduce some dispersion for the SMC SNRs unless all of them were placed in the same velocity component (which is not the case as we will see below) or the depth of the SMC would not be so large relative to the distance to the SMC. Thus favoring small depths (about 5 kpc) for the SMC.

We have searched for a correlation of this type for the 11 SNRs now known in the SMC and we find the same type of dependence:

$$\Sigma_{843MHz} = c^{te} D^{(-2.6 \pm 0.5)} Wm^{-2} Hz^{-1} sr^{-1}$$

From these data, Mathewson et al. (1983) derived a cumulative number-diameter, N-D, relation for SNRs in the MCs with D$\lesssim$50 pc:

$$N(D) = 0.36 D^{(1.0 \pm 0.2)}$$

From which they deduced that the SNRs in the MCs are in the free- expansion phase up to diameters of 50 pc. The physical reasons for such a long lasting phase are not clear and there are some other analysis which argue against the data treatment and the completeness of the data (Green, 1984).

Several authors have used other quantities, preferentially observed ones in individual SNRs, which could allow to deduce a trend in the evolution of the observed SNRs. For instance, Danziger and Leibowitz (1985), from optical spectroscopic work, have deduced a relation between the density measured from the [SII] line-ratios of SNRs in the LMC and their diameter which implies that D varies as $t^{\frac{2}{3}}$, Rosado (1986), from optical Fabry-Perot(FP) data on LMC SNRs, has found a relation between the expansion velocity and the diameter which gives a dependence of D on $t^{\frac{3}{5}}$. Note that the numerical value of the exponent is quite similar so that it is the same law. What is interesting to emphasize is that such a dependence is the one expected for wind blown bubbles in the energy conserving case(Weaver et al., 1977) so that even if some other possibilities occur, it could be possible that the optical diameters we use were the diameters of the bubbles formed prior to the SN explosion, probably by the precursor of the SN or the other members of an OB association. The predicted N-D relation in this case is N varying as $D^{1.67}$, much flatter than the adiabatic one but still steeper than that found by Mathewson et al. (1983). However these findings give a physical reason for a SNR which mimics the free-expansion phase: the very low density medium inside a wind blown bubble cannot slow-down the SN ejecta as the "normal" $1cm^{-3}$ uniform electron density medium currently assumed does.

5 Our Work

The team of interferometry of the Marseilles Observatory (P. Amram,, J. Boulesteix, Y.P. and Y.M. Georgelin, E. le Coarer, M. Marcelin and A. Viale) and A. Miranda and M. Rosado (IA-UNAM, Mexico) have a long term program for studying the kinematics of SNRs in the MCs. Till now we have used high spectral resolution observations by means of scanning Fabry-Perot (FP) interferometers with 6 and 16 kms^{-1} sampling velocity resolution. We have approached the problem in two ways: high spatial resolution observations (2".6/px) of selected SNRs in the LMC with the CIGALE equipment (Boulesteix et al., 1984) attached to the 1.52 m telescope of ESO which provide an angular field coverage of 7' and a modest spatial resolution observations (9"/px) with the $H\alpha$ SURVEY also installed at ESO (Amram et al., 1991, Le Coarer et al., 1992) which give a large field (39') sufficient enough to try to map the entire galaxies. Both approaches are complementary: in the first case we have the detailed velocity field of the SNR which allow us to know the particularities of its, in general, complicated kinematics; in the second case we have a global view of all the SNRs in a galaxy. High spatial resolution of SNRs with the CIGALE equipment have been performed in several SNRs in the LMC while the $H\alpha$ SURVEY has been completed allowing us to have kinematical information about all (but one) the SNRs in the SMC.

5.1 SNRs in the SMC

The Hα survey performed at ESO allows us to cover the kinematics of the main body of the SMC. Here, we show the results of the survey at the sampling resolution of $16kms^{-1}$. We have detected all but one of the SNRs in the SMC reported by Mathewson and co-workers. They are clearly recognized in the λ-maps, obtained from the FP scanning, because they have emission even at a wide range of velocities. The only non-detection was the SNR 0102-723. This is probably because of its small diameter or because of a small Hα emission.

Incidentally, there appear some other nebula which show an extended velocity range and may be related with nebulae formed by shocks (if not SNRs). Thus, the FP λ-maps are a powerful tool for detecting shocked nebulae even in complex ambients and may be applied in the search for SNRs in more distant galaxies. We have

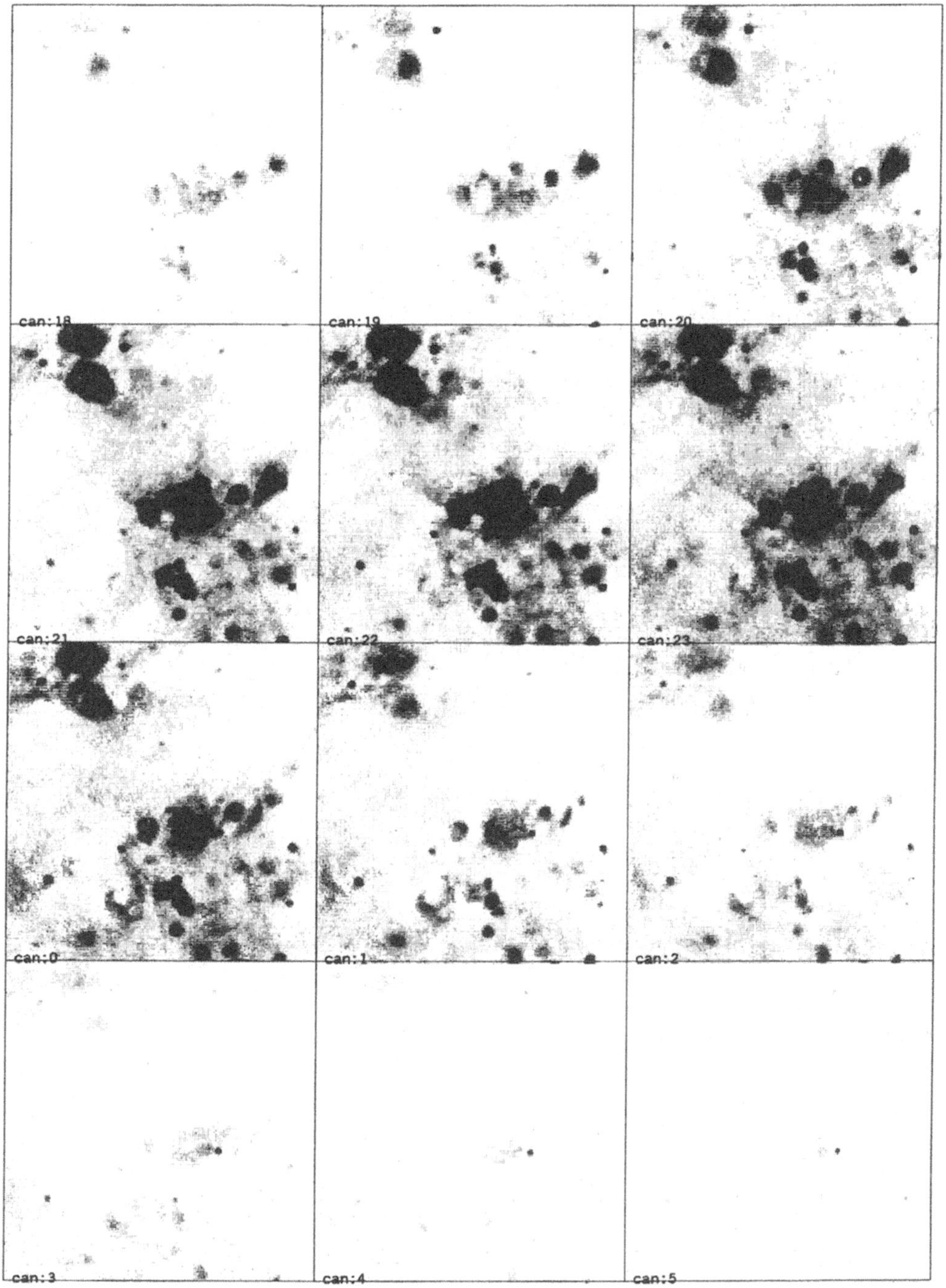

Fig. 1. FP λ-maps of the field where the SNR 0045-734 is located (marked by an arrow). The velocity difference between consecutive channels is of $16kms^{-1}$.

constructed a list of 11 candidates to SNRs or wind blown bubbles obtained with this instrument. The candidates are located inside the regions: DEM 7, 14, 24, 25 (N17), 31 (N19), 48, 56, 103 (N66), 118, 123 (N76A) and 128 (three of them have X-ray sources within their boundaries). We have obtained velocity profiles integrated over the whole reported area of the SNRs. The main characteristic is that all of them are complex, quite different to the profiles of the majority of the nebulae of the SMC. Another point is that almost all of them are contaminated by bright (or relatively bright) HII region emission. It is important to discriminate if the HII regions are physically associated with the SNRs or if they are only foreground or background nebulae. This would require the study of the detailed velocity field observed. Nonetheless, since 80% of the cases show either systemic velocities similar to HII region velocities or abnormal broadening of the profiles of the HII region component we can conclude that they are indeed associated with the SNRs. This work is in progress and the detailed velocity fields will be published elsewhere (Rosado et al., 1992b).

All the detected SNRs have complicated velocity profiles of several types: splitting in 2 components (S type), high velocity wings (W type) in both directions, high velocity wing in one direction-approaching or receding-, and "humps"(H type) in one direction (Fig. 2). This is interesting because these are the type of profiles we will obtain for SNRs in more distant galaxies.

Table 1 gives some preliminary results derived from the integrated profiles. Further work is in preparation based on the analysis of the whole velocity field of each SNR. The table contains the SNR name and diameter, the integrated profile type, the systemic velocity, V_{syst}, the velocity of the HII region, V_{HII}, the velocity difference between the red and blue components, ΔV, the velocity difference between the red or blue components and V_{HII}, $V_{B/R} - V_{HII}$, and the derived location of the contaminating HII region in one of the 4 regions of Martin et al. (1989). Under the assumption that the HII regions are associated with the SNRs, one can locate the positions of the SNRs relative to the 4 complexes of stars and HI found by Martin et al. (1989) as the main components of the SMC (VL, L, H and VH). These complexes, at different velocities, seem to be at different depths along the line of sight (being the VL component the closest and the VH component the farthest). There are SNRs associated with the L, H and VH components and no SNR is associated with the VL component; so that, we observe SNRs located at different depths in the SMC. Since there is no clear difference in the brightness of the optical and X- ray emissions of these SNRs, one can infer that the depth of the SMC must be not so large. Indeed, the SNR 0101-724 which is the only SNR associated with the VH component, and thus is more distant, has a high X-ray emission.

We have also searched for correlations between these quantities and the diameter, and find that $V_{HII} - V_{B/R}$ and ΔV are well correlated for those points where both quantities exist. On the other hand, there is a loose correlation between $V_{HII} - V_{B/R}$ and the diameter (the one quoted by Mathewson and co-workers) with an exponent of -0.44 ± 0.23. Thus, giving the same result as in the LMC SNRs (Rosado, 1986) which implies wind-bubbles evolution or marginally, free-expansion. Of course, these results are preliminary and much work must to be made in the radial velocity field as well in the diameter determination.

Table 1. Properties Derived From Integrated Velocity Profiles

Name	D(pc)	Profile Type	Vsyst	V_{HII}	ΔV	$V_{\mathrm{HII}} - V_{\mathrm{R/B}}$	Location
0045-734	26	W	146	141	121	65 -56	L
0046-735	26	W	—	148	—	-55	L
0049-736	48	H	—	160	—	50	H
0101-724	25	H	—	195	—	60	VH
0103-726	57	S	— (167)	142	— (50)	50	L
0047-735	51	W	165	163	72	38 -34	H
0050-728	58	S + H	127	133	91	39 -52	L
0056-725	56	H or W	—	(122)	—	(53)	L
0058-718	62	H	—	161	—	23	H
0104-723	29	W	178	177	91	46 -45	H

5.2 SNRs in the LMC

We have made observations, following the high spatial resolution approach, of several SNRs in the LMC:

N186D (Laval et al., 1989), N186E (Rosado et al., 1990) and N120 (Laval et al., 1992, Rosado et al., 1992a). We have in course of preparation: the SNRs N103A and N103B, the emission nebulae: N9 and N44, and several superbubbles.

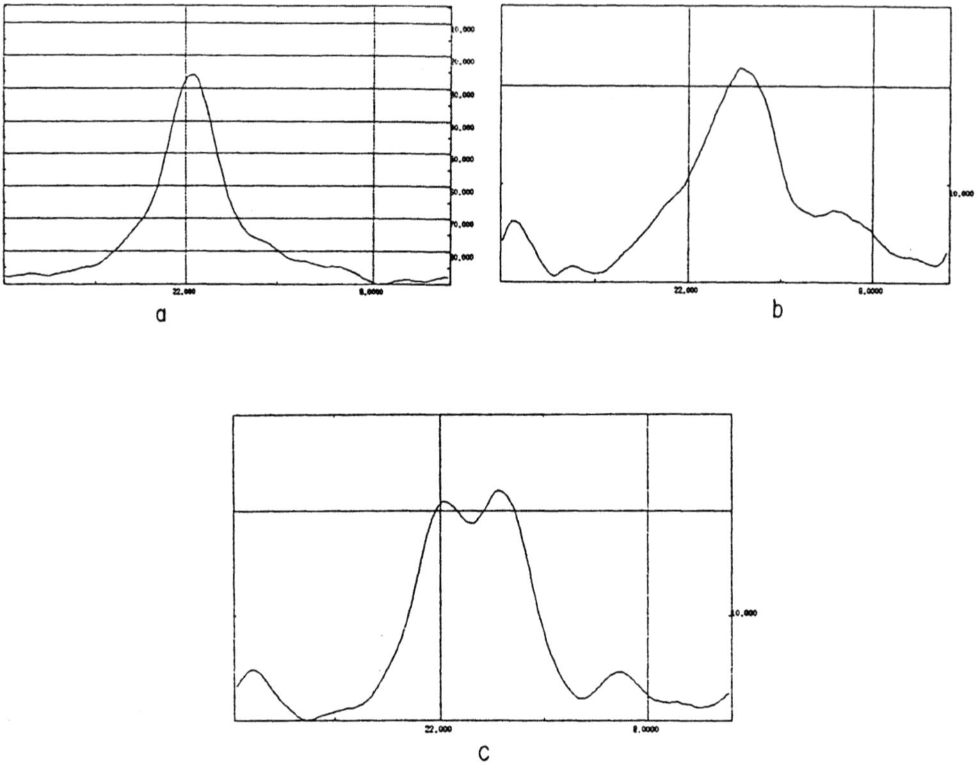

Fig. 2. Different types of FP velocity profiles found for the SMC SNRs: a) High velocity wings, b) Hump, c) Splitting

References

Amram ,P., Boulesteix, J., Georgelin, Y.P., Georgelin, Y.M., Laval, A., Le Coarer, E., Marcelin, M., Rosado, M.(1991): The Messenger **64** 44

Boulesteix,J., Georgelin, Y.P., Marcelin, M., Monnet, G.(1984): SPIE Conf. Instrum. Astron. **445** 37

Chevalier, R.A.(1990):In "Supernovae",ed. A.G. Petschek(Berlin:Springer)pp. 91

Chu, Y.H., Kennicutt, R.C.(1988): A.J.**95** 1111

Chu, Y.H., Mac Low, M.M.(1990): Ap.J.**365** 510

Chu, Y.H., Kennicutt,R.C., Schommer,R..A., Laff,J.(1992):A.J.**103** 1545

Danziger, I.J., Leibowitz, E.M.(1985): M.N.R.A.S.**216** 365

Franco, J., Tenorio-Tagle, G., Bodenheimer, P., Rozyczka, M.(1991): P.A.S.P.**103** 803

Georgelin, Y.M., Georgelin, Y.P., Laval, A., Monnet, G., Rosado, M.(1983): A&A Suppl. **54** 459

Goudis, C., Meaburn, J.(1984): A&A **137** 152

Green, D.A.(1984): M.N.R.A.S. **209** 449

Haynes, R., Milne, D.(1991): IAU Symp. No. 148, Kluwer, Dordrecht, Holland

Inoue, H., Koyama, K., Tanaka, Y.(1983): In "Supernova Remnants and their X-ray Emission", ed. by Danziger, I.J. and Gorenstein,P., Reidel, Dordrecht, Holland, pp. 535
Lasker, B.M.(1980): Ap.J. **239** 65
Laval, A., Rosado, M., Boulesteix, J., Georgelin, Y.P., Marcelin, M., Monnet, G., Le Coarer, E.(1989): A&A **208** 315
Laval, A., Rosado, M., Boulesteix, J., Georgelin, Y.P., Le Coarer, E., Marcelin, M., Viale, A.(1992): A&A **253** 213
Le Coarer, E., Rosado, M., Georgelin, Y.P., Viale,A. (1992): These proceedings
Long, K.S., Helfand, D.J., Grabelsky, D.A.(1981): Ap.J. **248** 925
Martin, N., Maurice, E., Lequeux, J.(1989): A&A **215** 219
Mathewson,D.S., Ford,V.L., Dopita,M.A., Tuohy, I.R., Long,K.S., Helfand, D.J.(1983): Ap.J.Suppl. **51** 345
Mathewson,D.S., Ford, V.L., Dopita, M.A., Tuohy, I.R., Mills, B.Y., Turtle,A.J.(1984): Ap.J.Suppl. **55** 189
Mathewson, D.S., Ford, V.L., Tuohy, I.R., Mills, B.Y., Turtle, A.J.(1985): Ap. J. Suppl. **58** 197
Meaburn, J., Terrett,D.L.(1980):A&A **89** 126
Oliva,E., Moorwood,A.F.M., Danziger,I.J.(1989):A&A **214** 307
Roger , R.S., Landecker, T.L.(1987): IAU Colloquium No. 101 Cambridge University Press, Cambridge
Rosado,M.(1986):A&A **160** 211
Rosado,M.(1989):Rev.Mex.Astr.Astrof. **18** 105
Rosado,M., Georgelin,Y.P., Georgelin,Y.M., Laval,A., Monnet,G.(1981):A&A **97** 342
Rosado,M., Georgelin,Y.P., Georgelin,Y.M., Laval,A., Monnet,G.(1982):A&A **115** 61
Rosado,M., Laval,A., Boulesteix,J., Georgelin,Y.P., Greve,A., Marcelin,M., Le Coarer,E., Viale,A.(1990):A&A **238** 315
Rosado,M., Laval,A., Le Coarer,E., Boulesteix,J., Georgelin,Y.P., Marcelin,M.(1992a):A&A submitted
Rosado,M., Le Coarer,E., Georgelin,Y.P., Viale,A.(1992b):In preparation
Rozyczka,M., Tenorio-Tagle,G., Franco,J., Bodenheimer,P.(1992): M.N.R.A.S. (submitted)
Seward,F.D.(1990): Ap.J. Suppl. **73** 781
Seward,F.D., Mitchell,M.(1981):Ap.J. **243** 736
Tenorio-Tagle,G., Bodenheimer,P., Franco,J., Rozyczka,M.(1990):M.N.R.A.S. **244** 563
Wang, Q., Helfand, D.J. (1991): Ap.J. **373** 497
Weaver,R., Mc Cray,R., Castor,J., Shapiro,P., Moore,R.(1977):Ap.J. **218** 377
Ye,T., Turtle,A.J., Kennicutt,R.C.(1991):M.N.R.A.S. **249** 722

Stars, Stellar Evolution

Bernhard Wolf, Michael W. Feast

Paolo Molaro, Paola Andreani, I. John Danziger

Cepheids,Miras and CH stars in the Magellanic Clouds

M.W.Feast

Institute of Astronomy, University of Cambridge, England

Abstract: Recent work on the use of Cepheids and Mira variables to investigate the distance and structure of the Magellanic Clouds is reviewed. This leads to a discussion of the nature of the CH stars in the LMC

1 Cepheids and the LMC Distance

For some long while now the most accurate distance to the LMC has been derived from Cepheids,with a zero point based on Cepheids in young open clusters in our own Galaxy. Feast and Walker (1987) derived in this way an LMC distance modulus of 18.47 ± 0.15. This modulus is ultimately based on the trigonometrical parallaxes of eleven main sequence stars using the Pleaides to define the shape of the main sequence (van Leeuwen 1983). The implied distance modulus of the Pleiades was 5.57 ± 0.08.

More recently, Feast (1991a) gave a revised LMC modulus of 18.55. This was based on a revision of LMC and galactic cluster data by Caldwell and Laney (1991), including revised metallicity corrections to the LMC Cepheid data, together with an adopted theoretical main sequence (vandenBerg and Poll 1989) which implies a Pleiades modulus of 5.60. A fit [Feast (1991a) following Sandage and Cacciari (1990)] of 33 main sequence stars with parallax probable errors $< 7\%$ [standard errors $<$ 10%] implies a somewhat greater Pleiades modulus,5.63 ± 0.04 or 5.68 ± 0.06 when a Lutz-Kelker correction is applied. The uncertainty of this last result is a rough estimate based on the work of Koen (1992) on the confidence limits of the Lutz-Kelker corrections.

A much improved calibration of the main sequence will no doubt eventually become available from HIPPARCOS. Until that time it is useful to look also at other methods of calibrating the Cepheid zero point.

At one time the Cepheid distance scale was based ultimately on the distance of the Hyades. There is recent work on the Hyades distance modulus which restores the Hyades to some of its former importance.

Table 1. Recent Distance Modulus Determinations for the Hyades

Method	Modulus	Ref.
Trig. parallaxes of cluster members	3.23 ± 0.06	[1]
Binary cluster members	3.35 ± 0.02	[2]
Proper motions	3.40 ± 0.04	[3]
Unweighted mean	3.33 ± 0.05	

[1] Patterson and Ianna (1991)
[2] Peterson and Solensky (1988)
[3] Schwan (1991)

The binary approach though of the highest internal accuracy is somewhat model dependent and an unweighted mean of the three determinations has been adopted as the best currently available modulus. A comparison of the Hyades and Pleiades main sequences together with a metallicity difference, Δ[Fe/H]=0.16 ± 0.03 (Boesgaard and Friel 1990), then gives a Pleiades modulus of 5.57 ± 0.06 [this estimate of the standard error includes the effects of uncertainty in the metallicity correction]. The mean of this result and 5.68 ± 0.06 from the trigonometrical parallaxes of the field stars is 5.62 ± 0.05. This is very close to the value of 5.60 adopted by Feast (1991a).

Table 2 lists the results of some further tests of the Cepheid distance scale. Laney and Stobie (1992a,b,c) have derived Baade-Wesselink radii for 51 galactic Cepheids from new JHK photometry and published radial velocities. Combining these results with model colours (Hindsley and Bell 1989,1990) they derive a Cepheid zero point which is 0.05 ± 0.04 [internal s.e.] brighter that that adopted by Feast (1991a). Baade-Wesselink results for Cepheids in the cluster NGC1866 in the LMC and statistical parallax work on galactic Cepheids both have significant uncertainties. The result of the IUE observations of companions to galactic Cepheids suggests a somewhat shorter distance scale.

Table 2. Tests of the Cepheid Scale

Method	ΔMod	Ref
Gal.Cep.,JHK,Baade-Wesselink	$-0.05[\pm 0.04$ int]	[1]
NGC1866[LMC] Cep.,Baade-Wesselink	-0.05 ± 0.30	[2]
Gal.Cep.,Statistical parallaxes	$+0.19 \pm 0.33$	[3]
Gal. Cep.,companions	$+0.26 \pm 0.12$	[4]

[1] Laney and Stobie (1992a,b,c)
[2] Coté et al (1991)
[3] Wilson et al (1991)
[4] Evans (1992)

The various results of this section suggest that the basic scale adopted by Feast (1991a) [equivalent to a Pleiades modulus of 5.60] is soundly based with an uncertainty of about 0.05. The most significant discrepancy is the result for the Cepheid companions which deviates from the adopted scale by twice its standard error.

The uncertainty in the distance of the LMC has been discussed by Feast and Walker (1987) and Feast (1991a). This included the uncertainties in the main sequence cluster fits in our Galaxy, as well as metallicity and reddening corrections for the clusters and for the LMC. A rough estimate of ± 0.15 was made. This may now be slightly reduced, due primarily to the improved basic calibration, but must still be about ± 0.10 or possibly slightly large.

The determination of an LMC modulus from the ring round SN1987A is particularly valuable being geometrical in nature. Panagia et al (1991) found 18.50 ± 0.13. However the result is somewhat model dependent and Dwek and Felten (1992) have suggested that the real uncertainty is significantly greater than the above value. Possibly therefore the Cepheid modulus is somewhat more accurate than that of the supernova, but more importantly it is very satisfactory that these two independent methods give closely the same result. For the present the Cepheid value of 18.55 ± 0.10 is adopted.

2 The Structure of the SMC from Cepheids

Work on the structure of the SMC as derived from Cepheids has recently been reviewed by Feast (1989,1991b) and by Caldwell and Laney (1991). Very recently Caldwell and Maeder (1992) have used 3-D visualization software to examine the Cepheid distribution in the SMC as viewed from different directions. They find that the elongated structure of the SMC gives the impression of being both bent and twisted. It can be roughly approximated by two discs bent about 30° out of alignment and with their planes twisted nearly at right angles to each other.

3 Mira Variables in the LMC

The LMC Mira PL and PLC relations derived from JHK photometry (Feast et al 1989) are potentially important for distance determinations. Unfortunately the data on which to base a galactic calibration of the zero points of these relations is quite poor (cf. Robertson and Feast 1981). Work is in progress on galactic Miras with physical companions which should improve this situation. In addition the precision of trigonometrical parallaxes even from the ground is rapidly improving and the situation will no doubt be transformed by the HIPPARCOS results. Gatewood (1992) has recently published what seems to be the most accurate trigonometrical parallax of a Mira so far. He found 8.3 ± 1.0mas for R Leo. This is a particularly interesting result because R Leo was earlier believed to be a member of a stellar moving group which implied that it had an absolute bolometric magnitude of -5.7. The new result shows that it is about one magnitude fainter than this and using the star to calibrate the M_{bol} - logP relation gives an LMC distance modulus of 18.61 ± 0.3, not significantly different from the Cepheid result.

It has been frequently pointed out (e.g. Wood 1991) that there may be differences between Mira PL relations in different systems due to possible metallicity differences at a given period. If this is so then this should also affect the period-colour relations. Evidence for this has so far been lacking (cf. Whitelock et al 1991,Feast 1992) but additional and extensive infrared photometry of Miras in the Magellanic Clouds and in our own Galaxy is desirable to carry out more stringent tests.

Mira variables are of importance for our understanding of the kinematics of the Clouds. Hughes et al (1991) showed that the shorter period Miras in the LMC [mean period 180days] have a radial velocity dispersion of 33 $\mathrm{km\,s^{-1}}$. This is greater than that of young objects [about 9 $\mathrm{km\,s^{-1}}$] or planetary nebulae [about 22 $\mathrm{km\,s^{-1}}$] but less than that expected for a halo-type population supported entirely by random motions, for which a dispersion of 40 or 50 $\mathrm{km\,s^{-1}}$ is anticipated. This result shows that these Miras are strikingly analogous in their kinematic behaviour to Miras of similar period in our own Galaxy which also have motions intermediate between those of halo and disc objects.

4 CH stars in the LMC

This last result for Miras leaves open the question of whether or not the LMC has a very old, pressure-supported halo or not. Hartwick and Cowley (1988,1991) and Cowley and Hartwick (1991) attempted to examine this problem using a group of stars that they had identified as having the spectroscopic characteristics of CH stars. In our own Galaxy the 'classical' CH stars define an extreme halo population but the Hartwick-Cowley stars have an overall velocity dispersion of only 24 ± 2 $\mathrm{km\,s^{-1}}$ leading Hartwick and Cowley to suppose that a kinematic halo is absent in the LMC.

However infrared JHK photometry of the Hartwick-Cowley stars (Feast and Whitelock 1992) shows that they are quite unlike 'classical' CH stars, but are very interesting objects in their own right. They are brighter than classical CH stars. their bolometric absolute magnitudes ranging up to -6 [see Fig.1]. In fact comparison with carbon stars in Magellanic Cloud clusters suggests that these CH stars are the most massive carbon stars in the Clouds with main sequence turn-off masses of about 3 solar masses or more. One would normally expect such stars to be relatively young [a few times 10^8 years old]. However the kinematic results of Hartwick and Cowley suggest an age of a few Gyr. This apparent contradiction can be resolved if the Hartwick-Cowley stars are the results of binary mergers [or extensive mass transfer] in an intermediate age population. That is, they are the evolutionary products of blue stragglers of such a population. Evidently these stars deserve further investigation. Cowley and Hartwick (1991) have suggested that their stars can be divided into an inner and an outer group which are kinematically distinct,the inner group being definitely old. Stars of the two groups are distinquished by different symbols in fig 1. No differences between the two groups are apparent in this diagram and in particular, very luminous CH stars are present in the inner group.

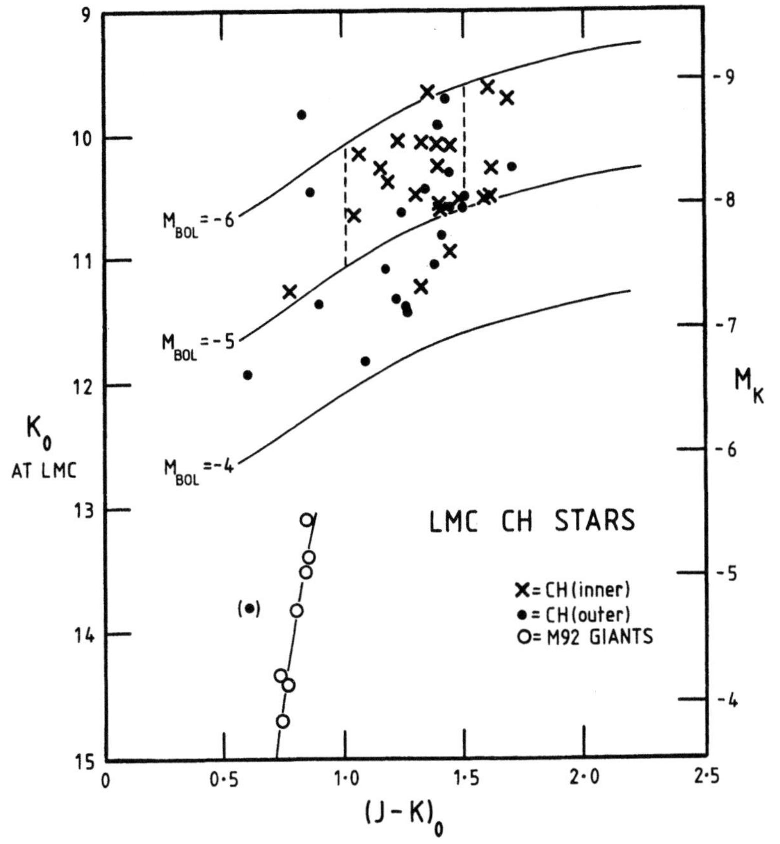

Fig. 1. CH stars in the LMC (from Feast and Whitelock 1992). The crosses and dots are for stars in the inner and outer regions of the LMC respectively.

5 Acknowledgements

I would like to thank Drs J.A.R.Caldwell and D.C.Maeder for allowing me to show their 3-D simulations of the SMC at the meeting. I am indebted to Drs C.Koen,C.D.Laney,R.S.Stobie and P.A.Whitelock for information in advance of publication. The hospitality of Prof.D. Lynden-Bell and the Institute of Astronomy, Cambridge as well as financial support from the Royal Society is gratefully acknowledged.

references

Boesgaard A.M., Friel E.D.(1990) Ap.J.**351**,467.

Caldwell J.A.R.,Laney C.D. (1991) in "The Magellanic Clouds" ed. R.Haynes and D.Milne [Kluwer:Dordrecht] p249.

Caldwell J.A.R.,Maeder D.C. (1992) in "Variable Stars and Galaxies" ed. B.Warner; ASP Conf.Series, in press.

Coté P.,Welch D.L.,Mateo M.,Fischer P.,Madore B.F. (1991) A.J.**101**,1681.

Cowley A.P.,Hartwick F.D.A. (1991) Ap.J.**373**,80.

Dwek E.,Felten J.E. (1992) Ap.J.**387**,551.

Evans N.R. (1992) Ap.J.**389**,657.

Feast M.W. (1989) in "Recent developments in MC research" ed K.S.de Boer,F.Spite and G.Stasinska [Obsevatoire de Paris] p75.

Feast M.W. (1991a) in "Observational Tests of Cosmological Inflation" ed. T.Shanks, A.J.Banday, R.S.Ellis, C.S.Frenk and A.W.Wolfendale [Kluwer:Dordrecht] p147

Feast M.W. (1991b) in "The Magellanic Clouds" ed. R.Haynes and D.Milne [Kluwer: Dordrecht] p1.

Feast M.W. (1992) in "Highlights of Astronomy" ed. J.Bergeron [Kluwer:Dordrecht] in press

Feast M.W.,Glass I.S.,Whitelock P.A.,Catchpole R.M.(1989) MNRAS **241**,379

Feast M.W.,Walker A.R. (1987) Ann.Rev.A.A. **25**,345.

Feast M.W.,Whitelock P.A. (1992) MNRAS in press

Gatewood G. (1992) PASP **104**, 23.

Hartwick F.D.A.,Cowley A.P. (1988) Ap.J. **334**, 135.

Hartwick F.D.A.,Cowley A.P. (1991) in "The Magellanic Clouds" ed R.Haynes and D.Milne [Kluwer:Dordrecht] p77.

Hindsley R.B.,Bell R.A. (1989) Ap.J. **341**,1004.

Hindsley R.B.,Bell R.A. (1990) Ap.J. **348**,673.

Hughes S.M.G.,Wood P.R.,Reid N. (1991) A.J. **101**, 1304

Koen C. (1992) MNRAS **256**,65.

Laney C.D.,Stobie R.S. (1992a) A.A.Sup. **93**,93.

Laney C.D.,Stobie R.S. (1992b) in "Variable Stars and Galaxies" ed B.Warner; ASP conf. series. in press

Laney C.D.,Stobie R.S. (1992c) to be published

Panagia N.,Gilmozzi R.,Macchetto F., Adorf H-M.,Kirshner R.P. (1991) Ap.J. **380**,L26.

Patterson R.J.,Ianna P.A. (1991) A.J. **102**,1091.

Peterson D.M.,Solensky R. (1988) Ap.J. **333**,256.

Robertson B.S.C.,Feast M.W. (1981) MNRAS **196**, 111.

Sandage A.R.,Cacciari C. (1990) Ap.J. **350**,645.

Schwan H. (1991) A.A. **243**,386.

VandenBerg D.A.,Poll H.E. (1989) A.J. **98**,1451.

van Leeuwen F. (1983) Ph.D. Thesis,Leiden University, The Netherlands.

Whitelock P.A.,Feast M.W.,Catchpole R.M. (1991) MNRAS **248**,276.

Wilson T.D.,Barnes T.G.,Hawley S.L.,Jefferys W.H. (1991) Ap.J. **378**,708.

Wood P. (1990) in "From Miras to Planetary Nebulae" ed. M.O.Mennessier and A.Omont [Editions Frontieres] p67.

The Most Massive Stars in the Magellanic Clouds

M. Heydari-Malayeri

European Southern Observatory

Abstract: We review our present knowledge about the mass of the most luminous or very massive stars in the Magellanic Clouds (MCs). Based on high resolution data gathered during past years, we propose that stars with a present mass over 100 $M_\odot$ do not seem to exist in the MCs, and stars with an initial mass higher than this limit may be extremely rare. It is shown that in many cases the upper mass limit is proportional to the "seeing" used! This 100 $M_\odot$ is not a magic number and should not be thought of a strict limit; it represents a working threshold for discussions. In view of the new observational results, models of massive star formation and evolution may be subject to revision. Future improvements in observational techniques and more elaborate theoretical models will undoubtedly provide a more precise value for the upper mass limit of stars.

1 Introduction

Massive stars, though they are relatively rare, play a key role in astrophysics. They are major sites of nucleosynthesis and, therefore, are mainly responsible for the chemical evolution of galaxies. They are progenitors of Wolf-Rayet stars and such rare and violent phenomena as supernova events. Because of their high luminosities they can be detected in faraway systems and so they tell us about the composition, distance and star formation activity in these systems. Due to their high ultraviolet flux and powerful stellar winds, they bring about interesting phenomena in the interstellar medium, like HII regions, shocks, bubbles, etc. In brief, the most massive stars are outstanding objects to study both for their extreme intrinsic properties and for their extreme energy input into the surroundings – and mass is is the fundamental parameter with which these characteristics scale. A precise knowledge of the masses is, therefore, important for theories of stellar evolution and internal structure, for example for tests of convective overshooting. In particular, the upper mass limit (UML), which also influences the derivation of the slope of the initial mass function (IMF), is a basic ingredient in star formation theories.

Since the concept of the UML was put forward (Ledoux 1941), there has been a conflict between theoretical calculations and observations. Models have often predicted very massive stars (VMSs), i.e. stars above $\sim$ 100 $M_\odot$, for which there has been no or only crude indirect observational support. However, high quality ground-based data obtained in recent years using the ESO facilities now show a clear trend towards smaller masses. In this review, we will try to examine the confrontation between theory and observation with emphasis on these new observational results. Our knowledge of the most massive stars a decade ago is reviewed in the proceedings of the ESO workshop on this subject (D'Odorico et al. 1981).

The Magellanic Clouds are favorite galaxies for this investigation since their distances are very well known and this facilitates reliable absolute magnitude estimates, which is of prime importance for the determination of stellar masses. Moreover, the MCs present unique opportunities for the investigation of massive star formation in metal deficient environments.

2 Theories concerning the UML

There are three main mechanisms that determine the UML in stars: pulsational instability, infall reversal during the protostar stage, and fragmentation of the molecular cloud.

2.1 Pulsational instability

Once a star is formed the nuclear energy generated at the center is continuously transformed into the mechanical energy of the pulsations. Ledoux (1941) was the first to show that massive stars should be pulsationally unstable above a certain critical mass, M_{crit}, or UML. Instability develops when the energizing mechanism in the core overcomes the damping mechanisms for pulsations. The damping, mainly caused by heat leakage, occurs predominantly in the outer envelope below the surface of the star where the radiative opacity is important. This means that the actual value of the UML is sensitive upon the opacity law used for the calculation. For a pure Thomson electron scattering, the maximum stable mass of a chemically homogeneous, non-rotating Population I star, based on a linear stability analysis, is $\sim 60\ M_{\odot}$ (Schwarzschild & Härm 1959, Boury et al. 1964), whereas using an opacity law dependent on temperature and density gives an UML of $\sim 90\ M_{\odot}$ (Stothers & Simon 1970, Ziebarth 1970). This is because the inclusion of atomic absorption in the total opacity leads to a larger central condensation of the star and hence to smaller pulsation amplitudes near the center as well as to the positive dissipation of pulsation energy in the envelope. Several authors have addressed the problem of the stable mass using various versions of the Los Alamos opacities generally with quasi-adiabatic treatment of the convection in the stellar core and nonadiabatic effects in the outer envelope. They have come up with a critical mass of 80–100 $M_{\odot}$ which is not a sensitive function of composition and, on the other hand, decreases with increasing Y at fixed Z; and for fixed Y, decreases slowly with decreasing Z.

More sophisticated opacities have increased the UML further. Recently, Stothers (1992) has used revised atomic opacities containing radiative contribution from many metallic absorption lines. The resulting critical masses are about ~ 45 % larger than the previous estimates (Stothers & Simon 1970). With the new opacities, the instability occurs above 120–150 $M_{\odot}$ for normal Galactic Population I chemical compositions, and above $\sim 90\ M_\odot$ for stars in metal-poor environments like the outer Galaxy and the SMC.

However, there are model calculations arriving at even higher UMLs. Non-linear analyses accounting for large velocity amplitudes (with respect to the velocity of sound) at the outermost layers of the star predict a much efficient damping of the pulsations through shock waves than allowed by the linear theory (Ziebarth 1970, Appenzeller 1970). Thus, pulsational mass loss is insufficient to completely disrupt stars of up to 200 $M_\odot$ during their main sequence life time (Appenzeller 1970). We can also mention theoretical works admitting stable stars of masses from 2000 to 10000 $M_\odot$ (Klapp 1983, Langer 1987).

2.2 Infall halt

Pulsational instability is not the only mechanism that imposes an upper limit to stellar masses. Several studies show that there are physical processes that at a much more earlier stage limit the mass of the protostar. Numerical calculations of the dynamical behaviour of collapsing interstellar clouds show that the central regions rapidly form a small hydrostatic core, while the surrounding dust-gas envelope continues to fall inward. If the core is massive enough, then hydrogen burning will begin in the interior while it continues to acquire mass by accreting the surrounding cloud (Appenzeller & Tscharnuter 1974). Larson & Starrfield (1971) and Kahn (1974) proposed that the interstellar grains in the cocoon absorb the ultraviolet and visible radiation from the star in a thin shell at the inner edge of the dust cocoon and re-emit in the infrared. In many cases the cocoon is also opaque to this re-emitted radiation and therefore a gradient of radiation pressure builds up. The pressure acting on grains in the infall will decelerate the infall of material. At some point in the inflow, the grains could become hot enough to be destroyed by sublimation. If the inward flow can be maintained through the grain destruction region, then accretion will continue since radiative pressure will no longer be able to halt the flow. If, on the other hand, the flow is stopped before grains can be destroyed, then accretion is halted and the core is considered to have reached its final mass.

Kahn's (1974) result was that when the core luminosity-to-mass ratio, L/M, exceeds $\sim 10^4$ ergs s^{-1} g^{-1} the inflow of material is reversed. This L/M ratio corresponds, in our Galaxy, to a maximum mass of $\sim 40\ M_\odot$. Kahn showed that the maximum L/M ratio, derived from his analysis, is inversely proportional to the grain opacity. Regions of lower grain-to-gas mass ratio would therefore form more massive stars.

Wolfire & Cassinelli (1987), on the basis of advances available in the understanding of interstellar grain properties (Mathis et al. 1877, hereafter MRN), investigated the effects of grain destruction in determining the maximum mass that a growing protostar may gather. They found that the maximum L/M ratio is proportional to T_{sub}^{-3}, where T_{sub} is the assumed temperature at which grains sublimate. Kahn's UML of 40 $M_\odot$ is based on a sublimation temperature of 3670 K for graphite. However, the laboratory data indicate that in the low-pressure environment of star formation graphite sublimates at much lower temperatures. If the sublimation temperature is decreased to 2000 K, the maximum L/M ratio derived from Kahn's analysis corresponds to $\sim 1000\ M_\odot$.

However, a more thorough treatment of the radiation field and grain opacity in an accretion flow carried out by Wolfire & Cassinelli (1987) showed surprizing results. They found that infall of a massive protostellar cloud will not occur for all plausible values of T_{sub} if the opacity is as large as that expected from the standard MRN mixture (graphite and silicate grains)! For massive stars to form, the total number of grains must be decreased by a factor of 4 or more from the standard Galactic number abundances, and the largest graphite grain size must be decreased. This means that special grain properties are required for massive star formation. These conditions are supposed to occur in the interstellar medium by shock fronts (Elmegreen & Lada 1977; Assousa & Herbst 1980), spiral density wave (Woodward 1976) or ionization fronts (Elmegreen & Lada 1977), from a nearly stellar association.

2.3 Fragmentation of the molecular cloud

An increasing body of data obtained in recent years using high resolution molecular line observations show that fragmentation may play an important role in collapsing molecular clouds. It is not yet fully understood how the fragmentation occurs and gives rise to the clumps in which the stars form. But, it seems that larger clumps subfragment into several yet denser clumps and subfragmentation becomes increasingly important for progressively larger clump masses (Larson 1991, Zinnecker et al. 1992). Apart from subfragmentation, there are other phenomena that may limit the amount of mass that can be accreted: magnetic fields, turbulence, and tidal effects. More progress in observation and theory are needed to clarify all these interesting questions.

3 Observations

It is very important in such a situation to have observational constraints that allow us to test theories. During recent years we have undertaken a program devoted to studying the formation and evolution of the massive stars in the MCs on the basis of high quality observations obtained during sub-arc seeing conditions at the ESO telescopes, especially NTT (Heydari-Malayeri et al. 1988, 1989; Heydari-Malayeri & Hutsmékers 1991a,b) and an efficient image restoration code using a maximum entropy method developed by P. Magain & M. Remy (see the above references).

First, we know that accurate measurement of the mass and radius of a star other than the Sun can only be determined directly if the star is located in a binary system in which both components are observable. The most reliably determined masses for O stars to date, based exclusively on very accurate data from detached, double lined eclipsing binaries, are recently reviewed by Andersen (1991). Masses and radii have been estimated with an accuracy better than 2%, and there are only four O type stars in this sample of 45 Galactic binaries (90 single stars). The highest measured mass is $22.89 \pm 0.21\ M_{\odot}$ corresponding to the primary of EM Carinae, which is a star of type O8 V with an absolute magnitude of $M_V = -4.56$.

However, in practice, masses of early type stars are inferred from observable (photometric or spectroscopic) parameters involving the use of theoretical calculations. This method depends strongly on the value of the bolometric magnitude used for the star. An important result obtained from high resolution observations of the most luminous stars in the MCs is that the majority of the stars previously considered as most luminous or most massive (Humphreys 1983) are in fact tight clusters of O type stars (Heydari-Malayeri & Hutsmékers 1991b, Heydari-Malayeri et al. 1992). This means that classical photometry using apertures and usually low spatial resolutions has overestimated the luminosities. These results cast real doubt about the existence of VMSs in the MCs. Moreover, if spatial resolution is already so crucial for our nearest companion galaxies it will be even more decisive for other members of the Local Group, not to mention galaxies at greater distances. For these reasons, a mass estimate of over 1000 $M_{\odot}$ for the progenitor of SN 1961v in NGC 1058 (Utrobin 1984) seems questionable.

4 Mass estimates

4.1 Present mass

The actual mass of a star is well below its ZAMS mass because the star loses mass due to stellar winds, pulsations, or other processes. The present mass (PM) can be derived from the following relation (see, e.g., de Jager 1980):

$$\log(M/M_{\odot}) = 12.50 - 0.4(V - 3.1E(B-V) - DM + BC) - 4\ \log T_{eff} + \log g$$

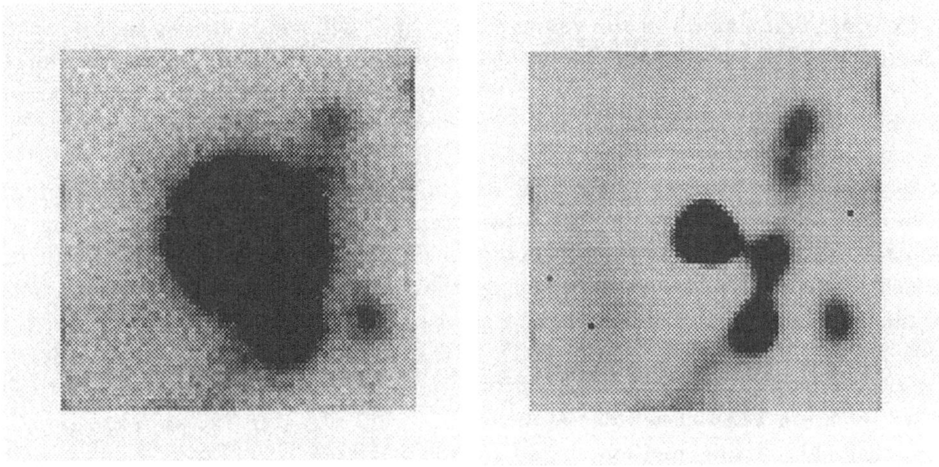

Fig. 1. Star #1 in the SMC NGC 346 observed with the ESO NTT through an R filter (left). Seeing 1″.0. Field 9″ × 9″. The same image after restoration by deconvolution (right). Resulting resolution 0″.55 (FWHM). North is at the top, east to the left.

The determination of temperature (T_{eff}) and gravity (g) has been the subject of controversy. They are usually derived from spectral types using an empirical calibration, or from fits to the profiles of H and He line via comparison with theoretical calculations (Kudritzki 1991). We have used a combination of both, i.e. we have taken T_{eff}s and gravities when available from the fitting method. Let us suppose that the errors for the distance moduli (DM), 18.5 and 18.9 for the Large and Small Clouds respectively (Westerlund 1990), are negligible. In order to show the sensitivity of mass estimates towards the M_V magnitudes used we consider two examples.

Table 1. Errors involved in mass estimates or how to create VMSs

Star	*Type*	V	$E(B-V)$	BC	$\log T_{eff}$ °K	$\log g$ cgs	PM $M_\odot$	L $L_\odot$	ZAMS M $M_\odot$
SMC									
NGC346#1a	O4 III(f)	12.43	0.08	-3.93	4.64	3.6	63	6.16	85
				5%			75	6.24	95
			100%				79	6.26	99
		0.3%					65	6.17	90
		0.3%	100%	5%			98	6.35	113
		0.3%	100%	5%	3%	25%	120	6.35	120
LMC									
HDE269676a	O5 If	12.2	0.11	-3.78	4.61	3.5	54	6.07	77
		6%					100	6.34	105
		6%	10%	5%	3%	25%	150	6.43	120

Consider first star #1 in the young cluster NGC 346 which excites the largest HII region in the SMC containing the hottest stars known in the SMC, including an O3 III(f) star. This star has been claimed to be the most massive star of the SMC with present and ZAMS masses 113^{+40}_{-29} and 130 ± 20 $M_\odot$ respectively (Kudritzki et al. 1989). High resolution observations have resolved star #1 into at least three components (Fig. 1); for details see Heydari-Malayeri & Hutsmékers 1991a. In Table 1, our estimates of the present masses (PMs) for each star, accurate within 30%, are given along with the parameters used. The figures accompanied with the "%" sign mean that if that particular parameter is overestimated by that amount, while all other parameters remain constant, the PM will get a new value, given in column 8. For instance, a 5% error in the bolometric correction (BC), in the case of NGC346#1a, is as important as a 100% error in $E(B-V)$. When all these overestimations conspire systematically, we will create a VMS of $\sim$ 120 $M_\odot$ instead of a 63 $M_\odot$ star, i.e. an overestimation of $\sim$ 90%. The fitting method, based on sophisticated NLTE treatment of stellar atmospheres, while it reproduces the line profiles perfectly, seems to overestimate the calculated fluxes, leading to higher values for L and M. At least in the case of the Galactic star HD 93129 the luminosity may be overestimated by about 60%. This is deduced from the fact that there is a good agreement between the radiation-driven wind method and the fitting method, and that the former overestimates M_V by 8% (Kudritzki & Hummer 1986).

HDE 269676 was believed to be one of the most massive stars of the LMC with a spectral type of O4–5 III(f). Here, the main reason for creating a VMS has been the multiplicity which has increased the V magnitude by 6%, thus enhancing the PM to 100 $M_\odot$. Fig. 2 shows that, from the viewpoint of photometry, HDE 269676 is resolved into 25 components! Spectroscopy has shown the presence of several O type stars in this cluster (Heydari-Malayeri & Hutsmékers 1991b), the most massive component being an O5 If. In Table 1, T_{eff} comes from Massey et al.'s (1989a) calibration and g is derived using the $\log g - \log T_{eff}$ diagram (Kudritzki 1991).

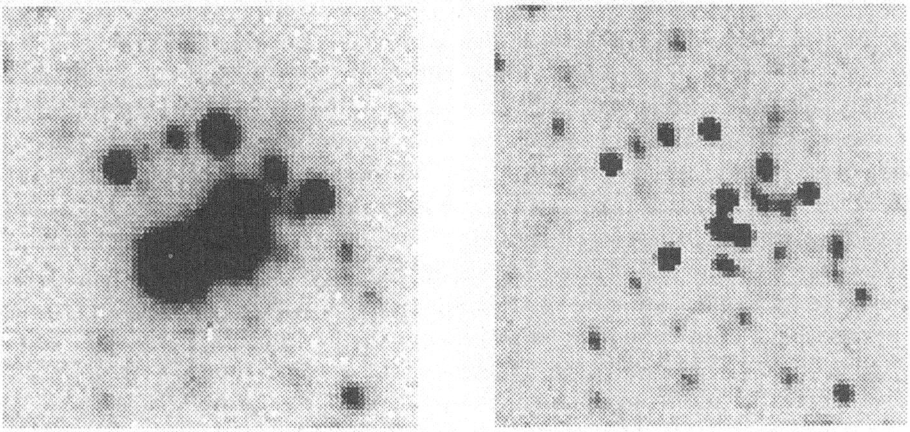

Fig. 2. HDE 269676 observed with the ESO NTT using an R filter (left). Seeing $0''.79$ (FWHM). Field $15'' \times 15''$. Restored image after deconvolution (right). Resulting resolution $0''.39$.

4.1.1 Bolometric corrections

BCs are a matter of controversy, and there are several attempts to calculate them. We have used the values given by Massey et al. (1989a) which are intermediate between those of Humphreys & McElroy (1984) and Schmidt-Kaler (1982) for hot stars. Accurate BCs should take into account the effects of wind blanketing, which heats the surface layers, and line blanketing, which causes them to cool. However, the size of these effects is not known at present (Kudritzki & Hummer 1986).

Even with Massey et al.'s BCs we may overestimate the masses. Let us apply the above-mentioned method to the four Galactic O stars in binary systems (EM Car, HD 97484, V 478 Cyg, HD 193611 of types O8 V and O9.5 V) which have the most accurately determined masses and for which accurate T_{eff} and g are available (Andersen 1991). Using Massey et al.'s BCs we get masses about 40–50% higher than those estimated on the basis of orbital elements! We trust the T_{eff} and g measurements; the discrepancy should therefore come from BCs. The Massey et al.'s BCs for these spectral types should be decrease by up to 15% in order to get the same results! This suggests that massive binaries are different with respect to isolated stars of the same spectral type, in particular they are less luminous.

4.2 Zero-age main sequence mass

The photometric and spectroscopic observations allow at the same time estimation of the ZAMS masses. This can be derived from T_{eff} and L:

$$\log(L/L_{\odot}) = 1.90 - 0.4(V - 3.1E(B - V) - DM + BC)$$

using the evolutionary tracks (Maeder 1990). Here also the use of accurate bolometric magnitudes is crucial. The results for the above-mentioned stars, 85 and 77 $M_{\odot}$ respectively, accurate within 15%, are also given in Table 1, along with values resulting from overestimations. We have used the new isochrones (Maeder 1990) calculated with metallicities (Z) appropriate for each Cloud, but we have not considered the involved inaccuracies. These models take into account Z-dependent mass loss and overshooting from convective cores. However, it should be underlined that the physical description of several key phenomena in stars (convection, mass loss, turbulent mixing, rotation, magnetic activity, etc.) are still highly simplified (Andersen 1991) and a physically acceptable theory of overshooting still does not exist (Trimble 1992, Baker & Kuhfuss 1987, Renzini 1987). Moreover, basic atomic data (nuclear reaction rates, opacities, detailed elemental and isotopic abundances) should be improved. It is known that overshooting increases the luminosity in model stars leading to higher evolved masses compared to models without overshooting.

4.3 Stellar wind method

An alternative method, based on the prediction of the theory of radiation driven winds, has recently been proposed by Kudritzki (1991). In this method the terminal velocity, v_{∞}, which is one of the most directly measurable quantities of the stellar winds, is used as an indicator for the mass. The basis for this procedure is the dependance of v_{∞} on T_{eff}, L, M and v_{esc} as predicted by the theory. Here also the magnitude from which L is derived should be well-determined. This method has given results comparable with those from model fittings (Kudritzki 1991). However, the predicted values of v_{∞} depend strongly upon one of the three force multiplier parameters, and using mean values is a severe oversimplification. Groenewegen et al. (1989) have compared the observed v_{∞}s with predicted values for radiation driven winds in which non-LTE effects and multiline scattering is taken into account. The predicted values are systematically higher than the observed ones by 40%, independent of luminosity class. Further confrontation of this method with observations of O type stars will undoubtedly teach us more about physical parameters of massive stars.

5 Massive star clustering

The resolution of the so-called VMSs into small clusters of lower mass stars indicate that massive stars tend to form in groups. The newly formed clusters are very tightly packed and later, due to dynamical interactions, the system disrupts. We can notice an age effect among the clusters so far detected. Sk–66°41 (Heydari-Malayeri et al. 1988) seems to be younger than HDE 269676 (Heydari-Malayeri & Hutsmékers 1991b) which itself is younger than HDE 269828 (Heydari-Malayeri et al. 1992). Spectroscopy of these clusters shows that they are composed of a large number of O type stars. HDE 269676 contains several O stars gathered in a region of $\sim$ 1 pc, while nine O stars have been detected towards HDE 269828. It is interesting that the latter cluster contains a number of Of and W–R stars. We can also mention the 30 Dor region where tens of hot O and W-R stars formed together ionize the core of this giant HII region.

These observational results are in line with numerical simulations carried out recently for explaining OB runaways as due to dynamical interaction of binaries (Leonard & Duncan 1989). The models suggest that O stars form preferentially in small groups of massive binaries containing $\leq$ 30 systems, and that the mass ratio in these systems is biased towards unity (Clarke & Pringle 1992). These authors argue that the IMF within these groups should be highly skewed towards the high mass compared to the field star IMF. Similarly, Larson (1990) and Zinnecker et al. (1992), on the basis of theoretical and observational considerations, suspect that massive s-tars can only form in a cluster. Note also that $\sim$ 40% of massive stars studied by Garmany et al. (1980) occur in binaries. The fraction will be higher if some of the single O stars are in fact runaways emanating from binary systems.

6 Extreme cases

There are a handful of cases where extremely high masses have been reported for stars. Speckle interferometry has resolved R136a, the central object of the giant HII region in the LMC once believed to have a mass of up to $\sim$ 2500 $M_\odot$ (Feitzinger et al. 1980, Cassinelli et al. 1981), into a dense cluster of at least eight stars packed within a diameter of $0''.7$ (Weigelt & Baier 1985). However, the most massive component of the group is suspected to have an upper limit of $\sim$ 250 $M_\odot$ (Walborn 1986). Recent high resolution observations of this object carried out using the FOC on board the Hubble Space Telescope have obtained spatial resolutions comparable with those reached by speckle techniques from the ground and have supported the previous mass estimate (Weigelt et al. 1991).

Another supermassive candidate is Melnick 42 lying in the R136 region less than $10''$ from the cluster center. Recently, based upon the GHRS observations on the Hubble Space Telescope and high resolution data, Heap et al. (1991) derived an effective temperature of 42500 °K, a present mass of $\sim$ 100 $M_\odot$ and an initial mass of $\sim$ 120 $M_\odot$ for this star to which a spectral type of O3f has been assigned (Melnick 1985). However, the models used are unable to reproduce the observed strengths and profiles of He II $\lambda\lambda$ 4686 and Hα. This is probably because, as the HST/GHRS observations by Walborn et al. (1992) have shown, Melnick 42 is in fact a hybrid O3If*/WN6-A object and therefore pure O-type models are unable to account for the observed features.

In any event, R136 is by far the most important site of violent starburst activity or collective star formation in the LMC. This means that R136a1 and Melnick 42 may be multiple stars. Let us suppose, for simplicity, that each of them is a binary system. Investigation of the orbital periods has shown that W-R binaries have mostly a period of several days (Moffat 1989 and references therein). If we assume an orbital period of 5 days and masses of 50 $M_{\odot}$ for each of the two stars constituting R136a1 or Melnick 42, the spatial separation (semi-major axis) between the components will be $\sim 5 \times 10^{-6}$ sec, i.e. out of reach of present and near future performances of the optical imaging. High resolution spectroscopic monitoring is therefore necessary in order to investigate the possible orbital motions in R136a1 and Melnick 42.

O3 stars are the most massive and hottest stars known. Only two O3 stars have been detected in the SMC and about 10 in the LMC. One of the SMC O3 stars lying in the NGC 346 starburst cluster is found to have a present mass of ~ 30 and a ZAMS mass of ~ 90 $M_{\odot}$ (Heydari-Malayeri & Hutsmékers 1991a). Likewise, the hottest Galactic stars are a number of O3 types gathered in the η Car starburst region. The O3 V component of HD 93205, which is the earliest known main sequence star in a binary system, was recently found to have a mass of ~ 70 $M_{\odot}$ (Moffat et al. 1990). Similarly, HD 93250, the hottest and the most massive star of this region, turns out to have a PM of ~ 75 $M_{\odot}$ using the observed physical parameters given by Kudritzki & Hummer (1986) and the *BC*s of Massey et al. (1989a).

7 Concluding remarks

The high quality observations obtained in recent years show a clear tendency towards smaller masses. In other words, we have witnessed an awful rate of mass loss for VMSs in these years! These observations have shown that the UML may be directly proportional to the "seeing" used! This trend has been confirmed by several independent research groups. For example, CCD photometry and spectroscopy of the LMC OB associations LH 9, 10, 117, 118 and the SMC NGC 346 have shown the absence of stars with an initial mass > 85 $M_{\odot}$ (Parker et al. 1992, Massey et al. 1989a,b). Note also that there is an apparent lack of stars with initial masses $> 60 M_{\odot}$ in M31 (Humphreys et al. 1990). The O3 V component of HD 93205 in the η Car starburst region, the earliest known main sequence star in a binary, is found to have a mass $\sim$ 70 $M_{\odot}$ (Moffat et al. 1990). Of course these observations do not exclude the existence of more massive stars, but underline their rarity, if they ever exist.

Should massive stars prefer to form in groups, as our observations show and theoretical calculations suggest, this will have serious implications for star formation models. We saw that massive stars in binaries seem to be less luminous and therefore less massive than their isolated counterparts. This seems to be due to an upper limit imposed on the mass of stars by fragmentation during the protostar stage. We may not be allowed to use for massive stars in groups the *BC*s derived for single stars. For these reasons, the current models devoted to formation and evolution of single stars may be subject to revision.

Recently, several attempts based on high quality CCD photometry have been devote to the investigation of the IMF in different regions of the MCs. Interestingly, these works show the same slope of -1.8 ± 0.2 for the OB associations LH 9, 117, 118, N159A and N59A in the LMC and NGC 346 in the SMC and a significantly different IMF for the LMC LH 10 (Massey et al. 1989a,b, Parker et al. 1992, Deharveng & Caplan 1992, Armand et al. 1992). These results would mean that the IMF is not universal and on the other hand it does not depend on metallicity.

The multiplicity of VMSs changes the IMF slightly, in the sense that previous estimates (Humphreys & McElroy 1984) have overestimated the number of stars with a mass $> 60M_\odot$ and at the same time have underestimated the stars in the interval $30-50M_\odot$. From the individual IMFs for the above-mentioned regions (except LH 10), we get a global, normalized IMF of $\xi(\log M) = 2.4(\mathrm{M}/\mathrm{M}_\odot)^{-1.8}$. In order to have an idea how multiplicity alters the slope, we assume that the above IMF overestimates the number of stars in the mass bins 50–70 and 70–90 $M_\odot$ by 10% and underestimates those in the range 30–50 $M_\odot$ by 10%. The result will be an IMF with a slope of about −2.0.

References

Andersen, J.: 1991, *A&AR* **3**, 91

Appenzeller, I.: 1970, *A&A* **9**, 216

Armand, C., Deharveng, L., Caplan, J.: 1992, *A&A*, in press

Assousa, G.E., Herbst, W.: 1980, in *Giant Molecular Clouds in the Galaxy*, ed. P.M. Solomon & M.G. Edmunds (New York: Pergamon), p. 275

Baker, N.H., Kuhfuss, R.: 1987, *A&A* **185**, 117

Boury, A., Gabriel, M., Ledoux, P.: 1964, *Ann. d'Astrophys.* **27**,

Cassinelli, J.P., Mathis, J.S., Savage, B.D.: 1981, *Science* **212**, 1497

Clarke, C.J., Pringle, J.E.: 1992, *MNRAS*, in press

Deharveng L., Caplan, J.: 1992 *A&A* **259**, 480

de Jager, C.: 1980, *The Brightest Stars*, Reidel, Dordrecht

D'Odorico, S., Baade, D., Kjär, K. (eds.): 1981, *The Most Massive Stars*, ESO Workshop, Garching, 23–25 Nov. 1981

Elmegreen, B.G., Lada, C.J.: 1977, *ApJ* **214**, 725

Feitzinger, J.V., Schlosser, W., Schmidt-Kaler, Th., Winkler, Ch.: 1980, *A&A* **40**, 50

Garmany, C.D., Conti, P.S.,Massey, P.: 1980, *ApJ* **242**, 1063

Groenewegen, M.A.T., Lamers, H.J.G.L.M., Pauldrach, A.W.A.: 1989, *A&A* **221**, 78

Heap, S.R. et al.: 1991, *ApJ* **377**, L29

Heydari-Malayeri, M., Magain, P., Remy, M.: 1988, *A&A* **201**, L41

Heydari-Malayeri, M., Magain, P., Remy, M.: 1989, *A&A* **222**, 41

Heydari-Malayeri, M., Hutsemékers, D.: 1991a, *A&A* **243**, 401

Heydari-Malayeri, M., Hutsemékers, D.: 1991b, *A&A* **244**, 64

Heydari-Malayeri, M., Melnick, J., Jorda, L.: 1992, *A&A*, in preparation

Humphreys, R.M.: 1983, *ApJ* **269**, 335

Humphreys, R.M., Massey, P., Freedman, W.L.: 1990, *AJ* **99**, 84

Humphreys, R.M., McElroy, D.B.: 1984, *ApJ* **284**, 565

Kahn, F.D.: 1974, *A&A* **37**, 149

Klapp, J.: 1983, *A Space Sci.* **93**, 313

Kudritzki, R.P., & Hummer, D.G.: 1986, *Luminous Stars and Association in Galaxies*, IAU Symp. **116**, eds. C.W.H. de Loore et al., Reidel, p. 3

Kudrtizki, R.P., Cabanne, M.L., Husfeld, D., Niemela, V.S., Groth, H.G., Puls, J., Herrero, A.: 1989, *A&A* **226**, 235

Kudritzki, R.P.: 1991, *Frontiers of Stellar Evolution*, McDonald Obs. 50th Anniversary, *ASP Conference Series* **20**, eds. D.L. Lambert, p.

Langer, N.: 1987, *Lecture Notes in Physics* **287**, eds. W. Hillebrandt et al., p. 180

Larson, R.B., Starrfield, S.: 1971, *A&A* **13**, 190

Larson, R.B.: 1990, in *Physical Processes in Fragmentation and Star Formation*, eds. R. Capuzzo-Dolcetta et al., Kluwer, p. 389

Larson, R.B.: 1991, in *Fragmentation of Molecular Clouds and Star Formation*, eds. E. Falgaronne et al., IAU Symp. 147, Kluwer, p. 261

Ledoux, P.: 1941, *ApJ* **94**, 537

Leonard P.J.T., Duncan, M.J.: 1989 *AJ* **99**, 608

Maeder, A.: 1990, *A&ASS* **84**, 139

Massey, P., Parker, J.W., Garmany, C.D.: 1989a, *AJ* **98**, 1305

Massey, P., Garmany, C.D., Silkey, M., Degioa-Eastwood, K.: 1989b, *AJ* **97**, 107

Mathis, J.S., Rumpl, W., Nordsieck, K.H.: 1977, *ApJ* **217**, 425

Melnick, J.: 1985, *A&A* **153**, 235

Moffat, A.F.J.: 1989, *ApJ* **347**, 373

Moffat, A.F.J., Drissen, L., Robert, C.: 1990, *Maasive Stars in Starbursts*, Workshop held at ST Sci, eds. C. Leitherer & N. Walborn, p. 40

Parker J. Wm., Garmany, C.D., Massey, P., Walborn, N.R.: 1992, *AJ* **103**, 1205

Renzini, A.: 1987, *A&A* **188**, 49

Schmidt-Kaler, T.: 1982, in *Landolt-Börnstein, New Series, Group VI*, vol. 26, eds. K. Schaifers & H.H. Voigt, Springer-Verlag, p.449

Schwarzschild, M., Härm, R.: 1959, *ApJ* **129**, 637

Stothers, R.: 1992, in press

Stothers, R., Simon, N.: 1970, *ApJ* **160**, 1019

Trimble, V.: 1992, *PASP* **104**, 1

Utrobin, V.P.: 1984, *A Space Sci.* **98**, 115

Walborn, N.R.: 1986, in *Luminous Stars and Associations in Galaxies*, IAU Symp. **116** eds. C.W.H. de Loore et al., Reidel, p. 185

Walborn, N.R., Ebbets, D.C., Parker J.Wm, Nichols-Bohlin, J., White, R.L.: 1992, *ApJ Letters*, in press

Weigelt, G., Baier, G.: 1985, *A&A*, **150**, L18

Weigelt, G. et al.: 1991, *ApJ* **378**, L21

Westerlund, B.E.: 1990, *A&AR* **2**, 29

Wolfire, M.G., Cassinelli, J.P.: 1987, *ApJ* **319**, 850

Woodward, P.R.: 1976, *ApJ* **207**, 484

Ziebarth, K.: 1970, *ApJ* **162**, 947

Zinnecker, H., McCaughrean, M.J., Wilking, B.A.: 1992, in *Protostars and Planets III*, eds. E.H. Levy et al., in press

Dusty B[e] Stars

F.-J. Zickgraf

Landessternwarte Königstuhl, D-6900 Heidelberg, Fed. Rep. of Germany

Abstract: Observations of the dusty B[e] stars in the Magellanic Clouds, the so-called B[e] supergiants, provide evidence that these stars are characterized by axially symmetric stellar winds. A two-component wind model was suggested consisting of a normal OB star wind in the polar region, and a cool and slowly expanding excretion disk in the equatorial region. The dust formation appears to take place in the cool and dense excretion disk.

1 Introduction

In the Magellanic Clouds (MCs) a surprisingly large number of early-type emission-line stars in the upper part of the H-R diagram show an infrared excess which due to its wavelength dependence is generally ascribed to the presence of hot circumstellar dust (Allen and Glass 1976, Glass 1977, Stahl et al. 1983, 1984, McGregor et al. 1988). Zickgraf et al. (1986) called these peculiar stars *B[e] supergiants.* The fact that the majority of early-type supergiants do not possess an IR excess caused by hot dust shows that the circumstellar environment of the B[e] supergiants provides the specific conditions with regard to temperature and density required for the formation of dust grains. Spectroscopically the B[e] supergiants are distinguished by extremely strong Balmer emission lines and the presence of permitted and forbidden emission lines of low-excitation species like FeII, [FeII], NaI, and in all known cases, [OI].

The analysis of the spectroscopic observations in the optical wavelength region and in the satellite UV lead to the so-called two-component stellar wind model (Zickgraf et al. 1985, 1986, 1989). In the following the observations suggesting this model will be summarized. The model will shortly be described and the physical conditions in the different wind components will be discussed.

2 IR Observations

The infrared excess provides an efficient tool to identify B[e] supergiants. In a (J-H)-(H-K) diagram they are clearly separated from normal OB supergiants and the related Luminous Blue Variables (LBVs) which show an IR excess due to $(f-f)$–$(f-b)$ emission from their stellar winds (Fig. 1). Binary systems with late-type companions are also well separated.

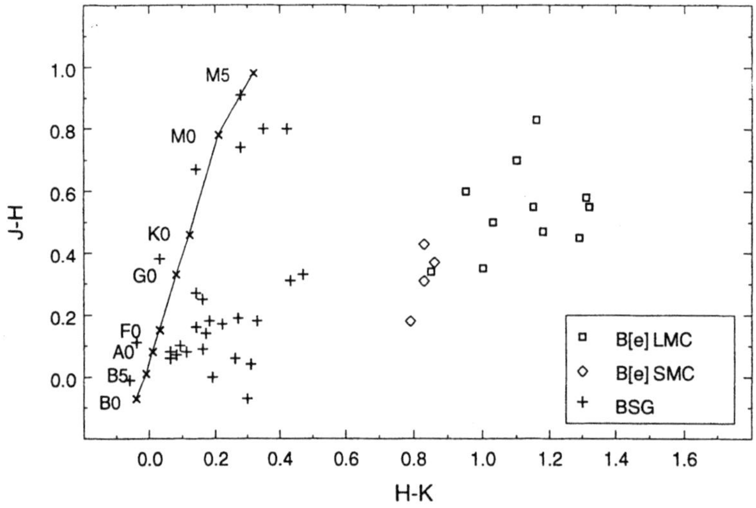

Fig. 1. (J-H)-(H-K) diagram showing the loci of the dusty B[e] supergiants. Also included are other emission-line stars in the MCs (BSG); the measurement were taken from Stahl et al. (1985). The stars in the lower left corner show an IR excess due to $(f-f)-(f-b)$ emission from stellar winds. In the upper left part stars with late-type companions are found. The solid line represents the locus of normal supergiants taken from Koornneef (1983).

The IR dust excess was used in September 1989 and November 1991 for a systematic search for B[e] supergiants in the SMC and LMC, respectively, in order to find all stars of this class in the two galaxies. The observations were carried out at ESO. We studied two samples of early-type stars with strong Hα emission lines collected from various emission-line star catalogs. The result of these IR surveys was the detection of one new object in the SMC (Zickgraf et al. 1992a) and four new B[e] supergiants in the LMC (Zickgraf et al. 1992b). Thus a total of 15 B[e] supergiants are now known in the MCs, 4 in the SMC and 11 in the LMC. Because the samples are fairly complete down to V $\approx$ 14 mag and 13 mag in the SMC and LMC, respectively, these 15 objects very likely represent the entire B[e] supergiant population in the MCs for the given brightness limits. Table 1 lists the 15 B[e] supergiants together with relevant references. More references can be found in the quoted literature.

Dust condensation is expected to result from the formation of molecules. IR spectroscopy in the 2.3μm region by McGregor et al. (1988) showed the presence of first overtone emission bands of CO molecules in several objects. Zickgraf et al. (1989) found a broad emission feature at 6159 Å which they ascribed to TiO molecules.

These observations show that in the circumstellar environment of the B[e] supergiants a region is present cool and dense enough to allow the formation of molecules and dust. This is not encountered in a normal OB type wind. For the CO emission region McGregor et al. (1988) estimated densities of $N_H \geq 10^{10}$ cm^{-3} at temperatures of $T_{ex} \approx 3000 - 5000$ K.

Table 1. B[e] supergiants in the Magellanic Clouds

Star	Reference	Star	Reference
R 4/SMC	2, 4	R 66/LMC	1, 4
R 50/SMC	4	R 82/LMC	4
Hen S18/SMC	5, 10	R 126/LMC	3
AV 172/SMC	6	Hen S12/LMC	4
Hen S22/LMC	4, 9	Hen S35/LMC	8, 12
Hen S59/LMC	8, 12	Hen S93/LMC	8, 12
Hen S111/LMC	7, 13	Hen S134/LMC	3, 11
Hen S137/LMC	8, 12		

References: (1) Stahl et al. (1983), (2) Stahl et al. (1984), (3) Zickgraf et al. (1985), (4) Zickgraf et al. (1986), (5) Zickgraf et al. (1989), (6) Zickgraf et al. (1992a), (7) McGregor et al. (1989) (8) Zickgraf et al. (1992b) (9) Bensammar et al. 1983, (10) Shore et al. (1987), (11) Shore, S.N. and Sanduleak, N. (1983), (12) Henize (1956), (13) Stahl et al. (1989)

3 The Two-Component Wind Model

3.1 Hybrid Spectra

A large fraction of the B[e] supergiants shows *hybrid* spectra. This term describes the simultaneous presence of narrow low-excitation emission lines and broad high-excitation absorption lines. An example is R126. The spectrum in the satellite UV is dominated by resonance absorption features of SiIV, CIV, and NV. The terminal wind velocity is about 1600 km s^{-1}. In the optical wavelength region very narrow low-excitation emission lines are present mainly of FeII, [FeII], NaI and [OI] with line widths of about 20 km s^{-1} (Zickgraf et al. 1985). Hen S134 and Hen S12 show similar spectral characteristics (Zickgraf et al. 1986). Hen S18 also shows very narrow emission lines of FeI, FeII, NaI, [FeII], and [OI] which are contrasted by P Cygni absorption components of HI and HeI lines indicating wind velocities of 750 km s^{-1} (Zickgraf et al. 1989).

These observations suggest that line forming regions with very different physical conditions must exist around the B[e] supergiants. The two-component wind model by

Zickgraf et al. (1985) can explain the hybrid nature of the spectra and also accounts for the observed infrared properties. The model consists of a hot and fast line-driven normal OB star wind in the polar region and a dense, cool and slowly expanding wind in the equatorial zone forming the so-called excretion disk. The high excitation lines originate in the polar wind. The low-excitation emission lines are formed in the excretion disk. The disk also is the site of dust formation (see below). Polarization measurements of B[e] supergiants in the MCs by Magalhaes (1992) confirmed the model. Also the polarization measurements of related stars in the Milky Way by Zickgraf and Schulte-Ladbeck (1989) are in agreement with the two-component model.

3.2 The Excretion Disk Wind

Hybrid spectra are only observed if the line-of-sight intersects the polar wind. This is the case if the particular star is viewed more or less pole-on. From a statistal point of view one would also expect a certain fraction of stars to show absorption features originating in the slow and cool excretion disk wind, the fraction depending on the extension of the proposed disk. From the number of stars showing edge-on and pole-on characteristics Zickgraf (1989) estimated the solid angle covered by the disk to about 20 to 30% of the full sphere, i.e. the opening angle is of the order of 20° to 40°.

Whereas for pole-on stars the wind velocity in the excretion disk can only be estimated from emission-line widths of the low-excitation species, UV spectroscopy of edge-on seen objects would allow to directly measure the expansion velocity of the slow equatorial wind. In December 1991 we obtained IUE high resolution spectra in the LWP range of three B[e] supergiants for which Zickgraf et al. (1986) had inferred from optical spectra that they are edge-on cases. We observed Hen S22 and R82 in the LMC, and R50 in the SMC. The LWP spectra are dominated by lines of singly ionized metals, mostly of FeII and CrII. MgII(1) shows strong emission components. Sections of the spectra are shown in Fig. 2. In Hen S22 the FeII lines exhibit pronounced P Cygni profiles. R 50 also shows FeII lines with P Cygni profiles, the absorption components, however, are less clear than in Hen S22. R82 shows FeII lines mainly in absorption. Only the lines of multiplets 62, 63 and 78 also show clearly visible emission components. The radial velocities measured from the lines centers and the blue absorption edges of the FeII lines indicate very low wind velocities. Mean expansion velocities of $v_{exp} = -58, -36$ and -27 km s^{-1} and terminal wind velocities of $v_\infty = -113, -77$, and -63 km s^{-1} were obtained for Hen S22, R82, and R50, respectively. Note, that the terminal wind velocities are about a factor of ten lower than expected from the spectral types which are B0 for Hen S22 and B2-3 for R82 and R50, respectively (Zickgraf et al. 1986).

Assuming a constant mass flux density over the entire surface of a B[e] star the velocity ratio of 10 between the equatorial and polar wind component leads to a density enhancement in the excretion disk of a factor of 10. Zickgraf et al. (1989) derived even higher density ratios of the order of 10^2 to 10^3 due to the larger mass flux densities they obtained for the equatorial disk. The densities derived for the disk are typically 10^9 to 10^{10} cm^{-3}. The temperature in the disk is about 5000 K (Zickgraf et al. 1986). These conditions are practically the same as those obtained by McGregor et al. (1988) from the CO emission. The excretion disk therefore seems to be the site of dust formation.

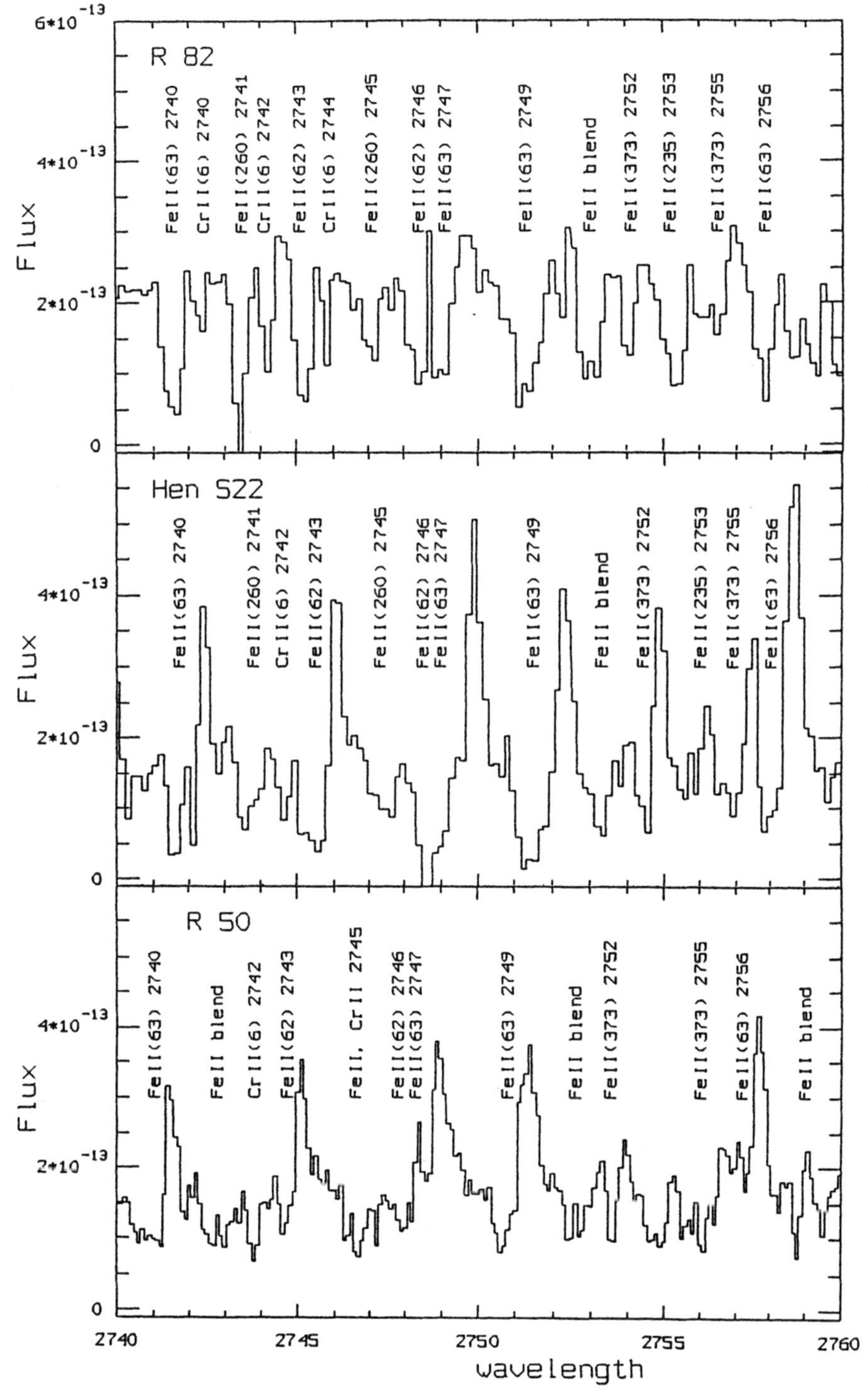

Fig. 2. Sections of the IUE LWP spectra of R82, Hen S22, and R50. These stars are most likely viewed edge-on. The spectra are dominated by lines of singly ionized iron. With -60 to -113 km s^{-1} the terminal wind velocities determined from the blue absorption edges are about 10 times lower than expected for normal early B-type supergiants.

4 Conclusions

The observations of the dusty B[e] stars in the MCs strongly suggest the presence of a two-component stellar wind. This model not only can explain the hybrid spectra observed in many B[e] supergiants, but also provides the setup for dust formation around these stars and is furthermore in agreement with polarization measurements. Whether stellar rotation is responsible for the formation of the excretion disk as suggested by Zickgraf et al. (1985) is not clear.

The recent IR surveys allow to estimate the duration of the B[e] phase by comparing the number of B[e] supergiants with the number of main-sequence stars in the same luminosity interval. Assuming that all massive O stars undergo a B[e] phase during their post-main sequence evolution Zickgraf et al. (1992a) obtained a lower limit for the duration of the B[e] phase of about 10^5 years. With mass loss rates of the order of 10^{-5} to 10^{-4} $M_{\odot}yr^{-1}$ as determined e.g. by Stahl et al. (1983), Shore and Sanduleak (1983), and Zickgraf et al. (1989) the total mass lost during the B[e] phase could be as much as several $M_{\odot}$. Such a large mass loss would strongly affect the post-main sequence evolution of massive stars.

References

Allen, D.A., Glass, I.S. (1976): ApJ 210, 666
Bensammar, S., Friedjung, M., Muratorio, G., Viotti, R. (1983): A&A 126, 427
Glass, I.S. (1977) MNRAS 178, 9P
Henize, K.G. (1956): ApJ Suppl. 2, 315
Koornneef, J. (1983) A&A 128, 84
Magalhaes, A.M. (1992): ApJ, in press
McGregor, P.J., Hillier, D.J., Hyland, A.R. (1988): ApJ 334, 639
Shore, S.N., Sanduleak, N. (1983): ApJ 273, 177
Shore, S.N., Sanduleak, N., Allen, D.A. (1987): A&A 176, 59
Stahl, O., Wolf, B., Zickgraf, F.-J., Bastian, U., de Groot, M., Leitherer, C. (1983) A&A 120, 287
Stahl, O., Leitherer, C., Wolf, B., Zickgraf, F.-J. (1984): A&A 131, L5
Stahl, O., Wolf, B., de Groot, M., Leitherer, C. (1985): A&A Suppl. 61, 237
Stahl, O., Smolinski, J., Wolf, B., Zickgraf, F.-J. (1989): in *Physics of Luminous Blue Variables*, eds. K. Davidson, A.F.J. Moffat, H.J.G.L.M. Lamers, Kluwer Academic Publishers, Dordrecht, p. 295
Zickgraf, F.-J. (1989): in *Physics of Luminous Blue Variables*, eds. K. Davidson, A.F.J. Moffat, H.J.G.L.M. Lamers, Kluwer Academic Publishers, Dordrecht, p.117
Zickgraf, F.-J., Schulte-Ladbeck, R.E. (1989) A&A 214, 274
Zickgraf, F.-J., Wolf, B., Stahl, O., Leitherer, C., Klare, G. (1985): A&A 143, 421
Zickgraf, F.-J., Wolf, B., Stahl, O., Leitherer, C., Appenzeller, I. (1986): A&A 163, 119
Zickgraf, F.-J., Wolf, B., Stahl, O., Humphreys, R.M. (1989): A&A 220, 206
Zickgraf, F.-J., Wolf, B., Stahl, O. (1992a): A&A 260,205
Zickgraf, F.-J., Wolf, B., Stahl, O. (1992b): A&A, in preparation

Luminous Blue Variables of the Magellanic Clouds

Otmar Stahl [1]

[1]Landessternwarte Heidelberg, Königstuhl

Abstract: We briefly review the properties of the LBVs of the MCs and discuss some new results about Luminous Blue Variables. Special emphasis is given to the recently discovered inverse P Cyg absorption in S Dor and R 127, the peculiar LBV R 110 of the LMC and to R 40, the first LBV discovered in the SMC.

1 Introduction

LBVs are the visually brightest stars in galaxies. They show outbursts with an amplitude of typically about 1 ... 2 mag in the visual range. Some stars have smaller or larger amplitudes. An extreme case is represented by η Car. The timescale of the outbursts is of the order of years or decades. In visual mimimum phase they are located close to the upper limit of the HRD, the Humphreys-Davidson limit (Humphreys and Davidson, 1979). During outburst, they normally show A-star spectra with indications for a very strong stellar wind with a mass-loss rate of several times $10^{-5} M_{\odot}$ yr^{-1} and a wind velocity of only about 100 km sec^{-1}. The luminosity in general does not change during the outburst, that means that the variations in the optical are due to changes in the bolometric correction.

Most LBVs are located above the upper luminosity limit for red supergiants, i.e. they probably do never evolve to red supergiants (RSG). The stars below this luminosity may be in a post-RSG phase, as indicated by their small masses. The LBVs probably are in an enhanced evolutionary state preceding the WR-phase. Some objects are even spectroscopically intermediate between Of and WR stars.

This is the standard picture of LBVs. For more details about LBVs see e.g. the Proceedings of IAU Coll. 113 on "Physics of Luminous Blue Variables", edited by Davidson, Moffat and Lamers (1989) or the recent reviews by Stahl (1990) and Wolf (1990). In the following, several new results, obtained by monitoring LBVs and candidate LBVs at ESO, will be discussed.

2 Inverse P Cygni absorption

Absorption redward of the emission peak was observed for the first time in an LBV by Stahl and Wolf (1982) in the case of S Dor. Later Wolf and Stahl (1990) reported the observation of a phase with particularly strong inverse P Cygni profiles in this star. Further monitoring has shown that this phase of strong inverse P Cygni absorption has lasted for about two years. Now S Dor is again in a normal state showing only P Cygni profiles, but no red-shifted absorption. The spectral evolution of S Dor in the last years can be seen in Fig. 1.

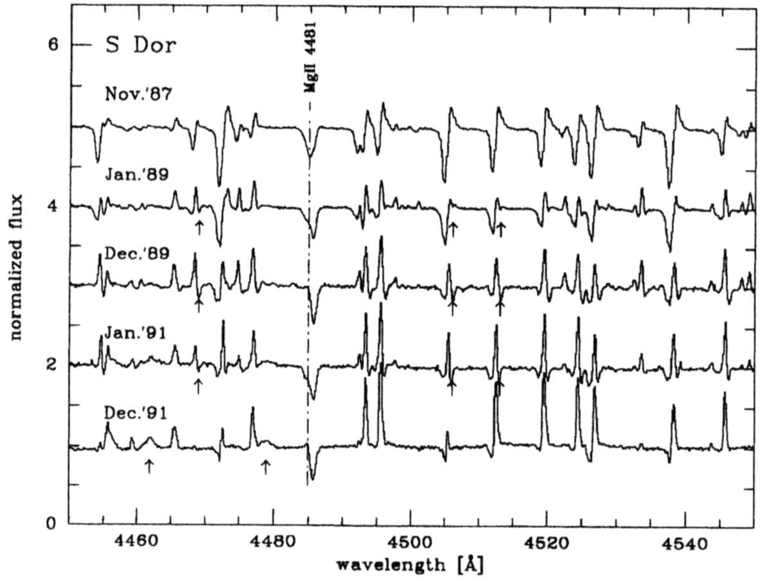

Fig. 1. Evolution of the spectrum of S Dor as observed with the ESO-CASPEC spectrograph during the last years. Note the inverse P Cygni absorption from Jan.'89 until Jan.'91. A few lines showing the inverse P Cygni profiles are indicated with arrows. Some [FeII]-lines appearing in Dec.'91 are also marked with arrows.

The inverse P Cyg absorption is most obvious in weak lines. Stronger lines sometimes show both inverse and normal (blue-shifted) P Cyg absorption components. Very strong lines show only the blue-shifted component.

In Fig. 2, we show the light curve, taken within the LTPV program at ESO, in the same time interval. It can be seen from this plot, that the inverse P Cygni components appear only during the fading of the star, after the brightest phase ever observed for S Dor.

We have now observed a similar episode in the LBV R 127 of the LMC. Inverse P Cyg absorption appeared in Jan. '91 (see Fig. 3) and disappeared again before Dec. '91. In this case, the timescale of the phenomenon is not well known, and the inverse absorption was less pronounced. However, if we compare the data with the

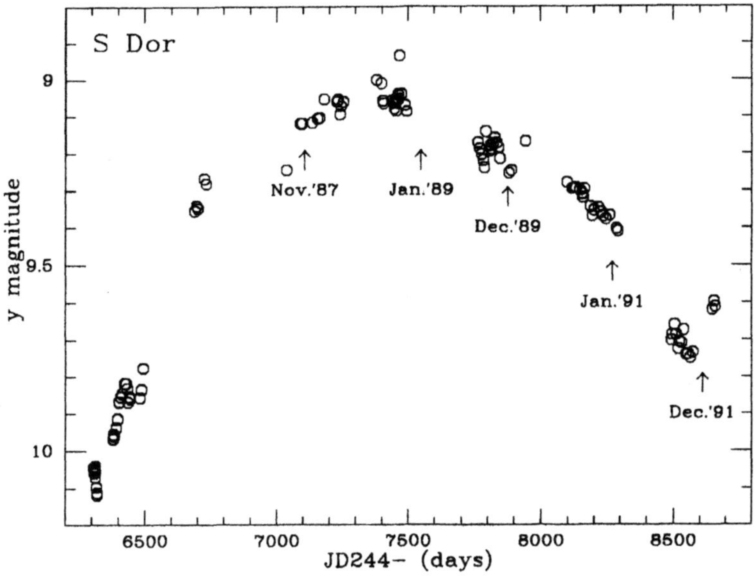

Fig. 2. The lightcurce of S Dor in the Strömgren y filter as obtained by the LTPV group at ESO, La Silla. Note the decreasing brightness during the phase of inverse P Cygni absorption.

LTPV photometry during the same interval (see Fig. 4), it appears that, as in the case of S Dor, the visual brightness was slightly decreasing during the phase where the inverse absorption appeared.

This suggests that the decreasing brightness and the appearance of inverse P Cyg absorption are related. A possible explanation for this relation could be, that after a large outburst the star can no longer support its stellar wind and part of the stellar wind is falling back to the star. Only the inner part of the wind, where the expansion velocity is below the escape velocity from the star, can fall back. This explains why mainly weak lines show the phenomenon. The stronger lines, which form in the outer part of the wind, show still the expansion of the wind.

Since the infall timescale is of the order of months only, it is difficult to explain in this picture, however, that the infall phase lasted for some two years in S Dor. Possibly more complicated (non-spherical) models have to be invoked to explain this behaviour.

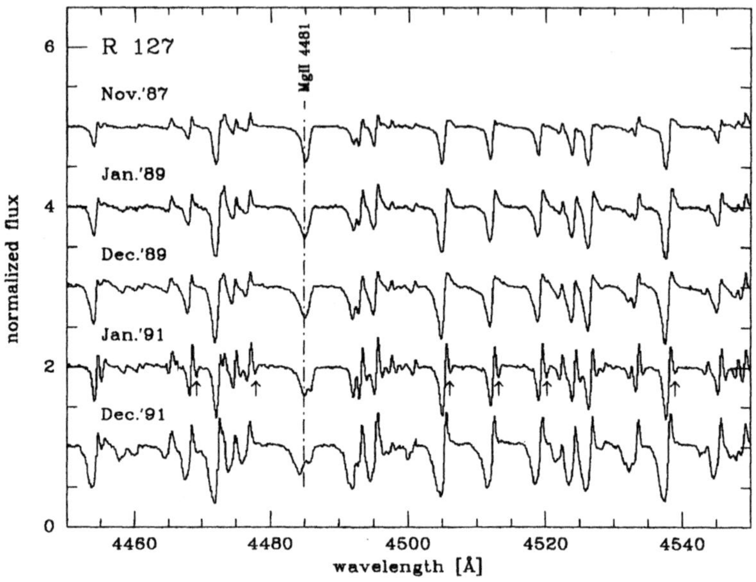

Fig. 3. Same as Fig. 1 but for R 127, the most luminous LBV of the LMC. Note the appearance of inverse P Cygni absorption in Jan.'91.

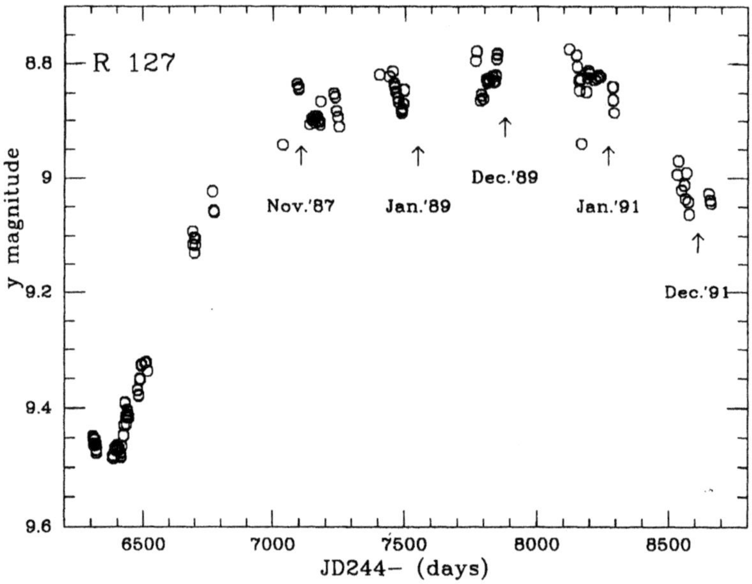

Fig. 4. Same as Fig. 2 but for R 127. Note the small decrease of brightness during the phase of the appearance of inverse P Cygni absorption.

3 The unique object R 110

R 110 was suspected of variability by Stahl et al. (1984) and Stahl et al. (1985). Later is was observed by Stahl et al. (1990) to be in a particularly cool state with a spectral type around F0. Possibly the strongest peculiarity of R 110 is, that its spectrum is so normal. It does not show signs of enhanced mass-loss or other strong spectral peculiarities.

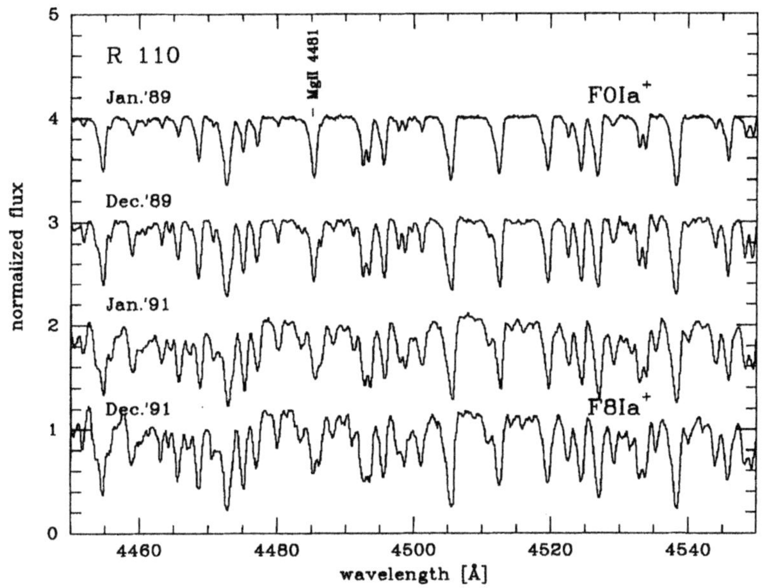

Fig. 5. Same as Fig. 1 but for R 110, the coolest LBV of the MCs.

From this time on, the star has been monitored by the LTPV group at ESO and it has been observed several times with CASPEC. The recent spectral evolution of R 110 can be seen from Fig. 5. During this time interval, R 110 evolved further to the red in the HRD. The LTPV photometry is shown in Fig. 6. As can be seen, the star was almost constant in the visual, but decreased strongly in the blue and UV. The brightness in the infrared increased during the same time. Spectroscopically, the star is now close to F8-G0, as can be seen by comparison with spectroscopic standards. Of course, it shows the spectroscopic signatures of very high luminosity and so does not match the MK standards, but it is virtually indistinguishable from the brightest yellow supergiants of the LMC (see Fig. 7). It should be noted also, that the luminosity of R 110 is even slightly below the luminosity of the brightest yellow supergiants.

R 110 is in a region of the HRD, where also the ρ Cas variables and the LBV Var. A in M 33 are found. These stars are variable yellow supergiants which show occasional excursions to the red. We compared R 110 with ρ Cas and the similar supergiant HR 8752 (see Fig. 8). The spectroscopic match is very good, but it should

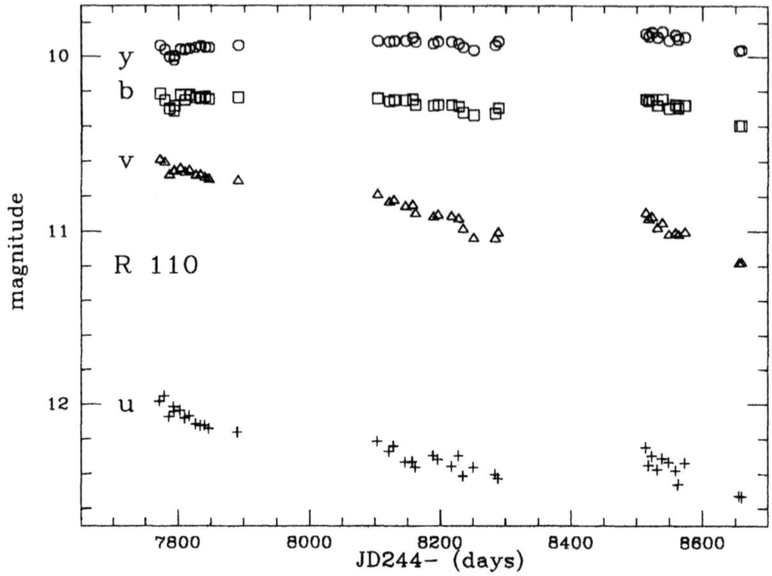

Fig. 6. The lightcurve of R 110 in the Strömgren *uvby* system. Note the strong reddening of R 110, the coolest LBV of the MCs

be emphasized that R 110 is the only one of these stars which was observed to be a blue star previously. It remains to be seen whether R 110 further evolves to the red.

R 110 is an important object for outburst models of LBVs. Since most LBVs stop the redward evolution at about 8000 K, this temperature has been considered as a limit for the outburst (see Davidson, 1987). The star R 110 shows that this limit can be crossed at least by some stars.

4 The LBV R 40 of the SMC

R 40 was suspected by Stahl et al. (1984) as a photometric and spectroscopic variable. It was subsequently monitored by the LTPV group at ESO and turned out to be the first LBV in the SMC (see Szeifert et al. in these proceedings).

R 40 does belong to the group of LBVs below the upper luminosity limit for RSGs. This group, although small, does contain very different objects. R 71, e.g., is very similar to the classical LBVs like S Dor, whereas e.g. R 110 seems more related to the ρ Cas variables. It is therefore very dangerous to attribute the properties of R 40 to the lower metallicity of the SMC. It is, however, already an interesting finding that the low metallity of the SMC does not prevent LBV outbursts. This is important for outburst models of LBVs. If opacity effects of metal lines play the dominant role, some effect of metallicity is to be expected.

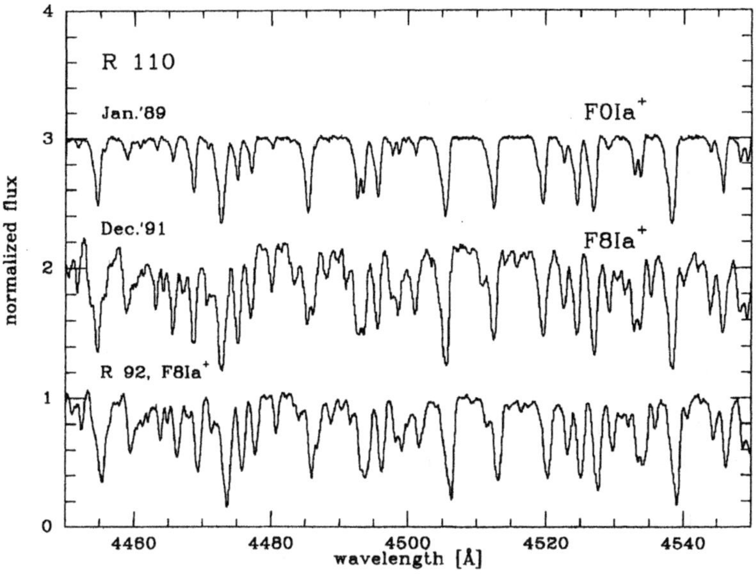

Fig. 7. Comparison of the spectrum of R 110 in two phases with the yellow hypergiant R 92, which is classified as F8Ia^{+}.

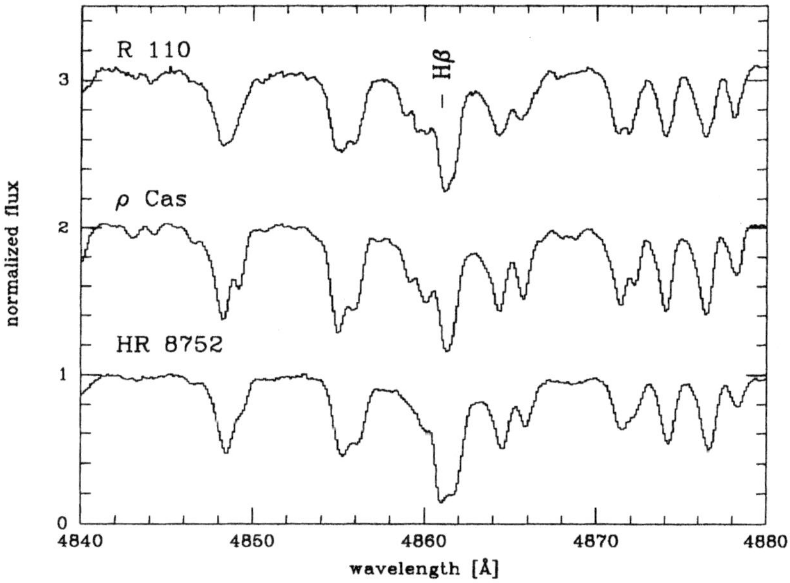

Fig. 8. Comparison of the spectrum of R 110 with ρ Cas and HR 8752. The spectra of ρ Cas and HR 8752 have been obtained at Heidelberg with our fiber-linked echelle spectrograph.

5 Conclusion

The observations discussed above show that LBVs, although better understood now than a few years ago, still offer many surprises. They also demonstrate the need for continuous monitoring of LBVs over long time-scales.

References

Davidson, K. (1987): *Astrophys. J.* **317**, 760

Davidson, K., Moffat, A.F.J., Lamers, H.J.G.L.M., eds. (1989): *Physics of Luminous Blue Variables*, Proc. IAU Col. 113, Kluwer

Humphreys, R. M., Davidson, K. (1979) *Astrophys. J.* **232**, 409

Stahl, O., Wolf, B. (1982): *Astron. Astrophys.* **110**, 272

Stahl, O., Wolf, B., Leitherer, C., Zickgraf, F.-J., Krautter, J., de Groot, M. (1984):, *Astron. Astrophys.* **140**, 459

Stahl, O., Wolf, B., de Groot, M., Leitherer, C. (1985): *Astron. Astrophys. Suppl.* **61**, 237

Stahl, O., Wolf, B., Klare, G., Jüttner, A., Cassatella, A. (1990): *Astron. Astrophys.* **228**, 379

Stahl, O. (1990): Reviews in Modern Astronomy, **3**, ed. G. Klare, Springer, p. 286

Wolf, B., Stahl, O. (1990): *Astron. Astrophys.* **235**, 340

Wolf, B. (1990): Evolution in Astrophysics, ESA SP-310, 33

Be STARS IN MAGELLANIC CLOUD CLUSTERS

Michael S. BESSELL and Peter R. WOOD
Mount Stromlo and Siding Spring Observatories
Private Bag, Weston Creek P.O.
Canberra, ACT 2611 Australia

Be stars have been identified in the Magellanic Cloud clusters NGC 330, NGC1818, NGC2004 and NGC 2100 by comparing H α and R images. When the colour H α - R is plotted against V-I, the Hα objects stand out very clearly with only a very small color term (Fig 1). Be stars had previously been discovered in NGC 330 by Feast (1972) and Grebel, Richtler and de Boer (1991). The Be stars tend to be significantly redder in V-I than the other B stars but there is little difference in B-V; in U-B the Be stars are bluer than normal. In order to investigate these effects we have synthesised the UBVI colors of Be stars in h and k Persei using their published energy distributions. It seems that the Be stars have additional light in both the U and I passbands due to hydrogen recombination radiation from an associated equatorial disk of material.

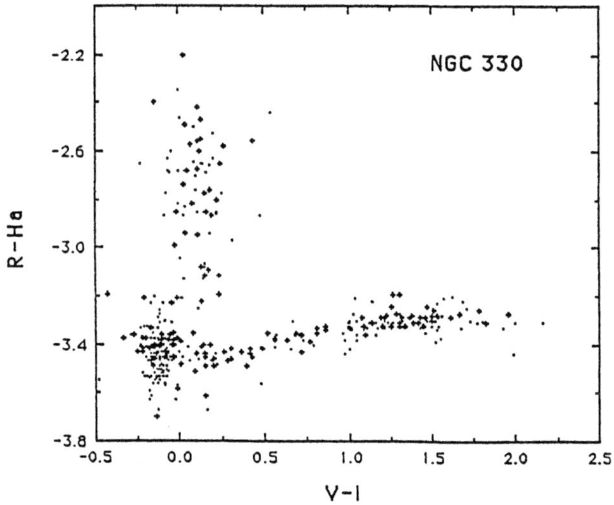

Fig 1. R-Hα versus V-I for the NGC 330 cluster and field. Dots are stars fainter than 16 mag.

We assume that Be stars are those stars in which rotationally induced mass-loss at the equator produces an extended disk of material giving rise to the Be features, namely Hα emission and the IR and UV excess. During hydrogen- burning main-sequence evolution of O and B stars, assuming conservation of angular momentum, the rotational velocity of the star initially increases (the star spins up), reaches a maximum, then decreases again before core collapse when there will be a final brief spin-up. Schild (1966) discusses such a scenario derived from work by Crampin and Hoyle (1960) for a 4M$_\odot$ star. The results should scale homologously to higher masses.

It follows in this scenario that if a group a stars arrives on the zero-age main-sequence with a distribution of rotational velocities, then as the stars evolve from the ZAMS we would expect the proportion of Be stars to increase as the stars evolve and spin-up. Beyond the point of maximum rotation, the Be star phenomena should vanish. The luminosity which marks the maximum in the proportion of Be stars will lie below the tip of the main-sequence turn-off. Indeed we find that in the four Cloud clusters considered (Fig 2) and in h and k Persei, Be stars occur at all luminosities below a certain absolute magnitude and that the proportion of Be stars increases with luminosity, reaching a maximum at about Mv = - 4. The main-sequence termination is about 1 magnitude brighter and the few blue stars brighter than -5 in these clusters we identify as He burning supergiants. The higher proportion of Be stars in the SMC cluster NGC 330 presumably reflects the higher rotation of its pre-cluster material.

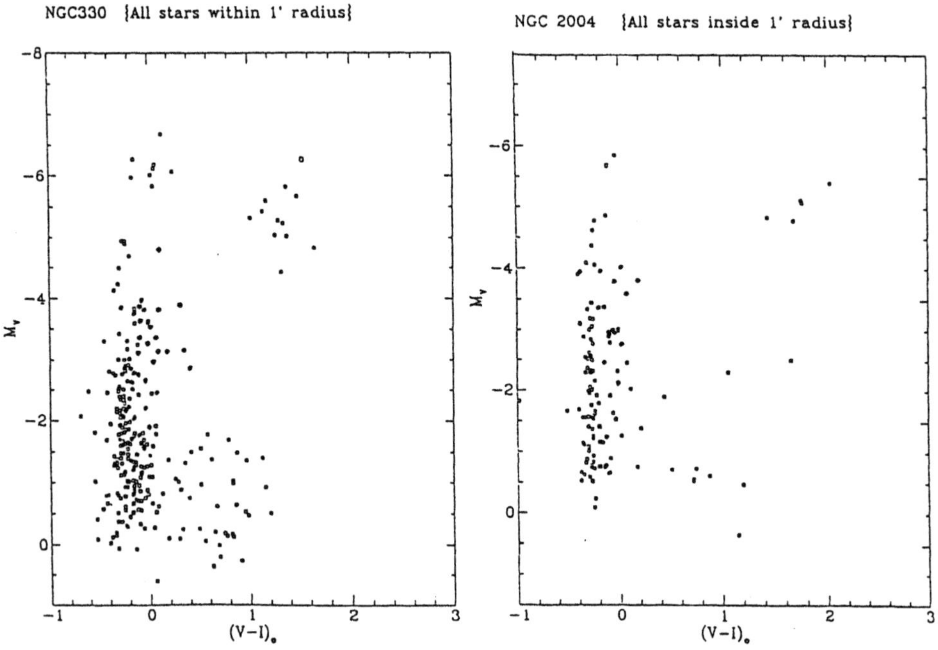

Fig 2. M_V versus V-I diagram for two of the clusters. Be stars are plotted as solid symbols.

Comparing this picture of the main-sequence with standard evolutionary tracks we find that the observed main-sequence turn-off is about 5000K cooler than predicted (20000K rather than 25000K). This has been known for some time, but has been ignored in an obfuscation of difficulties with reddening and association of the brightest main-sequence stars as He burning stars. Often a large range of masses and/or ages has been invoked to explain the large range in luminosities of the blue stars. A simpler explanation results from adoption of the Doom (1985) evolutionary tracks for stars of about 15$M_\odot$. These models are more centrally concentrated (thus aiding spin-up) and evolve toward higher luminosities and smaller effective temperatures than the standard models.

References

Crampin, J. & Hoyle, F. 1960 MNRAS 120, 33.
Doom, C. 1985 A&Ap 142, 143.
Feast, M.W. 1972 MNRAS 159, 113.
Grebel, E.K., Richtler, T. & de Boer, K.S. 1992 A&Ap 254, L5.
Schild, R. 1966 ApJ 146, 142.

Narrow band coronographic imaging of the bipolar nebula around the LBV R127

Mark Clampin [1 2], Antonella Nota [1 3], David A. Golimowski [2],

Claus Leitherer [1 3]

[1]Space Telescope Science Institute, operated by AURA, Inc. under contract with the National Aeronautics and Space Administration, Baltimore.
[2]Center for Astrophysical Sciences, The Johns Hopkins University, Baltimore.
[3]Affiliated with the Astrophysics Division, Space Science Department, ESA.

Abstract: New high resolution narrow band coronographic images of the R127 nebula have been recently obtained. The nebula displays a bipolar morphology and is similar in appearance to the nebula around AG Carinae. The observations improve the values for the linear dimensions (1.9x2.2 pc) and yield an estimated nebular mass of 3.1 $M_{\odot}$.

R127, like a number of other Luminous Blue Variables, shows strong, extended nebular emission lines suggesting the presence of an expanding circumstellar shell (Walborn 1982; Stahl and Wolf 1986). Stahl (1987) attempted direct imaging of the nebula by means of narrowband imaging and detected an excess in the stellar PSF at radii of 2 arcsec - 4 arcsec. Appenzeller *et al.*, (1987) also resolved nebular structure in the spectroscopic line profiles.

We obtained narrow band coronographic observations of R127 at the Las Campanas observatory on the 2.5 m DuPont telescope during the night of the 26th December 1991 using the Johns Hopkins Adaptive Optics Coronograph (AOC), in the attempt to resolve the nebular structures and to study their morphology. The AOC incorporates adaptive optics to improve angular resolution during an observation (Clampin *et al.* 1991; Golimowski *et al.* 1992), by means of wavefront tilt (image motion) compensation. Typically, the image quality is improved by factors of 1.2 – 2, depending on the telescope aperture and seeing conditions.

During this observing run, a Tektronix 512×512 CCD with 27 μm pixels gave an effective plate scale at the AOC's focal plane of 0.095 arcsec pixel^{-1} and a field size of 49×49 arcsec. One 1800 second Hα+[NII] ($\Delta\lambda$ = 110 Å) coronographic image was obtained with the central star, R127, occulted under a 2 arcsec diameter mask. The effective seeing measured from the average of the three brightest, unblended field stars is 0.7 arcsec.

In the Hα+[NII] image, the halo around the occulting mask appears asymmetric and extended with some structure resolved. Once subtracted the scattered light due to R127 itself and its bright companion (Figure 1), we see that the unprecedented signal-to-noise and angular resolution of the Hα+[NII] observation reveals the full extent of

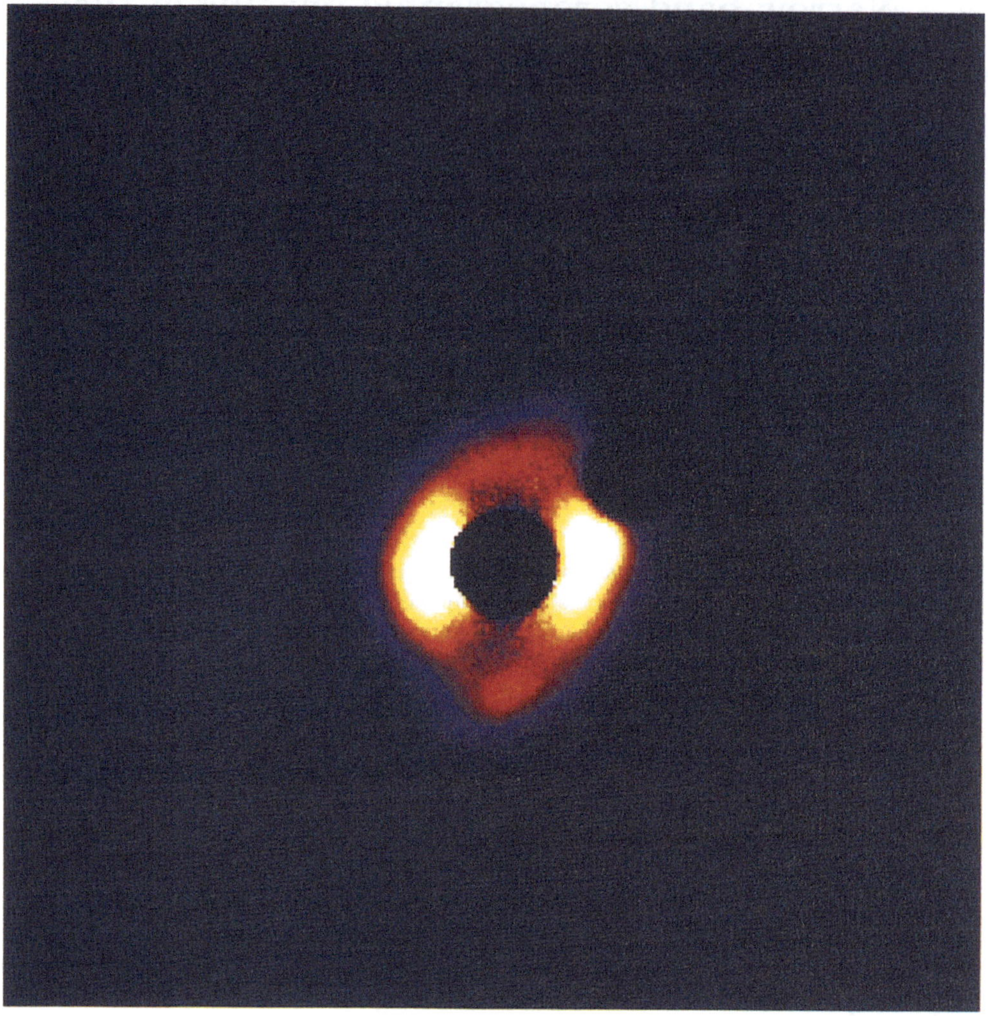

Fig. 1. A 1800 second coronographic exposure of the R127 nebula in the light of Hα+[NII]. The residual scattered light from R127 itself and of its NW companion have been subtracted out to enhance the contrast in the nebula, which displays a definite bipolar morphology. The field of view in the image is 15x15 arcsec, North is up and East to the left.

the R127 nebula and its morphology, which shows clear evidence of deviations from spherical symmetry. The nebula has a diamond shape, with its major axis lying at a position angle of 165°. The measured dimensions of 8.0 arcsec×9.0 arcsec yield a linear size of 1.9×2.2 pc, assuming a distance to the LMC of 50 kpc (Panagia *et al.* 1992). These dimensions are in conflict with the linear size of 0.8×1.1 pc previously published by Stahl (1987), although radial profile drawn from his data suggests that the nebula extends at least 4 arcsec from the star. Our dimensions are also inconsistent with the parameters derived from spectroscopic observations of Appenzeller *et al.* (1987), who also estimate that the nebula extends 2 arcsec from the star.

The nebula is reminiscent of the shell around the galactic LBV AG Carinae (Nota *et al.* 1992). There are large variations in the surface brightness of the nebula and an overall bipolar morphology. Two bright lobes lie along an axis with position angle (p.a.) 92° (≈EW) and an opening angle of ≃90°, possibly indicating gas density enhancements. The nebula is fainter along the major axis (≈NS) with the brightest regions lying farther (1 – 1.5 arcsec) from the central star than the brighter E–W lobes. In addition, their morphology is different, since they appear as partially resolved extensions of the brighter lobes, and are detached from the star. We can derive the ionized mass of the nebula from the integrated, extinction-free emission-line luminosity and the electron density using the relation $M \propto FD^2T_e^{0.85}n_e^{-1}$ (Pottasch 1980), where F is the emission line flux, D is distance, T_e electron temperature and n_e the electron density. The integrated Hα flux obtained from the Hα+[NII] coronagraphic image of the nebula is $3.3\times^{-12}$erg s^{-1}cm^{-2}, corrected for the contribution of the [NII] lines in the filter bandpass. This figure is in good agreement with Stahl (1987) who derived an Hα flux of 4.1×10^{-12}erg s^{-1}cm^{-2}, but did not correct for extinction. An extinction-corrected value of 4.0×10^{-12}erg s^{-1}cm^{-2} is derived, where E(B – V) has been given by Wolf (1989) as 0.2. In the assumption of $T_e = 7500$ K and $n_e = 10^3$ cm^{-3} (Stahl and Wolf 1986), we calculate a ionized gass mass of 3.1 $M_\odot$.

References

Appenzeller, I., Wolf, B., and Stahl, O. 1987, in *Instabilities in Early Type Stars*, ed. H. J. G. L. M. Lamers and C. W. H. de Loore, (D. Reidel), p241.

Clampin, M., Durrance, S. T., Golimowski, D. A. and Barkhouser, R. H., 1991 in *Active and Adaptive Optical Systems*, Proc. SPIE. Vol. 1542, ed. Mark A. Ealey, p. 165.

Golimowski, D. A., Clampin, M., Durrance, S. T. and Barkhouser, R. H., 1992, Applied Optics, **31**, *in press.*

Nota, A., Leitherer, C., Clampin, M., Greenfield, P. and Golimowski, D. A, 1992, Ap. J. in press.

Panagia, N., Gilmozzi, R., Macchetto, F., Adorf, H.-M., Kirshner, R. P., 1992, Ap. J. Letters 380, L23.

Pottasch, S. R. 1980, Astron. Astrophys. 89, 336.

Stahl, O. 1987, Astron. Astrophys. 182, 229.

Stahl, O. and Wolf, B. 1986, Astron. Astrophys. 158, 371.

Walborn, N. 1982, Ap.J. 256, 452.

THE ENVELOPES OF B[e] SUPERGIANTS IN THE MAGELLANIC CLOUDS

Antonio Mário Magalhães

Space Astronomy Laboratory, U. Wisconsin, Madison, WI 53705, USA
and
Instituto Astronômico e Geofísico, Universidade de São Paulo, Caixa Postal 9638, São Paulo, SP 01065-970, BRAZIL [1]

Abstract: We have obtained linear polarization data of B[e] supergiants in the Magellanic Clouds. Our analysis shows that they have non-spherically symmetric envelopes, with a range of intrinsic polarizations. Comparison of our data with spectroscopic data in the literature agrees with this interpretation. The polarization correlates best with the IR excess due to dust, suggesting dust scattering as the main polarizing mechanism. Comparison of our data with data uncovered from the literature for four stars indicate that the envelopes are stable.

Introduction

The B[e] supergiants populate the hot, luminous part of the H-R diagram, together with the Luminous Blue Variables and the Wolf-Rayet stars. While there is evidence that the LBV are progenitors of WR stars, the relation of B[e] stars to these classes and to the evolution of massive stars is not clear. Spectra of B[e] stars suggest (Zickgraf 1990) a hot, fast, radiatively driven polar wind together with a slow, dense, cool equatorial flow; in other words, their envelope could be very far from spherical symmetry. Non-spherically symmetric envelopes linearly polarize starlight. If measurable, polarization would hence potentially carry information on the B[e] star environment.

Observations

Optical linear polarization data have been obtained with the VATPOL polarimeter (Magalhães et al. 1984), at the 2.15m telescope at the CASLEO observatory, San Juan, Argentina. The data has been corrected for foreground (Galactic plus Magellanic) interstellar polarization (for details, see Magalhães 1992).

[1] Present address

Results

The main results may be summarized as follows:

a. Significant polarization has been detected in most of the objects. The B[e] star envelopes are hence notably non-spherically symmetric. Modelling shows that the histogram of observed polarizations is consistent with the Magellanic B[e] stars having a *distribution* of intrinsic polarizations.

b. The observed polarizations do not correlate to any statistically significant level with the near IR excess from the free-free radiation from the B[e] star envelopes. However, the observed polarizations do correlate significantly with the IR excess at longer wavelengths, due to dust, which forms in the outer parts of the B[e] star envelopes.

c.Data has been recovered from interestellar polarization surveys for 4 B[e] stars. A comparison with our data shows no statistical evidence for variability; this indicates that the envelopes are stable. More observations to stablish this on a firmer basis are still needed.

The full analysis and results are being published elsewhere (Magalhães 1992). They will include a discussion between the data and the detection of CO emission from the Magellanic B[e] stars and also a comparison with polarimetric data of Galactic B[e] stars.

Acknowledgements. This research has received support from grants by NASA (NAS5- 26777) and São Paulo state funding agency FAPESP (89/1670-9). His attending this meeting was supported by the Vatican Observatory and NASA grant NAG5-1463.

References

Magalhães, A.M. (1992): ApJ, 10 October, in press

Magalhães,A.M., Benedetti, E. and Roland, E.W. (1984): PASP **96**, 383

Zickgraf, F.-J. (1990): in "Angular Momentum and Mass Loss for Hot Stars", ed. by L.A. Willson and R. Stalio (Dordrecht: Kluwer), 245

A New Way to Calculate Doublet P-Cygni Profiles

Nadine Rons

Astrophysical Institute, Vrije Universiteit Brussel, B-1050 Brussels, Belgium
Bursaal I.W.O.N.L.

Abstract: A new way to calculate doublet P-Cygni profiles in comoving frame was used to calculate theoretical C IV profiles for early type stars in the Magellanic Clouds. The calculated profiles were compared to the profile fits presented by Patriarchi et al. (1992). Two examples are shown : HD 268605 and HD 169698.

1 Introduction

High resolution IUE spectra of early type stars in the Magellanic Clouds make it possible to study their stellar winds from the observed P-Cygni profiles.

Several methods have been developed to calculate theoretical profiles, i.e. the Sobolev method (Sobolev 1958), the Comoving Frame method (Mihalas et al. 1975) and the SEI method (Sobolev with Exact Integration) (Lamers et al. 1987).

A large number of these profiles originate from doublets, which are treated by these methods one component after the other, assuming the blue component's presence interferes with the absorption and emission by the red one, but not vice versa. This assumption does not always correspond to reality and should be treated with care when considering doublets with a small separation relative to the thermal (+ turbulent) velocity.

A new comoving frame doublet treatment was used here which considers the two doublet components together, forming a compound profile. Here the mutual influence of both components is taken into account.

A selection of C IV profile fits of presumed single stars from Patriarchi et al. were recalculated using the Comovong Frame method developed based upon the directions given by Mihals et al. and equipped with the new doublet treatment. The same parameterizations were considered as were presented in Lamers et al. and used in Patriarchi et al..

2 The calculated C IV profiles

In the figures below (emergent flux normalized to 1 at the continuum value versus wavelength), two C IV profiles computed with the SEI method by Patriarchi et al. (squares) are compared to those computed with the extended Comoving Frame method (full line). The larger the thermal + turbulent velocity, the more the components overlap. As expected, the best agreement is obtained for the star with the lowest $v_{th+turb}$.

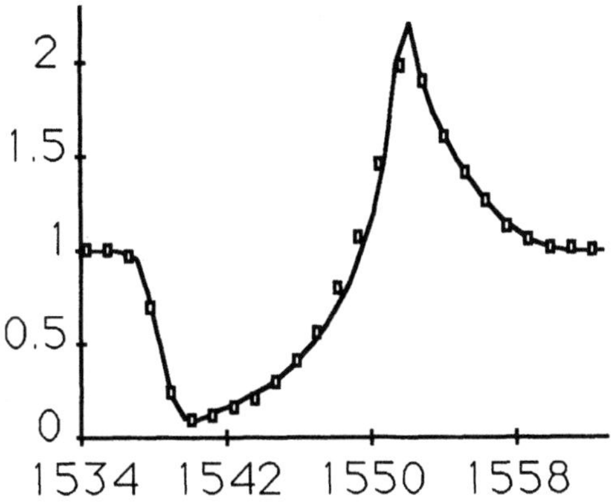

Fig. 1. HD 269698, $v_{th+turb} = 125\,km\,s^{-1}$

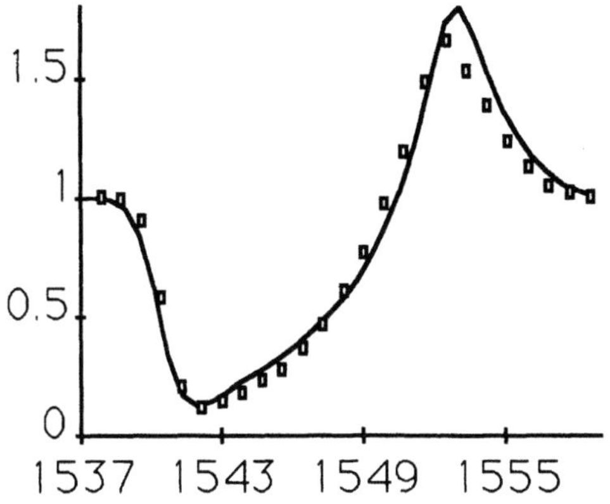

Fig. 2. HD 268605, $v_{th+turb} = 265\,km\,s^{-1}$

3 Future work

A theoretical investigation of the differences in the results obtained with both methods in comoving frame will make it possible to define the thermal + turbulent velocity range which allows a separate treatment of the doublet components. Profiles for which this velocity is found to be too large can be recalculated. A fair part of the C IV profiles shown in Groenewegen et al. (1989) e.g. are certain candidates. Finally, with new fits of C IV profiles for stars in the Magellanic Clouds and their Milky Way counterparts, a new estimate may be made for their relative mass loss.

References

Groenewegen, M.A.T., Lamers, H.J.G.L.M. (1989): A—&A Suppl. Ser. **79** 359

Lamers, H.J.G.L.M., Cerrutti-Sola, M., Perinotto, M. (1987): ApJ **314** 716

Mihalas. D., Kunasz, P.B., Hummer, D.G. (1975): ApJ **202** 465

Patriarchi, P., Perinotto, M. (1992): A—&A **258** 285

Sobolev, V.V. (1958): In *Theoretical Astrophysics*, ed. by V.A. Ambartsumian (Pergamon Press, New York)

R40: first Luminous Blue Variable in the Small Magellanic Cloud

Th. Szeifert [1], O. Stahl [1], B. Wolf [1], F.J. Zickgraf [1]

[1]Landessternwarte Königstuhl, Heidelberg, Fed. Rep. of Germany

R40 has been classified by Feast et al. (1960) as a B8Ie star of a visual magnitude of $V = 10.73$. Since 1985 it is being observed within Sterkens (1983) LTPV-campaign (see light curve in Fig.1). R40's present visual magnitude is about 10.3 with microvariations of a typical timescale of 120 days and an amplitude of 0.1 mag. A section of a CASPEC spectrum and of a high resolution IUE spectrum of R40 are shown in Figs. 2 and 3, respectively, along with a maximum spectrum of SDor and a spectrum of R45 (A0Ia-O). The photometric and spectroscopic observations show R40 to be an LBV presently at outburst, the first instance in the SMC. Since the metallicity in the SMC is down by a significant factor (−0.6 dex) compared to the Milky Way, R40 is a key object for investigating the metallicity dependence of the LBV phenomenon. A metallicity dependence of the outburst behaviour is to be expected from previous suggestions for the outburst mechanism , e.g. line opacity driving (Appenzeller, 1986, Lamers, 1986) or metallicity enhanced κ−mechanism (Moskalik and Dziembowski, 1992). From spectroscopic and photometric data and from a continuum fit (Fig.4) by use of ATLAS8 models we derived the stellar parameters for R40 in its present outburst phase (cf. also Fig. 5) and the mass-loss rate from the H_α-PCygni profile (see Fig.6):

$$T_{\rm eff} = 8700{\rm K} \qquad \log g = 0.75 \qquad M_{\rm bol} = -9.1$$

$$R = 247 \times {\rm R}_\odot \qquad M = 13 \times {\rm M}_\odot \qquad \dot{M} = 6 \times 10^{-6} \times \frac{\rm M_\odot}{\rm yr}$$

The location of R40 in the HRD is shown in Fig. 7 along with other LBVs. The estimated low mass provides evidence that R40 is in an evolved evolutionary stage, presumably in the post-red supergiant phase. The mass-loss rate of R40 is not overwhelmingly large but also not exceptionally small for an LBV of this luminosity if compared to R110 of the LMC which is located very close to R40 in the HRD. Likewise the amplitude of the eruption is small but similar to HD 160529 of the Milky Way and R110 of the LMC, which are of similar luminosity. From that it seems likely that metallicity is not the decisive parameter of the LBV phenomenon. (A more detailed paper is in preparation.)

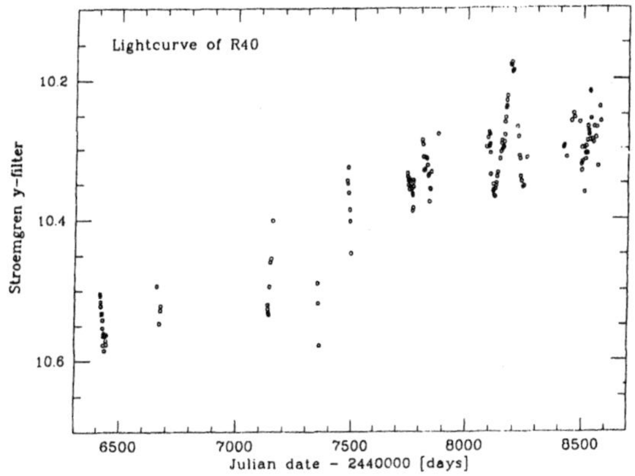

Fig. 1. Lightcurve of R40 which has recently been recognized within Sterken's LTPV campain at ESO as the first instance of an LBV in the SMC. As evidenced by the lightcurve, the visual brightness of R40 has increased by about 0.3 magnitudes during the past six years. Note that Feast et al (1960) give $V = 10.73$. The observed amplitude of eruption is about 0.5 mag. Superimposed to this general trend are microvariations with a quasi period of about 120 days. Such microvariations are typical for LBVs.

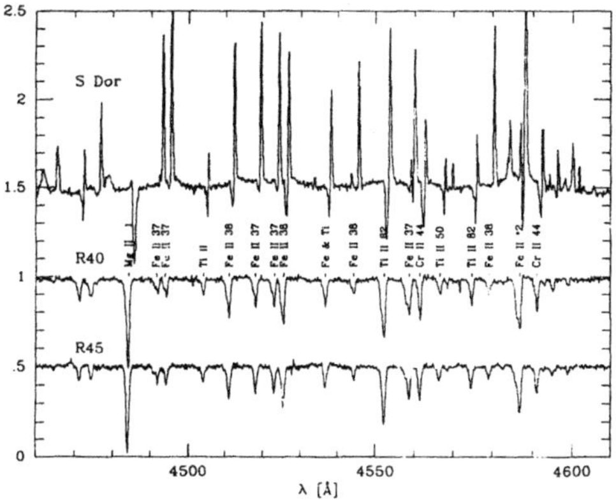

Fig. 2. Section of a CASPEC spectrum of the SMC LBV R40 along with a maximum spectrum of S Dor and of the most luminous A supergiant R45 (A0Ia-O) of the SMC. Although the same lines as in S Dor can be readily identified in R40, an expanding envelope is only indicated in this section by extended blue wings of singly ionized metallic lines which in contrast to H_α (see Fig. 6) do not show a P Cyg emission component.

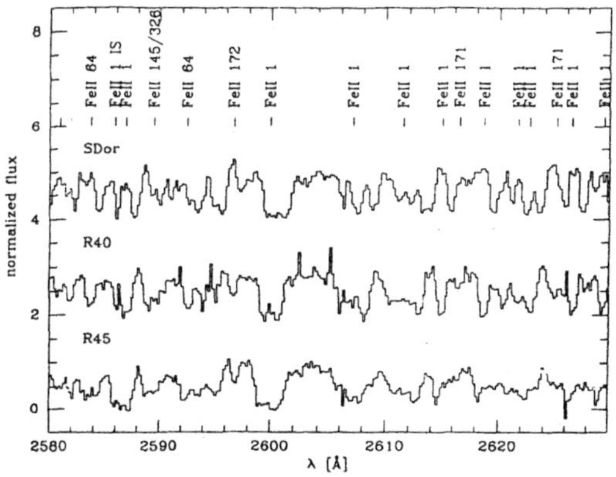

Fig. 3. Section of a high resolution IUE spectrum of R40 along with a spectrum of S Dor and a spectrum of the A0Ia-O star R45 in the SMC. The spectral appearance is very similar for all three stars. Particularly the close agreement of R40 and S Dor is surprising bearing in mind the striking differences in the optical range (cf. Fig. 2). Obviously the strong UV lines are saturated in both stars so that the metallicity difference between both stars is of minor importance here.

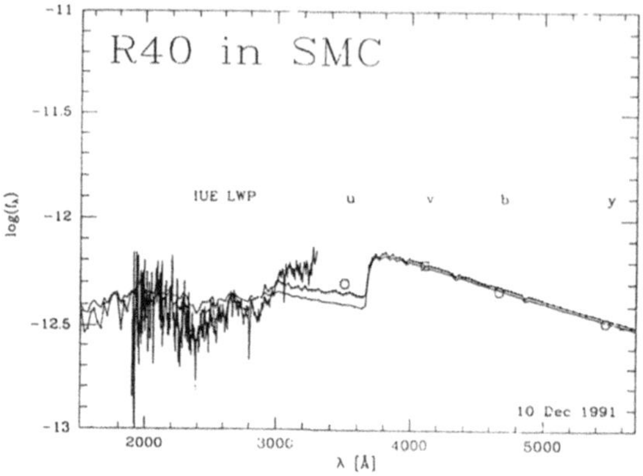

Fig. 4. Continuum energy distribution of R40. The IUE fluxes and Stroemgren observations have been dereddened by $E_{B-V} = 0.06$ using the galactic and SMC extinction laws (0.04 galactic, 0.02 SMC), respectively. The observations have been fitted by ATLAS models for solar (dotted line) and 1/10 solar (solid line) abundances , respectively. Surprisingly the model for solar abundance provides a better fit for the flux of the pseudo photosphere of this SMC LBV. A possible explanation is the enhancement of the line widths due to the expanding envelop which can mimic a higher metallicity.

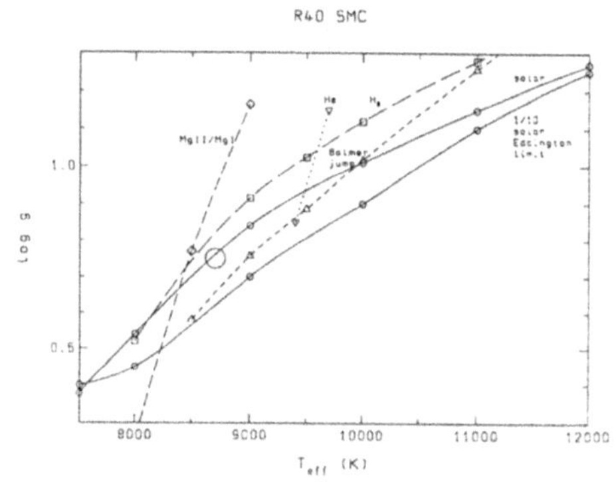

Fig. 5. $\log g - T_{\rm eff}$ diagram for a grid of ATLAS8 models. The dashed lines give the stellar parameters for the indicated observed features for models with a metallicity of 1/10 solar. "MgII/MgI" denotes the line for the magnesium ionisation equilibrium. "He" denotes the models which would yield $\log \epsilon = 11.0$ for helium. "H_8" represents the models reproducing the observed equivalent width of this Balmer line. The solid lines represent the Eddington limits for solar and 1/10 solar abundances. Helium can be overabundant in the atmosphere of this highly evolved star and moreover NLTE effects are expected to be important (see e.g. Lennon et al. 1991). We therefore adopted the stellar parameters indicated by the circle for R40 which are a good compromise for the MgII/MgI-ionisation equilibrium, the Balmer lines and the continuum energy distribution (see Fig.4). As expected R40 is close to the Eddington limit.

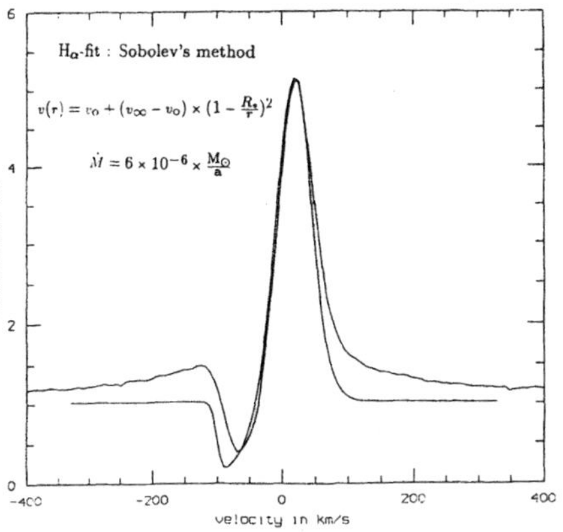

Fig. 6. Observed and computed H_α profile of R40 for the given wind parameters and the stellar parameters given in the text. The wings which are typical LBV features and which have been previously been interpreted by electron scattering of line photons could not be reproduced by our code. Another possibility is that the wings are due to NLTE effects as suggested by Hubeny and Leitherer (1989).

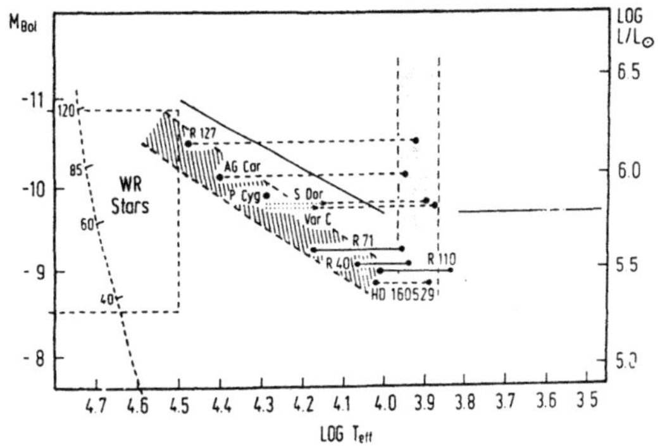

Fig. 7. HRD showing the position of R40 in the SMC along with other well studied LBVs of different galaxies. During minimum R40 is located in the inclined instability strip (Wolf, 1989). It is well below the Humphreys-Davidson limit (solid line) and belongs to the group of less luminous LBVs like R71, R110 and HD160529 which are supposedly post red supergiant objects.

References

Appenzeller, I.: 1986, in Luminous Stars and Associations in Galaxies, ed. by C.W.H. de Loore, A.J.

Willis, and P. Laskarides, Reidel, Dordrecht, p. 139

Feast, M.W., Thackeray, A.D., Wesselink, A.J.: 1960, MNRAS 121, 337

Hubeny, I., Leitherer, C.: 1989, in Physics of Luminous Blue Variables, eds. K. Davidson, A.F.J.

Moffat, and H.J.G.L.M. Lamers, Kluwer Academic Publishers, Dordrecht, p. 283

Lamers, H.J.G.L.M.: 1986, in Luminous Stars and Associations in Galaxies, eds C.W.H. de Loore,

A.J. Willis, and P. Laskarides, Reidel, Dordrecht, p. 157

Lennon, D.J., Kudritzki, R.P., Becker, S.T., Butler, K., Eber, F., Groth, H.G.,Kunze, D.: 1991, A&A 252, 498

Moskalik, P., Dziembowski, W.A.: 1992, A&A 256, L5

Sterken, C.: 1983, The Messenger 33, 10

Wolf, B.: 1989, A&A 217, 87

QUESTIONS RAISED BY MASSIVE STARS IN THE MAGELLANIC CLOUDS

André Maeder

Geneva Observatory CH-1290 SAUVERNY, Switzerland

Abstract: The differences between the properties of Wolf-Rayet (WR) stars in the Magellanic Clouds and in the Galaxy are shown and interpreted in terms of models of massive stars with different mass loss rates. About supergiants, we emphasize the remaining difficulties in explaining the properties of the precursor of SN 1987A with semiconvective models and suggest that one single diffusion mechanism should account for the various observational constraints. We show that models for the LMC with moderate overshooting predict that most blue supergiants are in a post-red supergiant stage with N-enhancements.

1. Introduction

Reviews on massive star evolution can be found elsewhere (cf. Maeder 1991a). Here we focus on a few specific topics concerning the Magellanic Clouds (MC): firstly the relative frequencies of WR stars in the Large Magellanic Clouds (LMC) and Small Magellanic Clouds (SMC), then the status of the discussions regarding the blue progenitor for SN 1987A and models for blue and red supergiants in the Magellanic Clouds.

2. WR stars in the Magellanic Clouds

2.1 The facts

WR stars are nowadays considered to be bare cores left over by the peeling-off due to stellar winds (cf. Lamers et al. 1991). The basic data on WR stars in the MC are from the works by Azzopardi & Breysacher (1979, 1985); Breysacher (1981); Smith (1968, 1988, 1991). The main characteristics of the WR distributions in the MC are the following ones.

2.1.1 WR/O number ratio

LMC:	lower than for the Galaxy (about $^1/_3$)
SMC:	much lower (about $^1/_{10}$ of the galactic ratio)

2.1.2 WN stars

All WN types are present. Relatively many WNE stars, among which many binaries (especially in the SMC). In the LMC, WNL stars are mainly located in the 30 Dor Nebula. No WNL in the SMC. (WNL means late WN, WN6-9; WNE means early WN, WN2-6).

2.1.3 WC stars

The WC/WN ratio is 0.14 in the SMC, 0.26 in the LMC compared to about 1.8 in the solar neighbourhood.

LMC:	all WC stars are WC5, except one or two;
SMC:	only one WC-WO, which is a binary.

It is interesting to notice that there is a general trend observed in galaxies of the Local Group (cf. Maeder et al. 1980; Smith 1988; Maeder 1991b), in the sense of a decrease of the WR/O number ratio with decreasing metallicity Z.

There is also a general diminution of the WC/WN ratio with decreasing Z. In addition, early WC types and WO stars are relatively more frequent at low Z. This trend is also found in our Galaxy (cf. Fig. 1, from Smith & Maeder 1991), where late WC types are mostly present in internal galactic regions, while the rare early WC types are found in external regions.

2.2 The interpretation

How does metallicity Z influence massive star evolution ? It is well known that metallicity effects generally enter the stellar structure mainly through the Z-dependence of the bound-free and line opacities. In the case of massive stars, such direct structural effects of metallicity are insignificant since electron scattering is the main opacity source in stellar interiors. Only in the external layers a metallicity effect influences the opacities, and thus the atmospheres and stellar winds. This is why metallicity effects mainly occur in massive stars through stellar winds. Wind models for O-stars by Abbott (1982) suggest a metallicity dependence of the mass loss rates $\dot{M}$ in the form $\dot{M} \sim Z^{\alpha}$, with $\alpha = 1.0$. Models by Kudritzki et al. (1987; cf. also Leitherer & Langer 1991) give a value $\alpha = 0.5$, applied in our model calculations.

For Wolf-Rayet stars, which are essentially He-CO stars, the initial metallicity is not expected to have great consequences on the effective mass loss rates. Recent works have shown that the mass loss rates $\dot{M}$ of the WR stars depend on their actual masses M with an exponent of about 2.5 (cf. Langer 1989), when hydrogen is no longer present. Such a relation is confirmed by several other works and accounted for in recent models. This $\dot{M}$ vs M relation leads to very low final masses for the WR

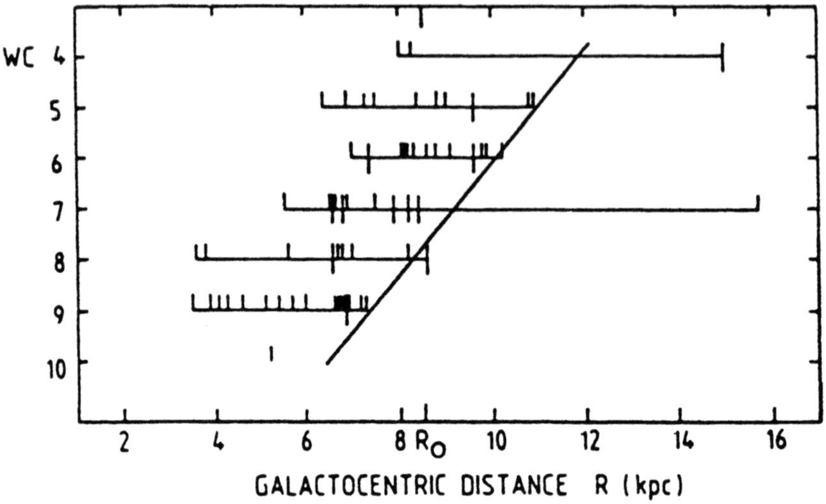

Fig. 1. Galactic distribution of WC subtypes from van der Hucht et al. (1988). We add the lines emphasizing the monotonic increase in the radius of the outer and inner boundaries.

stars, i.e. in the range 4 – 7 $M_\odot$. This corresponds finely to the observed WR masses and luminosities.

The observed number star ratios WR/O, WC/WR, WC/WN in the LMC, SMC and other galaxies have been compared to the theoretical predictions (cf. Maeder, 1991b). For a large galactic ring or area, the assumption of a constant star formation rate over the last few 10^7y is a reasonable one. In an HII region, or a starburst galaxy, this assumption is not valid. Initial mass functions $dN/dM = AM^{-(1+x)}$ with x=1.35 and 1.70 have been considered. The agreement between model predictions and observations is excellent. At high Z, the heavy mass loss rates lead to the formation of a lot of WR stars (bare cores) and high WC/WN number ratios. At low Z, like in the LMC and SMC, the lower mass loss rates produce lower WR/O and WC/WN ratios.

Let us recall that schematically WN stars exhibit products of H-burning, while WC stars show those of He-burning. Thus we also understand, why the ratio WC/WN is lower in the Magellanic Clouds. The lower Z, and thus lower mass loss rates, makes it more difficult to peel-off the star deep enough to see the products of He-burning.

The relative numbers of Wolf-Rayet stars with various WC subtypes are also analysed quantitatively. WC models are classified according to the (C+O)/He number ratios (cf. Smith & Hummer 1988; Smith & Maeder 1991).

WC9, (C+O)/He = 0.03 – 0.06	WC5, (C+O)/He = 0.55
WC8, (C+O)/He = 0.1	WC4, (C+O)/He = 0.7 – 1.0
WC7, (C+O)/He = 0.2	WO, (C+O)/He greater than 1.0
WC6, (C+O)/He = 0.3	

The models show that the entry points and lifetimes in the sequence WC9 → WC4, WO are extremely Z– and mass-dependent, as shown in Fig. 2 (cf. Smith & Maeder 1991). The entry into WC9 subtype occurs only for high initial Z and mass (i.e. Z $\gtrsim$ 0.02 and $M_\odot \geq 50\ M_\odot$) since the corresponding low (C+O)/He ratios can only occur when the previous mass loss rates have been sufficiently high. Lower initial masses and metallicities lead to an entry into the WC phase at an earlier subtype and it also gives shorter WC phases, because in those cases, – due to the lower mass loss rates – the products of He-burning are revealed at a more advanced stage of nuclear processing. After integration over the mass spectrum, we clearly find a dominance of late and intermediate WC types at Z $\gtrsim$ 0.02, while at lower Z, only early WC and WO stars are present and this in much lower numbers. Comparisons between the observed and predicted relative numbers of the stars of WC subtypes have been performed (cf. Smith & Maeder 1991) for Wolf-Rayet stars in galaxies of the Local Group, and the results again show good agreement.

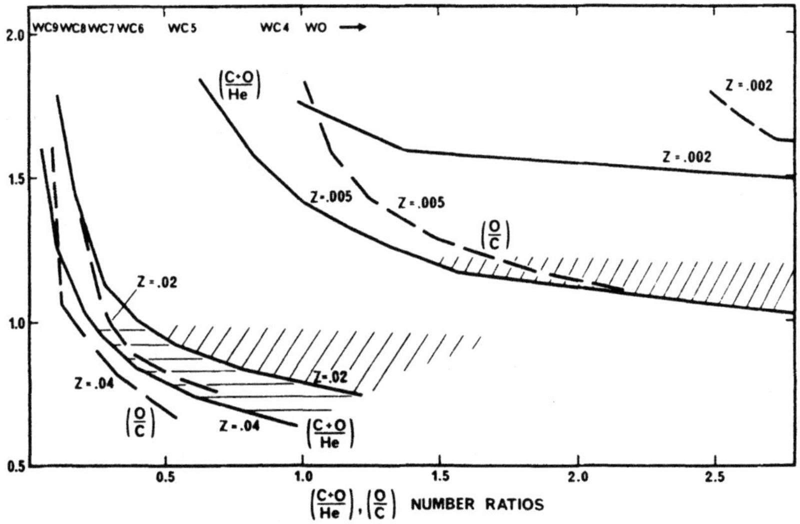

Fig. 2. The theoretical relation between log $M/M_\odot$ and abundance ratios for WC stars as a function of the initial Z.

3. Discussion on the models for the precursor of SN 1987A

Let us start by a short review of the models which have been proposed in order to explain the properties of the precursor of SN 1987A, in particular its blue colour and high N/C ratio (cf. Fransson et al. 1988).

- **Models with low Z, without mass loss:** such star models (cf. Hillebrandt et al. 1987) finish their life in the blue, however they produce no red supergiants and show no N/C-enhancements, in contradiction with observations.

- **Models with low Z, having mass loss:** may end their life in the blue as a result of mass loss (cf. Wood & Faulkner 1987; Maeder 1987) only if the envelope mass is smaller than 1 $M_\odot$, which is not supported by the interpretation of the light curve.
- **Models with the Ledoux criterion** for convection have been proposed by Woosley (1988) to account for the blue progenitor. However, when a moderate mass loss is taken into account, the final blue location occurs too early, i.e. during central He-burning, which is not in agreement with the suggestion that the precursor of SN 1987A was a red supergiant a few thousand years before the explosion.
- **Models with late mixing:** the possibility of mixing of the external layers of a red supergiant at the end of the He-burning phase has been envisaged by Saio et al. (1988). Our models also confirm that at this stage, the external and intermediate convective zones become extremely close to each other, so that any tiny convective extension is likely to bring the 2 zones into touch with one another. The resulting mixing produces simultaneously the He– and N/C-enrichment as well as the blue location of the SN progenitor and this makes the suggestion by Saio et al. (1988) an attractive one. In this case, standard models for the H– and He-burning are right and extra mixing comes only in a very late stage into play. Careful comparison with cluster sequences and abundance ratio will tell us whether this scenario is the right one or not, (see below).
- **Models with semiconvection:** semiconvection produces a mild mixing in zones with a μ-gradient. Models including this effect have been made by Langer (1991) and Arnett (1991), and are able to account for the blue progenitor. The physical connection, responsible for the blue final location (cf. Langer), is interesting and worth to be examined since it may have more general applications. Mixing (due to semiconvection or some other mixing process) reduces the He-content in the He-shell, which is therefore less efficient. Thus, at the time of CO core contraction (at the end of the He phase), the He-shell acts as a *weak mirror* and produces only a moderate expansion into the intershell region between the He and H-shells. Thus, the H-shell is not extinct and keeps very active. It acts as a very *active mirror* responding to the moderate intershell expansion by a strong contraction of the external envelope, therefore the blue supernova progenitor. The model results illustrate this effect very well. However, it is noticeable that some fine tuning is necessary to get the desired effect and that any changes in the parameters (Z, $\dot{M}$, opacity, etc.) may alter the conclusion.

However, like all models, models with semiconvection have to face several objections:

- The resulting distributions of supergiants differ from that observed by Fitzpatrick & Garmany (1990), but the real significance of this distribution is not indisputable (see § 4).
- The final blue loop was no longer found by El Eid & Höflich (1991) with new opacities at low T.
- The observed high N/C and N/O ratios are not explained by this mechanism and Langer is calling for turbulent diffusion in the external layers to explain the observed N-enrichment.

- The extensive studies of clusters made by the Geneva and Padova groups agree (at least !) on the fact that overshooting rather than semiconvection is required to account for the observations (cf. Meynet et al. 1992; Bressan et al., this meeting).
- On the theoretical side, overshooting appears to be called for (cf. Zahn 1991; Canuto 1992) and, as it is well known, this reduces the importance of semiconvection.
- At various Z, another difficulty is also visible in the models presented by Arnett (1991). His models with semiconvection predict no red supergiants at Z=0.002 (cf. SMC) and a lot at Z=0.010 or higher values. This is exactly the opposite of what has been observed by Meylan & Maeder (1982), who showed that the relative number of red supergiants, with respect to O-stars, is increasing at lower Z. This comparison shows that semiconvective models have to face a major difficulty.

My personal point of view on this matter is that the structural stellar property, i.e. the lower He content of the He-shell which by mirror effects makes the blue supernova progenitor as proposed by Langer (1991), is a very interesting possibility which has to be considered seriously. However, semiconvection is not necessarily the proper mixing mechanism for that, more especially as another mixing mechanism is needed to account for the high N/C, N/O ratios. Considering the number of hypotheses, it would be more economical to call for only one single mixing mechanism. At least, one ought to start by this approach.

In this context, it is interesting to notice that the diffusion coefficient D corresponding to semiconvection can, according to Langer, be written:

$$D = \frac{\alpha}{6} \; \frac{\nabla - \nabla_{ad}}{\nabla_L - \nabla} \; D_{rad} \;\; with \;\; D_{rad} = \frac{4\sigma T^3}{3\kappa\rho^2 C_p} \tag{1}$$

while the diffusion coefficient D corresponding to rotational mixing (cf. Zahn 1984) is

$$D = \frac{\Omega^2 r}{g} \; \frac{1}{\nabla - \nabla_{ad}} \; D_{rad} \; . \tag{2}$$

We notice that these two mixing processes have both a major dependence on D_{rad} (as it should be the case for all dissipative mixing processes). Thus, rather than using two partly different coefficients to explain the various properties of SN 1987A, it seems preferable to search for one single diffusion process, (necessarily scaled to D_{rad}), which in a unique way reproduces both the blue location of the SN 1987A progenitor and the nitrogen enrichment. Such an examination is now being undertaken in terms of models with rotational mixing.

4. Models for OB stars and supergiants at low Z

At present, there are a number of claims in favour of different treatment of convection and mixing in massive stars. We may identify the following treatments with the corresponding names of the authors:

- models with Schwarzschild criterion
- models with overshooting (Maeder; Chiosi)
- models with Ledoux criterion (Stothers; Woosley; Castellani)
- models with semiconvective diffusion (Langer; Arnett)
- models with rotational mixing (Maeder; Langer; Weiss)

The main common property of these models is that all of them are claimed by their authors to fit relatively well the observations. However, it is presently not quite clear what the observation is telling us. For example, the distribution of supergiants in the LMC by Fitzpatrick & Garmany (1990) does not at all give the same observational constraint as the colour magnitude diagram of some well studied LMC clusters like NGC 2004 (cf. Bencivenni et al. 1991; cf. also Meylan & Maeder 1982). Therefore, rather than using an ensemble of data of relatively poor quality, it is much more advisable to base the comparison on a set of well discussed clusters in the Magellanic Clouds and the Galaxy. Once this is done, there is a strong constraint in favour of models with overshooting for both, LMC clusters (cf. Bressan et al. this meeting) and galactic clusters (cf. Meynet et al., 1992). Ideally, one should also have enough observed data on the C/N, O/N, He/H ratios to establish the cartography of surface abundances in the HR diagram in order to compare it with model predictions. It is only in this way that we may be able to properly recognize the solution chosen by Nature in her wisdom.

Changes in the models may also affect some properties of WR stars (cf. § 3). But, as WR properties are mostly determined by mass loss effects, rather than by mixing processes, the main trends observed in WR distribution mainly result from mass loss effects.

Figure 3 shows new grids of models with moderate overshooting, mass loss, new Livermore opacities (cf. Rogers & Iglesias, 1992) and up-to-date nuclear cross-sections for the composition Z=0.008, appropriate to the LMC. Fig. 4 shows the lifetimes for the H- and He-burning phases for grids at various Z; grids of models at Z=0.004 and 0.040 will be added. Close comparisons with 30 galactic clusters have recently been made (cf. Meynet, Mermilliod, Maeder, 1992), they clearly favour models with overshooting from convective cores. Interestingly, the present models and the models at Z=0.001 from Schaller et al. (1992), both with mild overshooting parameterised by $d/H_p = 0.2$, lead to the same extensions of the blue loops as the models by Stothers & Chin (1992a) computed with the Ledoux criterion for convection at Z=0.004. From the presence of blue and red supergiants in two metal-poor clusters in the Small Magellanic Clouds (SMC), these authors confirm the upper limit of $d/H_p \lesssim 0.2$ using the Rogers & Iglesias (1992) opacities, and find *"..strong evidence in favour of the Ledoux criterion and against the Schwarzschild criterion"*, (Stothers & Chin, 1992b). For several reasons, such a conclusion seems premature, since from our models, using the Schwarzschild criterion and overshooting, we obtain both blue and red supergiants in the observed luminosity and T_{eff} domain.

Furthermore, other mechanisms for convective mixing as semiconvective diffusion have been proposed leading to similar results (cf. Langer, 1991). Such differences within the models probably reflect uncertainties in other ingredients which affect the extension of the loops.

Also, the inclusion of mass loss in models of massive stars at low Z favours the redwards evolution of models which would otherwise have stayed in the blue.

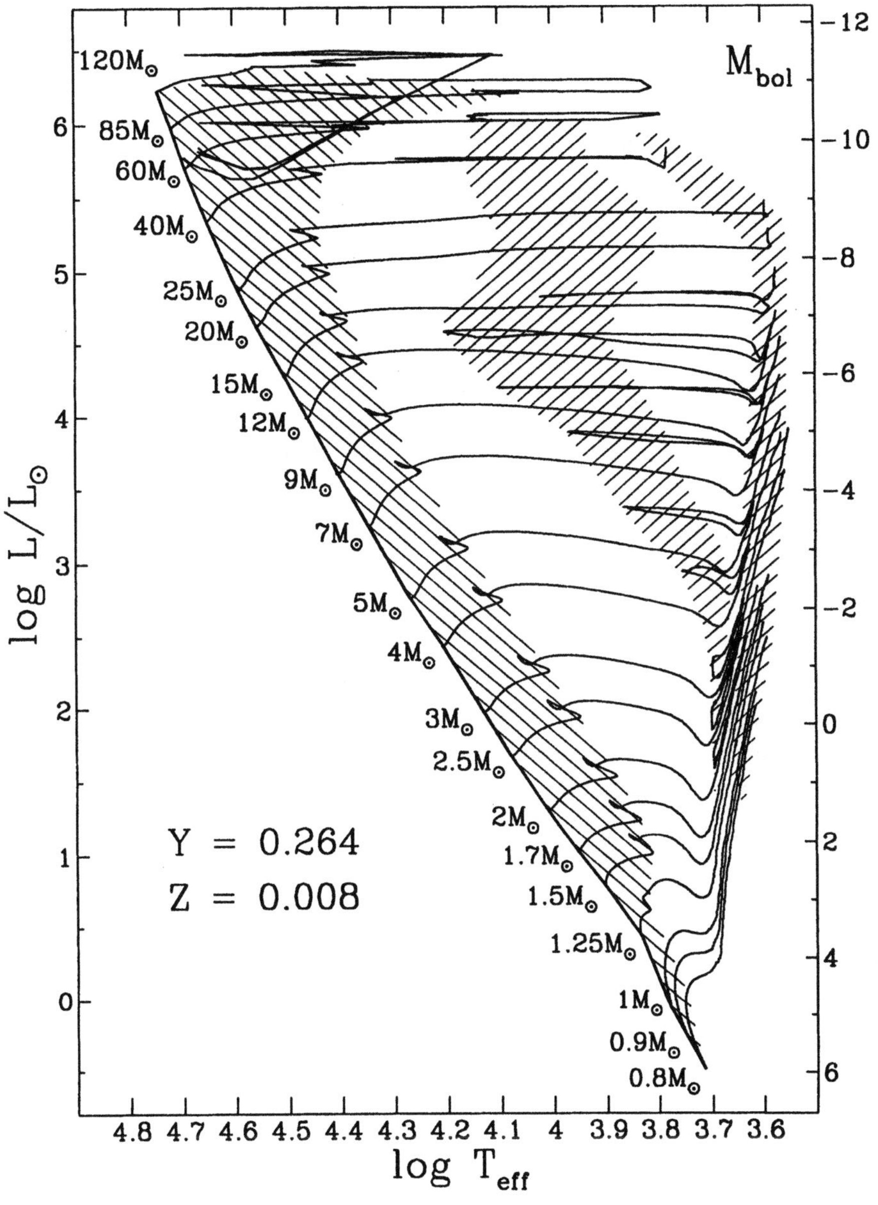

Fig. 3. New grids of models for the LMC at Z=0.008 (cf. Schaerer et al. 1992).

Let us also note that the models by Stothers & Chin (1992b), using the Ledoux criterion, predict unaltered N/C ratios while most other red supergiant models and the blue supergiant models evolving on a blue loop predict a convective dredge-up of H-burning products at the Hayashi-line.

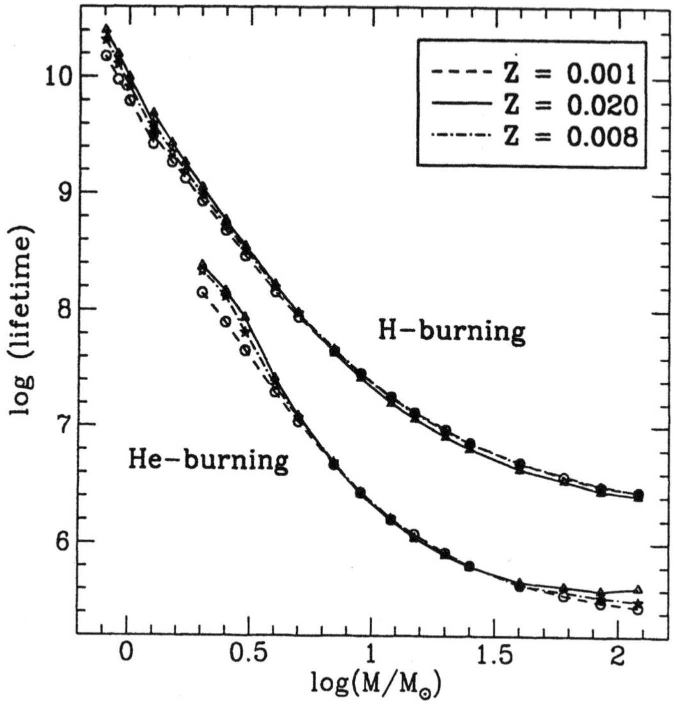

Fig. 4. The lifetimes as a function of the initial masses for models at Z=0.001, 0.008 and 0.020 (cf. Schaller et al. 1992; Schaerer et al. 1992).

Indeed, evidence for enhanced N/C ratios has been found in many of the studied blue supergiants of the LMC and SMC (cf. Kudritzki 1990; Reitermann et al. 1990). According to Fitzpatrick & Bohannan (this meeting), about 80% of the blue supergiants in the LMC show evidence of N-enhancement; this ratio seems to be lower in the Milky Way. Such a finding supports the view that many blue supergiants are in a post-red supergiant stage with N-enhancement, as found in the present models.

References

1. D. Abott: ApJ **259** 282 (1982)
2. D. Arnett: ApJ **383** 295 (1991)
3. M. Azzopardi, J. Breysacher: A&A **75** 120 (1979)
4. M. Azzopardi, J. Breysacher: A&A **149** 213 (1985)
5. D. Bencivenni, E. Brocato, R. Buonanno, V. Castellani: AJ **102** 137 (1991)
6. J. Breysacher: A&AS **43** 203 (1981)
7. V.M. Canuto: ApJ **393** 218 (1992)
8. M. El Eid, P. Höflich:6th Workshop on "Nuclear Astrophysics", Eds. W. Hillebrandt and E. Müller, Max Planck Institut für Astrophys. (1991)
9. E.L. Fitzpatrick, C.D. Garmany: ApJ **363** 119 (1990)
10. C. Fransson, Cassatella, R. Gilmozzi, N. Panagia, W. Wamsteker , R.P. Kirshner: ApJ **336** 429 (1989)
11. W. Hillebrandt, P. Höflich: Nature **327** 597 (1987)

12. K.A. van der Hucht, B. Hidayat, A.G. Admiranto, K.R. Supelli, C. Doom: A&A **199** 217 (1988)
13. R.P. Kudritzki, A. Pauldrach, J. Puls: A&A **173** 293 (1987)
14. H. Lamers, A. Maeder, W. Schmutz, J.P. Cassinelli: ApJ **368** 538 (1991)
15. N. Langer: A&A **220** 135 (1989)
16. N. Langer: A&A **253** 669 (1991)
17. L. Leitherer, N. Langer: in IAU Symp. 148, Eds. R. Haynes, D. Milne, Kluwer Acad. Publ. p. 480 (1991)
18. A. Maeder: in "SN 1987 A", ESO Workshop, Ed. I.J. Danziger, p. 251 (1987)
19. A. Maeder: QJRAS **32** 271 (1991a)
20. A. Maeder: A&A **242** 93 (1991b)
21. A. Maeder, J. Lequeux, M. Azzopardi: A&A **90** L17 (1980)
22. G. Meylan, A. Maeder: A&A **108** 148 (1982)
23 G. Meynet, J.C. Mermilliod, A. Maeder: A&AS in press
24 A. Reitermann, B. Baschek, O. Stahl, B. Wolf: A&A **234** 109 (1990)
25 F.J. Rogers, C.A. Iglesias: ApJS **79** 507 (1992)
26 H. Saio, M. Kato, K. Nomoto: ApJ **331** 388 (1988)
27 D. Schaerer, G. Meynet, A. Maeder, G. Schaller: A&AS in press
28 L.F. Smith: MNRAS **141** 317 (1968)
29 L. F. Smith: ApJ **327**, 128 (1988)
30 L. F. Smith: in the "Magellanic Clouds", IAU Symp. 148, Ed. R. Haynes, D. Milne, Kluwer Acad. Publ. p. 267-272 (1991)
31 L.F. Smith, D. Hummer: MNRAS **230** 511 (1988)
32 L.F.Smith, A. Maeder: A&A **241** 77 (1991)
33 R. Stothers, C.W. Chin: ApJ **390**, 136 (1992a)
34 R. Stothers, C.W. Chin: ApJ **390** L33 (1992b)
35 P.R. Wood, D.J. Faulkner: Proc. astr. Soc. Austr. **7** 75 (1987)
36 S.E. Woosley: ApJ **330** 218 (1988)
37 J.P. Zahn: in "Astrophysical Processes in Upper MS Stars", 13th Saas-Fee Course, Eds. B. Hauck, A. Maeder (1983)
38 J.P. Zahn: A&A **253** 179 (1991)

Implications of Processed Material in the Atmospheres of LMC B Supergiants

Bruce Bohannan [1], Edward L. Fitzpatrick [2]

[1]Kitt Peak National Observatory, P.O. Box 26732, Tucson, AZ 85726, U.S.A.
[2]Princeton University Observatory, Princeton, N.J. 08544, U.S.A.

Abstract: A high-dispersion study of 11 LMC B-type supergiants shows that the relatively rare nitrogen-weak stars (10-20% of LMC B-type supergiants) also have weaker helium lines and stronger oxygen lines than the more spectroscopically "typical" B supergiants. Such correlations in line strength are what would be expected from the contamination of the surfaces of the "typical" supergiants by material processed a hydrogen-burning core. The relatively rare "N-weak" stars likely have retained their original surface abundances.

Fitzpatrick (1991) found a wide range in the strength of N II lines of LMC B-type supergiants for a given spectral type. Among the early B sub-types, 10-20% of the stars had anomalously weak N II lines compared to the majority, the spectroscopically "typical" stars. To determine if this range of nitrogen strengths arises from surface contamination from core nuclear reactions, Fitzpatrick and Bohannan (1992) observed at high dispersion lines of helium, oxygen and carbon–elements which also participate in these reactions–in 11 LMC B supergiants from Fitzpatrick's sample.

As shown in Fig. 1, the "N-weak" stars are systematically slightly He-weak and slightly O-strong, both by about 0.05 dex relative to the equivalent widths of the "typical" stars, a result which strongly argues that there is indeed contamination of the surfaces of the spectroscopically "typical" supergiants in the LMC and that it is the relatively rare "N-weak" stars which most likely have retained their main sequence surface abundances. If the LMC "N-weak" stars are completely un-contaminated with processed material, the line strengths differences observed here suggest–assuming the lines are on the linear portion of the curve-of-growth–that "typical" LMC B supergiants have surface helium and nitrogen abundances enhanced by factors of 1.12 and 2-3 (+0.05 dex), respectively, and surface oxygen is deficient by a factor of about 0.89 (-0.05 dex) compared with their original surface elemental abundances.

With the majority, perhaps 80%, of the B supergiants in the LMC having processed material on their surfaces, peculiarities affecting a few stars cannot be invoked, rather this conclusion must explained in terms of a common evolutionary model for LMC supergiants. Four mechanisms may be considered: accretion in a binary system by blue supergiant secondaries of processed matter from initially more massive, now very evolved primaries; stripping of the original H envelope by mass loss; rotationally induced mixing during the main sequence phase; and convective mixing during an ear-

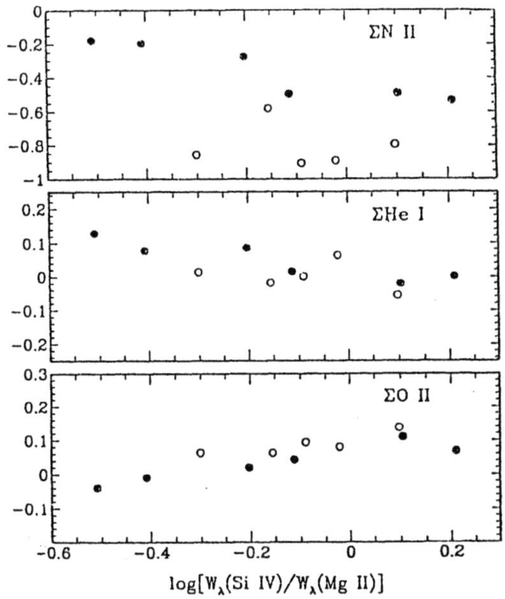

Fig. 1. Equivalent widths for the sums of several O II, He I, and N II lines plotted against a temperature indicator, the log of the equivalent widths of Si IV λ4089 to Mg II λ4481. Higher temperatures are to the right. The spectroscopically "typical" stars are plotted with filled circles and the "N-weak" stars are plotted with open circles. Vertical error bars are comparable to the symbol sizes.

lier supergiant phase. Only the previous red supergiant phase is considered plausible (Fitzpatrick and Bohannan 1992).

Interest in models of stellar evolution in which blue supergiants descend from red supergiants and have surface elemental abundance changes was revived by SN1987A. Langer (1991) has published models of evolutionary tracks which have "blue loops" for initial masses in the range we observe here. The blue supergiants show surface abundance changes of helium, nitrogen, oxygen, and carbon similar to what we estimate for the "typical" B supergiants in the LMC. However, his "blue loop" models do not extend far enough to high temperatures to produce the early B-type supergiants observed here. More theoretical calculations and diagnostic observations are needed before this situation will be resolved. Due to well-know observational advantages, the LMC will continue to be an important proving-ground for theories of stellar evolution.

References

Fitzpatrick, E.L. (1991): Publ. Astron. Soc.Pacific, 103,1123.

Fitzpatrick, E.L., and Bohannan, B. (1992): Astrohys. J., accepted for publication.

Langer, N. 1991: Astron. & Astrophys. 252,669.

THE INFLUENCE OF CLOSE BINARY EVOLUTION ON THE THEORETICALLY PREDICTED NUMBER DISTRIBUTION OF WR STARS IN THE GALAXY AND IN THE MAGELLANIC CLOUDS.

D. VANBEVEREN
Dept. of Physics, Vrije Universiteit Brussel, 1050 Brussels, Belgium

C. DE LOORE
Astrophys. Inst., Vrije Universiteit Brussel, 1050 Brussels, Belgium.

ABSTRACT Binary computations are used in combination with single star evolutionary calculations in order to study the theoretically predicted number distribution of WR stars in the Galaxy and in the Magellanic Clouds.

INTRODUCTION

In order to explain the observed WR/O and WC/WN number ratio gradient as a function of metallicity Z, three suggestions appeared in litterature, i.e.

a. $\dot{M}$ during the WR phase is a function of Z (Vanbeveren and Conti, 1980)

b. Different IMF for different Z (Garmany et al., 1982)

c. $\dot{M}$ during the RSG phase is a function of Z (Maeder et al. 1981).

Using single star evolutionary computations only, quantitative results for the ratio's have been discussed by Maeder (1991) using suggestion C and it was concluded that they match reasonably well the observations. In the present paper we will combine binary and single star evolutionary computations and rediscuss the predicted ratios. We will demonstrate that it is very unlikely that suggestion C is the (only) reason for the observed gradient.
For single stars we will use the evolutionary computations of Maeder (1991). During the RSG phase, M $Z^{0.5}$ whereas during the WR phase, M is independent from Z. The recent improvements of Schaller et al. (1992) lead to almost similar results. For massive close binaries we will use the Galactic results of Vanbeveren (1991). For the Magellanic Clouds we performed a complete set of binary computations which will be published in A.A. Supplements. The binary computations use the same physical ingredients as the single star results, i.e. the same M during core hydrogen burning, the same convective core overshooting (α = 0.25) and M during the WR phase is considered independent from Z. This set of single and binary computation will be called 'standard' evolution of massive stars.

THE RESULTS.

For the Galaxy and the MC's we adopt an IMF $\div M^{-\alpha}$ ($\alpha = 2.5 \pm 0.2$) and we assume that the IMF of primaries in massive close binaries is similar to that of single stars. Denoting by f the O + OB type binary frequency, it follows that the theoretically expected WR/O number ratio can be estimated as

$$\frac{WR}{O} = f\left(\frac{WR}{O}\right)_b + (1-f)\left(\frac{WR}{O}\right)_s$$

where the subscripts b and s stand for the predicted ratio for binaries, resp. single stars. It is obvious that the predicted ratio of WR single stars /WR binary components (with a normal companion) follows from

$$\frac{WR_s}{WR_b} = \frac{(1-f)\left(\frac{WR}{O}\right)_s}{f\left(\frac{WR}{O}\right)_b}$$

Using f = 0.35 (Garmany et al., 1980) for the Galaxy and for the MC's, the predicted ratios are given in table 1. The results are average values for $\alpha = 2.5$ and may differ by $\pm$ 10 % for α varying from 2.3 to 2.7.

TABLE 1 The theoretical WR distribution compared to the observed one; the first row (resp. the second row and the third row) holds for the Galaxy, resp. the LMC and the SMC.

WR/O				WR_s/WR_b	WC/WN	
single	binary	total	observed		theo	obs
0.072	0.12	0.09	0.116	1.12	1.3	
0.017	0.085	0.041	0.04	0.36	1.0	0.2
0.0027	0.045	0.018	0.015	0.11	1.3	0.06

The observed WR/O number ratio is given as well. We conclude

Standard evolution of single stars and close binaries predicts a WR/O number ratio for the Galaxy and for the MC's which corresponds to the observed values.

Standard evolution of single stars and close binaries predicts that there should be as many WR + OB close binaries as WR single stars in the Galaxy and that the number of WR + OB close binaries is at least 2.4 (resp. 8) times the number of WR single stars in the LMC (resp. the SMC), i.e.

Standard evolution of single stars and of close binaries predicts that the WR population in low metallicity environnements is largely dominated by close binaries.

In table 1 we also compare the theoretical WC/WN ratio predicted by standard evolution with the observed ratio. As can be noticed

Standard evolution of single stars and of close binaries predicts a WC/WN ratio which is far too large in the MC's compared to observations (i.e. a factor 5 in the LMC and a factor 20 in the SMC) and thus

Standard evolution of single stars and of close binaries does not reproduce the observed WR distribution in the Magellanic Clouds.

We have explored the model which has been proposed originally by Vanbeveren and Conti (1980) i.e. the $\dot{M}$ during the WR phase is a function of Z. The details of this model will be published in a forthcomming paper (Vanbeveren and de Loore, 1992). The quantitative results allow us to conclude that

only when the $\dot{M}$ during the WR phase is a function of Z the WC/WN number ratio predicted by stellar evolution corresponds to the observed value in the Galaxy and in the Magellanic Clouds.

REFERENCES.

Garmany, C.D., Conti, P.S., Massey, P.: 1980, Ap. J. **242**, 1063.
Garmany, C.D., Conti, P.S., Chiosi, C.: 1982, Ap. J. **263**, 777.
Maeder, A.: 1991, A.A. **242**, 93.
Maeder, A., Lequeux, J., Azzopardi, M.: 1980, A.A. **90**, L17.
Schaller, G., Schaerer, D., Meynet, G., Maeder, A.: 1992, A.A. Suppl. (in press).
Vanbeveren, D.: 1991, A.A. **252**, 159.
Vanbeveren, D., Conti, P.S.: 1980, A.A. **88**, 230.
Vanbeveren, D., de Loore, C.: 1992, A.A. submitted.

Clusters, Chemical Evolution

James Lequeux, Michael R. Rosa

Bengt E. Westerlund, Bernard E.J. Pagel, Michael W. Feast

Chemical Abundances of B-Type Stars in the Magellanic Clouds [1,2]

A. Jüttner

Landessternwarte Königstuhl, D-6900 Heidelberg 1, Germany

Abstract: Recent abundance determinations of B-type stars located in different parts of the Magellanic Clouds are presented and discussed.

1 Introduction

Recent abundance analyses of seven B-type giants carried out within the ESO key programme on "Coordinated Investigations of Selected Regions in the Magellanic Clouds" (cf. de Boer et al. 1989, 1991) by our group in Heidelberg will be presented and discussed. For homogeneity reasons the results of a reanalysis of two B-type giants published by Reitermann et al. (1990) are also presented. We compare our results with those for B main sequence stars in the MCs published by the Belfast group and with the results derived for B supergiants published by the Munich group. For a detailed comparison of the chemical abundances derived from B-type stars with the compositions derived from late type stars and from HII-regions reference is made to the contribution to this conference by Pagel (1992).

B-type giants and main sequence stars are supposed to represent in their atmospheres the present day abundance pattern of their host galaxies. Hence the major aim of abundance determinations from B-type stars in the Magellanic Clouds is to learn more about the star formation history of the MCs and about the detailed production mechanism for individual elements. Within the framework of the chemical evolution in the MCs it is also important to analyse both stars located in the general field and stars located in clusters. The latter are important observational targets because the simultaneous knowledge of both ages and abundances should provide insight in chemical enrichment mechanisms of the MCs.

[1] Based on observations collected at the European Southern Observatory at La Silla, Chile and on observations by the International Ultraviolet Explorer collected at the Villafranca Satellite Tracking Station of the European Space Agency, Spain

[2] Invited talk at the Conference on **New Aspects of Magellanic Cloud Research**, Heidelberg, June 15-17 1992, G. Klare(ed.), Springer-Verlag.

2 Observations

For the selection of the MC cluster stars we concentrated on the HR diagrams of NGC 330 (SMC), NGC 1818 (LMC), NGC 2004 (LMC) and NGC 2100 (LMC) published by Robertson (1974). The basis for the field stars were the catalogues published by Azzopardi et al. (1975) and Garmany & Conti (1987) and by Brunet et al. (1975) and Conti & Garmany (1986) for the SMC and LMC, respectively.

One drawback is that B-type stars are frequently fast rotators. Stars with projected rotational velocities $v \sin i \geq 75\,\mathrm{kms}^{-1}$ are not suitable for abundance analysis work. For this reason we first had to carry out a preselection program, using the ESO/MPI 2.2 m telescope and the B&C spectrograph at a resolution of 30Å/mm. The typical exposure times for stars with $V \approx 14$ mag are between one and two hours in order to reach sufficient signal-to-noise ratios.

Within three observing runs between Dec. 1989 and Dec. 1991, subsequent high resolution ($\lambda/\Delta\lambda \approx 20000$) and high signal-to-noise ($S/N \geq 75$) spectra covering wavelength ranges from 3940Å to 4940Å and 3870Å to 5250Å, respectively were obtained using the Cassegrain Echelle Spectrograph (CASPEC) attached the ESO 3.6 m telescope. A detailed journal of the CASPEC observations can be found in Jüttner et al. (1992).

In addition to the CASPEC data we obtained spectrophotometry (resolution $\approx$ 7Å) data in the satellite UV for each object using IUE in both the SWP range ($\lambda \approx 1150 - 2000$Å) and the LWP range ($\lambda \approx 1825 - 3300$Å). The flux distribution in the UV range was completed by photometry data (U,B,V,R,I) in the optical/visual range compiled from the literature.

3 Model Atmosphere Analyses

The analyses of our stars are based on LTE line blanketed model atmospheres calculated with the ATLAS 8 code (Kurucz, 1979 and literature quoted therein). The equivalent widths and line profiles are computed with a modified version of the BHT code (Baschek et al. 1966, Peytremann et al. 1967).

The surface gravity g is deduced from the fit of the Balmer line profiles Hβ, Hγ and Hδ using the broadening theory of Vidal, Cooper and Smith (Vidal et al. 1974). The temperature $T_{\rm eff}$ is derived using the ionization equilibria of SiII/SiIII and/or SiIII/SiIV. It should be noted that these equilibria depend via curve of growth effects significantly on the microturbulence and even on the abundances used in the model atmosphere calculations; both are initially unknown parameters. Therefore the ionization equilibrium calculations are done for a set of microturbulente velocities ranging from 5 to 20 kms^{-1} in steps of 5 kms^{-1}. Since we are performing the model atmosphere analyses for stars in the MCs we usually use a 1/3 solar abundance pattern in the model atmosphere code as an initial guess. The microturbulence velocity is finally derived in such a way that there is no obvious dependence of oxygen abundance on the line strength. The model atmosphere parameters which we finally derived from the procedure described above are shown in Table 1.

Table 1. Model atmosphere parameters of our target stars.

Object	$T_{\rm eff}$ [K]	$\log g$	$\xi_{\rm micro}$ [kms^{-1}]
NGC 330/B30	18800	2.91	10
NGC 330/B22	22700	3.22	10
AV 175	20500	2.84	20
NGC 2004/B30	23700	3.40	20
NGC 2004/B15	24400	3.72	15
NGC 1818/D12	19600	3.13	10
NGC 2100/C13	21000	2.78	20
Bru 217	21080	2.92	15
Bru 231	20000	2.63	15

The typical errors are $\Delta T_{\rm eff} = \pm 1000$ K, $\Delta \log g = \pm 0.15$ dex and $\Delta\xi_{\rm micro} = \pm 5$ kms^{-1}.

With the atmosphere parameters derived form the Balmer line fits and the silicon ionization equilibrium we are able to give reddening values for given reddening laws (see below) by comparing the observed flux distribution with the calculated continuum.

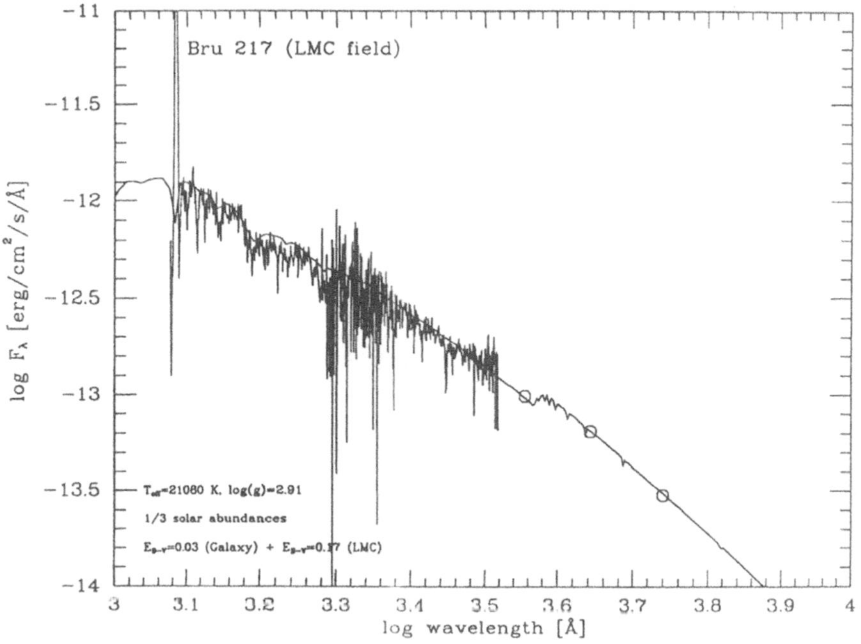

Fig. 1. Comparison of the observed continuum (IUE and UBV) and computed energy distribution of Bru 217

Since our target stars are members of the MCs we have a reliable knowledge of the distance and hence luminosity. A distance modulus of $m - M = 18.5 \pm 0.1$ for the LMC and $m - M = 18.85 \pm 0.1$ for the SMC has been adopted. The derived stellar parameters as well as the reddening values are presented in Table 2.

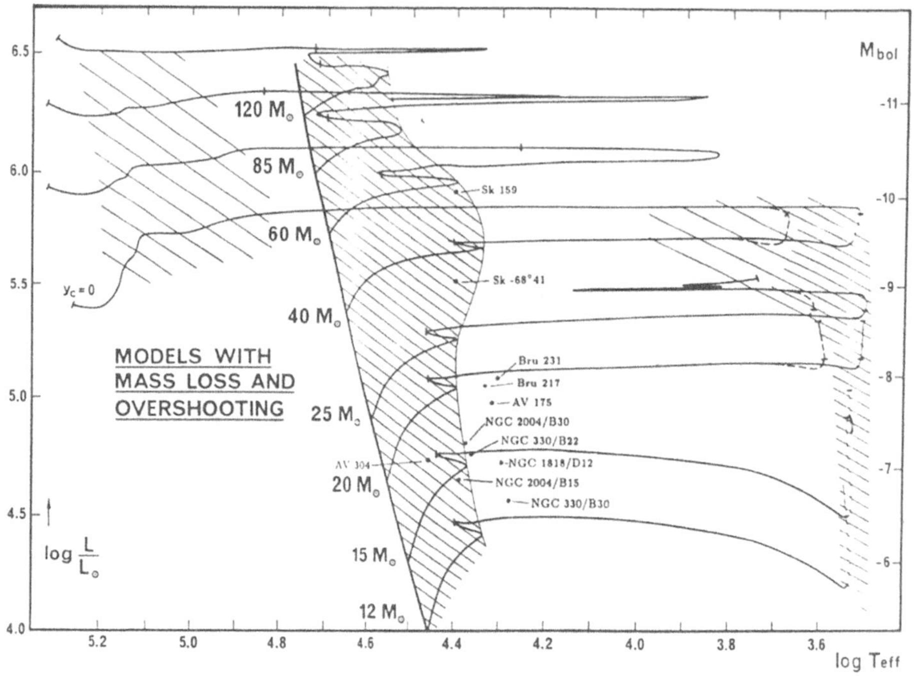

Fig. 2. Location of analysed objects in the HRD and evolutionary tracks of Maeder & Meynet (1987). Apart from our B giants the two B supergiants Sk 159 and Sk-68°41 (Lennon et al. 1991) as well as the SMC B main sequence star (Dufton et al. 1990) are indicated.

Table 2. Stellar parameters of the target stars.

Object	$M/M_\odot$	$R/R_\odot$	$M_{\rm bol}$	$M_{\rm V}$	$E_{\rm B-V}$ gal.	$E_{\rm B-V}$ MC
NGC 330/B30	9.6 ± 3	18.0	-6.66	-4.80	0.03	0.03
NGC 330/B22	14.6 ± 4	15.5	-7.15	-4.79	0.03	0.07
AV 175	15.3 ± 4	24.6	-7.70	-5.54	0.03	0.09
NGC 2004/B30	20.7 ± 6	15.1	-7.27	-4.91	0.05	0.01
NGC 2004/B15	26.5 ± 9	11.7	-6.86	-4.40	0.05	0.01
NGC 1818/D12	19.4 ± 5	19.8	-7.04	-5.17	0.05	0.05
Bru 217	19.1 ± 5	25.3	-7.89	-5.77	0.03	0.17
Bru 231	18.0 ± 8	34.4	-8.33	-6.25	0.03	0.20

The basis of the reddening values are the reddening laws published by Savage & Mathis (1979) for the galactic foreground, Howarth (1979) for the LMC and Prevot et al. (1983) for the SMC. Note that the total reddening is the sum of the last two columns in Table 2.

4 Abundances in B-type Giants

Using the model atmospheres presented in Table 1, the abundances of the metal lines were derived by comparing the calculated and the measured equivalent widths using the BHT code. For the diffuse helium lines the broadening theories of Barnard, Cooper and Smith (1974) and Gieske & Griem (1968) were used, while for the sharp helium lines that of Griem et al. (1966) was applied.

The Figures 3 to 12 present the differential element abundances ($\Delta \log \varepsilon = \log \varepsilon_{MC} - \log \varepsilon_{HR\,2618}$) versus the elements analysed in our target stars. For each object we also present a mean metal value indicating the mean metal abundance relative to HR 2618 for all elements heavier then oxygen. All error bars are the sample standard deviations.

So far three B-type giants in the SMC have been analysed. Two stars are supposed to be members of the prominent blue globular cluster NGC 330 (Robertson 1974, Carney 1985) and one star is located in the surrounding field of NGC 330. With respect to the galactic comparison star HR 2618, the two cluster stars exhibit a mean underabundance of 0.6 dex while the field star located at a distance of about 8′ away from the cluster shows a mean metal depletion of 0.4 dex.

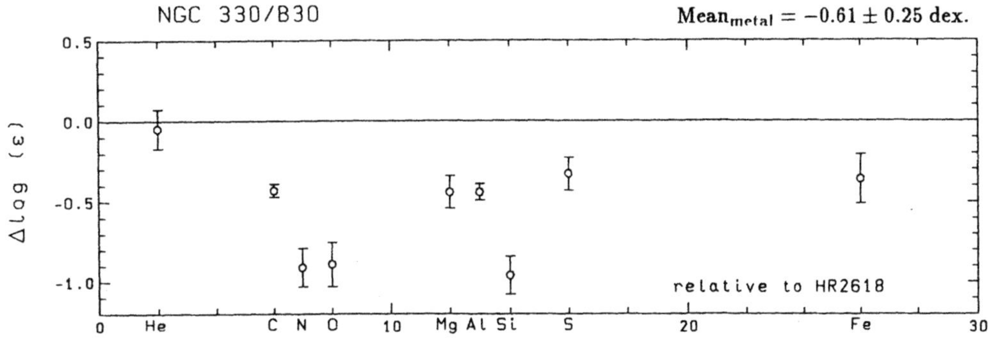

Fig. 3. Differential abundance pattern for NGC 330/B30.

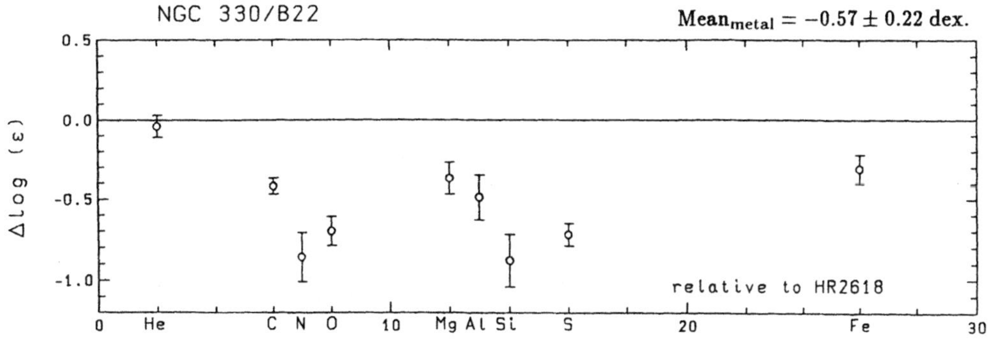

Fig. 4. Differential abundance pattern for NGC 330/B22.

Our sample in the LMC so far consists of one supposed member in NGC 1818, one in NGC 2100 and two in NGC 2004. The two field stars Bru 217 and Bru 231 are

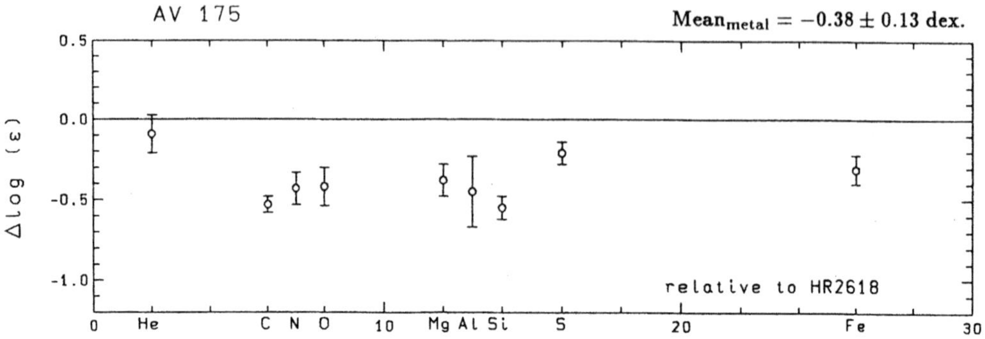

Fig. 5. Differential abundance pattern for the field star AV 175

located in the field around SN 1987A. The derived abundances of the cluster stars relative to HR 2618 are shown in the Figures 6 to 9.

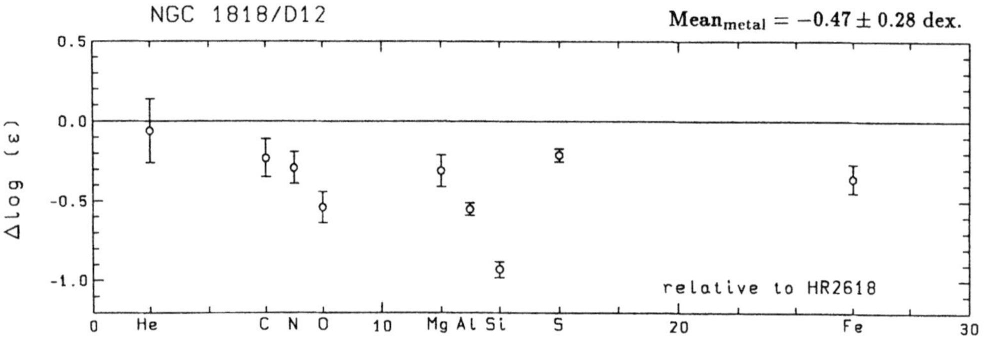

Fig. 6. Differential abundance pattern for NGC 1818/D12.

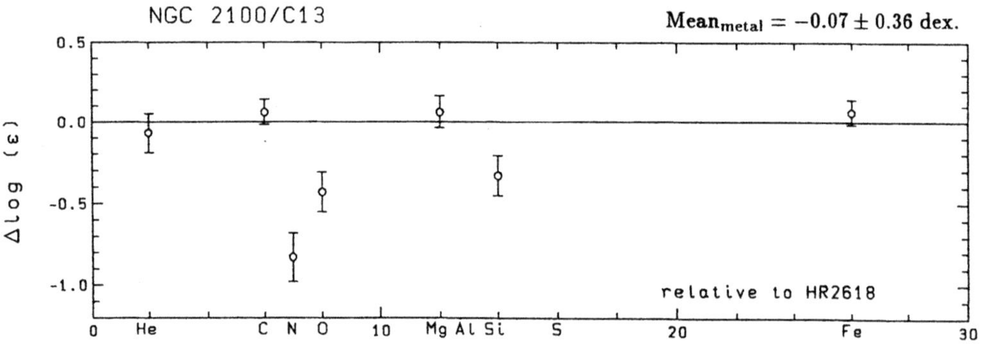

Fig. 7. Differential abundance pattern for NGC 2100/C13.

The results for NGC 2100/C13 should be treated with care because the preselection spectra provided by M. Bessell in 1989 showed narrow absorption lines while the spectra obtained in Dec. 1991 by our group exhibits broad lines. NGC 2100/C13 may hence be a binary system.

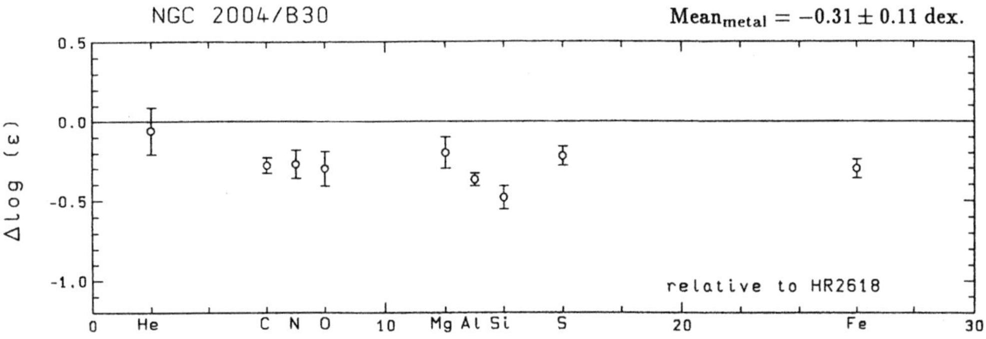

Fig. 8. Differential abundance pattern for NGC 2004/B30.

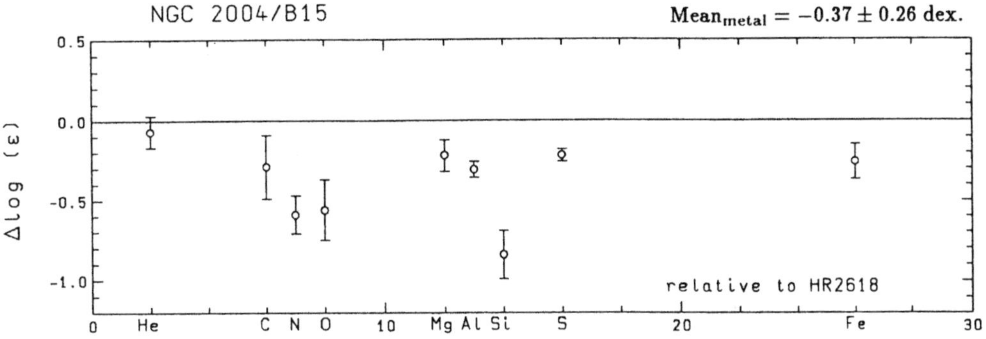

Fig. 9. Differential abundance pattern for NGC 2004/B15.

The cluster membership of NGC 2004/B15 may be doubted due to its 40 kms^{-1} higher radial velocity with respect to the cluster velocity. It is presumably a kind of "runaway" star. The derived abundances of the field stars relative to HR 2618 are presented in Figure 10 and 11, respectively.

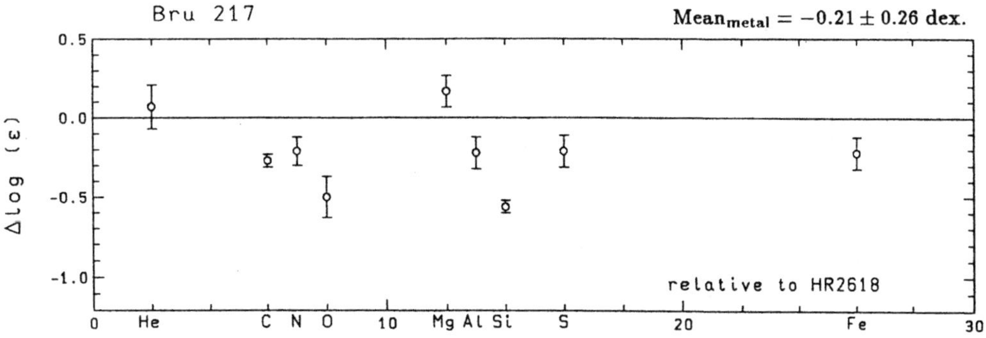

Fig. 10. Differential abundance pattern for Bru 217.

The data obtained from B-type giants located in the SMC indicates an abundance difference of approximately 0.2 dex between cluster and field stars. The abundance data of the LMC stars show almost no difference between the cluster and the field

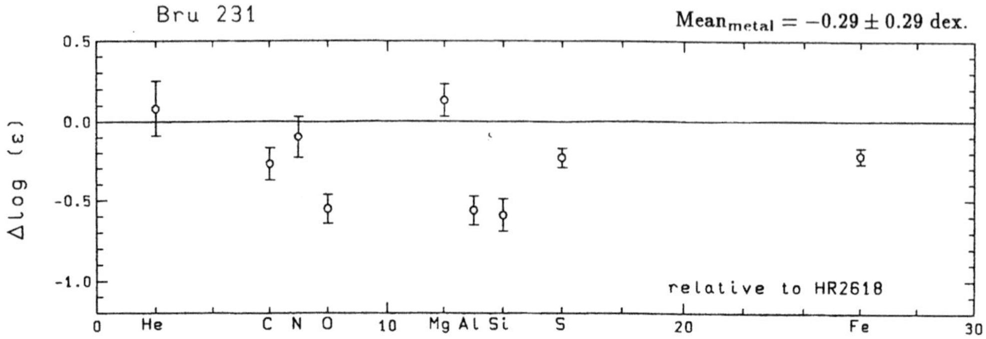

Fig. 11. Differential abundance pattern for Bru 231.

stars. Only NGC 1818/D12 exhibits an underabundance of about 0.2 dex if compared to Bru 217 and Bru 231.

Note that the galactic comparison star HR 2618 shows a slight underabundance with respect to the solar values (Grevesse & Anders 1988). The abundance pattern found for HR 2618 relative to the solar values is shown in Figure 12.

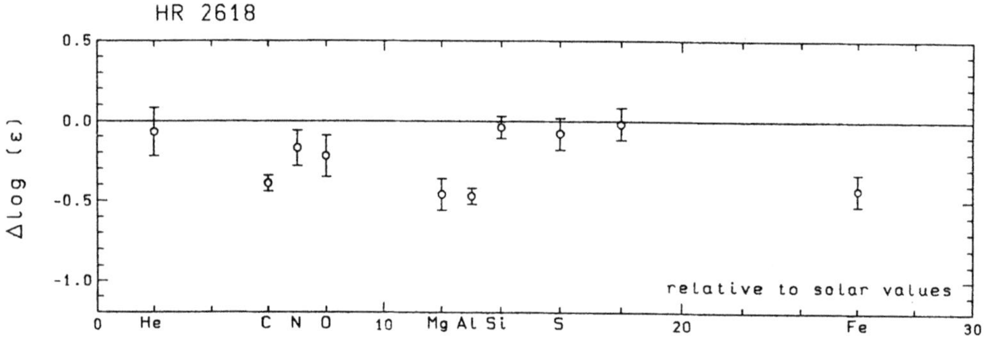

Fig. 12. Abundance pattern for HR 2618 relative to the solar values.

5 Abundances in B-type Supergiants

Recent NLTE analyses of MC supergiants are described in Lennon et al. (1991). They obtained high resolution ($\Delta\lambda \approx 0.25$Å) spectra of the B supergiants Sk-68°41 (LMC) and Sk 159 (SMC) using the ESO 3.6 m telescope and CASPEC. The abundance analyses are based on detailed NLTE model atmospheres and subsequent NLTE line formation calaculations for all elements of interest. The principle of the analysis is to use the profile fit of the Balmer lines (Hγ, Hδ) for the gravity determination and the profiles of HeI and HeII lines for the simultaneous determination of $T_{\rm eff}$ and the helium abundance ($y = N_{\rm He}/(N_{\rm H} + N_{\rm He})$).

The adopted stellar parameters ($T_{\rm eff}$, $\log g$), the helium abundances and gravitational masses (derived from $\log g$ and luminousity) and their dereddened magnitudes are given in Table 3.

Table 3. Adopted stellar parameters.

Object	V	$T_{\rm eff}$	$\log g$	y	$M_{\rm grav}/M_\odot$	$M_{\rm ZAMS}/M_\odot$
κ Ori	2.06	25000	2.70	0.20	14	25
Sk-68°41	12.01	25000	2.60	0.23	15	35
Sk 159	11.90	25000	2.55	0.35	29	55

Note that all objects indicate a helium enrichment in their atmospheres. A possible explanation for this, suggested by Lennon et al. (1991), is that these stars had evolved through a red supergiant phase, and had moved bluewards again in the HR-diagram to appear once again as blue supergiants. The helium overabundances could then be due to the influence of accumulated mass loss over the star's livetime and convective processes during the red supergiant phase of evolution.

The results of the abundance analyses relative to κ Orionis are summarized in Table 4. Note, that the NIII lines imply a nitrogen rich atmosphere while the NII lines show a nitrogen underabundance.

Table 4. Metal abundances in Sk-68°41 and Sk 159 relative to κ Ori.

Element	Sk-68°41 (LMC)	Sk 159 (SMC)
C	+0.06	-1.09
N (II)	-0.12	-1.00
N (III)	0.14	0.17
O	-0.33	-1.10
Mg	-0.78	-1.00
Si	-0.14	-0.75

Considering Sk 159 one can see that the general picture of the SMC being the more metal deficient of the MCs is confirmed by this NLTE analysis. Note that Sk 159 exhibts a much lower oxygen abundance than found in any other SMC star. The fact that Sk 159 is nitogen rich and carbon and oxygen poor is at least qualitatively consistent with the idea of CNO-cycled material having been mixed into its atmosphere.

Important improvements in analysing B supergiants' atmospheres are under way; more sophisticated NLTE "unified" expanding model atmospheres are being developed by the Munich group (cf. Puls 1991). However, apart from computational difficulties of deriving absolute abundances the finding that the atmospheres of B supergiants are contaminated by CNO-cycled material is even more serious. This makes B supergiants less suitable for deriving the abundances (He, C, N and O) characterizing the present evolutionary state of the MCs. To avoid these more principal difficulties analyses of less evolved stars are important.

6 Abundances of B Main Sequence Stars

A first differential abundance analysis of a main sequence B-type star located in the SMC was carried out by Dufton et al. (1990), preliminary results on a B main sequence star located in the LMC were presented by Howarth et al. (1991). They obtained high resolution spectra of AV 304 (SMC) and PS 34-16 (LMC) using the

University College London Echelle Spectrograph and the Image Photon Counting System (IPCS) attached to the 3.9 m Anglo Australian Telescope (AAT).

The model atmosphere parameters of AV 304 and PS 34-16 were deduced using line-blanketed LTE model atmospheres of Kurucz (1979). The effective temperature of AV 304 was deduced from the SiIII/SiIV ionization equilibrium while the temperature for PS 34-16 was derived from the Stroemgren *uvby* photometry. In both cases the surface gravity was derived from the fit of the Balmer line profiles.

The results of these differential analyses relative to the galactic comparison star τ Sco are shown in Table 6.

Table 6. Model atmosphere parameters and differential abundances for AV 304 and PS 34-16.

	AV 304 (SMC)	PS 34-16 (LMC)	PS 34-16 (LMC)
T_{eff}[K]	28000	22000	25000
$\log g$	4.1	4.0	4.2
ξ_{micro} [kms^{-}1]	5	5	5
Carbon	-0.8	-0.4	-0.3
Nitrogen	< -1.1	-0.2	-0.4
Oxygen	-0.4	+0.3	-0.2
Neon	-0.4	...	...
Magnesium	-0.6	-0.9	-0.6
Aluminium	< -0.7	...	...
Silicon	-0.7	-0.4	-0.7
Sulphur	-0.6	...	...

To demonstrate the sensitivity of the results to the adopted temperature, abundances for PS 34-16 have been also deduced for a model atmosphere with T_{eff}=25000 K and $\log g$=4.2. For this hotter model the helium abundance is consistent with that found in galactic B main sequence stars. The results agree qualitatively with those derived by us for B-type giants in the field and with those derived from analyses of e.g. late-type supergiants (Russell & Bessell 1989) or emission-line objects (Russell & Dopita 1990). The general picture that LMC and SMC are metal deficient compared with the Galaxy, with the SMC being more heavily depleted is confirmed.

7 Conclusions

Chemical abundances representing the present evolutionary state and the abundances of the ISM can be derived from B main sequence stars and B-type giants with fair or even with good accuracy. The results are not seriously affected by NLTE effects when working strictly differential.

From 6 B-type stars of the LMC both in the field and in clusters, we derive and underabundance of $\Delta \log \varepsilon \approx -0.3$ dex. (Only in the case of NGC 1818/D12 the composition seems to be slightly down by $\Delta \log \varepsilon \approx 0.2$ dex compared to the bulk of the stars). This underabundance is in reasonable agreement with the data derived from HII-regions and late type stars (cf. Pagel 1992).

The metallicity of the SMC B-type giant AV 175 is down by about -0.6 dex relative to the solar values. This result is in good agreement with the abundances derived by

Dufton et al. (1990) from the B main sequence star AV 304 and with the abundances derived from late type field stars and HII-regions (cf. Pagel 1992).

The metallicity in the young blue globular cluster NGC 330, however, appears to be below that of the surrounding field indicating a chemical inhomogeneity. Similar results are also found from photometric investigations by Grebel & Richtler (1992) and spectral analyses of late type stars (Spite & Spite 1992).

At variance with that, the B-type giants so far analysed in the LMC show almost no difference between cluster and field stars (apart from NGC 1818/D12, see above). However, one has to take into account that the two field stars Bru 217 and Bru 231 are not located in the surrounding fields of the clusters. We intend to observe further B-type giants around NGC 1818, NGC 2004 and NGC 2100 as well as NGC 330 in order to obtain more detailed information of the chemical compositions in cluster and field stars.

The abundance analyses of luminous supergiants in the MCs provide a good test of evolutionary scenarious of very massive stars located and formed in different physical and chemical environments compared to the Galaxy. Yet the so far derived absolute abundances are more problematic and have to be treated with care due to the complex structure of the atmospheres and the mixing with CNO-cycled material.

Acknowledgements: I am very grateful to Drs. B.Wolf and B.Baschek for stimulating discussions and for critically reading the manuscript. I would like to thank Dr. M. Bessell for providing us some of his spectroscopic data on B-type giants. This research was supported by the Deutsche Forschungsgemeinschaft (SFB 328).

References

Azzopardi M., Vigeneau J., Macquet M., 1975, A&AS 22, 285

Barnard A.J., Cooper J., Smith E.W., 1974, J. Quant. Spectros. Rad. Transf. 14, 1025

Baschek B., Holweger H., Traving G.: 1966, Abhandl. Hamburger Sternwarte, VIII, Nr.1, 26

de Boer K.S., Azzopardi M., Baschek B., et al., 1989 ESO Messenger 57, 27

de Boer K.S., Spite F., François P., et al., 1991 ESO Messenger 66, 14

Brunet J.P., Imbert M., Martin N., Mianes P., Prevot L., Rebeirot E., Rousseau J., 1975, A&AS 21, 109

Carney B.W., 1985, AJ 90, 1196

Conti P.S., Garmany C.D., 1986, AJ 92,48

Dufton P.L., Fitzsimmons A., Howarth I.D., 1990, ApJ 362, L59-L62

Garmany C.D., Conti P.S., Massey P., 1987, AJ 93, 1070

Gieske H.A., Griem H.R., 1968, ApJ 157, 963

Grebel E.K., Richtler T., 1992, A&A 253, 359

Grevesse N., Anders E., 1988, Solar-System Abundances of the Elements: A New Table. In: Waddington C.J.(ed.) AIP Conf. Proc. Cosmic Abundances of Matter, Minneapolis (1988)

Griem H.R., Baranger M., Kolb A.C., Oertel G., 1966, Phys. Rev. 125, 177

Howarth I.D., 1983, MNRAS 203, 301

Howarth I.D., Dufton P.L., Fitzsimmons A., Rolleston W.R.J., 1991, The Chemical Composition of Main Sequence Stars in the Magellanic Clouds. In: Heber U., Jeffery C.S.(eds) Proc. CCP7 Workshop on Atmospheres of Early Type Stars. Kiel(1991)

Jüttner A., Reitermann A., Stahl O., Wolf B., 1989, A&AS 81, 93
Jüttner A., Stahl O., Wolf B., Baschek B., 1991, In: The Magellanic Clouds (IAU Symp. No. 148), eds. R. Haynes and D. Milne, Kluwer, Dodrecht, p. 388
Jüttner A., Wolf B., 1991, Analyses of B-Type Stars in the Magellanic Clouds(Review). In: Heber U., Jeffery C.S.(eds) Proc. CCP7 Workshop on Atmospheres of Early Type Stars. Kiel(1991)
Jüttner A., Stahl, O., Wolf B., Baschek B., 1992, this volume
Kurucz R.L., 1979, ApJS 40, 1
Lennon D.J., Kudritzki R.-P., Becker S.R., Butler K., Eber F., Groth H.G. Kunze, D., 1991 A&A 252, 498
Pagel B.E.J., 1992, this volume
Peytremann E., Baschek B., Holweger H., Traving G., 1967 Un Programme FORTRAN d' Analyse Quantitative de Spectre Stellaires, Obs. Geneve
Prevot M.L., Lequeux J., Maurice E., Prevot L., Rocca-Volmerage B., 1984, A&A 132, 389
Puls J., 1991 A&A in press
Reitermann A., Baschek B., Stahl O., Wolf B., 1990, A&A 234, 109
Robertson J.W., 1974, ApJS 15, 261
Russell S.C., Bessell M.S., 1989, ApJS 70, 865
Russell S.C., Dopita M.A., 1990, ApJS 74, 93
Savage B.D., Mathis J.S., 1979, ARA&A 17, 73
Spite F., Spite M., 1992, this volume
Tammann G.A., 1987, In: Observational Cosmology (IAU Symp. No. 124), eds. A. Hewitt, G. Burbidge, Li Zhi Fang, Reidel, p. 151
Vidal C.R., Cooper J., Smith E.W., 1973, ApJS 25, 37

Analyses of F supergiant stars in the Magellanic Clouds

François Spite, Monique Spite

Observatoire de Paris-Meudon, F 92195 MEUDON cedex, France

Abstract: The metallicities found for the F stars in the Clouds depend strongly on the choice of the temperature of the model. The determinations of the different authors are in good agreement. A few more stars have been analyzed in the LMC. The difference of metallicities between field stars and stars in the young cluster NGC330 is a more difficult problem and obviously deserves further investigation. The relative abundances are more robust. The pattern of the abundances of the elements in the F supergiants of the Clouds is very similar to the (non-solar) pattern of the Galactic supergiant Canopus, except for the neutron capture elements heavier than barium: these elements are enhanced in the Clouds. The relative abundance of carbon is the same as the relative abundance in Canopus: this result desagrees with the overdeficiency of carbon found in the Clouds from the HII regions. The behavior of oxygen is similar to the Galactic one.

1 Introduction

Abundances are an important part of the information necessary for finding the history and the evolution of the Magellanic Clouds (see for example Russell,1991). The values of the stellar abundances are sensitive to the choice of the parameters of the model atmosphere which is best fitted to the observed star : the most critical parameter is the temperature, but the choice of the other physical parameters of the stellar atmosphere, such as the gravity and the microturbulence velocity parameter ξ, is also of some importance.

Moreover, in the Magellanic Clouds only supergiants are presently accessible to detailed analyses. In such stars, NLTE effects occur : the main effect is the overionisation, and it has to be taken into account. Finally, it is obvious that the analysis has to be conducted differentially, relative to a well analysed supergiant of our Galaxy : the different authors chose the F type supergiant Canopus (α Car), analysed in detail by Luck & Lambert (1985). A final warning : the relative abundances in Canopus show a pattern somewhat different from the solar pattern (*e.g.* Na, O). However, the metallicity (iron abundance) of Canopus is solar.

2 The field stars in the LMC

Unfortunately, no many new results are available since the last IAU Symposium; it is however worth discussing the coherence of the previous results, and trying to use at best the most recent ones. Russell & Bessell (1989) observed 8 F-type supergiants in the LMC (one is classified G0). They obtained spectra in the blue region and they determined the temperature from *uvby* photometry, using the calibration by Lester et al. (1986), based on the Kurucz's models. McWilliam and Williams (1991) observed 6 other supergiants, in the list of Grieve & Madore (1986a, 1986b), determining the temperature by a calibration of de-reddened *V–K, V–J, J–K* colors; the calibration is derived (McWilliam 1991) from the Infra-red Flux method of Blackwell (see Blackwell & Petford 1991 and references therein). The calibration by the Infrared Flux method is in principle the best one. Unfortunately, in the region of F type supergiants, the calibration is determined from only a small number of stars, and the McWilliam's formula providing T_{eff} as a function of *B–V* is perhaps not yet completely reliable, although it can be used in a differential way. Finally, François (1991) observed a few supergiants in the LMC, and, among them, one of the list of McWilliam and Williams and he found for this common star a fair agreement. Spite & Spite also observed one star. Russell & Bessell have also determined the $(B-V)_o$ colors for their 8 F-type stars. It is therefore possible to compute the temperature of the 8 stars of Russell & Bessell, using various calibrations: we will come back to this point about the discussion of the metallicity difference between field and clusters. Here we will only note that the formula of McWilliam gives temperatures higher than the temperatures found by Russell & Bessell. However, the temperature differences between the various authors are somewhat compensated by the choice of the other parameters: for example, McWilliam and Williams do not force the ionisation equilibrium. The final results are in rather good agreement.

The results of these spectroscopic investigations of field stars in the LMC may be summarised :

- a mean metallicity [Fe/H] = –0.30 dex for the 8 stars of Russell and Bessell; the results of François and Spite for three more stars are in good agreement. McWilliam and Williams analyzed 6 LMC stars without forcing the ionisation equilibrium. From the measurements of the FeII lines, they found [Fe/H] = –0.29 dex in perfect agreement with the preceding results, but [Fe/H] = –0.51 when the FeI lines are used.
- a carbon deficiency similar to the iron deficiency (Russell & Bessell, Spite & Spite 1991), a result contrasting with the previous analyses of HII regions (Dufour 1984)
- an enhancement of the elements heavier than barium (Russell & Bessell). Let us note however that this enhancement is not confirmed by McWilliam and Williams.

About the scatter of metallicity encountered from star to star let us note two points :

1) In collaboration with Tom Richtler (Richtler et al. 1989), a cool star (B12 following the designation of Robertson 1974) was observed in NGC 1818. The temperature is derived from *H–K* colors (interpolated using *B–V*) and leads to a metallicity as low as –0.8 dex or even –1 dex.

2) Recent observations of one star in the young association NGC 1948 provides a preliminary metallicity around –0.3 dex (see the Poster by Spite et al., this meeting).

The metallicity [Fe/H] is difficult to determine with accuracy in supergiant stars, the numerical value depends strongly on the choice of the temperature, and it happens that often only one type of photometry is available for the stars of the Magellanic Clouds (no independent check available); the reddening correction is not easy to determine and last but not least, the calibration of the colors of the supergiants is not so good as the calibration of more accessible types of stars. The relative abundances are more robust, and the fact that these relative abundances are slightly different in the Clouds and in the Galaxy is an important point for our knowledge of the Chemical Evolution of the Galaxies.

An example of the pattern of element abundances for a F supergiant of the LMC is compared to the pattern of Canopus in Fig. 1.

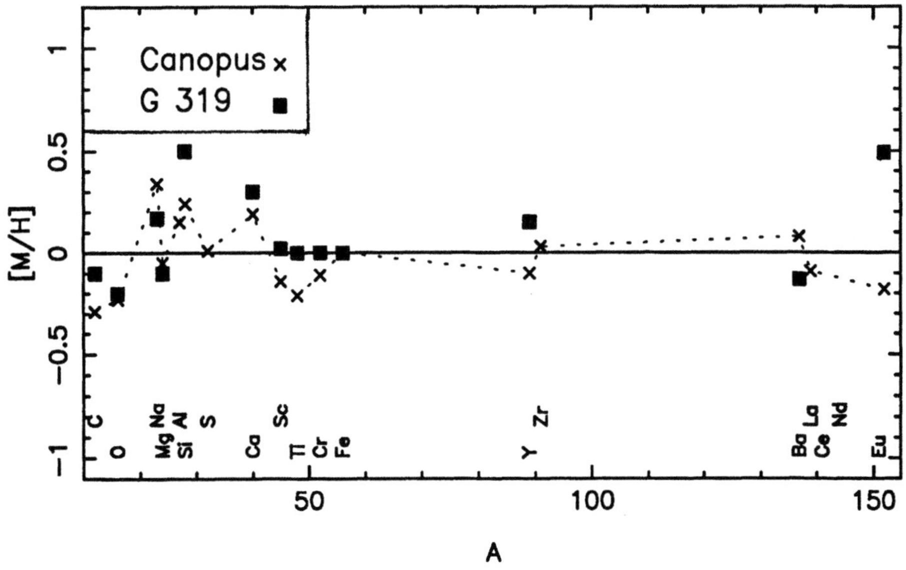

Fig. 1. Pattern of the element abundances for a typical LMC star G319 and for Canopus. The general trend is very similar. The only significative difference is the enhancement of Europium in the LMC star.

3 Field stars in the SMC

Spite et al. (1989) observed 3 field F-type supergiants, using Grieve and Madore 's color and reddening determinations. The reddening correction is rather high. We tried the temperature determination using the R–I colors and the calibration of Johnson (1966), but such temperatures were obviously much too high. We used the Schmidt-Kaler's calibration of B–V (Schmidt-Kaler 1982), which turned out to be in much better agreement with the excitation temperature. We are now able to check that our temperatures are only slightly higher than the temperatures computed from McWilliam's formula. Both our temperatures, and McWilliam's temperatures are higher than the Kurucz's temperatures.

Russell & Bessell (1989) observed also 8 field supergiants in the SMC. The temperature determination was made from the measured *b–y* color, following the same procedure as the one used for the LMC. Russell & Bessell determined moreover $(B-V)_o$ colors for these 8 stars. From this colors it may be noted that the temperatures of Russell & Bessell are similar to the temperatures of Kurucz, but lower, especially for the cooler stars.

It is clear that the Kurucz's calibration provides too low temperatures : it provides $T_{\rm eff} = 7100$ K for Canopus, although the detailed analysis by Luck & Lambert (1985) gives 7500 K, and any differential analysis relative to Canopus requires that we adopt the temperature of Luck & Lambert. Fortunately the Russell and Bessell calibration shows a trend relative to Kurucz's calibration, providing higher temperatures at lower $B-V$ values, correcting in this way, at least partly, the too low Kurucz's temperatures. The resulting abundances of the various authors are very similar. The mean metallicity of the SMC derived from the 3 stars analysed by Spite *et al.* (1989a) is [Fe/H]= -0.66 ± 0.1 dex, to be compared to the mean of the 8 stars of Russell & Bessell: -0.65 ± 0.2 dex . The near perfect numerical agreement is coincidental. The indicated errors reflect the scatter of the metallicity from star to star but does not take into account the systematic errors.

The pattern of the abundances of the elements is not solar, but very similar to the pattern found in Canopus (Fig. 2). The main difference with Canopus is the enhancement of the neutron capture elements, especially those heavier than Ba. The figure 18 of Russell & Bessell shows that this effect is more striking in the SMC than in the LMC.

The measurements of Russell & Bessell shows that the carbon deficiency is similar to the iron deficiency, again contrasting to the previous results derived from the HII regions (Dufour 1984). The same result is found quite independently from other lines of carbon in 3 SMC supergiants (Spite *et al.* 1989b). Moreover, it is also found (Spite *et al.* 1989b) that the ratio O/Fe is very nearly the same as in Canopus.

4 Comparison of the metallicity in field stars and in NGC 330

On photometric grounds (Richtler & Nelles 1983), it appears that the stars of the young globular cluster NGC 330 are somewhat more deficient than the supergiants of the field, in spite of a similar age. We are lead to a delicate problem: the comparison of the metallicity found in the field stars with the metallicity found in NGC 330, and as recalled hereabove, the metallicity depends strongly on the choice of the temperature. A spectroscopic analysis of 4 stars of NGC 330, made in collaboration with T. Richtler, shows such a difference (Spite *et al.* 1991). However, the uncertainties of the temperature determination are rather large, and it is worth examining if the difference (not larger than the error bar) is really established.

Among the four supergiants of NGC 330, two F stars were analysed (some more were observed but are not yet interpreted). The reddening is supposed to be low in the literature. Bessell (1991a,b) proposes a higher value of the reddening. The temperature

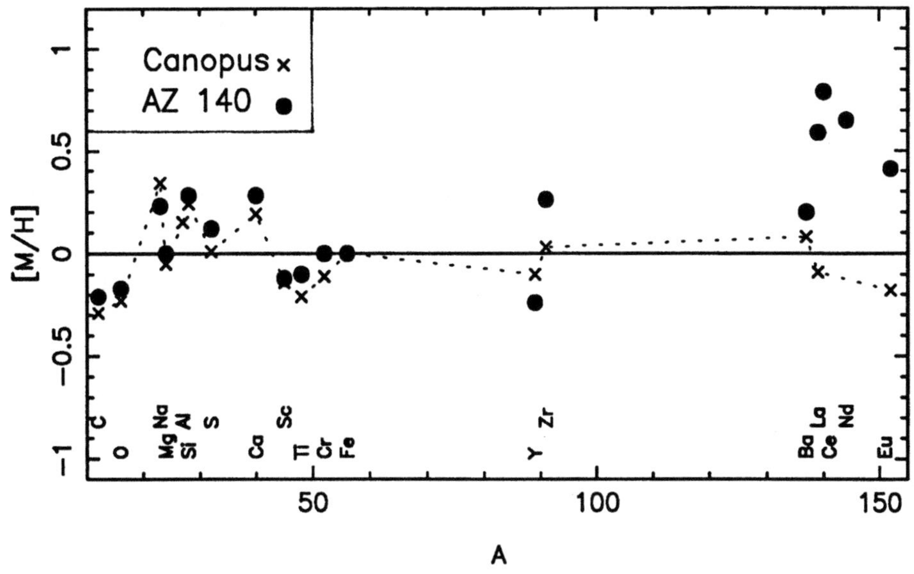

Fig. 2. Pattern of the element abundances for a typical SMC star AZ340 and for Canopus. Here again the general trend is very similar, but all the rare earths heavier than barium are significantly enhanced.

for our stars is approached in a first step from $B-V$, using the calibration of Kurucz (1979), assuming two different reddening values : a high one : $E(B-V) = 0.12$ (Bessell) and a low one : $E(B-V) = 0.03$. However the profiles of the Hα lines pushes clearly towards a higher temperature for the model, and we take this point into account. Our temperature determination is made for two different values of the reddening. As usual, the gravity determined from the evolutionary mass of the star is not able to provide the ionisation equilibrium. As usual, forcing the ionisation equilibrium provides a model with a "spectroscopic gravity" that cannot be taken at face value : the distortion of the gravity takes into account the main correction due to the departure of LTE in such stars : the overionisation. In this way, reliable metallicities are found, and it turns out that they are lower than the metallicities found in the field supergiants for the low value of the reddening, and hardly different for the high reddening.

It is somewhat surprising to find a difference of metallicities in stars of about the same age, but in his thesis, T. Richtler convincingly shows that a cluster can be chemically independent of its galaxy. We have here a very interesting problem. Additional information is provided by thc cool supergiants (see Bessell in this workshop). Additionnal information is also provided by the work of Grebel & Richtler (1992): see also their poster in this meeting. In order to solve the problem, a strictly identical procedure should be applied to field and cluster stars, preferably in the same region of the Small Cloud : we intend to do so in the near future. If additional colors such as $b-y$, $V-K$ *etc.* and reddenings were available for both field stars and cluster stars, this additional information would be very helpful.

5 Variation of the oxygen abundance in the Magellanic Clouds

Oxygen is an important element for stellar evolution and for models of the chemical evolution of the galaxies: it is one of the most abundant element in the Universe, a change of its abundance in the material which formed a star will change the stellar opacities and will increase the rate of the central H-burning (Simoda & Iben 1970) and as a consequence the age of the star. An important point is to compare the variation of [O/Fe] in our Galaxy and in the Magellanic Clouds. It has been suggested by McWilliam & Williams (1991) that this relation was steeper in the Magellanic Clouds. In Fig.3 we plot the ratio [O/Fe] versus [Fe/H] for all the stars we studied in the SMC and the LMC (field stars and one star in NGC330), and we compare to the galactic stars. No significant difference is apparent between the trend of [O/Fe] in the Galaxy and in the Magellanic Clouds.(For the Magellanic stars, the [O/Fe] ratio is not relative to the Sun but to the Galactic supergiant Canopus.) One of the star (G274), has been independently measured by François (1991) and by McWilliam & Williams (1991). It seems that there is a systematic difference in the determination, of the oxygen abundance but the paper of McWilliam & Williams is not enough detailed to allow us to explain this disagreement. The O/Fe ratios found in the Magellanic Clouds are in qualitative agreement with the model of Matteucci & Brocato (1990); see also Matteucci & François (1992).

6 Conclusion

The analysis of F type supergiants bring some useful information: a confirmation of the metallicities of the Clouds, as already obtained from HII regions, with however a difference for the carbon abundance. A few new determinations confirm the previous results. Oxygen is in reasonable agreement with the predictions of the chemical evolution. Some neutron capture elements are enhanced. The more delicate question of the metallicity difference between cluster and field stars, suggested by photometry and by a first spectroscopic analysis, will require a strictly differential analysis of field and cluster stars: some additional photometry and reddening determinations would help. A coordination will be maintained with the other analyses made in other institutes, especially those which begun a collaboration within the ESO Key program on coordinated research in the Magellanic Clouds.

Aknowledgments: A large part of our observations was made inside the ESO Key-program and we are gratefull to Klaas de Boer for his kind and efficient coordination.

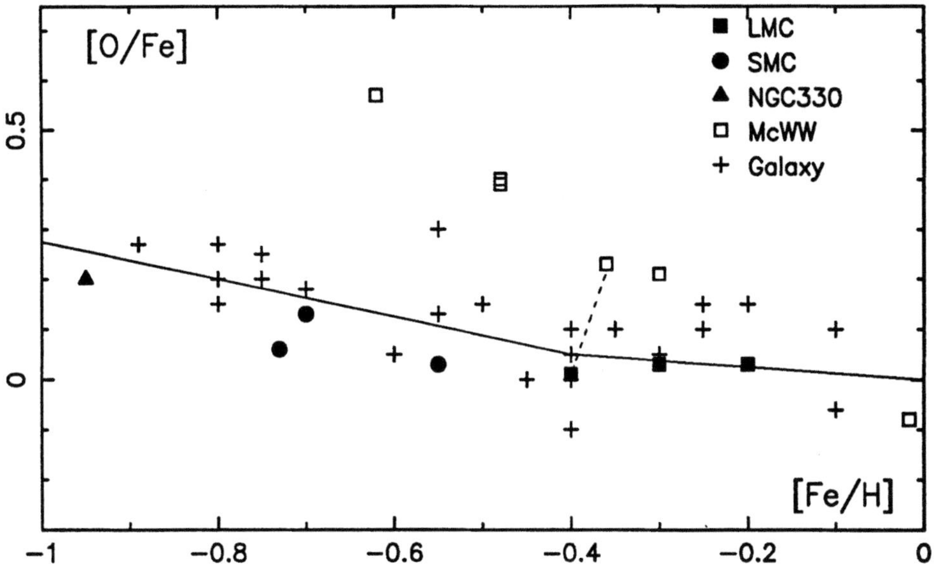

Fig. 3. Comparison of the trend of oxygen abundance in the Magellanic Clouds and in the Galaxy. The (relative) abundance of oxygene of field dwarfs and giants in the Galaxy is plotted versus the metallicity. The data (from the literature) are represented by the symbol +. Oxygen abundance measurements by Spite *et al.*(1989), Spite & Spite (1991) and François(1991) are represented by filled symbols. They define a trend similar to the Galactic trend with a very slightly lower oxygen abundance. The open symbols are determinations in the LMC by McWilliam & Williams (1991): their trend seem to reveal the presence of an unidentified systematic effect. (The dotted line joins the two determinations of the abundance ratios in G274.)

References

Bessell M. 1991a, in The Magellanic Clouds IAU symp. 148, eds. R.Haynes and D. Milne. Kluwer, Dordrecht p.273

Bessell M. S. 1991b, A&A 242, L17

Blackwell D. E., Petford A. D. 1991, A&A 250, 459

Dufour R.J. 1984, in Proceedings of IAU symposium 108 "Structure and evolution of the Magellanic Clouds" S. Van den Bergh and K.S. de Boer eds., Reidel, Dordrecht, p. 157

François P. 1991, Communication at the Workshop of the ESO Key-program, unpublished.

Grieve,G.R., Madore,B.F.: 1986 a, Astrophys. J. Suppl. Ser. ,62 ,427

Grieve,G.R., Madore,B.F.: 1986 b, Astrophys. J. Suppl. Ser. ,62 ,451

Grebel E. K., Richtler T. 1992, A&A 253, 359

Johnson H. L. 1966, ARA&A 4, 193

Kurucz R. L. 1979, ApJ Sup 40, 1

Lester J.B., Gray R. O., Kurucz R. L. 1986, ApJ Sup 61, 509

Luck R.E., Lambert D. L. 1985,ApJ 298, 782

Matteucci F., Brocato E. 1990, ApJ 365, 539

Matteucci F., François P. 1992, A&A *in press*

McWilliam A. 1991, A.J., 101, 1065

McWilliam A., Williams R. E. 1991, in : IAU Sympos. 148, R. Haynes & D. Milne, eds., Kluwer, p. 391
Richtler T., & Nelles B. 1983, A&A, 119, 75
Richtler T., Spite M., & Spite F. 1989, A&A, 225, 351
Robertson J. W.1974, A&AS 15, 261
Russell S. C., Bessell M. S., 1989, ApJ Sup 70, 865
Russell S. C. 1991, in : IAU Sympos. 148, R. Haynes & D. Milne, eds., Kluwer, p. 367
Schmidt-Kaler Th. 1982 in : Landolt -Börnstein Numerical data and functional Relationship in Science and Technology, New series, Group VI, Vol 2b, eds. K. Schaifers & H.H. Voigt, Springer, p. 21
Simoda M., Iben I., 1970, ApJS 22, 81
Spite F., Spite M., François P. 1989a, A&A 210, 25
Spite M., Barbuy B., Spite F. 1989b, A&A 222, 35
Spite M.,Spite F.,in : IAU Sympos. 148, R. Haynes & D. Milne, eds., Kluwer, p. 372
Spite F., Richtler T., Spite M. 1991, A&A 252, 557
Spite M., Spite F. 1992 (in preparation).

Abundances of KM Supergiants in Magellanic Cloud Clusters

Michael S. Bessell

Mount Stromlo and Siding Spring Observatories
Private Bag, Weston Creek P.O.
Canberra, ACT 2611 Australia

Abstract: Examination of the stellar content of YPC and the neighbouring fields indicates that the young stars appear to have been formed at the same time and place and therefore should have the same abundances. Those analyses which have come to a different conclusion have been carefully examined and explanations are given for discounting that work. Problems with model atmospheres and the importance of closely differential analyses are addressed and a method of easily assessing the temperature of K stars from a line ratio near the NaD lines is shown. Using these techniques, the abundances of the KM supergiants in the Cloud YPC and field are found to be the same and similar to the abundances of the FG supergiants.

1 Why YPC and Field Stars should be Similar

The Kinematics of the young stars and gas in the Clouds are well established due to the high accuracy radial velocities of late type supergiants in the LMC published by Prevot et al. (1985) and similar data for young stars and HII regions in the SMC by Maurice et al. (1986, 1989). The young supergiants in the LMC are distributed in super-associations of between 800 and 2400 pc in diameter which are confined to a disk and participate in a well established rotation along with the nearby gas. The internal velocity dispersion of these stellar associations average about 6.6$\pm$1.3 kms^{-1} which is similar to that of the HII regions and close to the velocity dispersions of CO clouds and HI in some external galaxies (Martin & Prevot 1989). The Young Populous Clusters (YPC) NGC 1818, NGC 2004 and NGC 2100 are embedded within the LMC associations B1-3, SHIII and 30 Dor 3 respectively and have the same mean radial velocity as the associated supergiants. The velocity dispersions of the cluster red giant stars within 15 pc (1 arcmin) are between 1 and 2.5 kms^{-1} (NGC 1866: Fischer et al. 1991; NGC 2004; Andersen 1992) while the velocity dispersion of the red giants in the surrounding 300 pc diameter field is about 3-4 kms^{-1}. This is reminiscent of the double cluster h and χ Persei within the Per OB1 association, where nine innermost red giants around χ Persei have a dispersion of 3.6 kms^{-1} and the 24 outer giants within a diameter of 170 pc have a dispersion of 6 kms^{-1}.

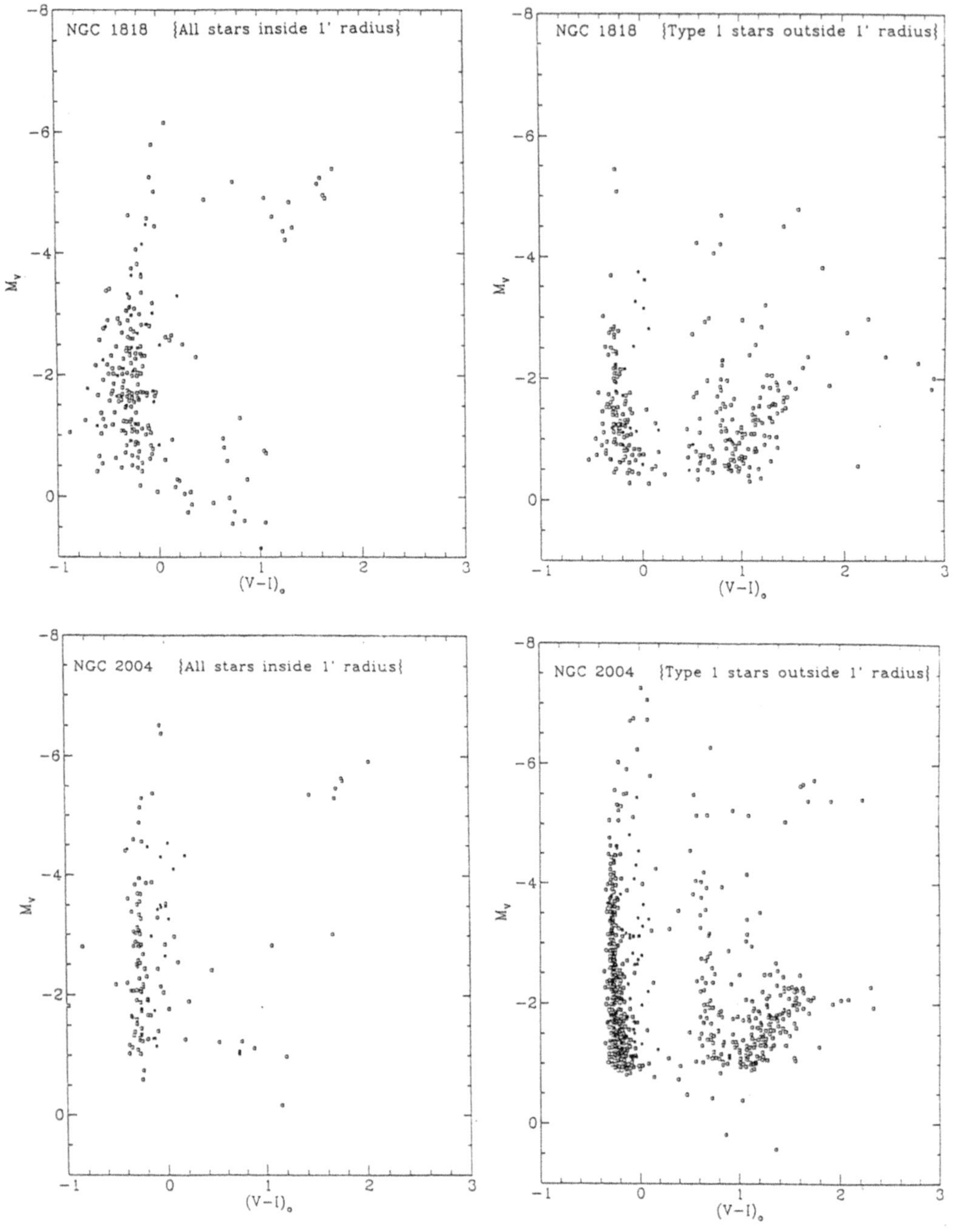

Fig. 1a. V versus V-I diagrams of the inner cluster region and outer 5′ field of the clusters NGC 1818 and NGC 2004 in the LMC.

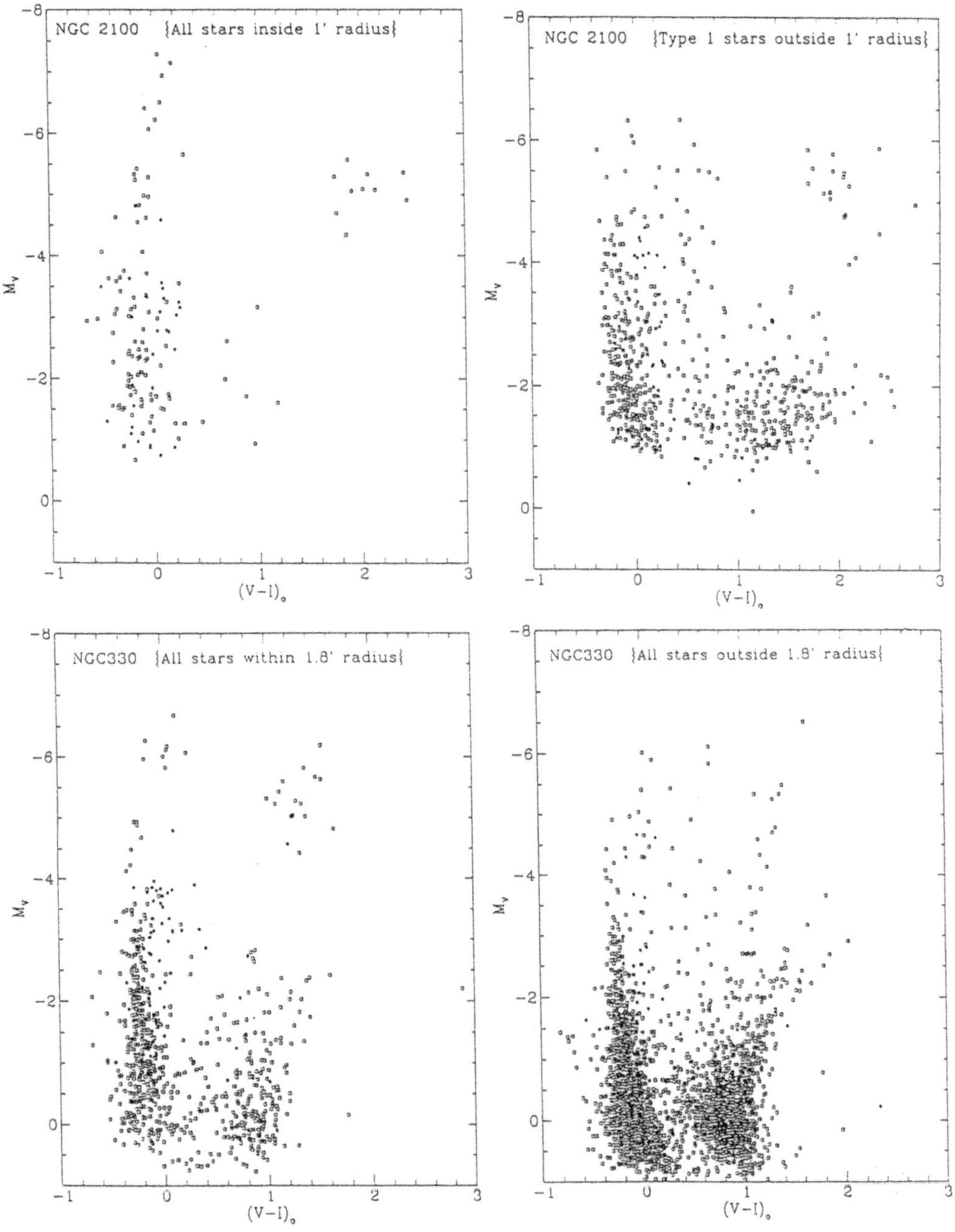

Fig. 1b. V versus V-I diagrams of the inner cluster region and outer 5′ of the cluster NGC 2100 in the LMC and NGC 330 in the SMC.

Star counts and light profiles measured by Elson, Fall & Freeman (1989) for YPCs indicate a power-law stellar density profile and shows that the clusters are not tidally bound. It appears that more than half of the mass of these clusters is in unbound halos and stars associated with the clusters extend up to 5 arcmin. Bessell and Wood (1992) have obtained VI photometry of 10′ fields containing the YPCs in the Clouds and the color magnitude diagram of the fields around the clusters (Fig. 1) and around h and χ Persei, likely reflecting the greater mass density of the gas clouds from which the clusters formed in the Clouds compared to the gas clouds in the Galaxy.

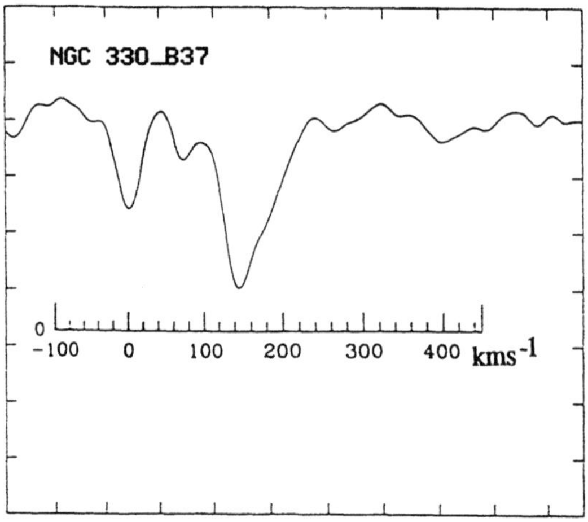

Fig. 2. The interstellar CaK line complex in star B37 in the SMC cluster NGC 330.

The distribution of young stars in the SMC is more complex than that of the LMC. About 2/3rds of the young stars and HII regions in the SMC are associated with two main HI components with sub-component velocities near 114, 132 (VL and L), 170 and 193 kms^{-1} (H and VH) (eg Martin, Maurice & Lequeux 1989). The low velocity HI is in front of the higher velocity clouds and molecular gas has only been detected in the nearer main complex. Stars associated with the higher velocity complexes at 170 and 193 kms^{-1} have reddenings on the average E(B-V) = 0.05 mag larger than those associated with the lower velocity HI complexes. NGC 330 has an intermediate velocity of 153.6±1.3 kms^{-1} and is associated with a group of supergiants within a 40′ diameter field with velocities between 143 and 160 kms^{-1}. The interstellar CaK line seen in the spectra of the cluster stars is quite complex with a strong galactic component, a maximum absorption at the velocity of the cluster but with components extending to at least 200 kms^{-1}. This high velocity interstellar Ca gas seems to be in front of the main HI complexes. Bessell (1991a,b) shows that the E(B-V) reddening of NGC 330 is about 0.10 mag (0.05 mag foreground and the remainder internal SMC reddening), much higher than previously accepted, but this is not unreasonable given that NGC 330 is behind the bulk of the HI gas seen in the low velocity complexes. The color magnitude diagrams of the fields surrounding NGC 330 are also very similar to that of the cluster (Fig. 1).

What this discussion has attempted to show is that there are many a priori reasons why the stars in YPCs and young stars in the surrounding fields should have very similar ages and metallicities and it was very surprising that significantly different abundances between cluster supergiants and field supergiants have been claimed by several authors (eg Spite et al 1986, 1989, 1991; Grebel & Richtler 1992). It is clearly very important to re-examine these claims and see whether we can understand how they arose, in particular, because Bessell (1991a, 1992) finds no difference between field and cluster K supergiants.

2 Observations of Differences between YPC and Field Stars

2.1 The Stömgren Photometry of NGC 330

First we will examine the CCD Strömgren photometry of Grebel & Richtler (1992) (GR) which has been interpreted as showing a clear difference in metallicity between stars within 40 arcsec and stars outside 1 arcmin. The photometry actually shows that a group of stars near the cluster centre have bluer *v-b* colors compared to more distant stars with the same *b-y* color. This effect, rather than resulting from lower metal-line blanketing in the centrally located stars, is more likely caused by contamination of the *v* magnitude (4100Å) of the red stars by blue light from the bright blue cluster core background or from nearby blue stars. Because of obvious contamination the colors for several red giants were excluded from the cluster average, but the possibility of contamination at a lower level in the remaining stars used was not considered. We have obtained BVI magnitudes for the cluster and surrounding field in conditions of good seeing ($\sim$ 1 arcsec) and examination of the B-V versus V-I colors (Fig. 3) shows less contamination in B (4300Å) for two of the red stars rejected by GR (due to our data being obtained in obviously better seeing) but clear contamination for one of the included stars. The B-V and V-I colors for most of the inner "cluster" stars and the outer "field" stars are indistinguishable within the scatter. We suggest that the results of GR should be restated as - the further one is from the cluster core the less likely the chance of blue contamination (hence apparent low metallicity), especially under poor seeing conditions.

2.2 Problems with Fine-Analysis of K Supergiants

As suggested previously (Bessell 1991a, 1992) the low metallicities derived in the previously published fine analyses of K supergiants in Magellanic Cloud clusters (Spite et al 1986, 1989, 1991) result from a combination of three effects. Firstly, before CCDs, the B-V colors of the cluster supergiants were difficult to measure accurately and B-V colors are still difficult to interpret due to reddening and uncertain empirical and theoretical calibration. It appears that in the analyses, temperatures too low by at least 200 K were used. Secondly, when one uses solar derived gf values, because the model atmospheres of the Sun from the grids of Bell et al. (1975) and Kurucz (1979) differ from the actual Sun's atmosphere and from the empirical solar model of Holweger and Müller (1974), it matters which model is used. Abundance differences of up to 0.3 dex can occur in the analyses of K stars depending on which solar model is used. Thirdly, the models for cooler stars like the K and M supergiants are not in good agreement with real stellar atmospheres, as can be seen (Fig. 4) in the systematic

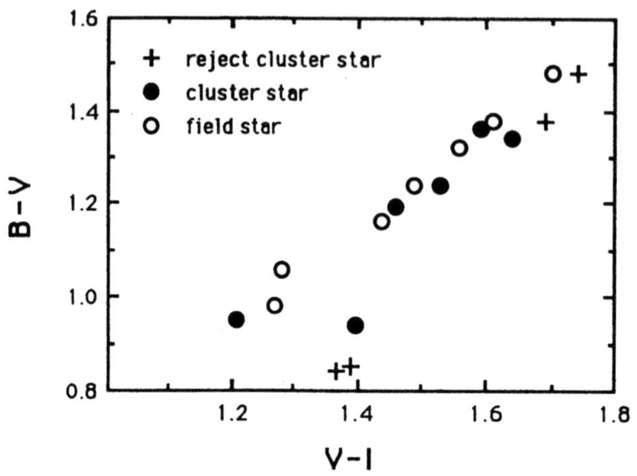

Fig. 3. The B-V versus V-I diagram for stars within a 4′ diameter field centered on NGC 330.

differences in the derived abundances of G K and M supergiants analysed by Luck (1982) and from the abundance of -0.5 dex relative to the Sun derived by Vieira (1986) for α Ori, a very young M supergiant. Clearly, the abundances of supergiants should not depend on their temperature.

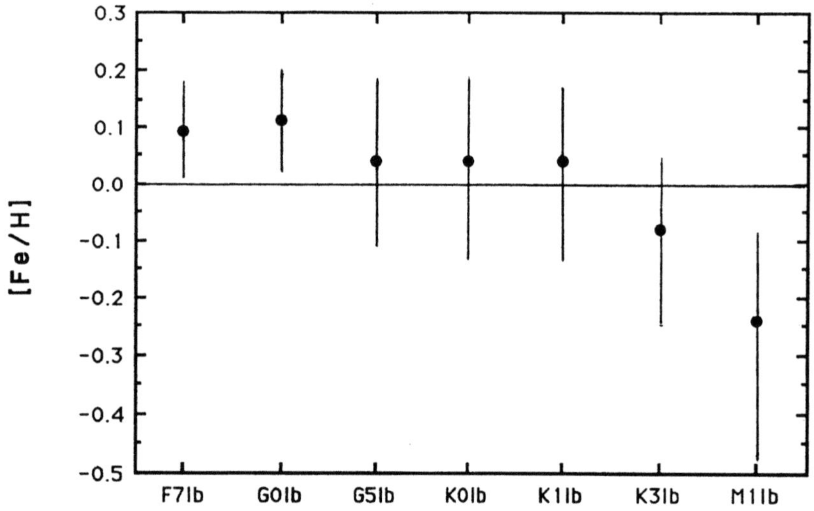

Fig. 4. Abundances for galactic supergiants (Luck 1982) binned by spectral-type.

3 Techniques of Differential Analysis

3.1 Comparison Stars

The way to overcome these difficulties is to do a closely differential analysis using a star very similar in temperature to the Cloud K supergiant stars and/or to calibrate out any systematic abundance differences with temperature and luminosity, by analysing stars with a large temperature range in one galactic cluster, moving group or association. Arcturus was chosen as the primary standard as it is a very bright, well-known star with a well-established temperature close to that of the hotter K supergiants; in addition, red-giants with a range in spectral types were observed in the clusters M67 and 47 Tuc, in the Hyades, Arcturus and W630 moving groups together with bright K and M supergiants in galactic associations in order to explore the systematic changes in the derived abundances with temperature and luminosity.

3.2 A Useful Line-Ratio for Assessment of K Star Temperatures

During this investigation it was discovered that a pair of TiI (0.02 eV) and FeI (4.61 eV) lines near the NaD lines were naturally very sensitive to temperature and that by inspection of this line ratio for stars between spectral types K0 and M2 one could accurately rank stars in temperature or even assign an accurate temperature without knowledge of the stars intrinsic colors. Some sample spectra of K stars are shown in Figure 5.

3.3 The Correlation between Line-Width and Microturbulence

Another useful discovery made was that at a resolution of about 50000, the actual half width of weak lines was found to correlate very well with the microturbulent velocity determined in the fine-analysis. Galactic globular cluster stars and old disk stars have sharp virtually unresolved lines at this resolution and microturbulent velocities of about 1.5-1.75 kms^{-1}, while the massive supergiants have significantly broader lines and microturbulent velocities up to 5 kms^{-1}. Having a good estimate of the microturbulence independently of the abundance analysis greatly strengthens the abundance results.

3.4 New Model Atmospheres

It is important that the systematic problems of the model atmospheres outlined above be overcome and perhaps the new MARCS models from the Uppsala group and the new Kurucz (1992) models will be successful in this regard; however, it is still possible to obtain reliable abundance analyses and the old models provided that a closely differential analyses is made.

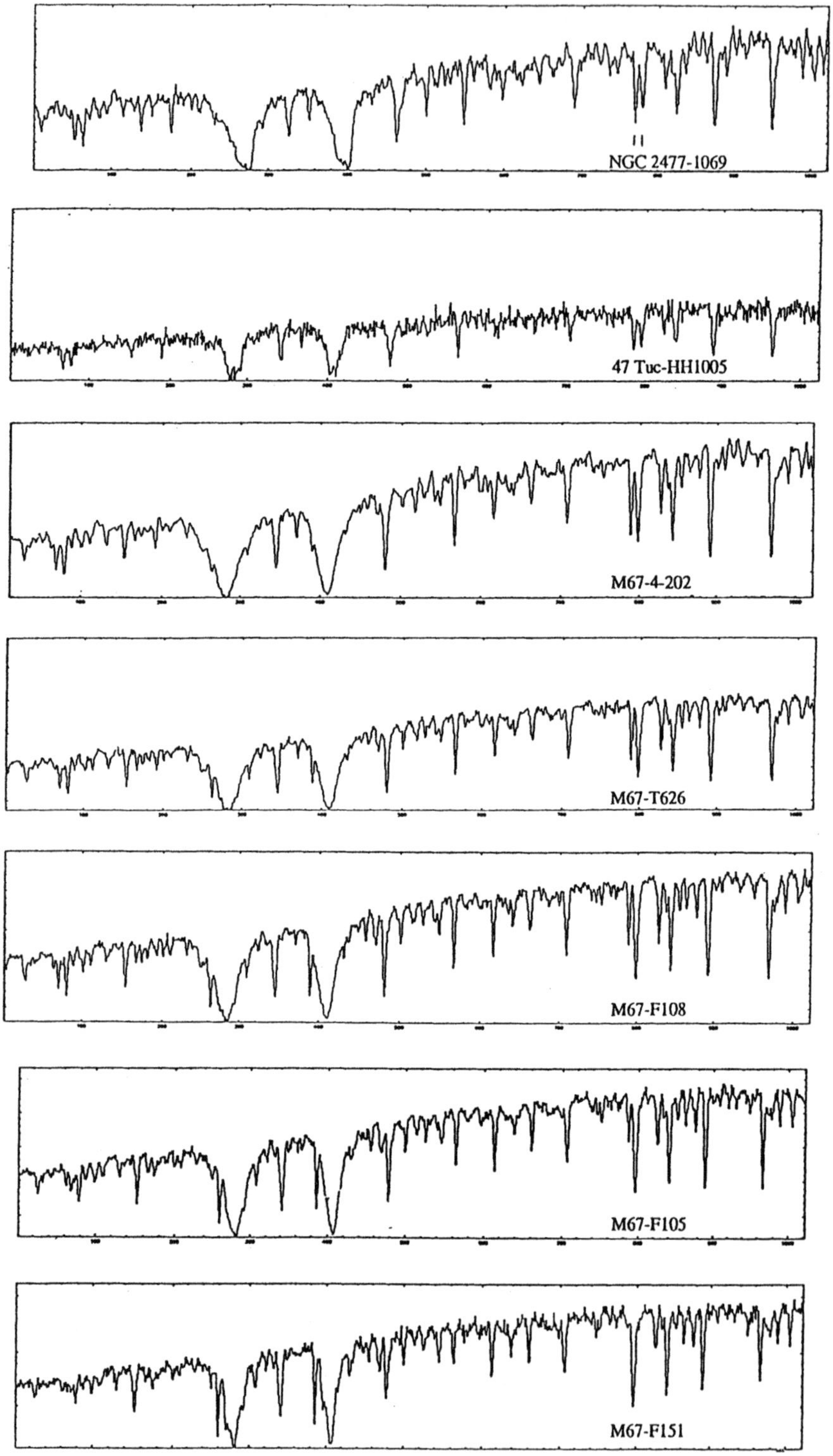

Fig. 5. A selection of echelle spectra for K giants with different spectral-types showing the temperature sensitive line pair (directly above the star name).

4 V, V-I and V, V-K Color-Magnitude Diagrams for YPCS

Bessell and Wood (1992) have obtained V versus V-I diagrams of the 4 YPCs using a CCD. Comparison of this data indicates a systematic difference in the V-I color of the red giant branches as one goes in the sequence NGC 330, NGC 1818, NGC 2004 and NGC 2100. We assume that as the blue stars in the clusters have very similar luminosities and temperatures, the temperature of the red giant branch reflects the different abundances of the clusters. The mean V-I colors of the red giants in the surrounding fields are similar to the cluster stars, further supporting the similar metallicities of cluster and field. We have also obtained J, H and K images of a 2.5′ diameter field centered on the clusters using a mercatel array on the AAT. These IR images enable the complete sampling of the large numbers of red giants in the clusters and clearly show that there are at least 3 times as many red as blue supergiants. The V-K versus V-I relation defined by the stars is sensibly linear indicating that V-I is as good as V-K as a temperature indicator for the KM supergiants.

References

Bell, R.A., Eriksson, K., Gustafsson, B. & Nordland, A. 1976 A&AS 23, 37.
Bessell, M.S. 1991a IAU Symposium 148 The Magellanic Clouds (Eds R. Haynes & D. Milne), (Kluwer, Dordrecht) p 273.
Bessell, M.S. 1991b A&A 242, L17
Bessell, M.S. & Wood, P.R. 1992 in preparation.
Elson, R.A.W., Fall, S.M. & Freeman, K.C. 1989 ApJ 336, 734
Fischer et al. 1991 IAU Symposium 148 The Magellanic Clouds (Eds R. Haynes & D. Milne), Kluwer, Dordrecht) p 200.
Grebel, E.K. & Richtler, T. 1992 A&A 253, 359.
Holweger, H. & Müller, E.A. 1974 Solar Phys. 39, 19.
Kurucz, R.L. 1979 ApJS 40, 1.
Kurucz, R.L. 1992 Private communication.
Luck, E. 1982 ApJ 256, 177.
Martin N. & Prevot, L. 1989 in Recent Developments of Magellanic Cloud Research (Eds K.S. de Boer, F. Spite, G. Stasinska), (Observatoire de Paris) p 155.
Maurice et al. 1986 A&AS 67, 423
Maurice et al. 1989 A&AS 78, 445
Prevot et al. 1985 A&AS 62, 23.
Spite, M., Cayrel, R., Francois, P., Richtler, T. Spite, M. 1986 A&A 168, 197.
Spite, F., Richtler, T. & Spite, M. 1991 A&A 252, 557.
Vieira 1986 Thesis, University of Uppsala.

Stellar vs. Interstellar Abundances in the Magellanic Clouds[1]

B. E. J. Pagel

NORDITA, Blegdamsvej 17, 2100 Copenhagen Ø, Denmark

Abstract: Results of abundance determinations in the interstellar gas and young stars of the Magellanic Clouds, and corresponding Galactic standards, are intercompared. Given the uncertainties in both sets of sources, and the great variety of ways in which the intercomparisons can be carried out, the agreement is about as good as could be expected.

1 Introduction

The article "Trouble in the Magellanic Clouds" (de Boer *et al.* 1991) reminded me of an old BBC programme called "So you think you've got troubles!" In the realm of abundances, we already have discrepancies between the large–scale abundance gradients in our own Galaxy deduced from HII regions (Shaver *et al.* 1983) and from hot stars (Lennon *et al.* 1990; Fitzsimmons *et al.* 1990) which would be expected to represent identical populations, and the apparent oxygen deficit in the Orion Nebula (Baldwin *et al.* 1991) relative to the Sun (Grevesse & Anders 1989). One consequence of this is that somewhat different conclusions can be arrived at according as one compares MC abundances with the Sun, Galactic B-stars, F-supergiants, HII regions, supernova remnants (SNR) etc., and whether one compares element:hydrogen or element:element ratios. There is a large number of ways in which the comparisons can be made, all giving slightly different results, and the consequence is that both trends and discrepancies in abundances tend to be, like beauty, in the eye of the beholder, and, again like beauty, often only skin-deep. In this review I consider chiefly elements that are detected both in young stars and in the interstellar medium (ISM), namely C, N, O, S, Fe, Zn, and emphasize comparisons between similar objects.

[1]Invited talk at Conference on **New Aspects of Magellanic Cloud Research**, Heidelberg, June 15–17 1992, G. Klare (ed.), Springer-Verlag.

2 The Galactic standards

Table 1 shows abundances adopted in 10 (!) different Galactic standards on the usual logarithmic scale 12+log(X/H). Solar abundances of these elements (Grevesse & Anders 1989) are believed now to be accurate to within ±0.1 dex, except for iron which has subsequently been revised down to 7.5 (Grevesse 1992).

Table 1: Reference Abundances

	C	N	O	S	Fe	Zn	
Sun	8.6	8.0	8.9	7.2	(7.7)	4.6	GA 89
Canopus	8.3		8.8		(7.7)		SBS 89
"	8.3			7.2	(7.6)		RB 89
Mean Gal. supergi	8.4	(8.4)	8.7	7.3			BSSM 91
HR 2618 (B2 II)	8.2	7.8	8.6	7.1	7.3		JSWB 92
κ Ori (B0.5 Ia)	(7.5)		8.8				Len, *et al.* 91
Solar vicinity	8.3	(7.6)	8.7	7.1	7.4	4.7	RD 92
Orion neb.	8.4	(7.5)	8.6	7.1			Dufour 90
Local Gal. HII		(7.6)	8.7	7.1			Shaver *et al.* 83
Cygnus loop	8.0	8.2	8.8	7.2			RD 90
Diffuse clouds	8.3	7.8	8.6	7.2	(6.1)	4.5	de Boer 91
Range	0.6	0.6	0.3	0.2	0.2	0.2	

The two sets of abundances for Canopus (the main standard for F supergiants), based on a comparison with the Sun, are in good agreement but carbon is somewhat lower than solar, and the nitrogen abundance adopted for the mean Galactic supergiant is high, both effects being possible consequences of mixing in the course of stellar evolution. The iron abundances are presumably affected by the correction to the solar iron abundance, which is taken into account in the last row. The B stars are in fair agreement with the Sun, apart from a uniform deficiency of about 0.2 dex in HR 2618 (ϵ CMa) and a low carbon abundance in the supergiant κ Ori. The nitrogen abundance for the Orion nebula and local Galactic HII regions is probably too low by nearly a factor of 2 (Rubin *et al.* 1988), and this in turn affects the value adopted by Russell & Dopita for the Solar vicinity.

The Cygnus loop (standard for SNR) again shows signs (real or spurious) of CN processing, but its oxygen and sulphur abundances agree well with solar, while in diffuse clouds sulphur and zinc agree well with solar values whereas other elements seem to be more or less affected by depletion on dust grains, iron especially so.

The bottom line of the table shows the range in abundances derived from different sources, not counting the more dubious values. The ranges in C and N are very large, probably owing to CN processing, those in S, Fe and Zn are satisfactorily small and oxygen is slightly worse.

3 The LMC

Table 2 shows various abundance determinations and comparisons for the LMC.

Table 2: Abundances in LMC

	C	N	O	S	Fe	Zn	
ceph./Sun					-0.1 to -0.4		Feast 91
NGC 1866/Sun					(≥-0.3:)		" "
cF/Sun	-0.8		-0.3		0 to -0.6		McW. W. 91
" "	-0.4		-0.5	-0.2			Sp. Sp. 91
cF/cF(Gal.)	-0.2		-0.3	-0.3			" "
" "	-0.3		-0.4	-0.4	-0.2	-0.4	RB 89; RD 92
av. like/like	-0.3		-0.3	-0.4	-0.2	-0.4	
B 0.5 Ia/κ Ori	(+0.1)	(0.0)	-0.3				Lenn.*et al.* 91
Br217,231/HR2618	-0.2	-0.2	-0.6	-0.2	-0.2		JSWB 92
N2004 B15,30/HR	-0.2	-0.4	-0.4	-0.2	-0.2		" "
N1812 D12/HR	-0.2	-0.3	-0.6	-0.2	-0.3		" "
N2100 C13/HR	+0.1	-0.9	-0.5		+0.1		" "
HII/Sun	-0.6	-1.1	-0.5	-0.4			Dufour 90
" "		-1.0	-0.6	-0.3			RD. 90
HII/HII(Gal.)	-0.5	(-0.6)	-0.3	-0.3			" "
SNR/Sun	-0.9	-0.6	-0.7	-0.6	-0.4		" "
SNR/SNR(Gal.)	-0.3	-0.7	-0.6	-0.6			" "
Diff.cl./Sun			-0.4	-0.4:		-0.6	de Boer 91
Dif.cl./Dif.cl.(Gal.)			-0.1	-0.4:		-0.5	" "
av. like/like	-0.4	-0.7	-0.4	-0.4		-0.5	

The first two rows are based on general data — colours and HR diagrams — and indicate a modest deficiency in metallicity. The next 4 rows refer to F supergiants in which McWilliam & Williams (1991) find rather a large range in iron abundance (from FeII) but a much smaller one in oxygen abundance, the latter result being in agreement with what was found from HII regions (Pagel *et al.* 1978); but McWilliam & Williams have given hardly any details about their analysis. Russell & Bessell also find a range, somewhat smaller than that of McWilliam & Williams, but deem it insignificant compared to an error bar of ±0.2 dex. Compared to Galactic F supergiants, the data are consistent with a uniform deficiency of 0.3 dex for C, O, S, Fe and Zn.

The same result is derived by Lennon *et al.* (1991) for oxygen in a non-LTE analysis of the B-supergiant Sk -68° 41 relative to κ Ori. The next row refers to LTE analyses of two field B-stars of considerably lower luminosity by Jüttner *et al.* (1992) relative to the comparable Galactic star ϵ CMa; given the slight underabundances found in the latter, the results for C, S and Fe are remarkably close to those found for F-supergiants; oxygen looks substantially lower, but I wonder how seriously to take this.

Under these are three B-stars in clusters, also analyzed relative to ϵ CMa by Jüttner *et al.*. Apart from the third one, which had less good quality spectra, the results are indistinguishable from those for the field stars.

The remaining entries refer to the ISM, which has its own problems, e.g. possible effects of temperature fluctuations and depletion on dust grains, both of which may depress abundances relative to solar below the true values. Comparing like with like, as on the bottom line, one finds quite good agreement at the level of ± 0.1 dex with the results from stars, with the possible exception of nitrogen which is rather a dicey element in B-stars. (The discrepancy between optical and infra-red determinations of the N/O ratio which affects Orion and some other Galactic HII regions appears to be reassuringly absent in 30 Doradus (Lester *et al.* 1987)). Overall, there seem to be no outstanding discrepancies, either between stars and the ISM or between field stars and cluster members.

A deficiency of 0.3 ± 0.1 dex relative to similar Galactic objects appears to be an adequate description of the C, O, S, Fe and Zn abundances in young objects of the LMC, with nitrogen probably more deficient than that.

4 The SMC

My compilation of abundance determinations in the SMC is given in Table 3.

Table 3: Abundances in SMC

	C	N	O	S	Fe	Zn	
ceph/Sun					-0.6		
cF/Sun	-0.9		-0.8	-0.6	-0.7		SSF 89; SS 91
cF/cF	-0.7		-0.6	-0.6	-0.7		" "
"	-0.5		-0.7	-0.5	-0.6		RB 89; RD 92
av. like/like	-0.6		-0.6	-0.6	-0.6		
N330cF,K/Sun	-0.9	-1.1	-1.0	-1.0	-1.0		SBS89;BSSM91;SRS91
AV304/τ Sco	-0.8	$\leq$-1.4	-0.4	-0.6	-0.6		DFH 90
AV175/HR	-0.5	-0.4	-0.4	-0.2	-0.3		JSWB 92
N330 B22/HR	-0.4	-0.8	-0.7	-0.7	-0.3		" "
N330 B30/HR	-0.4	-0.9	-0.9	-0.3	-0.3		" "
HII/Sun	-1.3	-1.6	-0.9	-0.7			Dufour 90
" "		-1.5	-0.8	(-0.4)			RD 90
HII/HII(Gal.)	-1.2	(-1.1)	-0.6	-0.6			" "
SNR/Sun		-1.2	-1.0	-0.8	-0.7		" "
SNR/SNR(Gal.)		-1.3	-0.9	-0.7			" "
Dif.cl./Sun				-1.2		-1.1	de Boer 91
av. like/like	-1.2	-1.3	-0.8	-0.8		-1.0	

F supergiants in the field display a uniform deficiency of 0.6 dex, for the elements listed, compared to their Galactic counterparts. Field B stars are in fair agreement with this, bearing in mind the slight deficiency of HR 2618. The cluster NGC 330 shows a slight deficiency of O, S and Fe, relative to field stars, averaging 0.3 dex in the cool stars, but these are affected by a controversy about the reddening (Bessell 1991), and different reddenings have been adopted by different authors (cf. Reitermann *et al.* 1990; Barbuy *et al.* 1991). The results of the Strömgren photometry (Grebel & Richtler 1992) do not completely convince me and the B-stars do not show a consistent cluster-field difference, maybe because the uncertainties are too large. At the present time, I have to regard the suggested deficiency of NGC 330 with respect to neighbouring field stars as being within the noise.

Coming to the ISM determinations, there is remarkable agreement among all objects (except diffuse clouds) in the sulphur abundance, which is one of those least sensitive to how one chooses to make the comparison and comes out close to the canonical value of -0.6. Nitrogen is undoubtedly more deficient and oxygen has a barely significant scatter between values quoted for field F-supergiants and NGC 330. Carbon is apparently more deficient in HII regions than in field supergiants. In absolute terms (i.e. relative to the Sun) the discrepancy is 0.4 dex; this, however, is reduced to 0.3 dex when one considers the C/O ratio, and that amount is not very large in relation to likely errors. Assuming the right answer to be somewhere in between, one derives log C/O = -0.6 in the SMC, which is quite a typical value for low-metallicity HII regions (cf. Pagel 1990). The deficiency in oxygen itself, relative to the Sun, is the same in HII regions as in F-supergiants, and the larger discrepancy in "like/like" comparisons of C/O or C/H can be traced back to a slightly low carbon abundance (possibly due to CN processing) in Canopus!

There are thus no conclusive indications of real discrepancies between stellar and interstellar, or between stellar and other stellar, abundances among young objects in either of the Magellanic Clouds. At the same time, more accurate abundances are obviously needed and the way to get these is to obtain more accurate spectra of less luminous stars. Bring on the VLT!

References

Baldwin, J.A., Ferland, G.J., Martin, P.G., Corbin, M.R., Cota, S.A., Peterson, B.M. & Slettebak, A. 1991, *Astrophys. J.*, **374**, 580.

Barbuy, B., Spite, M., Spite, F. & Milone, A. 1991, *Astr. Astrophys.*, **247**, 15.

Bessell, M.S. 1991, *Astr. Astrophys.*, **242**, L17.

de Boer, K.S. 1991, *in* IAU Symp. no. 148: **The Magellanic Clouds**, R. Haynes & D. Milne (eds), Kluwer, p. 401.

de Boer, K.S. *et al.* 1991, ESO *Messenger*, no. 66, p. 14.

Dufour, R.J. 1990, *in* Int. Symp. **Evolution in Astrophysics**, ESA SP-310, p. 117.

Dufton, P.L., Fitzsimmons, A. & Howarth, I.D. 1990, *Astrophys. J. Let.*, **362**, L59.

Feast, M.W. 1991, *in* IAU Symp. no. 148: **The Magellanic Clouds**, R. Haynes & D. Milne (eds), Kluwer, p. 1.

Fitzsimmons, A., Brown, P.J.F., Dufton, P.L. & Lennon, D.J. 1990, *Astr. Astrophys.*, **232**, 437.

Grebel, E.K. & Richtler, T. 1992, *Astr. Astrophys.*, **253**, 359.

Grevesse, N. 1992, *in* **Origin and Evolution of the Elements**, Int. Symp. in honour of H. Reeves, M. Cassé, N. Prantzos & E. Vangioni-Flam (eds), in press.

Grevesse, N. & Anders, E. 1989, in **Cosmic Abundances of Matter**, J. Waddington (ed.), New York: Amer. Inst. Phys., p. 1.

Jüttner, A., Stahl, O., Wolf, B. & Baschek, B. 1992, in these proceedings.

Lennon, D.J., Dufton, P.L., Fitzsimmons, A., Gehren, T. & Nissen, P.E. 1990, *Astr. Astrophys.*, **240**, 349.

Lennon, D.J., Kudritzki, R.-P., Becker, S.T., Butler, K., Eber, F.,

Groth, H.G. & Kunze, D. 1991, *Astr. Astrophys.*, **252**, 498.
Lester, D.F., Dinerstein, H.L., Werner, M.W., Watson, D.M., Genzel, R. & Storey, J.W.V. 1987, *Astrophys. J.*, **320**, 573.
McWilliam, A. & Williams, R.E. 1991, *in* IAU Symp. no. 148: **The Magellanic Clouds**, R. Haynes & D. Milne (eds), Kluwer, p. 391.
Pagel, B.E.J. 1990, in **Dynamical and Chemical Evolution of Galaxies**, F. Ferrini, J. Franco & F. Matteucci (eds), Pisa: Giardini Editore, p. 229.
Pagel, B.E.J., Edmunds, M.G., Fosbury, R.A.E. & Webster, B.L. 1978, *Mon. Not. R. astr. Soc.*, **184**, 569.
Reitermann, A., Baschek, B., Stahl, O. & Wolf, B. 1990, *Astr. Astrophys.*, **234**, 109.
Rubin, R.H., Simpson, J.P., Erickson, F. & Haas, M.R. 1988, *Astrophys. J.*, **327**, 377.
Russell, S.C. & Bessell, M.S. 1989, *Astrophys. J. Suppl.*, **70**, 865.
Russell, S.C. & Dopita, M.S. 1990, *Astrophys. J. Suppl.*, **74**, 93.
Russell, S.C. & Dopita, M.S. 1992, *Astrophys. J.*, **384**, 508.
Shaver, P.A., McGee, R.X., Danks, A.C. & Pottasch, S.R. 1983, *Mon. Not. R. astr. Soc.*, **204**, 53.
Spite, F., Richtler, T. & Spite, M. 1991, *Astr. Astrophys.*, **252**, 557.
Spite, F. & Spite, M. 1991, *in* IAU Symp. no. 149: **The Magellanic Clouds**, R. Haynes & D. Milne (eds), Kluwer, p. 243.
Spite, F., Spite, M. & François, P. 1989, *Astr. Astrophys.*, **210**, 25.
Spite, M., Barbuy, B. & Spite, F. 1989, *Astr. Astrophys.*, **222**, 35.

New Abundance Analyses of B-Type Giants in the Magellanic Clouds [1]

A. Jüttner [1], O. Stahl [1], B. Wolf [1], B. Baschek [2]

[1]Landessternwarte Königstuhl, D-6900 Heidelberg 1, Germany
[2]Institut für Theoretische Astrophysik der Universität Heidelberg
Im Neuenheimer Feld 561, D-6900 Heidelberg 1, Germany

1 Introduction

Our nearest galactic neighbours, the Magellanic Clouds provide an excellent opportunity to investigate problems like the initial mass function and star formation rates, dust compositions and interstellar extinction, and the evolution and mass loss of individual stars. All these problems are connected with the chemical composition of the MCs and with the question of how homogeneous are the MCs in metal content.

2 Observations

Within the ESO key programme on "Coordinated Investigation of Selected Regions of the Magellanic Clouds" (cf. de Boer et al. 1989, 1991) we concentrated our efforts on early B-type giants located both in young blue globular clusters and in the field.

We obtained so far for 6 stars in the LMC and for 3 stars in the SMC high resolution ($\lambda/\Delta\lambda \approx 20000$) and high signal-to-noise ($S/N \geq 75$) spectra with CASPEC at the ESO 3.6 m telescope. In addition to these high resolution spectra in the optical range, we obtained low resolution spectrophotometry with IUE in the satellite UV range.

Table 1. Journal of CASPEC observations in the SMC.

Object	V[mag]	Exp.time[h]	S/N	Date	v_r[kms^{-1}]
NGC 330/B30	14.22	2.0+2.0	100:1	Nov 1987	158
NGC 330/B22	14.34	3.5	95:1	Dec 1991	151
AV 175	13.65	2.0+2.5	90:1	Dec 1989	191

[1] Based on observations collected at the European Southern Observatory at La Silla, Chile and on observations by the International Ultraviolet Explorer collected at the Villafranca Satellite Tracking Station of the European Space Agency, Spain

Table 2. Journal of CASPEC observations in the LMC.

Object	V[mag]	Exp.time[h]	S/N	Date	v_r[kms^{-1}]
NGC 2004/B30	13.78	2.0+2.0+2.0	110:1	Nov 1987	317
NGC 2004/B15	14.29	3.0+3.0	75:1	Dec 1989	357*
NGC 1818/D12	13.65	1.5+1.5+1.5	110:1	Nov 1987	302
NGC 2100/C13	13.85	3.5+3.0	100:1	Dec 1991	254
Bru 217	13.40	2.0+2.5	100:1	Jan 1991	269
Bru 231	13.10	2.0	90:1	Jan 1991	271

* membership doubtful due to $v_{\rm rad}$

3 Analyses and Results

The aim of these observations was to derive chemical abundances for the elements He, C, N, O, Mg, Al, Si, S and Fe. The abundance determinations were carried out differentially to the galactic B-type giant HR 2618 which shows intermediate atmospheric parameters relative to the coolest and hottest stars of our sample.

The low resolution IUE spectrophotometry is used for an additional determination of the model atmosphere parameters (mainly $T_{\rm eff}$) by fitting the continuum energy distribution in the satellite UV range. In addition our model atmospheres are used in connection with the IUE data to study the interstellar reddening in the MCs.

A detailed description of the analysis methods can be found in Jüttner & Wolf (1991) and in Jüttner (1992) elsewhere in this volume. The model atmosphere parameters which we finally derived are presented in Table 3.

Table 3. Model atmosphere parameters of the target stars.

Object	$T_{\rm eff}$ [K]	$\log g$
NGC 2004/B30	23700	3.40
NGC 2004/B15	24400	3.72
NGC 1818/D12	19600	3.13
NGC 2100/C13	21000	2.78
Bru 217	21080	2.92
Bru 231	20000	2.63

Object	$T_{\rm eff}$ [K]	$\log g$
NGC 330/B30	18800	2.91
NGC 330/B22	22700	3.22
AV 175	20500	2.84
HR 2618	21080	3.16

In Table 4 we present new results of differential abundance analyses for six LMC stars and three SMC stars. For reasons of homogeneity we reanalysed the two stars published by Reitermann et al. (1990), namely NGC 330/B30 and NGC 1818/D12.

Table 4. Logarithmic element abundances relative to HR 2618.

	$\Delta \log \varepsilon = \log \varepsilon_{MC} - \log \varepsilon_{HR\,2618}$								
Object	He	C	N	O	Mg	Al	Si	S	Fe
NGC 330/B30	-0.05	-0.43	-0.91	-0.89	-0.44	-0.44	-0.96	-0.33	-0.36
NGC 330/B22	-0.04	-0.42	-0.86	-0.70	-0.37	-0.49	-0.88	-0.72	-0.37
AV 175	-0.09	-0.53	-0.43	-0.42	-0.38	-0.45	-0.55	-0.21	-0.31
NGC 2004/B30	-0.06	-0.28	-0.27	-0.30	-0.20	-0.37	-0.48	-0.22	-0.30
NGC 2004/B15	-0.07	-0.29	-0.59	-0.56	-0.22	-0.31	-0.84	-0.22	-0.26
NGC 1818/D12	-0.06	-0.23	-0.29	-0.54	-0.31	-0.55	-0.93	-0.21	-0.36
NGC 2100/C13	-0.07	0.06	-0.83	-0.43	0.06		-0.33		0.06
Bru 217	0.07	-0.26	-0.21	-0.50	0.17	-0.22	-0.56	-0.21	-0.22
Bru 231	0.08	-0.27	-0.10	-0.55	0.13	-0.56	-0.59	-0.23	-0.22

The abundance pattern found for the galactic comparison star with respect to the solar values (Grevesse & Andres 1988) is shown in Table 5.

Table 5. Logarithmic element abundances for HR 2618 relative to the solar values.

	$\Delta \log \varepsilon = \log \varepsilon_{HR\,2618} - \log \varepsilon_{solar}$								
Object	He	C	N	O	Mg	Al	Si	S	Fe
HR 2618	-0.07	-0.39	-0.17	-0.22	-0.46	-0.47	-0.04	-0.08	-0.43

Acknowledgements: This research was supported by the Deutsche Forschungsgemeinschaft (SFB 328).

References

de Boer K.S., Azzopardi M., Baschek B., et al., 1989, ESO Messenger 57, 27

de Boer, K.S., Spite F., François P., et al., 1991, ESO Messenger 66, 14

Grevesse N., Anders E., 1988, Solar-System Abundances of the Elements: A new Table. In:Waddington C.J.(ed.), AIP Conf. Proc. Cosmic Abundances of Matter, Minneapolis (1988)

Jüttner A., Wolf B., 1991, Analyses of B-Type Stars in the Magellanic Clouds (Review). In: Heber U., Jeffery C.S. (eds) Proc. CCP7 Workshop, Atmopheres of Early Type Stars. Kiel(1991)

Jüttner A., 1992, this volume

Reitermann A., Baschek B., Stahl O., Wolf B., 1990, A&A 234, 109

Observation of Stars in NGC1948

Monique Spite, François Spite, Patrick François, Suzanne Huille

Observatoire de Paris-Meudon, F92195 MEUDON Cedex, France

1 Introduction

Spectroscopic abundance determinations are still lacking for many Magellanic Clouds clusters, despite the importance of their role in the age-metallicity and distance calibrations. In the framework of the ESO key program on "Coordinated Investigations of Selected Regions in the Magellanic Clouds" we have initiated a program to determine the abundance ratios in a few clusters: NGC330, NGC1818, and NGC1948. NGC 1948 is a young cluster located in the "Selected Region E" of the LMC.

The *(B-V)* color of several stars has been measured in this region (Bomans *et al.* 1991, Vallenari *et al.* 1992, Bomans 1992) and we tried to analyse the chemical composition of the cool stars.

2 Observations

Three stars (Fig. 1) were observed with EMMI at the NTT with a resolution $R \approx 30000$. At the present time this resolution is the highest possible for faint stars at La Silla.

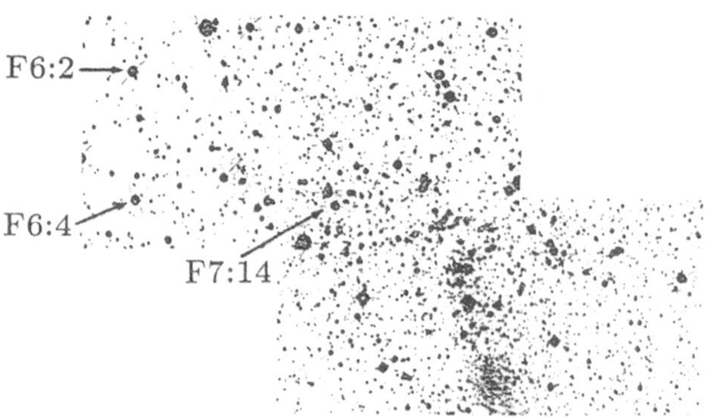

Fig. 1. The three stars observed in NGC1948

The S/N ratio of the spectra was about 50 and in a first step we could estimate the radial velocity of the cool stars with a precision of about 2 km s^{-1} (Table 2).

Table 1. Characteristics of the stars observed in the region of NGC1948 (colors following Bomans 1992)

name	V	B-V	Vr (km s^{-1})
F6:2	14.27	1.08	27
F6:4	14.41	1.63	304
F7:14	14.08	0.76	(315)

From these measurements it is clear that F6:2 is a galactic star. F7:14 is a hot emission line star probably reddened by dust and F6:4 a K star. Part of the spectrum of F6:4 in the Hα region is displayed in Fig. 2.

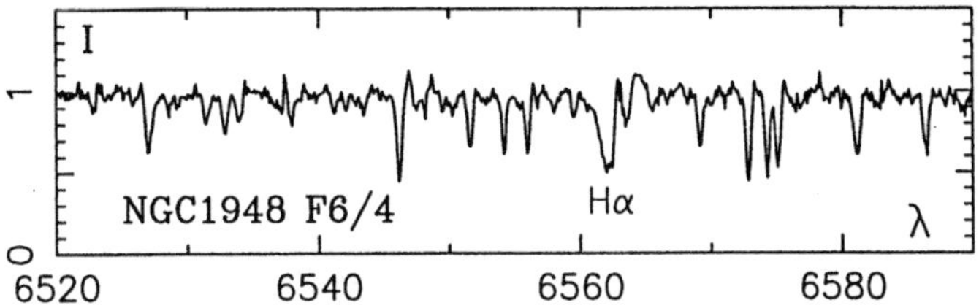

Fig.2. Spectrum of NGC1948 F6:4 in the region of the Hα line.

3 Analysis and Conclusion

A classical model atmosphere analysis has been performed for the star F6:4. More than 250 lines have been measured from λ 500 nm to 720 nm. We found for this star a temperature of about 4300 K, a gravity log g $\approx$ 1 and a microturbulent velocity $v_t \approx 2.2$ km s^{-1}.

The mean metallicity of NGC1948 F6:4 (relative to the Sun) is: [M/H]=–0.3.

Table 2. Abundances of the elements in NGC1948 F6:4

	[M/H]	n		[M/H]	n		[M/H]	n		[M/H]	n
O I	–0.3	2	Si I	–0.2	12	Ti II	+0.1	5	Ba II	0.0	3
Na I	–0.1	4	Ca I	–0.2	20	Fe I	–0.3	148	La II	+0.2	2
Al I	–0.4	4	Ti I	–0.1	38	Fe II	–0.2	13	Eu II	+0.2	1

The abundances of the elements (Table 2) in F6:4 are typical of a LMC star: mean deficiency by a factor of $\approx$ 2 and slight relative overabundance of the rare earths (especially those heavier than Ba).

References

Bomans, D.J., Vallenari, A., Seggewiss, W., Richtler, T., Grebel, E.K., de Boer, K.S. (1991) in :"The stellar populations of galaxies" IAU symp. 149, eds. B.Barbuy and A.Renzini p.393

Bomans, D.J. (1992): private communication

Vallenari, A., Bomans, D.J., de Boer, K.S. (1992) : A&A, *in press*

New Aspects of Young Magellanic Cloud Clusters[1]

Tom Richtler

Sternwarte der Universität Bonn, W-5300 Bonn 1, Auf dem Hügel 71, F.R.Germany

Abstract: New photometric metallicities are presented for the young globular clusters NGC 1818, NGC 1850, NGC 2004. These results confirm the earlier observed trend of surprisingly low metallicities. The example of NGC 1818 shows that the interpretation of Strömgren colours as a general metallicity indicator becomes questionable, when applied to evolved supergiants. Recent detections of Be stars and shortperiod variables in some young clusters have interesting consequences for the interpretation of those phenomena. The knowledge about the Initial Mass Functions of young clusters is briefly reviewed and an explanation for the apparent universality of stellar IMFs is proposed.

1 Introduction: Cluster metallicities as a basic topic

Among the young stellar population of the Magellanic Clouds, the young globular clusters always attracted particular attention (see the review by van den Bergh 1990 and the numerous articles refering to them in IAU Symp 148). Richly populated star clusters are laboratories for various directions of astrophysical investigations and in most cases the knowledge of metal abundance is the basis for all further interpretation. Through metallicity, star clusters are linked with the physics of galaxy evolution: Because age and abundance can be determined simultaneously, star clusters are probes of galaxy-wide chemical evolution.

Metallicity also provides the connection with stellar physics: Stellar winds, pulsations, lifetimes, colours, and luminosities all depend on metal abundance. Moreover, Magellanic Cloud clusters help to calibrate integrated metallicity indicators which are needed in extragalactic work.

Given the importance of the parameter "metallicity" we have to admit, unfortunately, that the metallicities of young clusters in the Magellanic Clouds are not easy to determine. Because integrated parameters fail, one has to investigate individual stars, but high dispersion analyses are tedious and have become available only for stars in

[1] Based in part on observations obtained in time of the MPI für Astronomie, Heidelberg, with the 2.2m telescope on La Silla

NGC 330 and NGC 1818 (e.g. Richtler et al. 1989, Spite et al. 1991, Reitermann et al. 1990). Most of the prominent bright young clusters are still unexplored with respect to their metallicities.

CCD photometry in a suitable system often means an estimation rather than a determination of metallicity, but has the advantage of measuring many stars simultaneously with high S/N even in crowded fields. The Strömgren system proved to provide good sensitivity for metallicities of late-type giants and supergiants and has already been succesfully applied to the young SMC cluster NGC 330 (Grebel and Richtler 1992).

Here I want to report about new Strömgren photometry of LMC clusters (in collaboration with Eva Grebel), before I address the topic of Be stars and short period variables with these metallicities in mind.

2 More about photometrically derived metallicities of young clusters

In December 1990, E. Grebel obtained CCD Strömgren data with the 2.2m telescope at La Silla on time granted by the Max-Planck Institut für Astronomie, Heidelberg. Objects were the LMC clusters NGC 1818, 1850, 1866, 2004, 2214. We used the 1m in January 1992 to measure local zero-point stars in some clusters to allow for an absolute calibration. Here I want to give preliminary results for the clusters NGC 1818, 1850, and 2004.

NGC 1818, a cluster with an age between 20 and 40 Myr, dependent on the stellar models used (Will 1992), was a target of the ESO key programme on "Coordinated Investigations of Selected Regions in the Magellanic Clouds". Previous spectroscopic investigations (Richtler et al. 1989, Reitermann et al. 1990) resulted in a surprisingly low metallicity of about -0.9 dex, thus being well below the value which has been found for LMC field supergiants (Russell and Bessell 1989). The Strömgren photometry was therefore of high interest.

Fig.1 shows the dereddened (we adopted a reddening of $E_{B-V} = 0\overset{m}{.}09$) b-y, m1 diagram for selected supergiants in this cluster. The selection was done according to photometric quality, i.e. stars suffering apparently from crowding problems have been omitted. All stars are probable cluster members, as the nearby field population shows no red supergiants at all. The star which we analysed already spectroscopically (B26 in the notation of Robertson 1974) is marked and the photometric metallicity agrees excellently with the spectroscopic value of -0.9 dex. However, a disturbing feature is the group of somewhat hotter supergiants that scatter around the solar value. I can find no reasonable explanation for the location of these stars except that they are evolved supergiants with strong CN-bands which can feign a high metal abundance in the Strömgren system. It is known (Bell and Gustafsson 1979) that the Strömgren system reacts sensitively to the nitrogen abundance via the CN-band at 4216 A. Note also that the analyses of Jüttner et al. (this conference) support the overall low metallicity of NGC N1818 once again. If that is true, NGC 1818 would be a unique object for studying differentially the change in surface abundance during the evolution of supergiant stars.

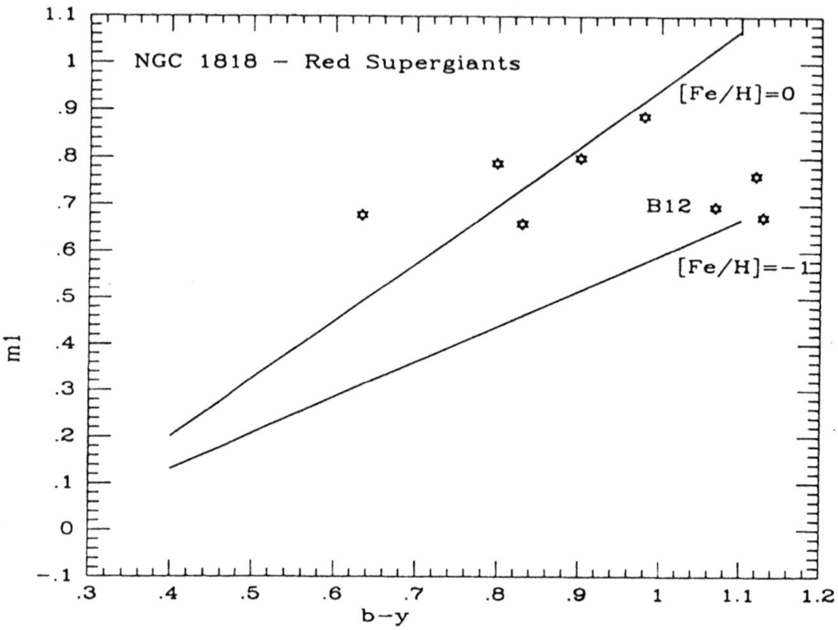

Fig. 1. The b-y, m1 diagram for selected supergiants in NGC 1818. Indicated are two lines of constant metallicity according to the calibration of Grebel and Richtler (1992). Stars have been selected according to presumably good photometry. Two groups are recognisable: red supergiants at a metallicity of roughly -1 dex and a group scattering at about solar metallicity. It is possible that these latter stars are evolved supergiants with strong CN-bands. The star B12 has been analysed by Richtler et al. (1989).

The cluster NGC 1850 is also interesting. It is located in the bar of the LMC in a very crowded region, thus uncertainties concerning cluster membership are expected when discussing individual stars. Fig.2 gives the dereddened b-y, m1 diagram (again we adopted $E_{B-V} = 0\overset{m}{.}09$) for selected supergiants (according to photometric quality). In lack of a better criterion for cluster membership, we considered all stars closer than 50" to the cluster centre as members and mark them "*". All stars within this region lie more or less on a line of constant metallicity, which corresponds to about -0.7 dex. It is striking that the other stars, which in majority presumably do not belong to the cluster, scatter over a large range of metallicities. Perhaps some of the apparently metal deficient stars are binaries, where a bluer component cause a higher flux in the v filter, but the global pattern is similar to what Luck and Lambert (1992) in their recent analyses of LMC cepheids and supergiants have found: The young stars in the LMC exhibit no uniform metallicity, but scatter over an appreciable range of abundances.

The third cluster for which we have calibrated Strömgren data, is NGC 2004. For reasons of space, we give no b-y, m1 diagram, but state simply that the location of

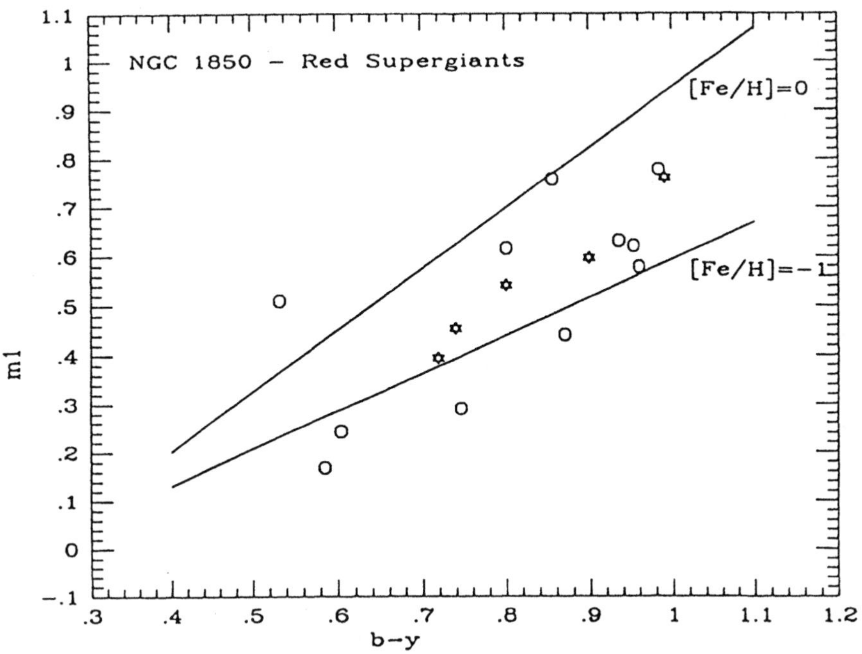

Fig. 2. The b-y m1 diagram for selected supergiants in NGC 1850. Indicated are two lines of constant metallicity according to the calibration of Grebel and Richtler (1992). Only stars with good photometric quality are plotted. The stars closer to the cluster centre than 50" are marked with "*". The mean metallicity is about -0.7 dex. Note the scatter of the stars presumably being field stars

the red supergiants hint to a metallicity of about -0.8 to -0.9 dex. This agrees well with the spectroscopic result by Jüttner et al. (this conference).

Summarising we find again quite low metallicities for young LMC clusters, which at face value do not really fit into a smooth age-metallicity relation (e.g. Da Costa 1991). They also do not fit to the higher metallicities of somewhat older clusters given by Olszewski et al. (1991). However, there is still the question what Strömgren photometry actually measures. As the puzzling pattern in NGC 1818 indicates, it must not be primarily the iron abundance, which determines the colour. Bell and Gustafsson (1979) already noted a sensitive reaction of the Strömgren colours to N abundance. On the other hand, our metallicity calibration has been performed in terms of iron abundance. Thus, it is questionable to what degree the metallicities derived from Strömgren photometry resemble those derived by Olszewski et al. (1991), who measured the strength of the infrared Ca triplet. A systematically low CN abundance in Magellanic Cloud clusters may be a possible reason for this discrepancy, but unfortunately, there are yet no objects in common to investigate this matter in more detail.

3 Emission-line stars and Variables

Metal abundances of young Magellanic Cloud clusters gain a new importance with the interpretation of the occurence of emission-line stars and variables among B main sequence stars. Grebel et al. (1991) found from CCD photometry (Strömgren b,y, and $H\alpha$) in the young SMC cluster NGC 330 that the majority of B main sequence stars in this cluster shows $H\alpha$ emission and are in fact Be stars (adopting a simplified view of the Be star phenomenon, see Baade 1992 for a more thoughtful definition; see also Muratorio and Azzopardi, this conference). Such a large fraction of Be stars was not expected and has no known counterpart in galactic clusters. Bessell and Wood (this conference) performed V,I photometry in NGC 330. They find that the $H\alpha$ emission is clearly coupled with an excess in I, a fact that is already known from galactic Be stars. Lamers (1986) cites evidence that the infra-red emitting region is the same that produces the $H\alpha$ emission and that the infrared radiation is free-free emission from an equatorial shell rather than dust emission. This is also plausible in view of the low metal abundance of NGC 330. It would be very interesting to monitor NGC 330 simultaneously in infrared bands and in $H\alpha$.

Be stars have been found also in other Magellanic Cloud clusters. Bessell and Wood (this conference) and Grebel et al. (this conference) report further $H\alpha$ photometry for the LMC clusters NGC 2004 and 2100. Whether Be stars in these objects are as abundant as in NGC 330, remains to be shown, but again their number is strikingly high.

What causes the high number of Be stars in young Magellanic Cloud clusters? Galactic Be stars show rapid rotation (e.g., Balona 1990). Is there any mechanisms perhaps dependent on metallicity which causes high rotation velocities for more stars than in galactic clusters? It would be very interesting to measure $v \cdot sini$ for Be stars and normal B stars in Magellanic Cloud clusters, which should be feasible for the brighter stars.

Balona (1992a), surveyed NGC 330 for variables stars. He detected 19 variables with periods between 0.5 and 0.9 days and amplitudes of typically 0.05 mag. Some of them are proved Be stars according to Feast (1972) and Grebel et al. (1991) and it may be suspected that the majority of them are periodic Be stars, i.e. λ Eri stars.

Balona (1992a) pointed out that the fact that so many Be stars and variable Be stars occur in a metal-poor cluster, bears consequences for the understanding of the Be star phenomenon in the sense that pulsation excited by opacity via the classical κ mechanism cannot longer be considered the plausible mechanism causing the periodic variability of Be stars. He argued (Balona 1990) that for galactic Be stars the rotation period is directly connected with the period of variability so the light curve may reflect the rotation of the star. However, the debate about this is still going on (see for example Gies 1991 who advocates nonradial pulsations).

The recent revision of metal opacities (Rozsnayi 1989, Iglesias and Rogers 1991) makes pulsation attractive for explaining the variability of β Cephei stars (Cox and al. 1992). These new opacities provide a solution by their very strong dependence on temperature via iron absorption at about 150 000 K. While it is unclear how a low metallicity would favour the occurence of Be stars, one would expect a paucity of β Cephei variables as a general effect. Indeed, Balona (1992a,1992b) did not find a single β Cephei star, (i.e. a star with a period about 0.2 days which must be due to pulsation) with an amplitude exceeding $0\overset{m}{.}02$ among many variables in NGC 330, NGC 2004, and NGC 2100, although these stars are quite common among early type stars in galactic clusters.

Among the variables in those clusters are also a few eclipsing binaries (Fig.3). This is a remarkable result since it shows that we have to expect a significant fraction of close binaries in these young, metal-poor populations. A large initial population of binaries in globular clusters is particularly interesting in view of the various binary manifestations in galactic globular clusters and of the importance of binaries for cluster dynamics (e.g. McMillan 1990).

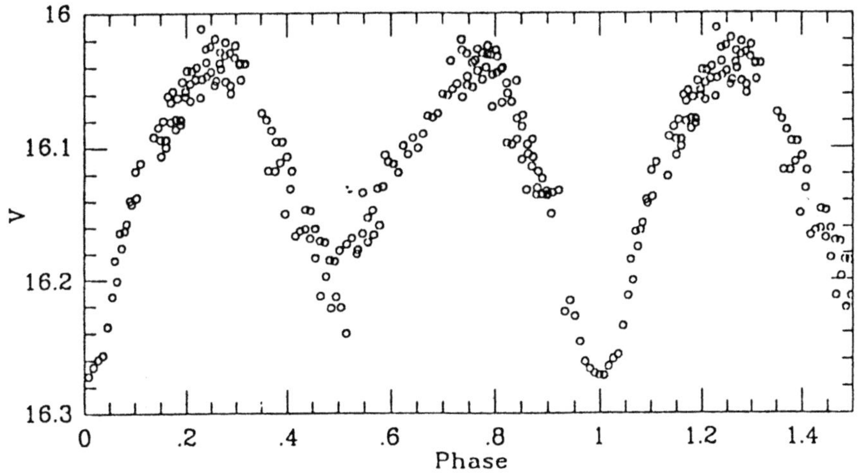

Fig. 3. This is the light curve of an eclipsing binary in NGC 2004 (taken from Balona 1992b, star 87) The period is 1.664 days.

4 Initial Mass Functions of young clusters

The question whether the Initial Mass Function (IMF) of stars is universal or not, is a central question in the general framework of star formation (see the reviews of Scalo 1985, Cayrel 1990, and Zinnecker et al. 1992). The young globular clusters in the Magellanic Clouds are ideal objects for studying the IMF in the mass range 1 to 10 solar masses: They provide a good statistic due to their richness in stars and they are young enough that their mass spectra have not yet undergone dynamical modifications. Moreover, they can serve as test objects whether the IMF is a function of metallicity or not. The main observational difficulty is the immense crowding of stars which could not be overcome in the photographic era of astronomy. With the advent of linear detectors, it has become feasible to perform crowded field photometry and to quantify completeness factors of star counts which is a necessary prerequisite for obtaining luminosity functions in crowded fields. The transformation of luminosity functions in mass functions is now possible on the basis of a large set of evolutionary tracks and isochrones (e.g., Bertelli et al. 1990, Schaller et al. 1992).

Determining the IMF of young Magellanic Cloud clusters is therefore a particularly interesting aspect in Magellanic Cloud research. In what follows we shall use the description of IMFs as power laws $dN/dm \sim m^{-x}$, where dN is the number of stars in the mass range m-dm/2 ,m+dm/2. The famous "Salpeter-value" is then x = 2.35.

The first papers which addressed this topic seriously give still contradicting results (Elson et al. 1989, Mateo 1988). However, Sagar and Richtler (1991) found for 5 young clusters x in the range 2.1 to 2.5 with an error of + 0.2. Further support for Salpeter-like mass spectra came from Fischer et al. (1991) who got x = 2.3 for NGC 1866, from Bencivenni et al. (1990) who found x = 2.2 for NGC 2004 and from Cayrel et al. (1989) who got x = 2.3 for NGC 330. Elson (1991, 1992) measured light profiles for a sample of clusters and found the trend that core radii increase with age. If that effect is due to core expansion caused by stellar evolution, again Salpeter-like mass spectra are inferred.

In the frame of the ESO Key Programme we obtained B,V,R -photometry for NGC 1818. Will (1992) used the isochrones of Schaller et al. to convert the observed luminosity function into a mass function. The result is x = 2.2 for a mass range of 2 - 10 solar masses, fitting perfectly to the cited work above. This value is characteristic not only for these young clusters, but similar mass functions have been found also for associations like NGC 1948 (Seggewiss et al., this conference, Vallenari et al. 1992) and even for associations between the Magellanic Clouds (Grondin et al. 1992).

Taken all the available evidence, we have good reason to believe that the IMF is an (almost?) universal feature of stellar aggregates.

5 Formation scenarios and Initial Mass Functions

The fundamental question connected with the occurence of young globular clusters in the Magellanic Clouds is of course their formation. Why do these objects occur in the Magellanic Clouds (and in other galaxies like M33, NGC 2403 etc.) and not in the Galaxy or in M31? Several theoretical scenarios for globular cluster formation have been proposed (e.g., Murray and Lin 1990, Shapiro et al. 1992) and there is no space to detail their relevance for Magellanic Cloud clusters.

The scenario I favour, is described in a poster paper (Richtler and Fichtner, this conference) and again it is the metallicity which is the key property. Rather than to go into theoretical details, I want to outline a simple application of this scenario on the problem of the IMF.

Starting point is the fragmentation scheme of Murray and Lin (1989a). These authors showed that fragmentation in a collapsing protocluster occurs via thermal instabilities which develop later on into gravitational instabilities (Murray and Lin

1990b). Whatever the details, the relevant property of thermal instabilities in our context is that they have no preferred length scale, contrary to gravitational instabilities where we have the Jeans-length. We may thus assume that within some reasonable limits R_{min} and R_{max}, each fragment radius R occurs with the same probability. If we further assume that the density is constant, we may write for the mass of each fragment $M \sim R^3$ and the corresponding mass spectrum is $dN/dM \sim M^{2/3}$, if dN/dR = const.

If we require that the fragments may not overlap then it is obvious that this mass spectrum is realised only in an infinite volume. In a finite volume, such as a collapsing protocluster, the largest possible R_{max} for each fragment is the distance to the nearest neighbour. The free space between the fragments decreases with the fragments filling their available volume more and more. Therefore, if the filling factor approaches 1, the resulting distribution is biased towards smaller masses. What is the shape of the resulting mass spectrum?

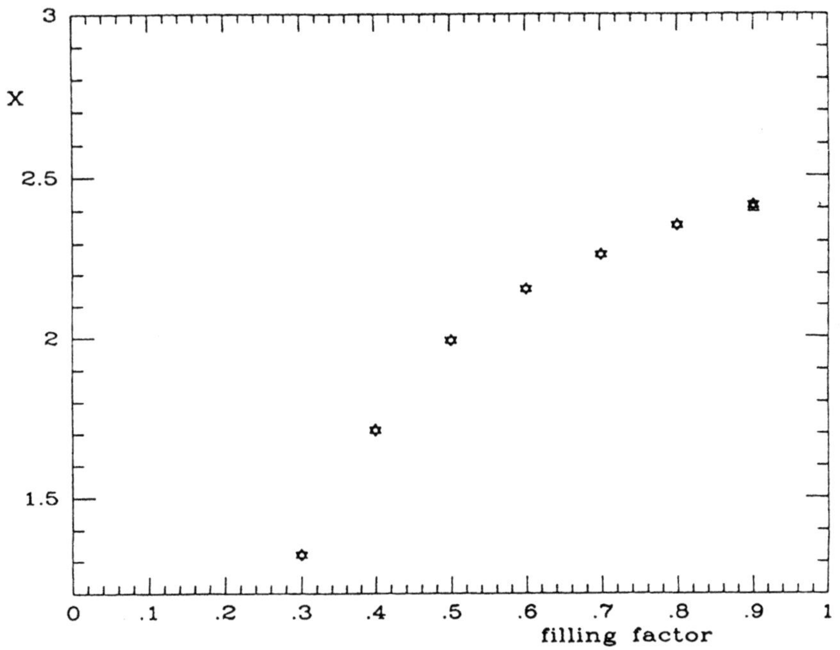

Fig. 4. The exponent of the power law of the mass distribution of fragments is plotted against the filling factor, following from our numerical experiments . Even while the filling factor is still moderate, the exponent is close to what is actually observed.

We performed numerical experiments to investigate this problem. The principle is simple: Fill a volume (parent cloud, radius 1) with smaller subvolumes (=fragments) whose radius distribution follow dN/dR = constant. The locations coordinates and the radii are chosen randomly. If a newly chosen subvolume overlaps with a neighbour, a new radius in the range between R_{min} and the distance to this neighbour is chosen.

Fig. 4 shows the resulting mass spectrum slope in dependence of the filling factor. Each point is the mean of 10 single experiments on the average. Typical scatter is 0.2 in x. We have arbitrarily assumed that a volume of radius 1 should correspond to 5000 solar masses, so $R_{min} = 0.02$ and the initial $R_{max} = 0.27$, corresponding to 0.1 and 50 solar masses, respectively. The distribution converges (statistically) against $dN/dM \sim M^{-2.4}$. The dependence on the actual R_{min} and initial R_{max} is small. It is amazing that with this simple assumptions, the main characteristics of the observed IMF, the power law and even the correct exponent, can be duplicated.

Is this the physical reason for a universal Salpeter-like mass spectrum? One may object that the assumptions are partly questionable, in particular that the final stellar mass is determined by the initial mass of the thermally unstable region. It remains to be seen whether this simple geometric concept persists in a more realistic framework of cluster formation.

6 Concluding remarks

Almost 40 years after it has been recognised that the Magellanic Clouds host a type of star cluster unknown in the Milky Way, these objects still present a challenging case. Neither fundamental properties like metallicities nor their formation history have been ultimately explored. They still offer new and surprising aspects for investigation as the Be stars in NGC 330 and other clusters show. There are many more interesting phenomena which could not be addressed due to space limitations, for example binarity among Magellanic Cloud clusters (e.g, Bhatia et al. 1991) or the detection of substructure in the light profile of young clusters (Elson 1991), which perhaps hint to merger processes of smaller entities.

Whatever future investigations will reveal, we can be sure that young Magellanic Cloud clusters will continue to serve as excellent test laboratories for a tremendous variety of astrophysical branches.

I would like to thank Eva Grebel for a fruitful collaboration and Klaas S. de Boer for discussions and a critical reading of the manuscript. I am also indebted to Dietrich Baade for helpful comments on Be stars. We thank the directorate of the Max Planck Institut für Astronomie, Heidelberg, for generous allocation of observing time.

References

Cayrel, R., Tarrab, I., Richtler, T., 1988, The Messenger, 54, 59

Cayrel, R., 1990, "Physical Processes in Fragmentation and Star Formation", Astr. Sp. Sc. Library Vol. 162, p. 343, eds. R. Capuzzo-Dolcetta, C. Chiosi, A. Di Fazio, Kluwer Academic Publishers, Dordrecht

Cox, A.N., Siobahn, M.M., Rogers, F.J., Iglesias, C.A., 1992, ApJ, 393 , 272

Baade, D., 1992, in "Evolutionary Processe in Interacting Binary Stars", IAU Symp. 151, eds. Y. Kondo, P. Sistero, R.S. Polidan, Kluwer Academic Publishers, Dordrecht, in press (ESO Preprint 792).

Balona, L.A., 1990, MNRAS, 245, 92

Balona, L.A., 1992a, MNRAS, 256, 425
Balona, L.A., 1992b, MNRAS, in press
Bell, R.A., Gustafsson, B., 1979, A&A, 74, 313
Bencivenni, D., Brocato, E., Buonanno, R., Castellani, V., 1991, AJ, 102, 137
Bertelli, G., Betto, R., Bressan, A.G., Chiosi, C., Nasi, E., Vallenari, A. , 1990, A&AS, 85, 845
Bhatia, R.K., Read, M.A., Hatzidimitriou, D., Tritton, S., 1991, A&AS, 87, 335
Da Costa, G.S., 1991, in "The Magellanic Clouds", IAU Symp. 148, eds. R. Haynes and D. Milne, Kluwer Academic Publishers, Dordrecht, p. 183
Elson, R.A.W., Fall, S.M., Freeman, K.C., 1989, ApJ, 336, 734
Elson, R.A.W., 1991, ApJS, 76, 185
Elson, R.A.W., 1992, MNRAS, 256, 515
Feast, M.W., 1972, MNRAS, 159, 113
Fischer, P., Welch, D.L., Cote, P., Mateo, M., Madore, B.F., 1992, AJ, 103, 857
Gies, D.R., 1991, in "Rapid Variability of OB-Stars: Nature and Diagnostic Value", ed. D. Baade, ESO Conference and Workshop Proceedings No. 36, p. 229
Grebel, E.K., Richtler, T., de Boer, K.S., 1992, A&A, 254, L5
Grebel, E.K., Richtler, T., 1992, A&A, 253, 359
Grondin, L., Demers, S., Kunkel, W.E., 1992, AJ, 103, 1234
Iglesias, C., Rogers, F.J., 1991, ApJ, 371, L73
Lamers, H.J.G.L.M., in "Physics of Be Stars", IAU Coll. 92, eds. A. Slettebak, T.P. Snow, Cambridge University Press, p. 219
Luck, R.E., Lambert, D.L., 1992, ApJS, 79, 303
Mateo, M., 1988, ApJ, 331, 261
McMillan, S.L.W., 1991, in "The Formation and Evolution of Star Clusters", p. 324, ed. K. Janes, A.S.P. Conf. Ser. 13, BookCrafters, Inc., San Francisco
Murray, S.D., Lin, D.N.C., 1989a, ApJ, 339, 933
Murray, S.D., Lin, D.N.C., 1989b, ApJ, 346, 155
Murray, S.D., Lin, D.N.C., 1990, ApJ, 357, 105
Olszewski, E.W., Schommer, R. A., Suntzeff, N.B., Harris, H.C., 1991, AJ, 101, 515
Reitermann, A., Baschek, B., Stahl, O., Wolf, B., 1990, A&A, 234, 109
Richtler, T., Spite, M., Spite, F., 1989, A&A, 225, 351
Robertson, J.W., 1974, A&AS, 15, 261
Rozsnayi, B.F., 1989, ApJ, 341, 414
Russell, S., Bessell, M., 1989, ApJS, 70, 865
Sagar, R., Richtler, T., 1991, A&A, 250, 324
Schaller, G., Schaerer, D., Meynet, G., Maeder, A., 1992, A&AS, in press
Scalo, J.M., 1986, Fund. Cosm. Phys., 11, 1
Shapiro, P.R., Clocchiatti, A., Kang, H., 1992, ApJ, 389, 269
Spite, F., Richtler, T., Spite, M., 1991, A&A, 252, 557
Stothers, R.B., Chin, C., 1992, ApJ, 390, 136
Vallenari, A., Bomans, D.J., de Boer, K.S., 1992, A&A, in press
van den Bergh, S., 1991, ApJ, 369, 1
Will, J.-M., 1992, Diploma Thesis, Bonn University
Zinnecker, H., McCaughrean, M.J., Wilking, B.A., 1992, in "Protostars and Planets III" , eds. E.H.Levy, J.I.Lunine, M.S. Matthews, University of Arizona Press, in press

Evolutionary Tracks and Color Magnitude Diagrams

A. Bressan [1], G. Bertelli [2], C. Chiosi [3]

[1]Astronomical Observatory, Padova, Italy
[2]National Council of Research CNR Padova Italy
[3]Department of Astronomy, Padova, Italy

Abstract: In this paper we discuss the effect of the new radiative opacities by Iglesias et al. (1992) on the properties of stellar models in relation to the problem of the type of mixing (classical semiconvection versus overshoot) in stellar interiors. Models with the opacities are briefly presented and the constraints imposed by the integrated luminosity function (ILF) of main sequence stars and the mass discrepancy problem of the Cepheids are outlined. In addition to this, we present the Johnson-Cousins integrated magnitudes and colors of theoretical star clusters calculated with aid of the last generation of stellar models and the library of synthetic spectra by Kurucz (1992). The theoretical colors are used to discuss the family of clusters of the LMC with the aim of understanding the history of cluster formation in this galaxy. Finally, we summarize recent work on the history of star formation in this galaxy derived from the study of the color-magnitude diagrams (CMDs) of field stars.

1 Introduction

The study of the CMD of cluster and field stars is the primary step toward understanding stellar structure and evolution and the past history of stellar populations and galaxies as a whole. In this paper we will touch upon various topics that are deeply related, and which ultimately depend on the adequacy and accuracy of stellar models. Particular emphasis will be given to the effects of the new radiative opacities by Iglesias et al. (1992) on stellar models and on the related subject of the extension of convective regions.

The radiative opacity is basic to the calculation of stellar evolutionary models because it affects the luminosity, the radius, and the total lifetime of a star. The recent advancements in the calculation of astrophysical radiative opacities (Iglesias et al. 1992:OPAL) showed that two significant enhancements are present with respect to the classical tabulations of the Los Alamos Opacity Library (Huebner et al. 1977: LAOL). The first one, more important (about a factor of 3), occurs at temperatures of a few hundred thousand degrees, whereas the second one, less pronounced (about 20%), occurs at temperatures of about one a million degrees. The low temperature enhancement in the opacity bears on the pulsational properties of Beat and Bump Cepheids, removing for the first time the disagreement between the observed and predicted period ratios (Moskalik et al. 1992, Antonello 1992, and Kanbur & Simon 1992). Because this enhancement is confined in regions of relatively low temperatures (Log T $\simeq$ 5.3), it is expected to have little influence on the global structure of a star (hence lifetimes). On the contrary, the second enhancement, which is located at Log T $\simeq$ 6.2, is expected to more strongly affect the structure and the evolution of stellar models, because a significant fraction of the star mass is involved. As nowadays, there is no systematic comparison between models based on the LAOL and models based on the OPAL. Stothers & Chin (1991) computed stellar models with the OPAL and compared them with previous evolutionary tracks that were obtained with the old Cox & Stewart (1970) opacities. Because these latter are significantly different from the LAOL, great part of the effect discussed by Stothers & Chin (1991) is due the use of old opacities and not to the difference OPAL and LAOL. It is often claimed that the use of *modern* opacities would rule out the need of overshoot from the convective core. See the discussions by Brocato et al. (1990) in the case of LAOL and Cox & Stewart (1970) opacities, and by Stothers & Chin (1991) in case of OPAL and Cox & Stewart (1970). However no comparison has been standing on physically homogeneous stellar models and on homogeneous observational data. To cast light on this topic, we computed new sets of stellar models (Bertelli et al. this meeting) with chemical composition suitable for the solar vicinity and the LMC, in which the OPAL tabulations and moderate overshoot (Alongi et al. 1992) were adopted. In the case of the LMC composition, we also computed models with the classical semiconvective mixing during the central He-burning phase. The main reason for it, was to clarify that homogeneity in the input physics is the main requirement for a meaningful comparison between theory and observations. Various tests can be devised to disentangle which of the mixing schemes yields stellar models in better agreement with the observational data. In the following we will concentrate on two of those. The ILF for main sequence stars normalized to the number of evolved stars in a cluster is a powerful tool (Chiosi et al. 1989), because it reflects the ratio between the central H- and He-burning lifetimes, which in turn depend on the adopted mixing scheme. Another signature of the type of mixing is given by the mass-luminosity relation of He-burning stars, because they are brighter by about Δ Log $L/L_{\odot}$ $\simeq$ 0.3 or 0.5 in the case of mild or full overshoot, respectively (Chiosi 1990). Since the mass (both evolutionary and pulsational) of the cepheid stars depend on the underlying mass-luminosity relationship, the young star clusters of the LMC containing Cepheids are ideal laboratories in which the internal structure of the stars, hence type of mixing, can be tested. The lack of satisfactory theory of mixing in stellar interiors (see Chiosi et al. 1992a for a detailed discussion of this subject) forces on us an empirical approach to determine the amount of overshoot during the main evolutionary phases (core H- and He-burnings).

These new stellar models are used to derive the integrated magnitudes and colors of single stellar populations (clusters) following the method described by Chiosi et al. (1988). This allows us to study those clusters for which CMDs and luminosity functions are not yet available and only integrated magnitudes and colors are known. The analysis is made limited to the family of clusters of the LMC. On the one hand the reconstruction of the integrated colors and magnitudes as a function of the age and metallicity casts light on contribution from particular groups of stars, such as red giant branch (RGB) and asymptotic giant branch (AGB) stars, whose effect has long been debated (see Chiosi et al 1988 and 1992b for discussion of this topic). On the other hand, ranking clusters as a function of age and chemical composition, and knowing how the intrinsic colors of any individual cluster evolves with time, would allow us to understand the physical cause of the color distribution of the whole family of clusters, and in principle to infer something about the combined effect of cluster formation and disruption rates (see Chiosi et al 1988). The final goal of this being to identify the signatures of the history of cluster formation and the possible correlation with the dynamical evolution of the parent galaxy (LMC in our case) as a whole.

Finally, constraints on the history of star formation in LMC are derived from the analysis of the CMDs and luminosity functions of field stars made by Bertelli et al. (1992).

2 Stellar Evolution with OPAL

Two large grids of evolutionary tracks, one with a composition suitable for the LMC (Z=0.008, Y=0.250) and another with a solar like composition (Z=0.020, Y=0.280), were computed by Bertelli at al. (this volume) with the aid of the most recent OPAL tabulation, which adopts the fractional abundances by Grevesse (1991). The initial masses of the models range from 0.6 $M_{\odot}$ to 100 $M_{\odot}$ and the sequences terminate either at the stage of carbon ignition, or at the beginning of the Thermally Pulsing AGB phase (TP-AGB). All the models are calculated allowing for both core and envelope overshoot according to the prescription by Alongi et al. (1992).

In Figure 1 we compare two evolutionary sequences for a 5 $M_{\odot}$ star with LMC composition calculated using both OPAL and LAOL. With the new opacities (OPAL), the main sequence phase is slightly cooler, the loop during the central He-burning phase is shorter and less luminous, and the Hayashy line is slightly hotter than with the old opacity (LAOL). The last effect is caused by the larger value of the mixing length in the outer convection that is required to calibrate the our evolutionary sequences on the Sun. In Figure 2 we show the run of the OPAL with the interior mass for a red giant star model taken from the sequence of Fig.1 labelled OPAL and the corresponding LAOL based on the Ross-Aller mixture and calculated for the same temperatures and densities. Figure 2 clearly illustrates the large enhancement in the OPAL, which in the H-rich envelope extends up to temperatures above 10^6 K. In the central, He-rich regions of the star, OPAL gets lower than LAOL by about $\simeq$ 15%. This is due to the different interpolation scheme allowed by the finer grid in the OPAL tabulation. In the case of the solar composition, the effects of the new opacities are more pronounced, in particular during the central He-burning phase, where the loop is less luminous (by half a magnitude) and much shorter. Finally, let us recall that the enhancement in the opacity decreases at increasing density (for a given temperature) so that it will vanish going toward lower stellar masses.

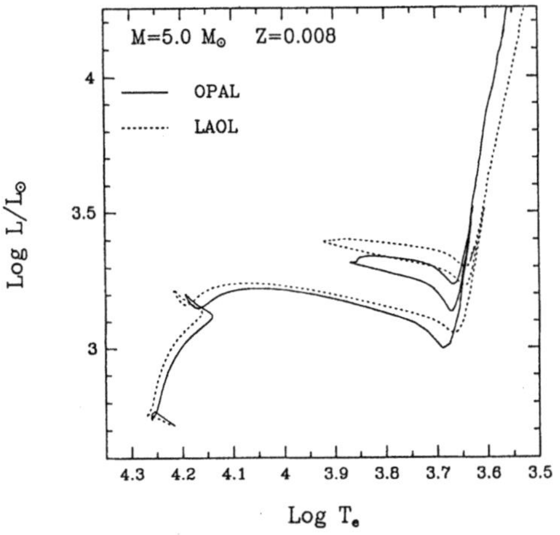

Fig. 1. The evolutionary tracks of a 5 $M_{\odot}$ model computed with the OPAL and the LAOL opacity tables respectively. The adopted chemical composition is suitable for young clusters of LMC

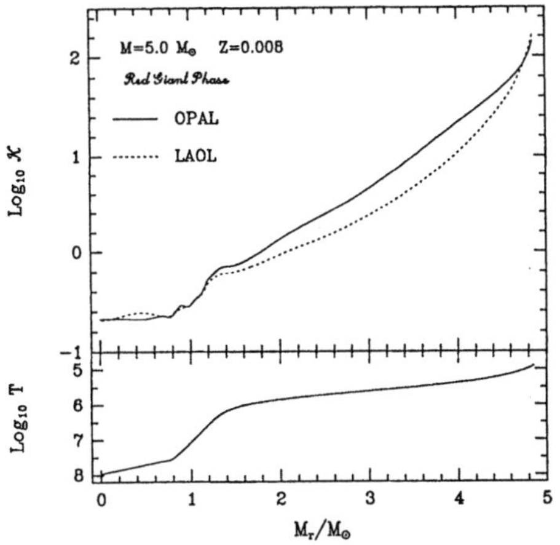

Fig. 2. The run of the opacity inside a red giant model of the sequence of Fig.1

3 Star Clusters and Integrated Luminosity Functions

The star clusters of the LMC are powerful workbench to test the input physics of stellar models, because their richness implies that even the short lived evolutionary phases are well populated. Among others, convective mixing during the major phases of nuclear burning is one of the most uncertain physical aspects of stellar structure, which is often parameterized in terms of the amount of extra-mixing (overshoot) beyond the region determined the Schwarzschild condition. The amount of overshoot during the H-burning phase significantly alters the subsequent evolution in a way that has been analyzed by several authors and recently summarized by Chiosi et al. (1992a). In short, at given initial mass, models with overshoot have a larger core mass than the corresponding standard models, so that they possess a wider and more luminous main sequence band, a longer core H-burning lifetime, a more luminous but shorter loop in the CMD, and finally, a shorter core He-burning lifetime. Moreover, the critical masses separating low, intermediate, and high mass stars are also affected and shifted toward lower values. Beginning with Becker and Mathews (1983), who first produced a synthetic simulation of the LMC cluster NGC 1866, it was clear that the relative number of stars in the He- and H-burning phases predicted by the theory was at variance with the observations, for any reasonable slope of the initial mass function (IMF). Chiosi et al. (1989) were able to quantify the discrepancy by introducing the ILF of the main sequence stars normalized to the core He-burning stars. They noticed that first it was not possible to discriminate between classical and overshoot models on the basis of the CMD and the differential luminosity function, second that the ratio between the central He- and H-burning lifetimes resulting from the different mixing schemes yields very different values for the expected number of main sequence stars (brighter than a certain magnitude) per red giant star, or equivalently yields different normalized ILFs. In the case of NGC 1866, the ratio in question derived from models with full overshoot and the Cox & Stewart (1970) opacities and evaluated at the magnitude where the completeness of star counts was 90%, turned out to be more than a factor of two larger than in the case of classical models, as suggested by the observational star counts. Similar analysis on the same cluster was performed by Brocato et al. (1990) who concluded that classical models are to be preferred when the LAOL opacities are used, with the obvious implication that the need of core overshoot simply masked the effect of adopting older (lower) opacities (Cox & Stewart 1970). However, Bressan (1990) showed that even with the LAOL opacities the discrepancy between theory and observation still remains, and that Brocato's et al. (1990) conclusion was entirely due to different star counts obtained for the same region of the CMD. Vallenari et al. (this conference) present two LMC clusters, i.e. NGC 2249 and NGC 2134, for which they have obtained good fits of the CMDs adopting models calculated with overshoot and the new opacities (OPAL). Furthermore, comparing the observed ILFs of the two clusters with those obtained from their synthetic CMDs for three different values of the exponent of the IMF, x=0.35, x=1.35 (Salpeter), x=2.35 we get the following results. In the case of NGC 2134 (age $\simeq 1.9\times10^8$yr) the observed ILF agrees with the theoretical one for the Salpeter slope, whereas in the case of NGC 2249(age $\simeq 5.5\times10^8$yr) the observational ILF runs steeper than that obtained with the higher IMF exponent. Although the correspondent analysis with the classical models is not yet available, simple considerations allow us to say that they would not be able to fit those data. In fact, the theoretical ILFs at the magnitude level where the main sequence is not affected by evolution, are simply proportional to the lifetime ratios t_H/t_{He} (Chiosi et al. 1989). These are given in Table 1 for a few selected models computed both with overshoot and semiconvection. The data of Table 1 clearly show that the ILFs of classical models would underestimate the observational ILFs by a significant factor.

Table 1 H- and He-burning lifetimes (10^6 yr.) and lifetime ratios (Z=0.008)

	Overshoot			Semiconvection		
$M/M_\odot$	t_H	t_{He}	t_H/t_{He}	t_H	t_{He}	t_H/t_{He}
3	331.09	87.22	3.85	245.57	152.31	1.61
4	171.00	24.94	6.25	130.48	48.06	2.70
5	105.55	10.67	10.00	82.20	21.17	3.85
7	52.71	3.73	14.29	42.91	6.59	6.67

4 LMC Clusters with Cepheid Stars

Among the many problems that can be tackled by studying the star clusters of the LMC, the unsettled question of the Cepheid mass discrepancy is of particular interest. The advantage with the LMC clusters containing cepheids is that much more numerous stars are present (cepheids included), membership of the cepheids is not a problem, accurate CMDs are available, and finally the evolutionary and pulsational properties of the cepheid stars can be estimated in a coherent fashion. For the galactic cepheids it has been shown that the mass derived from their pulsational properties is significantly smaller (about 40%) than the mass inferred from their evolutionary status by means of a suitable mass-luminosity relation (Iben & Tuggle 1975). Several ways out have been proposed (Cox 1980) none of which is fully satisfactory. Using the observations of NGC 2157, NGC 1866 and NGC 2031 (Mateo et al. 1990a,b; Welch et al. 1990) for which both good CCD photometry of the whole stellar content and accurate determinations of the fundamental parameters of the cepheid stars were given, Chiosi et al. (1992c) and Bertelli et al. (1992b) were able to compare the mass-luminosity relation given by the CMD fit with that derived from the pulsation properties of the Cepheids. The former turns out to depend on the amount of overshoot and, to a lesser extent, on the adopted distance modulus, while the latter turns out to be much more sensitive to the adopted distance modulus (see also Iben & Tuggle 1975). This is shown in Figure 3, where we plot the evolutionary and pulsational mass of the three Cepheids of NGC 2157 as a function of the distance modulus of the cluster and the underlying evolutionary scheme.

While both classical and overshoot models can lead to the formal solution of the mass discrepancy problem, the results displayed in Figure 3 and the discussion in the Chiosi et al. (1992c) paper clarify that only models with convective overshoot yield a good fit of the CMD and a distance modulus in agreement with other independent estimates, e.g. the value of $(m\text{-}M)_o$=18.5 obtained from the Supernova 1987A (Panagia et al. 1991). The same conclusions hold for the clusters NGC 1866 and NGC 2031 analyzed by Bertelli et al. (1992), in which the dependence of the distance modulus inferred from the solution of the mass discrepancy problem on the metallicity is investigated.

Those analyses were based on the pulsational models of Cepheid stars calculated by Chiosi et al. (1992d) who adopted the LAOL opacities. Fortunately, new opacities (OPAL) do not modify the above conclusions, because the variation of the fundamental period at a given mass, luminosity, and effective temperature, is less than a few percent (Figure 4). This result does not contradict the finding that the new opacities resolve the mass discrepancy problem of the Beat and Bump Cepheid stars (Moskalik et al. 1992, Kanbur & Simon 1992), for which another kind of mass determination is possible. Moreover, as shown by Moskalik et al. (1992), Kanbur & Simon (1992), and Antonello (1992), the mass discrepancy of the Bump Cepheids can be solved only by simultaneously adopting the new opacities and the mass-luminosity relation of models with overshoot (contrary to what often claimed in recent literature), while for the Beat Cepheids the effects of the new opacities overwhelm those of a different mass-luminosity relation.

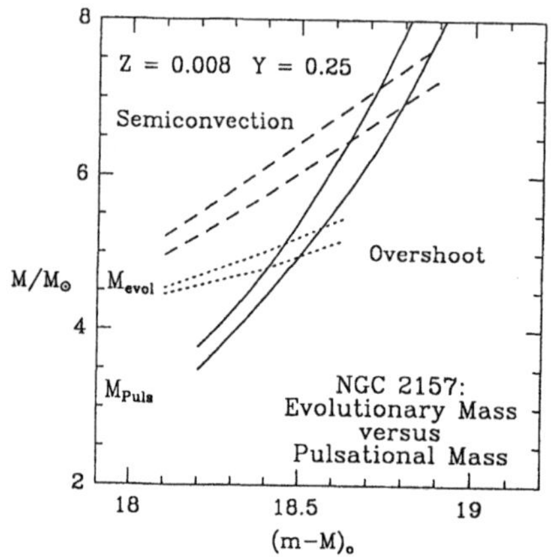

Fig. 3. The pulsational and evolutionary masses of the three Cepheids in NGC 2157 as a function of the distance modulus

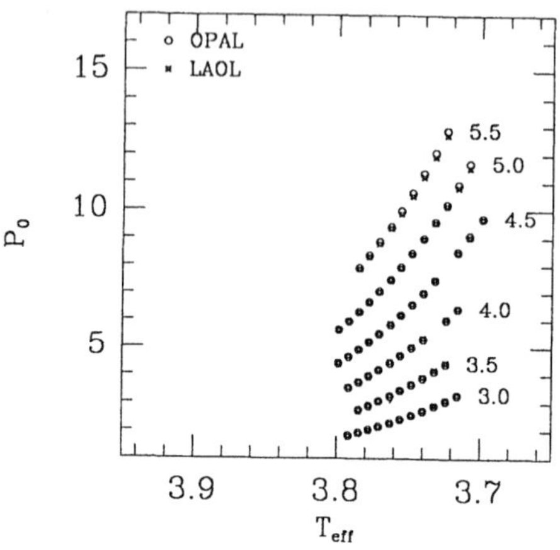

Fig. 4. Fundamental periods of Cepheid models calculated with the LAOL (Chiosi et al. 1992d) and OPAL (Antonello 1992) opacities

5 Integrated Magnitudes and Colors of Star Clusters

The above discussion has been restricted to those LMC clusters possessing a good photometry of individual stars. However, much more numerous are the clusters for which only integrated magnitudes and colors are known. Bica (1992) gives integrated UBV photometry for 624 clusters, a substantial fraction of the total population in LMC estimated by Hodge (1988), which amounts to about 4000 clusters. This larger sample allows a verification of the reliability of the predicted integrated colors and magnitudes of single stellar populations (SSP) as a function of the age and composition. To construct these relations we proceed as follows. The integrated monochromatic flux of a SSP with age τ and metallicity Z is

$$F_\lambda(\tau, Z) = \int_{M_L}^{M_U(\tau)} \phi(m)\ f_\lambda[g(m,\tau), T_{eff}(m,\tau), Z]\, dm$$

where $f_\lambda[g(m,\tau), T_{eff}(m,\tau), Z]$ is the monochromatic flux of a star of mass m, gravity $g(m,\tau)$, effective temperature $T_{eff}(m,\tau)$, and composition Z. The integral is performed along the isochrone of age τ and the dependence on the mass and the time of the various terms is shown explicitly. The quantity $f_\lambda[g, T_{eff}, Z]$ is taken from the Kurucz (1992) library of stellar spectra while the isochrones are calculated following the procedure described in Bertelli at al. (1990) with the IMF law given by $\phi(m) = A \times\ m^{-2.35}$ and A=1. The integrated spectrum is then convolved with the response of the filters in the UBVRIJKL system to obtain the integrated magnitudes and colors as a function of the age and composition. The comparison of the theoretical integrated colors (U-B) and (B-V) obtained for three different chemical compositions

(Z=0.02, Y=0.28; Z=0.008, Y=0.25; Z=0.001, Y=0.23) with the data for 624 LMC clusters by Bica (1992) shows that the trend in the observed colors is well reproduced (Figure 5).

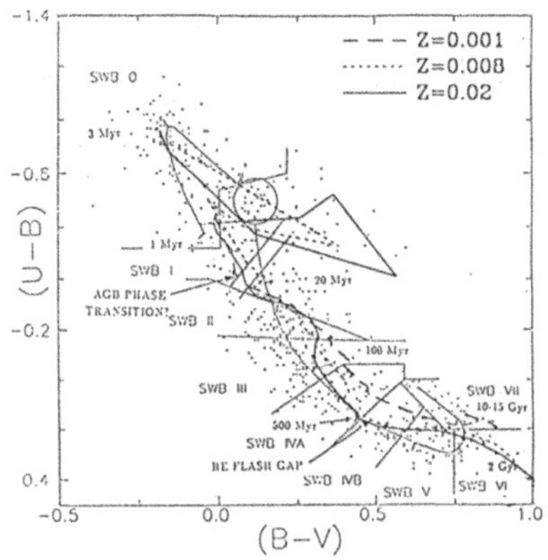

Fig. 5. (U-B)–(B-V) two color diagram for the 624 LMC clusters from Bica (1992). Superposed are the theoretical relationships with an average reddening of E(B-V)=0.07 and E(U-B)=0.05

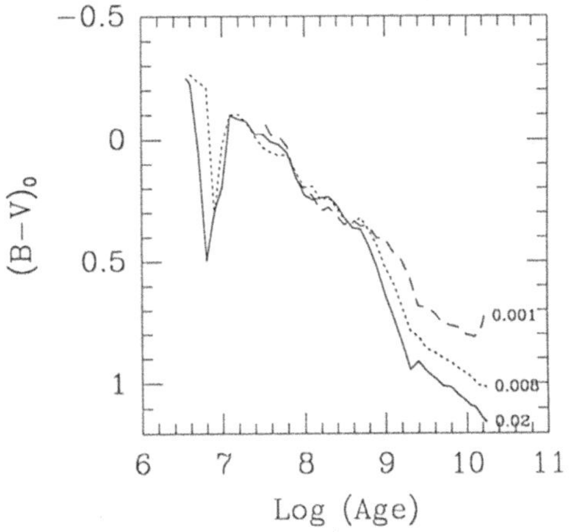

Fig. 6. $(B\text{-}V)_0$–age relationships for the three different chemical compositions

The observational data scatter around the theoretical curves. This can be explained as due to effects of different chemical compositions, and stochastic fluctuations in the number of bright stars in each cluster (Chiosi et al. 1988). Other than the fast change in colors due to the appearance of the red supergiants at age of a few Myr (Figure 6), there is no evidence of "phase transitions" in the visible colors caused by the sudden appearance of the AGB or the RGB stars (see also Chiosi et al. 1988). A transition of this kind is visible in the infrared colors (V-I, V-K, etc...) at the age of about 10^8 yr, which is clearly related to the appearance of the AGB stars. However, as already shown by Chiosi et al. (1992b), this powerful indicator of the epoch of the galaxy formation is wiped out by cosmological effects. Chiosi et al.(1988) analysing a much smaller sample of clusters concluded that the gap in the number of clusters at (B-V) $\simeq$ 0.4–0.6 (van den Bergh, 1981) is due to the concomitant effect of the intrinsic color evolution and age distribution of clusters. The bimodal distribution of the color could be understood without invoking strong discontinuities in the rate of cluster formation even if these could not be excluded. The gap in the (B-V) color is still present in the more numerous sample of Bica (1992). The present analysis confirms that the gap cannot be caused by discontinuities in the colors related to the sudden appearance of RGB or AGB stars. Whether it reflects a true gap in the star formation history of the LMC is less evident. Alongi & Chiosi (1991) argued for a real age gap.

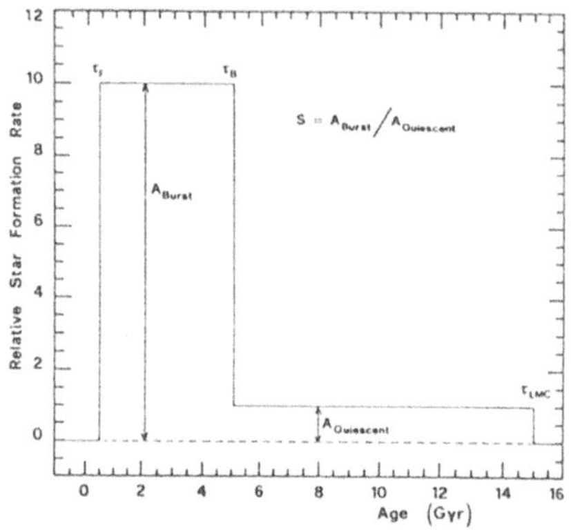

Fig. 7. The simple model describing the star formation history of LMC

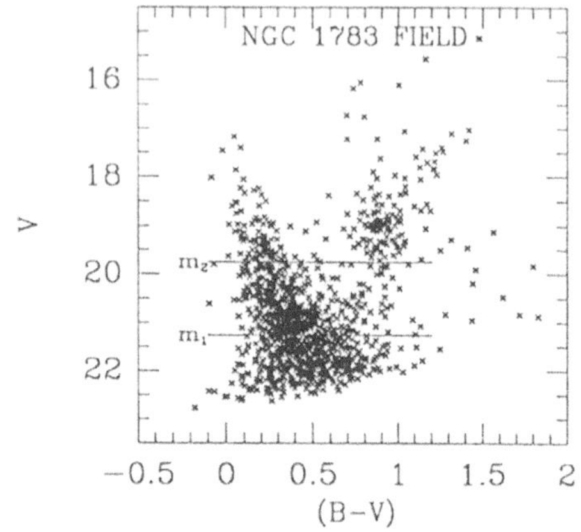

Fig. 8. The CMD of the stars in the field nearby NGC 1783. The horizontal lines labeled m_1 and m_2 delimit the areas selected for determining the star count ratios

6 The Star Formation History of LMC

Bertelli et al. (1992a) tried to reconstruct the history of star formation in the LMC by studying the CMDs of field stars in three selected areas named from the nearest cluster, i.e. NGC 1783, NCG 1866, NCG 2155. The simplest model of star formation compatible with the observational CMDs and luminosity function is characterized by a prominent burst that initiated τ_B and ended τ_F years ago, respectively, superposed to a quiescent star formation activity all over the galaxian lifetime τ_{LMC} (Figure 7). τ_F, τ_B, the ratio S between the strength of the star formation rate (SFR) before and during the burst, the slope of the IMF, and the age of LMC τ_{LMC} were free parameters adjusted in such a way that both the CMDs and suitable star counts were matched. The galaxian age was assumed to be $\tau_{LMC}=15\times10^9$ yr and τ_F was evaluated from the magnitude of the main sequence termination point of the CMD of each field. Figure 8 shows the CMD of the stars in the NGC 1783 field and the areas selected for the star counts. Three ratios of star counts were constructed: R_1 the ratio between the number of main sequence stars and that of the evolved stars both brighter than m_1, R_2 the ratio of the evolved stars brighter than m_2 to those fainter than this and brighter than m_1, and finally R_3, the ratio of main sequence stars between m_1 and m_2 to those brighter than m_2. By demanding that the simulations match all the above ratios at the same time Bertelli et al. (1992b) showed that the three fields likely underwent a burst of star formation with strength from 5 to 10 times higher than that of the previous quiescent activity. The episode begun 3 or 4 Gyr ago in all the three fields (the precise value depending on the underlying set of evolutionary models -with or without overshoot from the convective core-, the adopted distance modulus, and other details) and lasted until a few 10^8 yr ago. This analysis confirmed the result obtained long ago by Butcher (1977). What triggered this sudden increase in the SFR remains unclear. To map this process in more detail we are analyzing several other fields in LMC.

References

Alongi, M., Chiosi, C., 1991, in *The Magellanic Clouds* , ed. R. Haynes and D. Milne, Kluwer Academic Publ., p. 193

Alongi, M., Bertelli, G., Bressan, A., Chiosi, C., Fagotto, F., Nasi, E., Greggio, L., 1992, A&AS in press

Antonello, E., 1992, preprint

Anders, E., Grevesse, N., 1989, Geochim. Cosmochim. Acta 53,197

Bertelli, G., Betto, R., Bressan, A., Chiosi, C., Nasi, E., Vallenari, A., 1990, A&AS 85,845

Becker, S. A., Mathews, G. J., 1983, ApJ 270, 155

Bertelli, G., Mateo, M., Chiosi, C., Bressan, A., 1992a, ApJ 388,400

Bertelli, G., Bressan, A., Chiosi, C., Mateo, M., Wood, P. R., 1992b, ApJ submitted

Bica, E., 1992, in *The Stellar Populations of Galaxies* , ed. B. Barbuy and A. Renzini, Kluwer Academic Publ., p. 215.

Bressan, A., 1990, in *Chemical and Dynamical Evolution of Galaxies*, ed. F. Ferrini, P. Franco, F. Matteucci. Pisa: Giardina

Brocato, E., Buonanno, R. Castellani, V., Walker, A. R., 1990, ApJS 71, 25

Butcher, H., 1977, ApJ, 216, 327

Chiosi, C., 1990, in *Confrontation between Stellar Pulsation and Evolution*, ed C. Cacciari and G. Clementini, P.A.S.P. Conference Series, vol. 11, p. 158

Chiosi, C., Bertelli, G., Bressan, A., 1988, A&A 196,84

Chiosi, C., Bertelli, G., Meylan, G., Ortolani, S., 1989, A&A 219,167

Chiosi, C., Bertelli, G., Bressan, A., 1992a, ARA&A in press

Chiosi, C., Bertelli, G., Bressan, A., 1992b, in *The Stellar Populations of Galaxies*, ed. B. Barbuy and A. Renzini, Kluwer Academic Publ., p. 321.

Chiosi, C., Wood, P. R., Bertelli, G., Bressan, A., Mateo, M., 1992c, ApJ 385, 205

Chiosi, C., Wood, P. R., Capitanio, N., 1992d, ApJS submitted

Cox, A. N., 1980, ARA&A, 18, 15

Cox, A.N., Stewart, J.N., 1970, ApJS 19, 243

Grevesse, N., 1991, A&A 242, 488

Hodge, P., 1988, PASP 100, 1051

Huebner, W.F., Merts, A.L., Magee, N.H., Argo, M.F., 1977, Los Alamos Sci. Lab. Rep. LA-6760-M

Iben, I. Tuggle, R.S., 1975, ApJ 197, 39

Kanbur, M.S., Simon N.R., 1992, preprint

Kurucz, R. L., 1992, private communication

Langer, N., 1986, A&A 210, 93

Mateo, M., Olszewski, E., Madore, B.F., 1990, ApJ 107, 203

Moskalik, P., Buchler, R.J., Marom, A., 1992, ApJ 385,685

Stothers, R.B., Chin, C., 1991, ApJ 381,L67

Panagia, N., Gilmozzi, R., Macchetto, F., Adorf, H. M., Kirshner, R. P., 1991, ApJ 380, L23

van den Bergh, S., 1981, A&AS 46, 79

Welch, D. L., Mateo, M., Cote', P., Fisher, P., Madore, B., 1991, AJ 101, 490

New Sets of Evolutionary Tracks

G. Bertelli [1 3], A. Bressan [2], C. Chiosi [3] F. Fagotto [3]

[1]National Council of Research, Italy
[2]Astronomical Observatory, Padova, Italy
[3]Department of Astronomy, Padova, Italy

Abstract: We present four large grids of stellar models covering the main evolutionary phases. Owing to the homogeneity of their updated input physics, they constitute a useful data base to be used in many astrophysical applications.

1 Introduction

The new stellar opacities computed at the Livermore Laboratories by Iglesias, Rogers and Wilson (1992) (OPAL) present a large enhancement at temperature near $3 \cdot 10^5$ K when compared with the Los Alamos tables of Huebner et al. (1977) (LAOL). Because the opacity has a great impact on stellar models, we have computed new sets of evolutionary tracks with the most recent OPAL tables.

2 The Stellar Models

The following abundances were selected: Z=0.001, Y=0.230; Z=0.008, Y=0.250; Z=0.020, Y=0.280, and finally Z=0.100, Y=0.425. The models account for convective overshoot both from the core and the envelope. In progress are the computations with the classical semiconvective scheme. More details can be found in Alongi et al. (1992). The initial masses span the range from 0.6 $M_\odot$ to 100 $M_\odot$. The evolution is followed from the main sequence till either the stage of central carbon ignition or the beginning of the thermally pulsing AGB phase, according to the initial mass of the star. The nuclear reaction network contains sixteen elements; the reaction rates are taken from Caughlan and Fowler (1988); the neutrino emissions are from Munakata et al. (1985). OPAL provides the radiative opacities for the Grevesse (1991) mixture of elements (Iglesias, Rogers and Wilson 1992), while the molecular contribution to the opacity is from Bessel et al. (1989, 1991). The calibrated mixing length parameter of the outermost convective layers is set $1.63H_p$. Models of massive stars (M $\geq 12M_\odot$) take into account the effect of mass loss by stellar wind following the De Jager et al. (1988) prescription. The mass-loss rate increases to 10^{-3} $M_\odot$ yr^{-1} when the star appears as a LBV, or follows Langer (1989) when the star is a WR. On the contrary, low and intermediate mass stars are calculated at constant mass, because mass loss during the RGB and AGB phases can be easily taken into account following the standard analytical procedure (Iben and Renzini, 1983). Figure 1 shows the HR diagrams for the composition Z=0.008, Y=0.250.

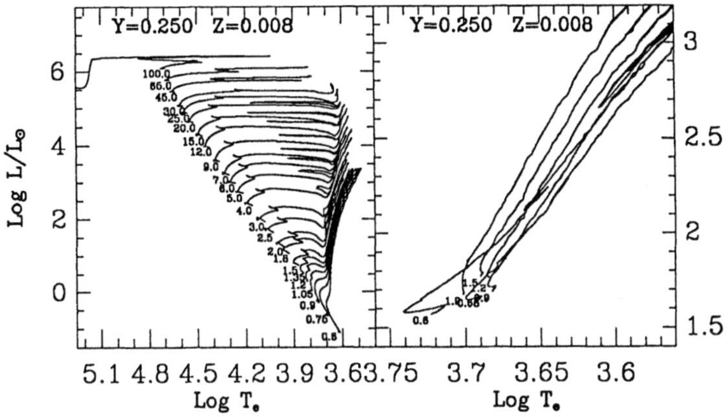

Fig. 1. HR diagrams of the evolutionary tracks with Z=0.008 and Y=0.250

References

Alongi, M., Bertelli, G., Bressan, A., Chiosi, C., Fagotto, F., Nasi, E., Greggio, L., 1992, A&AS in press

Anders, E., Grevesse, N.,1989, Geochim. Cosmochim. Acta 53,197

Bessel, M.S., Brett, J.M., Scholz, M., Wood, P.R., 1989, A&AS 771,1

Bessel, M.S., Brett, J.M., Scholz, M., Wood, P.R., 1991, A&AS Erratum,621

Caughlan, G.R., Fowler, W.A.,1988, Atomic Data Nuc. Data Tables 40,283

de Jager, C., Nieuwenhuijzen, H., van der Hucht, K.A., 1988, A&AS 72,259

Grevesse, N., 1991, A&A 242,488

Huebner, W.F., Merts, A.L., Magee, N.H., Argo, M.F., 1977, Los Alamos Sci. Lab. Rep. LA-6760-M

Iben, I.Jr, Renzini, A., 1983, ARA&A 21,271

Iglesias, C.A., Rogers, F.J., Wilson, B.C., 1992, ApJ submitted

Langer, N., 1986, A&A 210,93

Munakata, H., Kohyama, Y., Itoh, N., 1985, ApJ 296,197

CCD Photometry Of The LMC Clusters NGC 2134 And NGC 2249: The Effect Of The Opacities

A. Vallenari [1], A. Aparicio [2], F. Fagotto [1],

G. Bertelli [3], A. Bressan [4], C. Chiosi [1]

[1]Dip. Astronomia Università di Padova, I-35122 Padova, Italy
[2]Instituto de Astrofísica de Canarias, E-38200 La Laguna, Spain
[3]Consiglio Nazionale delle Ricerche, I-35122 Padova, Italy
[4]Osservatorio di Padova, I-35122 Padova, Italy

Abstract: We present CCD BV photometry of two LMC globular cluster (NGC 2134 and NGC 2249). The (B–V)-V diagrams of this clusters have been used to test three sets of stellar evolutionary models, calculated assuming the overshooting and the semiconvection scenarios for the treatment of the convective core and using two different libraries of opacities. The age and the reddening of the clusters have been also determined.

1 The stellar evolution scenario.

We can reasonably assert that there are two main debated questions which nowadays theoretical stellar models suffer. On one hand there is a considerable effort in investigating the effects of changes in the opacity input. In fact the new stellar opacities computed at the Livermore Laboratories by Iglesias, Rogers and Wilson (1992 and references therein) with the solar meteoritic abundance for the Fe (OPAL), present a large opacity enhancement for temperature near 3×10^5 K when compared with the Los Alamos tables of Huebner et al (1977) (LAOL).

On the other hand the efficiency of the mixing processes occurring in real stars is a topic that since long is under vivid debate. There are two main evolutionary scenarios according to which a mixing scheme can be adopted, i.e. either the classical or the overshoot treatment of the extension of convective regions (for a recent review on these topics, see Chiosi et al. 1992).

To carry on the proposed tests, we can rely on stellar models calculated in an homogeneous fashion differing only in the mixing prescription and/or in the opacity data base. Therefore the investigations are not biased by the usage of evolutionary models calculated by different authors. We will use three sets of models: A (using overshooting plus OPAL), B (using overshooting plus LAOL) and C (using semiconvection plus LAOL). The grids of stellar evolution models used are the same discussed by Bertelli et al (1992), with a metallicity of 0.008.

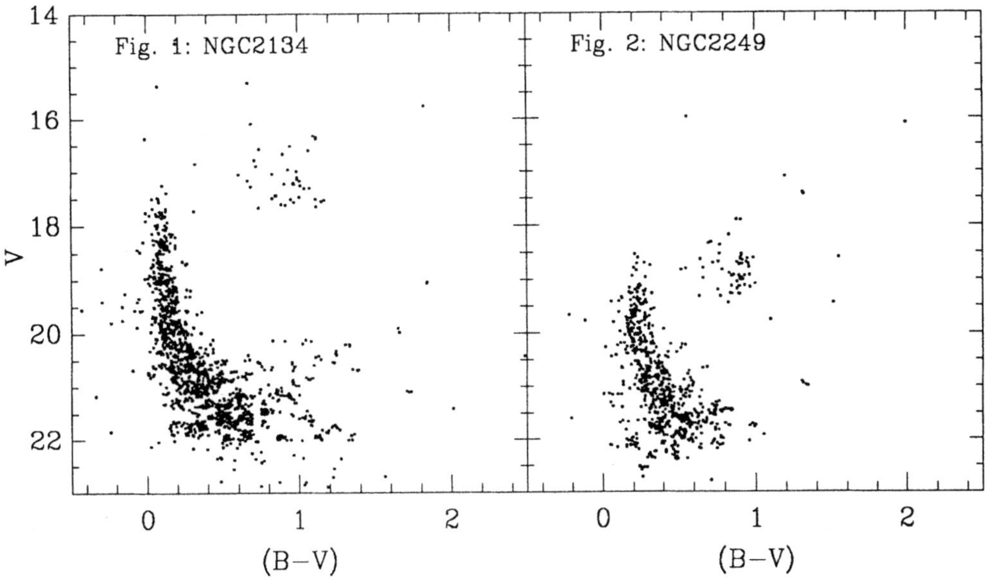

2 Results.

Figures 1 and 2 respectively show the colour-magnitude (CM) diagrams of NGC 2134 and NGC 2249 after correction for uncompleteness and for contamination by field stars have been done. Synthetic CM diagrams have been constructed using the stellar evolutionary models of each set (A, B and C) and according to the procedure described by Chiosi et al.(1989). A Salpeter IMF have been used assuming a percentage (25%) of unresolved binary stars with mass ratio spreading from 0.8 to 1.

We have found that the best overall fitting is obtained in both cases with the synthetic diagrams constructed using set A. In particular, after fitting the upper part of the MS, the luminosity predicted by the set C and B for the red giants clump of NGC 2134 and NGC 2249 is too high and the colour turns out to be too red in comparison with the experimental data. The set A that uses OPAL opacities well reproduces both the luminosity and the colour of the red giant clump. We have found a reddening for NGC 2134 of E(B–V)=0.22 from the comparison with the three sets of evolutionary models. About its age, set A and B give 1.9×10^8 years and set C gives 1.1×10^8 years. For NGC 2249 we have obtained E(B–V)=0.22 from set A and 0.25 from sets B and C. Its age turns to be 5.5×10^8 years if models of sets A or B are used and 3.5×10^8 years if set C is used.

References

Bertelli, G., Bressan, A., Chiosi, C., Fagotto, F., 1992, in this book
Chiosi, C., Bertelli, G., Meyland, G., Ortolani, S., 1989, A&A, 219, 167
Chiosi, C., Bertelli, G., Bressan, A., 1992a, ARA&A in press
Huebner, W.F., Merts, A.L., Magee, N.H., Argo, M.F., 1977, Los Alamos Scientific Laboratory Report LA-6760-M
Iglesias, C.A., Rogers, F.J., Wilson, B.C., 1992, ApJ submitted

New Strömgren and Hα Photometry of Young LMC Clusters

Eva K. Grebel, Tom Richtler, Klaas S. de Boer

Sternwarte der Universität Bonn, F.R.Germany

We present new Strömgren photometry (v, b, y) for the young LMC clusters NGC 1818, 1850 and 2004 in order to estimate their metallicities. For NGC 2004, we performed additional $H\alpha$ photometry to search for emission line stars. All Strömgren data were observed with the 2.2m telescope at La Silla in December 1990. The Hα observations were made at the NTT in February 1992. A preliminary photometric calibration in the Strömgren system was carried out using three local standards in each cluster, measured with the 1m telescope on La Silla. This work continues earlier work by Grebel & Richtler (1992), where also the metallicity calibration is explained. It must be noted that the calibration refers to iron abundance, but the Strömgren system is also particularly sensitive to CN abundance.

1. NGC 1818

This cluster presents an interesting case. Earlier spectroscopy (Richtler et al. 1989) indicated a metallicity of −0.9 *dex*. The Strömgren colours of th respective star are in agreement with that value. However, there are data points indicating the existence of supergiants with distinctly larger metallicities, which may be evolved supergiants with strong CN bands (see Richtler, this conference).

2. NGC 1850

Three overlapping fields were observed in the crowded region of this young cluster. The $(b-y), m_1$ diagram for photometrically selected red supergiants in the cluster and the surrounding field is shown by Richtler (this conference). Stars in the inner cluster region lie reasonably well on a line of constant metallicity. The suspected nonmembers at larger distances from the cluster scatter over a large range of metallicities. Such a scatter is also indicated among the abundances recently derived for LMC Luck and Lambert (1992). From our Strömgren photometry we estimate the metallicity of NGC 1850 to be around −0.7 *dex*.

3. NGC 2004

We observed one field centered on NGC 2004. Only a few red supergiants are present (Fig. 1). The metallicity is around −1 *dex*. Once again, it should b emphasized that the Strömgren system is sensitive not only to iron, but also to N abundances. The CNO abundances in NGC 2004 are very low (Jüttner et al., this conference).

$H\alpha$ photometry in NGC 2004 revealed the presence of numerous emission line stars. In Fig. 1 all stars are plotted that were found simultaneously on the b, y, and $H\alpha$ frames. All blue stars that appear bright in $H\alpha$ are marked with an asterisk. Given their brightness and $(b - y)$ colour, they lie in the B star range and thus are Be stars. This number of Be stars is rather a lower limit since the completeness of the $H\alpha$ data is lower at fainter magnitudes than that of the b and y data. Moreover, Be stars have variable $H\alpha$ emission, so that the ones in an $H\alpha$ minimum were not detected.

In addition to NGC 330 (Grebel et al. 1992) and NGC 2100 (Bessell and Wood, this conference) the third young Magellanic Cloud cluster with a strikingly high content of Be stars.

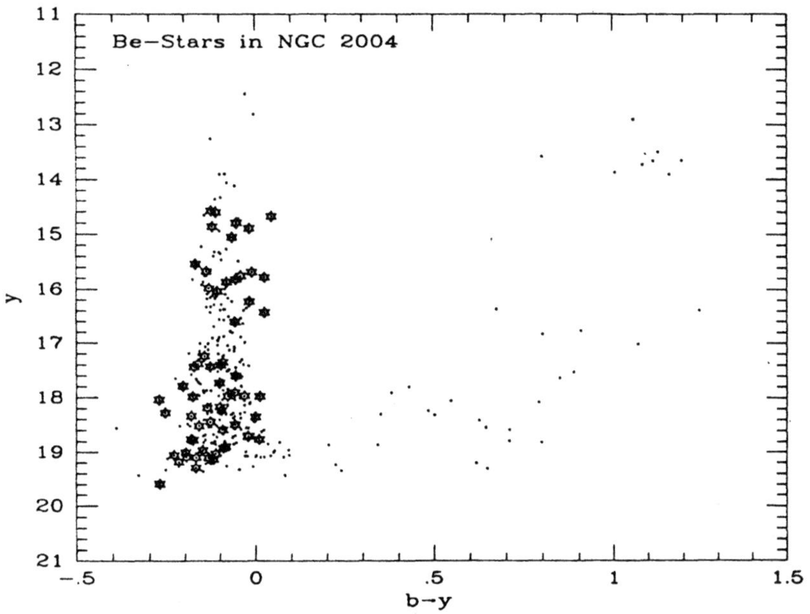

Fig. 1. The Strömgren colour-magnitude diagram of NGC 2004. All blue stars that appear bright in $H\alpha$ are marked by an asterisk. Note that this diagram probably does not reflect the true content in Be stars, but rather gives a lower limit. The metallicity of NGC 2004 is about -1 *dex*, according to our Strömgren photometry.

References

Grebel, E.K., Richtler, T., 1992, A&A, 253, 359
Grebel, E.K., Richtler, T., de Boer, K.S., 1992, A&A, 254, L5
Luck, R.E., Lambert, D.E., 1992, AJ, 103, 1234

Search for Mass Segregation in NGC2098 in the Large Magellanic Cloud

M.Kontizas [1], D.Hatzidimitriou [2], Y.Bellas-Velidis [3], R.D.Cannon [4], E.Kontizas [3], and A. Dapergolas [3]

[1]National University of Athens, Section of Astrophysics, Greece

[2]Institute of Astronomy, University of Cambridge, UK

[3]Astronomical Institute, National Observatory of Athens, Greece

[4]Anglo-Australian Observatory, Australia

1 Introductory remarks

The spatial distribution of stars with different masses in star clusters is very important both in the context of the dynamical processes taking place in the stellar system and of the star-formation mechanism. Mass segregation is a well known phenomenon in galactic globular clusters, the stellar-evolutionary age (τ_e) of which is much larger than their 2-body relaxation time (τ_r) in the central regions. Little information is retained in such systems about the original location within the cluster of stars with different masses.

In the Magellanic Clouds, there are numerous star clusters with $\tau_e << \tau_r$. In such clusters, mass segregation -if present- can be due to either some different mechanism of dynamical relaxation or to the way stars are formed in the protocluster.

2 Observations

Two very young LMC clusters, NGC2098 and SL666, were selected for the present study. The analysis of the data for both clusters is still in progress and a detailed analysis will be published elsewhere. Here, we are presenting the first results on NGC2098.

Both clusters were imaged in B and V during two service-observing runs in 1990 and 1991, with the 3.9m Anglo-Australian Telescope, using the RCA CCD at prime focus. The seeing was only moderate (2-2.5arcsec) during the observations of NGC2098, but significantly better (1arcsec) for SL666. The reductions were carried out using standard procedures in IRAF/DAOPHOT.

Two adjacent frames were observed in each case (one including the cluster), as well as a field region several tens of arcmins away from the clusters.

Using the Meylan and Maeder (1992) isochrones and assuming solar abundance, an age of $6 \times 10^7 yr$ was derived for NGC2098. However, the reddening and metallicity are poorly known and this number may well be revised. Nonetheless, it is confirmed that NGC2098 is a very young system.

3 Search for Mass Segregation in NGC2098

Kontizas *et al.* (1987) found on the basis of star-counts and spectral classification on UKST photographic plates that NGC2098 (as well as SL666) appeared to be extended beyond 1-2arcmin and had the brightest early-type stars concentrated in the innermost region.

The present data confirm that the brighest main sequence stars of NGC2098 are concentrated in the central 1.5 arcmin.

The question now arises whether the cluster itself is indeed extended beyond 1.5 arcmin. If it is, then we have observed mass-segregation. The bright massive stars near the centre were either formed preferentially there, or some form of dynamical relaxation other than two-body encounters is at play. If the cluster is not really extended beyond the innermost region then we have to understand another fact: the overall density of the field seems to keep falling radially from the centre of the cluster, well beyond the inner region. This decrease involves only the younger stars in the field, as was found by performing the following test: the stars in the red horizontal branch/clump region correspond mostly to old field stars, which cannot be related to the cluster. The main sequence stars (V=14-19mag) on the other hand belong generally to younger populations. The number of clump stars is constant (within the errors) in all three frames observed, while the number of main sequence stars falls significantly with distance from the centre of NGC2098. This can be interpreted as either showing that there are still cluster stars well beyond the radius of 1.5 arcmin or that there is an independent underlying population older than the cluster by a few $\times 10^7$ yrs, which is not uniformly distributed across the region of our three CCD fields. A crucial test is to find whether this population has the same density centre as the cluster. Extensive star counts are now being performed in all directions and at large radii around NGC2098, in order to investigate this possibility.

References

Kontizas E., Kontizas, M., Xiradaki, E. (1987) in "Evolution of Galaxies" ed. J.Palous, p177.
Maeder and Meylan (1992) preprint.

On the Formation of Young Globular Clusters in the Magellanic Clouds

Tom Richtler [1], Horst Fichtner [2]

[1]Sternwarte der Universität Bonn,
[2]Institut für Astrophysik and extraterrestrische Forschung der Universität Bonn, W-5300 Bonn, Fed.Rep. of Germany

1 Foundations

We propose a scenario for the formation of young globular clusters (GCs) in the Magellanic Clouds which describes GC formation as a trait of cluster formation in a metal-poor environment. Our scenario fits also well to a variety of observations of globular cluster systems (GCSs) in other galaxies. The characteristic property of GCs with respect to other types of stellar aggregates is their strong concentration, i.e. their large binding energy $-M^2/R$. Starting with a moderately bound collapsing protocluster, this requires a possibly small value for R and only little mass loss from the protocluster by the influence of massive stars. These requirements can be met in the framework of the fragmentation scheme of Murray and Lin (1989) who investigated the nature of fragmentation in a collapsing cloud. According to their results, fragments develop via the onset of thermal instabilities rather than by gravitational instabilities.

2 The role of metallicity

Thermal instabilities grow exponentially on a cooling time scale. Since metals are the main cooling agens under the physical conditions of interest, a low metal abundance means a long cooling time. This means that a low-metallicity protocluster undergoes a long "dormant" period during which star formation does not take place and the protocluster contracts towards a state of high binding energy by energy loss due to radiation. When the first massive stars form they more probably form in a denser environment in a metal-deficient than in a metal-rich configuration. Our point is now that massive star formation in a high background density leads to a significantly smaller mass loss than at low density levels. We estimated this effect by calculating the ionisation in the protocluster taking into account the effects of stellar winds. Under the assumption that the period of star formation takes place at a certain density value, we let stars form randomly according to a Salpeter mass spectrum until all matter is transformed into stars or has been ionised. The mass which can be ionised by an UV-emitting star is inversely proportional to the density of the ambient material (classical Strömgren sphere). Additionally, stellar winds create shells which in case of a sufficiently high density quickly trap the ionisation front and protect the surrounding from being ionised. More mass is then available for star formation and the star formation efficiency is very high, ending up with a strongly bound cluster. In a metal-rich environment star formation commences early and massive stars can disrupt the cluster.

3 Metallicities and IMFs of Magellanic Cloud Clusters

In the context of our scenario, we call attention to the low metallicities of young Magellanic Cloud clusters which emerges from both spectroscopic (e.g. Reitermann et al. 1990, Spite et al. 1991) and photometric work (Grebel and Richtler 1992, Richtler, this conference). GC formation can thus be understood as a trait of star formation in galaxies, which are not well mixed and where star forming regions can be metal-poor as well as metal-rich. Apart from Magellanic Cloud clusters, we also note the low metallicities of the young clusters in M33 (Brodie and Huchra 1991) and the fact that the mean colurs of GCSs around early type galaxies are bluer (i.e., presumably metal poorer) than the light of their host galaxies (e.g., Harris 1991). Concerning galactic GCs, the metallicity scale for suspected metal-rich objects seems not well established. There are indications for some of the disk clusters to be distinctly metal-poorer than previously thought (Richtler et al. 1990, Francois 1991, Friel and Geisler 1991), but no certain statements can be made.

Young GCs in the Magellanic Clouds have mass spectra similar to a Salpeter mass function (Sagar and Richtler 1991, Will 1992). If thermal instabilities initiate fragmentation in the protocluster, we have an interesting argument to explain the observed mass spectra theoretically. In contrast to gravitational instabilities, thermal instabilities have no preferred length scale, so we may assume that the radii of thermally instable fragments are equally distributed. Filling up a volume with fragments, whose radii are equally distributed between some R_{min} and R_{max}, and requiring that the fragments may not overlap, leads to a modified distribution, because the possible R_{max} is the distance to the closest neighbour and decreases with increasing filling factor. Numerical experiments show that the corresponding mass spectrum converges against a power law with an exponent around -2.4.

References

Brodie, J., Huchra, J., 1991, ApJ, 379, 157
Francois, P., 1991, A&A, 27, 56
Friel, E.D., Geisler, D., 1991, AJ, 101, 1388
Grebel, E.K., Richtler, T., 1991, A&A, 253, 359
Harris, W.E., 1991, ARAA, 29, 543
Murray, S.D., Lin, D.N.C., 1989, ApJ, 339, 933
Reitermann, A., Baschek, B., Stahl, O., Wolf, B., 1990, A&A, 234, 109
Richtler, T., Grebel, E.K., Seggewiss, W., 1992, "Stellar Populations", IAU Symp. 148, eds. B. Barbuy & E. Bica, Kluwer, Dordrecht, in press
Sagar, R., Richtler, T., 1991, A&A, 250, 324
Spite, F., Richtler, T., Spite, M., 1991, A&A , 252, 557
Will, J.-M., 1992, Diploma Thesis, Bonn University

How to Model the Chemical Evolution of the Magellanic Clouds

Joachim Köppen

Institut für Theoretische Astrophysik, Heidelberg, F.R.G.
present address: Institut für Theoretische Physik und Sternwarte, Olshausenstr 40, D-W-2300 Kiel, F.R.G.

1 The First Glimpse: The Simple Model

For a first interpretation of the comparison of observational data, the crude "Simple Model" of chemical evolution is quite useful. Since it has well been described in the literature (e.g. Pagel and Patchett 1975, Tinsley 1980), let us here just review the assumptions and whether they are satisfied in the Magellanic Clouds:

(1) The galaxy is a closed system, with no exchange of matter with its surroundings: This is most certainly wrong, because of the presence of the Inter-Cloud Region and the Magellanic Stream. Also, it is most likely that the Clouds have interacted with each other and the Galaxy as well (cf. e.g. Westerlund 1990).
(2) It initially consists entirely of gas (without loss of generality of primordial composition): This is also good approximation for models with gas infall, as long as the infall occurs with a time scale shorter than the star formation time scale.
(3) The metal production of the average stellar generation (the yield y) is constant with time: in the absence of any observational evidence or theoretical constraint, it is reasonable to make this assumption. For tables of the oxygen yield see Köppen and Arimoto (1991).
(4) The metal rich gas ejected by the stars is instantly and completely mixed with the ambient gas: This "instantaneous recycling approximation" is appropriate for elements synthesized in stars whose life time is much shorter than the star formation time scale, such as oxygen, neon, sulphur, and argon.
(3) The gas is well mixed at all times: This may well be true, since the oxygen abundances of the H II regions show a rather small dispersion (±0.08 dex, Dufour 1984). However, Russell and Bessell (1989) note a "suggestion of a scatter ... of perhaps $\sim$ 0.2 dex in addition to the observational uncertainties".

Then the metallicity Z of a primarily produced element (such as O, Ne, S, Ar) at every time t is a simple function of the gas fraction (by mass) f_g at that moment:

$$Z(t) = y \ln(1/f_g(t))$$

Note that the metallicity does not depend on the actual temporal history of the star formation rate (which could be very simple or very complicated) or on its dependence on the gas fraction. Also, the metallicity is always proportional to the yield of the stellar population, thus the 'astration parameter' $\ln(1/f_g)$ fully describes the chemical evolution of this system.

Pagel et al. (1978) determine the elemental abundances of H II regions in the Magellanic Clouds. Plotting the oxygen abundances against gas fraction, they find that the average abundances of both Clouds can well be explained with the simple model with the same metal yield $y = 0.003 = 0.15\,Z_\odot$ (the same model is also compatible with regions in M 33 and M 101). However there remains an ambiguity: the more gas-rich SMC could be a galaxy genuinely younger than the LMC, but with a similar star formation time scale. On the other hand, it could be of the same age, but with a lower star formation time scale. From chemical evolution alone, one cannot distinguish between these two possibilities.

2 A Broader Perspective: Irregular Galaxies

Though the Magellanic Clouds can be investigated at great detail, it seems worthwile to look at other irregular galaxies to try to find out more of the general properties of this type of galaxy. Lequeux et al. (1979) study 6 irregular and blue compact galaxies, and find that the chemical compositions and gas fractions can very well be described by a single simple chemical evolution model with constant yield $y = 0.004$, i.e. also in good agreement with the findings of Pagel et al. (1978) for the Magellanic Clouds.

Chiosi and Matteucci (1982) construct full numerical chemical evolution models (taking into account the detailed stellar nucleosynthesis prescriptions and the finite stellar life times) for the Magellanic Clouds and two other Irr galaxies. They assume infall of metall-free gas (exponentially decreasing with time), and star formation rates similar to that of the solar neighbourhood (LMC), and lower (for the SMC). Because of the dilution by the infalling gas, their choice of the IMF gives a higher yield $y = 0.008$. Since gas infall is assumed to take place with a time scale (7 Gyr) comparable to that for star formation, the gas density initially increases up to a maximum and then declines. Thus the star formation rate is more constant in time than in the closed box Simple Model.

In Fig. 1 we show the chemical evolution for the closed-box Simple Model and models with moderate and slow exponential gas infall. Despite the rather different behaviour of the gas *density*, and this the star formation rate ($\propto g$), all models are almost indistinguishable in the $\lg Z$ vs. $\lg \ln(1/f_g)$ plane. This means that the effective yield

$$y_{\rm eff} = Z / \ln(1/f_g)$$

is close to the yield y of the stellar population, and remains constant with time. Thus, even slow infall models may be called "Almost Simple Models".

In a sample of galaxies ranging from blue compacts to magellanic irregulars, Matteucci and Chiosi (1983) find that the metallicities of most objects are *smaller* than expected from the prediction from the Simple Model which explains the Magellanic Clouds. Thus a lower effective yield would be necessary. The authors discuss

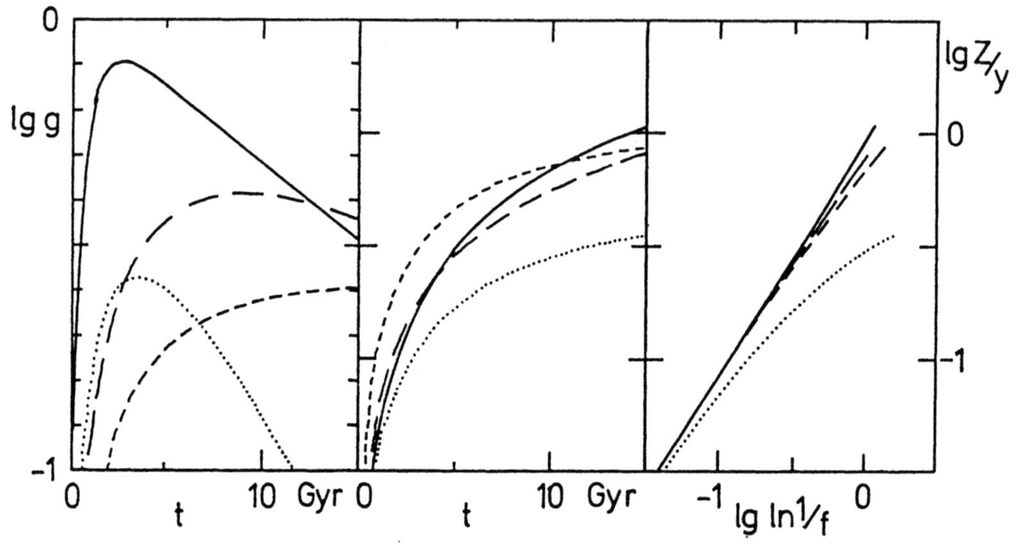

Fig. 1. Chemical evolution of Simple Model (full line) and infall models: time evolution of gas density g (arbitrary units), metallicity Z, and the lg Z vs. lg ln$(1/f_g)$ diagram for models with yield 1, a star formation rate proportional to g, and an approximative treatment of finite stellar lifetimes. Infall time scales are 0.7 τ_{SFR} (long dashes) and 50 τ_{SFR} (short dashes). Also shown (dots) is a model with continuous galactic wind (mass loss rate = 3∗SFR).

metallicity-dependent yields, gas infall (with the infall rate proportional to the instantaneous star formation rate), and galactic winds (mass loss rate proportional to star formation rate, see also the dotted curves in Fig. 1). They show that if the Magellanic Clouds are explained by a Simple Model (with $y = 0.0032$), the data of the other galaxies can well be explained by additional infall or strong galactic winds, which always decrease the yield.

This picture is confirmed by Gallagher and Hunter (1984): The metallicities are distributed over a rather wide range, but below of what is inferred from the Magellanic Clouds(their Fig. 3). No systematic differences between different types of irregulars are noted, with the possible exception of luminous and blue compact irregulars ($y \approx$ 0.0004).

However, in this comparison, one has to keep in mind two major problems: Firstly, the determination of oxygen abundance in distant systems is rather uncertain. Since the weak [O III] λ4363 Å — which is necessary to measure the electron temperature of the gas in the H II region — is often not observable, the abundance determination must be based on rather coarse, empirical schemes are used (cf. Hunter and Gallagher 1985). The other problem is the determination of the gas fraction. Often, one uses as the total mass of the galaxy which takes part in the chemical evolution the dynamical mass, as estimated from the velocity dispersion seen in the widths of the H I 21 cm line. Ideally, it should be the total mass in gas and stars. The latter is obtained from the absolute luminosity of the stars by using stellar population models, thus it is model-dependent. Gallagher and Hunter (1984) use this approach.

From this comparison of the Magellanic Clouds with other irregulars, the following questions must be considered as still being tantalizingly unanswered:

(a) Are the Magellanic Clouds typical irregular galaxies or are they somewhat special (e.g. because they happen to lack a galactic wind)?
(b) What is the physical explanation for the scatter in the irregulars (if it is genuine)?

3 Photometric Colours

The basic weakness of pure chemical evolution models shows up in the fact that in the Simple Model the metallicity is independent of the actual star formation history. The relevant parameter is not the actual age, but the 'chemical' age, i.e. the integrated star formation rate or the age measured in star formation time scales. Considering the evolution of the photometric properties of galaxies provides independent information.

First calculations of the photometric evolution of late-type galaxies (Searle et al. 1973) showed that the colours of irregular galaxies can be understood, if the galaxies are old (10 Gyr), have initial mass functions (IMF) as steep as Salpeter's IMF, and if the star formation rate declines rather slowly with time. Searle et al. interpret that the range of observed colours is due to a range of star formation rates, and that the bluest galaxies undergo intermittent and strong star formation bursts.

Olson and Peña (1976) construct single-zone chemical and photometric evolution models with gas infall a lower mass limit for the stars varying with time, and periodic star formation bursts (duration 0.1 Gyr, period 1 Gyr). The observational constraints are the present gas fraction and metallicity, and an age of 12 Gyr is assumed. For the LMC models without starbursts (or rather weak ones) can match the observed B–V colour, if a weak dependence of the star formation rate on gas density (SFR $\propto g^{0.25}$) is adopted. In contrast, the SMC can only be matched by a model with strong star bursts.

In their studies of the UV colours of the Clouds, Rocca-Volmerange et al. (1981) emphasize that for a given metallicity and IMF the colours are only sensitive to the ratio of present and past integrated star formation rate. This rate is found to have been constant in the LMC, whereas for the SMC it declines exponentially with time (time scale 6.7 Gyr).

To explain the chemical and photometric properties of irregular galaxies, Arimoto and Tarrab (1990) computed the chemical and photometric evolution for single-zone closed-box model. With this type of model, they find it difficult to reconcile the the low metallicity and high gas content of these galaxies — indicative of steep IMF and low average SFR — with their blue photometric colours — for which flatter IMF and high present SFR would be favourable.

Using the results of their grid of models, we may estimate the colours of Simple Models for the Clouds (with the data from Westerlund (1990)):

- LMC: from metallicity ($\lg(Z/Z_\odot) = -0.2$dex) and the ratio of H I mass to blue luminosity ($\lg M_{\rm HI}/L_{\rm B} = -0.38$ in solar units) one gets the slope of the IMF as $\mu = 1.55$ (Salpeter 1.35), and the constant ν_0 of the star formation rate (linear dependence on gas density) in units of $\nu_s = 0.19$ Gyr^{-1}, as $\nu_0/\nu_s = 1$. This gives colours (B–V) = +0.60 (obs.: +0.55) and (U–B) = −0.02 (obs.: 0.0).

- SMC: $\lg(Z/Z_\odot) = -0.6$dex and $\lg M_{\rm HI}/L_{\rm B} = +0.36$ yield $\mu = 1.55$ and $\nu_0/\nu_s = 0.25$. This gives colours (B–V) = +0.47 (obs.: +0.50) and (U–B) = −0.12 (obs.: −0.2).

Thus, the model colours for both LMC and SMC are in good agreement with the observed colours, i.e. within the relative accuracy of ±0.05 mag of the photometric model code.

4 Star Formation Histories

4.1 The Magellanic Clouds: Late Bloomers

The star formation history can be deduced from comparing indicators for stellar populations of different ages. Gallagher et al. (1984) obtain

(1) the integrated SFR from the mass fraction of stars in the galaxy,
(2) the SFR a few Gyr ago from the blue luminosity,
(3) the present SFR from the Hα luminosity of the H II regions ionized by the short-lived hot stars.

In their paper, they find that most irregulars have a rather constant star formation history. However, their Fig. 1 clearly shows that both Magellanic Clouds have a current star formation rate about 3 times higher than in the past.

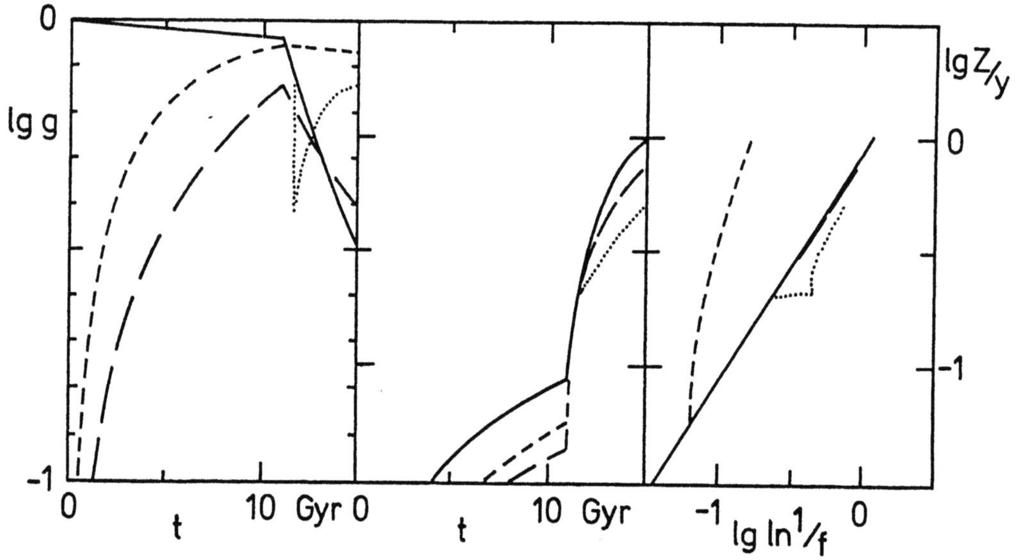

Fig. 2. Chemical evolution of models with enhanced star formation in the recent past: at 11 Gyr the SFR is increased by a factor 30 in a closed-box model (full line) and a slow infall model (dashed); simultaneous increase of SFR (*3) and yield (*10); SFR increase (*30) followed by a loss of half of the gas mass at 11.5 Gyr and subsequent flow back of this gas with a time scale of 1 Gyr (dots).

Evidence for a large population of field stars with ages about 3 Gyr old, present in both Clouds, is summarized by Hyland (1991): In the Bar region of the LMC there seems to have occurred a major star burst at that time, and continued at a lower rate until 0.1 Gyr ago. The star clusters give further clues on the star forming history: Firstly, the small number of very old clusters may well indicate that the star formation some 10 Gyr ago was rather low (Hyland 1991). Secondly, the age-metallicity relation of the SMC clusters is quite different from that determined in the solar neighbourhood (Twarog 1980) or expected for simple models of chemical evolution (Da Costa 1991): Between ages of 12 and 4 Gyr, [Fe/H] is rather constant at -1.4, and thereafter rises to a current value of -0.2 dex, which is compatible with the H II region abundances. Thus, the rate of metal enrichment has been enhanced over the last 3 Gyr. In view of this, one also wonders whether the large number of LMC clusters with ages up to 4 Gyr, and the paucity of clusters older than that, indicates an enhanced cluster formation rate during this time. A low level of star formation in the LMC at ages between 4 and 14 Gyr cannot be excluded or confirmed with the presently available data (cf. Da Costa 1991).

With deep photometry of field stars, interpreted with synthetic colour-magnitude diagrams and luminosity functions, Bertelli et al. (1992) show that in the LMC the level of star formation had been enhanced by as much as a factor of 10 since a time about 3–4 Gyr ago, lasting until about 1 Gyr ago.

In Fig. 2 we show a few examples of chemical evolution models with low initial star formation activity, followed by enhanced SFR after 11 Gyr age. The evolution is not much different for closed-box or slow infall models, The jump in metallicity can also be produced by an increase of the yield, but then the number of stars formed in this second phase remains small (gas density does not change much). Finally, a model which later also experiences a loss of half of its gas mass is shown. Note that except for the change in yield and the massive gas loss all models behave very closely like a Simple Model!

It is worth noting that for the model shown with solid lines in Fig. 2 the star formation time scale is 3 Gyr. With a H I gas mass of $5\,10^8 M_\odot$ (Westerlund 1990) one gets a current star formation rate of 0.7 $M_\odot yr^{-1} M_{\rm H\,I}^{-1}$ in good agreement with the rate deduced from observations (Hyland 1990).

The photometric colours of such a model would be bluer than one in which star formation had been going on all the time and thus had created a strong population of old red stars.

4.2 Bursts and SSPSF

That star formation must be a simple, continuous function of time is not a necessity of the physics of galaxies, but rather a simplifying assumption to aid the computation of models. That short bursts of star fomation are helpful in explaining the photometric colours of the irregulars and especially the blue compact galaxies, has been recognized by Searle et al. (1973), Olson and Peña (1976) and others. In connection with irregular galaxies the idea of stochastical self-propagating star formation (SSPSF, Gerola et al. 1980) has frequently been discussed, which considers that star formation may be induced in the neighbourhood of previous star formation activity. This can give rise to a strongly fluctuating star formation rate. Similar fluctuations are also possible as a consequence of the nonlinear behaviour of the equations for the evolution of interstellar clouds by collisional coalescence, disruption, energy dissipation, and star formation (Struck-Marcell and Scalo 1987), or the mass and energy balance in a multi-phase model of the interstellar medium which incorporates all relevant interactions between the stars and the gas phases (Theis et al. 1992).

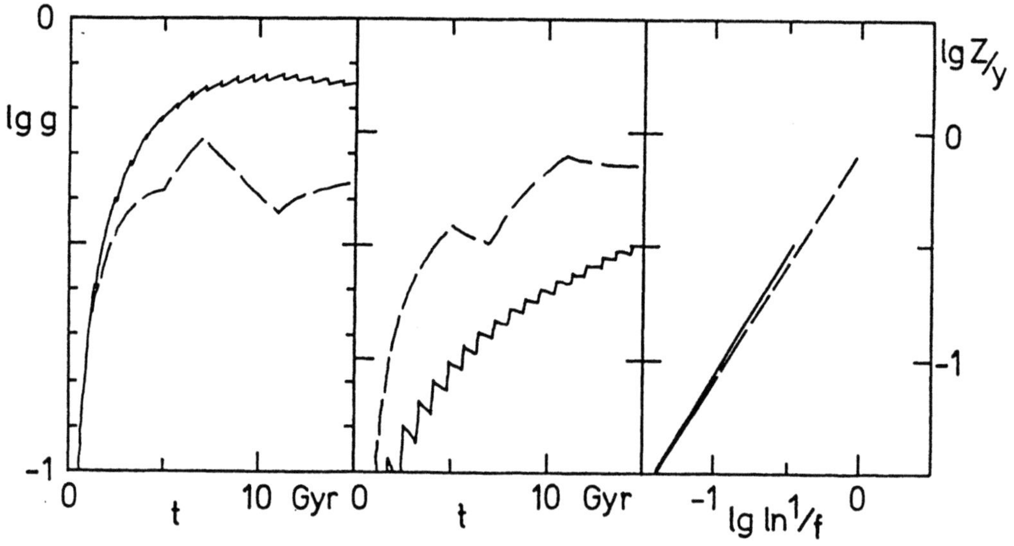

Fig. 3. Chemical evolution of models with very short star formation bursts every 1 Gyr (full line) and long ones (from 1 ... 5 and 7 ... 11 Gyr; dashed line).

Let us consider first how starbursts affect the chemical evolution of the galaxy: Matteucci and Chiosi (1983) simulated the SSPSF process by computing the chemical evolution of models with rapidly fluctuating star formation rate. Since perfect mixing in the interstellar medium is assumed, one expects that the abundance of elements which are produced in short-lived stars (such as oxygen) is still closely linked to the gas fraction as in models with a slowly varying star formation rate. Thus it is not surprising that the relationship between metallicity and the logarithmic gas fraction remains a straight line whose slope is determined *essentially* by the yield (see our Fig. 3 and their Fig. 8), which is found as $y = 0.003$.

However, these simulations are rather simplified: They do not take into account the multiple phase structure of the ISM (and thus the various delay times of mixing and enrichment). Also, they assume that chemical enrichment is a purely local process: gas flows in e.g. the hot phase, distribution of metals over a finite volume by galactic fountains, etc. are ignored). Exemplary calculations of the evolution with incomplete mixing (Wilmes and Köppen 1991) show that non-local effects do not affect the average abundance values, but only the correlation between secondary and primary elements (e.g. N/O ratio). These processes must be properly simulated, before we know whether deviations from the Simple Model are significant.

The photometric colours are quite strongly affected by the occurrence of star bursts in the recent past: depending on the intensity of the burst, the light from the massive, hot stars born during the burst will shift the colours more or less strongly to the blue, as long as these stars live. When they move off the main sequence, the amount of blueing will diminish, and the colours will eventually fade towards their pre-burst values. That bursts of enhanced star formation would be helpful in explaining the blue colours of blue irregulars, had been recognized by Searle et al. (1972). As mentioned earlier, Olson and Peña (1976) are able to explain the colours of the SMC, only if they assume strong starbursts. Effects of an oscillating star formation rate on the photometric colors are discussed by Yoshii and Arimoto (1991). Arimoto and Jablonka (1992) propose that star formation in irregulars takes place essentially in closely-spaced short bursts (interval 0.5 to 1 Gyr), so that the colours always remain sufficiently blue while the gas consumption rate is sufficiently low.

It is obvious that the time after a burst is quite important for the colours, and deviations from the pre-burst colours are strong only for 1 Gyr after the burst. Thus an explanation of the colours might essentially give information about the last star burst, which may not tell very much about the evolution of the galaxy as a whole.

5 Effects of interactions

The Magellanic Clouds are not isolated: They are connected with the Inter-Cloud Region consisting of gas and embedded stars, which contains as much H I gas as the SMC. The Magellanic Stream forms a long, gaseous tail to the Clouds, and is regarded as an indication of global interaction in the Magellanic system. Wayte (1990) lists other evidence for interaction as: velocity-distance measurements, the presence of young stars in the Inter-Cloud Region, and the velocity dispersion in the Magellanic Stream indicate that the splitting of the Clouds occurred some 0.1 Gyr ago. Computations of the orbital motions of the three-body system (LMC-SMC-Galaxy) also show that a collision of LMC and SMC within the past 1 Gyr seems difficult to avoid.

What could happen during an interaction? While we do not claim to produce a physically consistent list of processes and event, let us consider the effects of the following items on chemical evolution (as shown in Fig. 4):

1. Sudden gas loss by tidal interaction: Within a few Gyr the gas that had been bound in stars of intermediate life time (and low metal production) is released and — depending on the IMF — a fair fraction of the lost gas is replenished, the metallicity drops. If star formation continues, the system recovers to an almost normal Simple Model after few τ_{SFR} (full lines).
2. Sudden inflow of gas or loss of stars: Both processes increase the gas fraction for some time. After few τ_{SFR} the model evolves again as a Simple Model (dashed lines).
3. Partial exchange of gas: This would be a combination of the previous two events.
4. Star formation burst: The effects can be seen in Fig. 3: a quick increase of the metallicity followed by a decrease when the intermediate mass stars die. Essentially like a Simple Model.
5. SN-driven galactic wind following an intense starburst: The dotted curves in Fig. 4 show that the evolution is dominated by the gas loss effects.

All these models — done just to illustrate possible effects — show that one must have really drastic changes in the parameters, if substantial (factor 2 or more) deviations from the Simple Model are to be achieved.

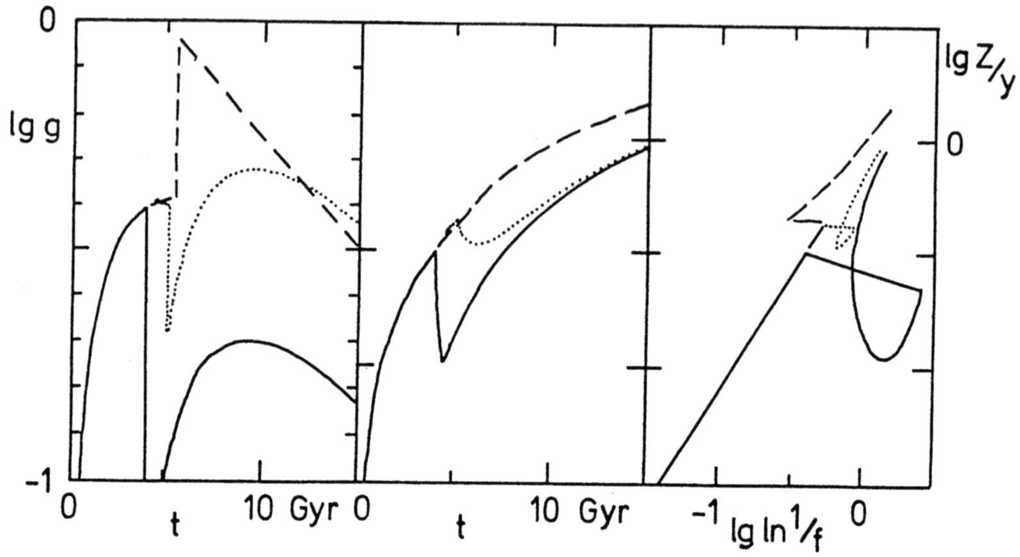

Fig. 4. Effects of possible processes caused by galactic interactions: sudden 99 percent gas loss at 4 Gyr (full lines), sudden gas inflow (of additional 50 percent mass) at 6 Gyr (dashes), star burst (3*SFR) at 4.5 Gyr and 50 percent gas loss at 5 Gyr followed by this gas flowing back with time scale 1 Gyr (dots).

6 So What Can We Learn from the Magellanic Clouds?

The Magellanic Clouds – especially the LMC – have global chemical and photometric properties which are fairly close to what is expected from a (nearly) Simple Model. This is quite fortunate for studies of stellar nucleosynthesis (e.g. Russell and Dopita 1992), since the abundance ratios can be expected to be insensitive to the details of galactic evolution (infall, galactic winds, etc.) and star formation history. On the other hand, this is a disadvantage, because separate studies of the star formation history are necessary (via the age-metallicity-relation, star clusters, synthetic colour-magnitude diagrams). From the experience with the solar neighbourhood, one should emphasize that *stellar metallicity distribution functions* (the G-dwarf-"problem") are extremely useful as constraints to chemical evolution models. Finally, the effects of galactic interaction may not be too apparent in the Magellanic Clouds. But here we can study them in much better detail than anywhere else. Freeman (1984) noted that "magellanic systems frequently come in pairs". It may well be that the Magellanic Clouds are representative for a type of interacting galaxies right, at our door-step.

Acknowledgements: I thank H.Meusinger and N.Arimoto for helpful discussions. Financial support from the Deutsche Forschungsgemeinschaft (SFB 328) is gratefully acknowleged.

References

Arimoto N., Tarrab I.: 1990, *Astron.Astrophys.* **228**, 6
Arimoto N., Jablonka P.: 1992, *Astron.Astrophys.* , submitted
Bertelli G., Mateo M., Chiosi C., Bressan A.: 1992, *Astrophys.J.* **388**, 400
Chiosi C., Matteucci F.: 1982, *Astron.Astrophys.* **110**, 54
Da Costa G.S.: 1991, *IAU Symposium* **148**, 183
Dufour R.J.: 1984, *IAU Symposium* **108**, 353
Freeman K.C.: 1984, *IAU Symposium* **108**, 107
Gallagher J.S., Hunter D.A.: 1984, *Ann.Rev.Astron.Astrophys.* **22**, 37
Gallagher J.S, Hunter D.A., Tutukov A.V.: 1984, *Astrophys.J.* **284**, 544
Gerola H., Seiden P.E., Schulman L.S.: 1980, *Astrophys.J.* **242**, 517
Hyland A.R.: 1991, *IAU Symposium* **148**, 125
Hunter D.A., Gallagher J.S.: 1985, *Astrophys.J.Suppl.* **58**, 533
Köppen J., Arimoto N.: 1991, *Astron.Astrophys.Suppl.* **87**, 109
Erratum: *Astron.Astrophys.Suppl.* **89**, 420
Lequeux J., Peimbert M., Rayo J.F., Serrano A., Torres-Peimbert S.: 1979, *Astron.Astrophys.* **80**, 155
Matteucci F., Chiosi C.: 1983, *Astron.Astrophys.* **123**, 121
Olson G.L., Peña J.H.: 1991, *Astrophys.J.* **205**, 527
Pagel B.E.J., Patchett B.E.: 1975, *Mon.Notices Roy.Astron.Soc.* **172**, 13
Pagel B.E.J., Edmunds M.G., Fosbury R.A.E., Webster B.L.: 1978, *Mon.Notices Roy.-Astron.Soc.* **184**, 569
Rocca-Volmerange B., Lequeux J., Maucherat-Joubert M.: 1981, *Astron.Astrophys.* **104**, 177
Russell S.C., Bessell M.S.: 1989, *Astrophys.J.Suppl.* **70**, 865
Russell S.C., Dopita M.A.: 1992, *Astrophys.J.* **384**, 508
Searle L., Sargent W.L.W., Bagnuolo W.G.: 1973, *Astrophys.J.* **179**, 427
Struck-Marcell C., Scalo J.M.: 1987, *Astrophys.J.* **64**, 39
Theis C., Burkert A., Hensler G.: 1992, *Astron.Astrophys.* , submitted
Tinsley B.M.: 1980, *Fund.Cos.Phys.* **5**, 287
Twarog B.A.: 1980, *Astrophys.J.* **242**, 242
Wayte S.R.: 1991, *IAU Symposium* **148**, 447
Westerlund B.E.: 1990, *Astron.Astrophys.Rev.* **2**, 29
Wilmes M., Köppen J.: 1991, *Evolution of Interstellar Matter and Dynamics of Galaxies*, CTS Workshop 1, Prague, ed. J.Palouš, Cambridge University Press, Cambridge, in press
Yoshii Y., Arimoto N.: 1991, *Astron.Astrophys.* **248**, 30

Nomenclature

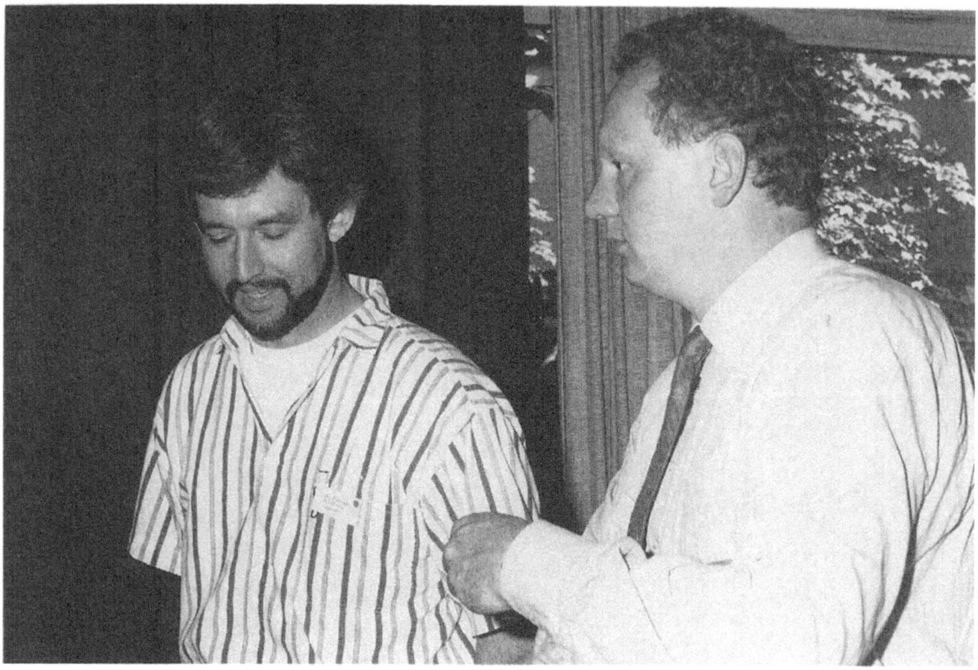

Andreas Jüttner, Josef Gochermann

Alessandro Bressan, Robin E.S. Clegg, Gianpaolo Bertelli

Where is SMC 1 ?

P. Dubois [1], M.-C Lortet [2]

[1]Observatoire de Strasbourg. F-67000 Strasbourg, France
[2]Observatoire de Paris. F-92195 Meudon, France

Introduction

With the examples given below, we want to draw attention to the necessity of using unambiguous names for astronomical objects. This should specially be the case in crowded regions like the Magellanic Clouds.

As a help, IAU Commission 5 gave some recommendations. (1990, IAU Transactions, vol XXB, pp S46-S47 ; 1990, Pub. Astron. Soc. Pacific **102** 1231-1233 ; 1992, Astron. and Astrophys. Special Issue, March, A11-A13).

Where is "H1-36 Arae" ?

reference : 1992, Astrophys. J. **387** L77.
comment : H1-36 could be "Haro 2-36", but this object is in Scorpius. Fortunately, the author gives the position which is that of "Haro 2-36".
main recommendation :
Object listing should contain, next to the acronym, a **second** designation or positional information.

Where is "RX 05402-6928" ?

reference : 1991, Nature **349** 578
comment : The position is not $5h\,40.2m\; -69°\,28'$ (1950) because the equinox is 2000. A proper designation would be "RX J05402-6928"
recommendation :
If coordinates, in any form, are used to encode an object, coordinates shall be preceded by a code for the reference frame, especially "J" for 2000 equatorial coordinates.
Example : QSO in 1991, Astrophys.J.Suppl. **77** 203
QSO 0000-398 = QSO J0003-3932

Where is "C313841313" ?

reference : 1987, Astron. and Astrophys. Suppl. Ser. **67** 447.
comment : the position is $0h\,31m\,38s\;+41°\,31.3'$.
A better designation would be "C 003138+4131.3".
recommendation :
If coordinates, in any form, are used to encode an object, coordinates shall contain leading zeroes, if necessary, and the plus or minus sign.

Where is SMC 1 ?

reference : 1992, The Strasbourg-ESO Catalogue of Galactic Planetary Nebulae.
comment : It is a planetary nebula in the LMC (!) which has no name in the original publication (1982, Observatory **102** 229).
recommendation :
Give yourself an acronym to your object, but **don't use an existing one.**

General recommendation : on-line access to the "Dictionary".

To avoid ambiguity, consult the "Dictionary of Acronyms" and its supplements. It is also a tool to decode the names of objects found in the literature.

1. e-mail : BORDE@FRMEU52 , IAPOBS::BORDE , MESIOB::BORDE
2. on-line access 1: (SPAN or internet) in the database SIMBAD, type "info -help" or one of the commands: "info cati XX..., info -l cata AA..., ...
3. on-line access 2: (SPAN) type "EDI MESIOB::USER1:[BORDE]ACRONYM. DIC" for the first file :
   a) ACRONYM.DIC : Alphabetic list and information on the acronyms
   b) MEANING.DIC : Meaning of the acronyms
   c) AUT.DIC : Authors
   d) REF1.DIC : References, years 1784-1962
   e) REF2.DIC : References, years 1963-1973
   f) REF3.DIC : References, years 1974-1992
   g) REMARK.DIC : Remarks on the acronym (not yet available)
   h) ACROMAGELLAN.DIC, MEANMAGELLAN.DIC and REFMAGELLAN. DIC (for the Magellanic Clouds, also included in the general files)
4. FAX : (33) 1 40 51 22 32 , Suzanne BORDE
5. printed version of the "Dictionary" :
   1983, Astron.and Astrophys. Suppl. Ser. **52** No.4
   1986, Astron.and Astrophys. Suppl. Ser. **64** 329-390
   a new version is in preparation for the end of 1992

Summary Talks

Jocelyne Silvy, Despina Hatzidimitriou, Dominik J. Bomans, Hildegard Domgörgen

Joachim Krautter, Sumner Starrfield

A Catalogue with Accurate Positions and a Grid of Magellanic Cloud Selected Areas

Klaas S. de Boer

Sternwarte University of Bonn, Auf dem Hügel 71, D-5300 Bonn

Abstract: In order to help structure research of the Magellanic Clouds two proposals are made.

1. The construction of a catalogue of all stars brighter than V=15 in the region of the Magellanic Clouds.
2. The definition of a grid of Magellanic Cloud Selected Areas.

Introduction

The growing importance of the Magellanic Clouds (MCs) for astrophysics is based on much improved optical astronomical techniques in the recent years. It has also led to a higher frequency of conferences with these galaxies as main topic. The 1960's saw three conferences on the MCs, the IAU Symposium 20 on "The Galaxy and the Magellanic Clouds" in Sydney in 1963 (Kerr & Rodgers 1964) followed by smaller conferences on only the MCs at Mount Stromlo (Hindman & Westerlund 1965) and in Santiago (Muller 1971). Two conferences were held in the early 1980s, in Strasbourg on the Small Magellanic Cloud (Dubois, Florsch & Jaschek, 1982) followed by the second IAU Symposium on the MCs in Tübingen in 1983 (van den Bergh & de Boer 1984). Activities increased recently with a european meeting in 1989 in Paris (de Boer, Spite & Stasinska 1989) preceding the 3rd IAU Symposium on the MCs of 1990 in Sydney (Haynes & Milne 1991). Heidelberg now hosts the 8th conference on the Magellanic Clouds.

To the major steps forward in the technical area count the coming into use of the 3.6m, 3.9m and 3.5m-telescopes at the ESO and the AAO, as well as the now very general use of the CCD as photon detector. These improvements led to the advantage of larger light collecting power and higher intrinsic spatial resolution, but the disadvantage of reducing the size of the field imaged onto the detectors. On the other hand, the high sensitivity of the CCD enables the detection of ever fainter stars.

There are on average on the order of 10^4 stars per square arcminute in the LMC, based on a projected size of 40 square degrees and using the estimate of its mass of 10^{10} $M_{\odot}$. Of these roughly half have a mass larger than that of the Sun, thus a visual magnitude $V<23$ and will therefore be visible in deeper CCD exposures. Giving these all names for identification purposes is not realistic and one has to rely, as in the past, on finding charts. However, the detail needed in such maps runs counter to the coarseness of many of the existing photographic representations of the MCs and their surroundings.

A dense grid of positional reference stars is highly desirable to match with the smallness and the large magnification of the CCD fields common today and in the future. Given the fact that a CCD field measures, with the current telescopes, somewhere between $2' \times 2'$ and $5' \times 5'$ on the sky, the ideal situation would be to have 5 to 15 reference stars of sufficient brightness per CCD frame.

A catalogue for all MC stars with V< 15

At the Sternwarte in Bonn we now have started to work to come to a catalogue of stars brighter than V=15 of the Magellanic Clouds, along the lines indicated before (de Boer 1989). The positions of the stars will be derived from scans of ESO Schmidt plates obtained with limiting magnitude of V=20 and will be tied to the PPM-Catalogue (Bastian et al 1991) including the reference catalogue of Périé et al (1991), the details being under discussion. The equinox will be J2000, the epochs around 1989-1990. Proper motions will not be available.

The name of the objects listed has to have a format which is unique, is easy to sort, and allows easy identification of the object in existing charts. The BD has names for objects as running number within a declination zone; the disadvantage of that system is, that the zones have different lengths and thus contain widely differing numbers of stars so that the running number has little meaning in terms of right ascension. The HD has all stars ordered according to right ascension so that a HD number has no meaning in terms of declination. Even though the MCs cover only a small portion of the sky, it is unrealistic to expect that a star name based on position is enough to find the star in the field. It is unavoidable to rely on maps for ultimate identification. It seems therefore worthwhile to have names which are based on a mix of the BD and HD scheme. We propose therefore a name which is based on approximate positional information followed by a running number.

The format of the names of the stars in the catalogue (and of names to be given to stars later) will be

MX hhmm-ddd:nn

with the following explanation. MX will be ML for the LMC, MS for the SMC and MB for the Bridge between the MCs; hhmm-ddd will be the RA-DEC position of the NW corner of a box in which the stars will be identified by the running number nn. The declination will be given in degrees with one decimal (without dot), the right ascension in steps of 1^m. The size of the box in which the running number holds is therefore $6'$ in declination and 1^m in right ascension. The structure of the format is conform to the rules agreed by the IAU (see, e.g. the instructions with the 1991 Index Supplementary Issue of A&A, p A11). A general description with considerations on how to create names is given by Dickel, Lortet & de Boer (1987). For further and more detailed lists of stars, names can then be given by using the same format but with at nn a new (extended) range of sequential numbers.

A grid of selected areas

Coordinating Magellanic Cloud reasearch has been the major goal of the research in the visual with the ESO key-program on "Coordinated Investigations of Selected Regions in the Magellanic Clouds". The main purpose of this international programme has been realised. In the MCs, six Regions had been defined in which spectroscopic surveys, photometry, and stellar, interstellar, and emission-line spectroscopy is being carried out (de Boer et al 1989). First results have become available on the study of the age structure, star formation history, and chemical evolution of some of the Regions in the MCs (de Boer et al 1991; and see research contributions to this conference).

Stellar statistics in relation with the overall structure of the MCs has not been covered in the above mentioned programme. A programme to investigate the star density in the entire region of the MCs would be most interesting, in particular if good colour information would become available as well.

One significant result of photometry in such a grid could be the determination of the distribution of, e.g., F main-sequence stars in the MCs. These objects would delineate the spatial extent of the stellar population with ages of near 10^9 years. Also one could determine the actual age of the population in each Selected Area from a comparison with isochrones. Another project could be to obtain field spectroscopy of most objects within the Selected Areas to get radial velocities for stars, to study the rotational behaviour of the stellar body of each of the MCs as well as of the Bridge region.

The statistics of the spatial distribution of stars of various kinds will be of great importance for the study of the extent, the star-formation history, and the dynamical evolution of the Magellanic Clouds. The speed and accuracy of present day CCD detectors allows it to obtain, with a moderate amount of telescope time, the data required for such studies. For such a programme it is not useful to measure randomly distributed fields in the MCs. In the same vein as the proposal to coordinate such measurements for galactic research by Kapteyn (1906) I propose here a grid of positions at which over the coming years photometry should be performed.

The grid of Selected Areas is defined as boxes on the sky of $6' \times 1^m$, with corner points at full degrees of declination and every 15 minutes of right ascension. Outside the more limited regions of the LMC and the SMC the Selected Areas are defined in a more coarse grid. Figure 1 shows the layout of the grid. In addition, in some specific directions further grid points have been defined. The CCD observations should be carried out with as a minimum one CCD field in the NW corner of the box. In case of local dense clusters or hampering emission nebulae a slightly deviant position is acceptable.

The CCD photometry should be carried out in the Johnson B,V,R system, and reach stellar brightnesses of V=21 with good accuracy. Inside the data x,y positions on the frame shall be given togetherOB with a tie to the positional grid of reference stars, defined by the catalogue.

The positions of the Magellanic Cloud Selected Areas are given in the Table.

LMC / SMC Field

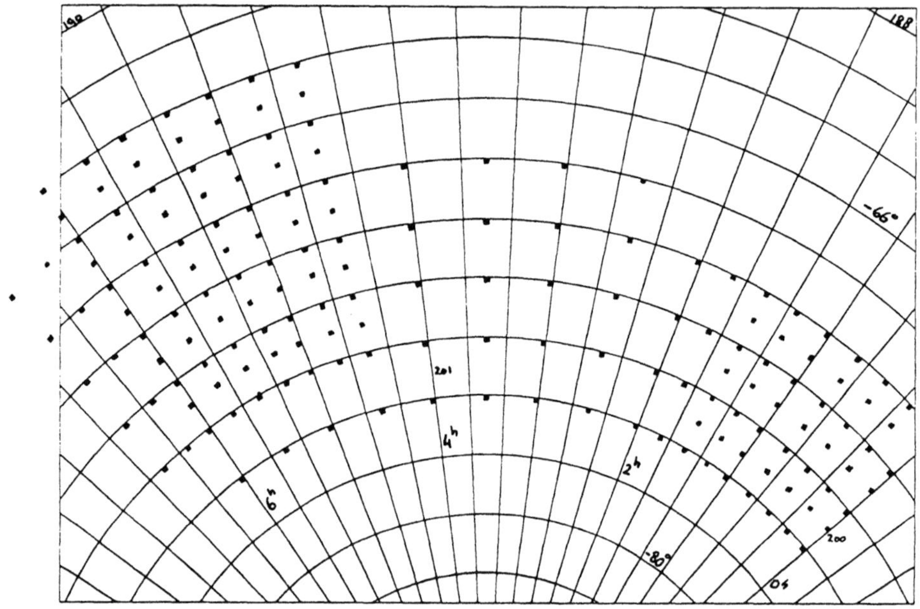

Fig. 1. Distribution of the Magellanic Cloud Selected Areas over the sky. The positions of the NWcorners of each MCSA are given in the Table. There are 4 Kapteyn Selected Areas in this field given by their number (188, 190, 200, and 201).

References

Bastian, U., Röser, S., Nesterov, V.V., Polozhentsev, D.D., Potter, Kh.I., Wielen, R., Yagudin, L.I., 1991, A&AS 87, 159

de Boer, K.S., Azzopardi, M., et al., 1989, ESO Messenger 57, 27

de Boer, K.S., Spite, F., et al., 1991, ESO Messenger 66, 14

de Boer, K.S., Spite, F., Stasinska, G., Editors, 1989, "Recent Developments of Magellanic Cloud Research", Obs. de Paris

Dickel, H.R., Lortet, M.-C., de Boer, K.S., 1987, A&AS 68, 75

Dubois, P., Florsch, A., Jaschek, M., Editors, 1982, "La Structure du Petit Nuage Magellan", Obs. de Strasbourg

Haynes, R., Milne, D.K., Editors, 1991, "The Magellanic Clouds", IAU Symp. 148, Kluwer, Dordrecht

Hindman, J.W., Westerlund, B.E., Editors, 1966, "Symposium on the Magellanic Clouds", Mount Stromlo Obs.

Kapteyn, J.C., 1906, "Plan of Selected Areas", Groningen

Kerr, F.J., Rodgers, A.W., Editors, 1964, "The Galaxy and the Magellanic Clouds", IAU Symp. 20, Austral. Acad. Sci., Canberra

Muller, A.B., Editor, 1971, "The Magellanic Clouds", Reidel, Dordrecht
Périé, J., Prévot, L., Rousseau, M., Peyrin, Y., Robin, A., 1991, A&AS 90, 1
van den Bergh, S., de Boer, K.S., Editors, 1984, "Structure and Evolution of the Magellanic Clouds", IAU Symp. 108, Reidel, Dordrecht

Data for the Grid of Magellanic Cloud Selected Areas (J2000)

MCSA	R.A. h m	Dec. deg
SMC01	0 0	-72
SMC02	0 0	-73
SMC03	0 0	-74
SMC04	0 0	-75
SMC05	0 0	-76
SMC06	0 15	-70
SMC07	0 15	-71
SMC08	0 15	-72
SMC09	0 15	-73
SMC10	0 15	-74
SMC11	0 15	-75
SMC12	0 15	-76
SMC13	0 30	-70
SMC14	0 30	-71
SMC15	0 30	-72
SMC16	0 30	-73
SMC17	0 30	-74
SMC18	0 30	-75
SMC19	0 30	-76
SMC20	0 45	-70
SMC21	0 45	-71
SMC22	0 45	-72
SMC23	0 45	-73
SMC24	0 45	-74
SMC25	0 45	-75
SMC26	0 45	-76
SMC27	1 0	-70
SMC28	1 0	-71
SMC29	1 0	-72
SMC30	1 0	-73
SMC31	1 0	-74
SMC32	1 0	-75
SMC33	1 0	-76
SMC34	1 15	-70
SMC35	1 15	-71
SMC36	1 15	-72
SMC37	1 15	-73
SMC38	1 15	-74
SMC39	1 15	-75
SMC40	1 15	-76

MCSA	R.A. h m	Dec. deg
SMC41	1 30	-70
SMC42	1 30	-71
SMC43	1 30	-72
SMC44	1 30	-73
SMC45	1 30	-74
SMC46	1 30	-75
SMC47	1 30	-76
BRI01	1 45	-70
BRI02	1 45	-72
BRI03	1 45	-74
BRI04	1 45	-76
BRI05	2 0	-70
BRI06	2 0	-72
BRI07	2 0	-74
BRI08	2 0	-76
BRI09	2 30	-68
BRI10	2 30	-70
BRI11	2 30	-72
BRI12	2 30	-74
BRI13	2 30	-76
BRI14	3 0	-68
BRI15	3 0	-70
BRI16	3 0	-72
BRI17	3 0	-74
BRI18	3 0	-76
BRI19	3 30	-68
BRI20	3 30	-70
BRI21	3 30	-72
BRI22	3 30	-74
BRI23	3 30	-76
BRI24	4 0	-68
BRI25	4 0	-70
BRI26	4 0	-72
BRI27	4 0	-74
BRI28	4 0	-76

MCSA	R.A. h m	Deg. deg
LMC01	4 30	-64
LMC02	4 30	-65
LMC03	4 30	-66
LMC04	4 30	-67
LMC05	4 30	-68
LMC06	4 30	-69
LMC07	4 30	-70
LMC08	4 30	-71
LMC09	4 30	-72
LMC10	4 30	-73
LMC11	4 30	-74
LMC12	4 30	-76
LMC13	4 45	-64
LMC14	4 45	-65
LMC15	4 45	-66
LMC16	4 45	-67
LMC17	4 45	-68
LMC18	4 45	-69
LMC19	4 45	-70
LMC20	4 45	-71
LMC21	4 45	-72
LMC22	4 45	-73
LMC23	4 45	-74
LMC24	5 0	-64
LMC25	5 0	-65
LMC26	5 0	-66
LMC27	5 0	-67
LMC28	5 0	-68
LMC29	5 0	-64
LMC30	5 0	-70
LMC31	5 0	-71
LMC32	5 0	-72
LMC33	5 0	-73
LMC34	5 0	-74
LMC36	5 0	-76
LMC37	5 15	-64
LMC38	5 15	-65
LMC39	5 15	-66
LMC40	5 15	-67
LMC41	5 15	-68
LMC42	5 15	-69
LMC43	5 15	-70
LMC44	5 15	-71
LMC45	5 15	-72
LMC46	5 15	-73
LMC47	5 15	-74

MCSA	R.A. h m	Dec. deg
LMC48	5 30	-64
LMC49	5 30	-65
LMC50	5 30	-66
LMC51	5 30	-67
LMC52	5 30	-68
LMC53	5 30	-69
LMC54	5 30	-70
LMC55	5 30	-71
LMC56	5 30	-72
LMC57	5 30	-73
LMC58	5 30	-74
LMC59	5 30	-76
LMC60	5 45	-64
LMC61	5 45	-65
LMC62	5 45	-66
LMC63	5 45	-67
LMC64	5 45	-68
LMC65	5 45	-69
LMC66	5 45	-70
LMC67	5 45	-71
LMC68	5 45	-72
LMC69	5 45	-73
LMC70	5 45	-74
LMC71	6 0	-64
LMC72	6 0	-65
LMC73	6 0	-66
LMC74	6 0	-67
LMC75	6 0	-68
LMC76	6 0	-69
LMC77	6 0	-70
LMC78	6 0	-71
LMC79	6 0	-72
LMC80	6 0	-73
LMC81	6 0	-74
LMC82	6 0	-76
LMC83	6 15	-66
LMC84	6 15	-68
LMC85	6 15	-70
LMC86	6 15	-72
LMC86	6 15	-74
LMC88	6 30	-66
LMC89	6 30	-68
LMC90	6 30	-70
LMC91	6 30	-72
LMC92	6 30	-74

Conclusions and Perspectives

J. Lequeux

Observatoire de Paris-Meudon, 92195 Meudon Cedex, France

I have been struck by the considerable advances made in our knowledge of the Magellanic Clouds since the IAU Symposium in Sydney, only two years ago. It is not an easy task to summarize those advances, and what I will say is certainly biased by my own interests. Consequently, I will rather concentrate on what I think we should do in the future.

1 Structure of the Magellanic Clouds.

It was known already for a long time that the SMC has a very complex structure which is not completely understood yet; in particular the problem of the extent in depth is still not really solved. I believe that the answer will come from the cepheids: infrared photometry gives accurate distances (at least relative, and this is what counts here) and accurate mean radial velocities will give the appartenance of the cepheids to one of the systems of gas and young stars which are the characteristic features of the SMC: this will yield the relative distances to those systems, and even possibly their own 3-D configuration. The necessary data are unfortunately only partly available. Other interesting data concerning the structure and kinematics of the SMC are coming from interstellar absorption-line studies and from the Marseille-ESO Hα survey, but the latter has not yet been exploited.

The LMC used to look simpler but this appears now a false impression: there are two different systems of globular clusters with different space distribution and kinematics, while radio polarization and HI observations reported at this meeting reveals important structures out of the plane.

All this seems to suggest the role of violent interactions between the two Clouds and perhaps the Galaxy. Did these interactions produce a mixing of populations? Can this explain the anomalous abundances of some clusters like NGC 330 (if they indeed exist!)? This is an interesting topic for future dynamical studies.

The bridge between the two Clouds and the Magellanic stream are certainly interesting places to look at in this context; they may contain the key of the above problem. It will be very interesting to obtain the abundances of heavy elements and of dust which will tell much about their origin. The Stream could not be detected by IRAS but the limit obtained on the amount of dust is not incompatible with a LMC and even Galactic dust-to-gas ratio. The stream and the bridge should both be targets of high priority for ISO.

2 Interstellar matter

Much new has been presented at this meeting. The Hα survey and the Molonglo 843 MHz survey are both gold mines that are just starting to be exploited, in particular in searches for supernova remnants. The Hα survey shows a substantial amount of diffuse ionized gas which deserves further detailed studies. In particular similar surveys in the Hβ line will allow to map extinction by comparison to Hα, and surveys in the lines of [SII], [OI] and [OIII] will provide constraints on the physical state and ionization processes in this medium.

The CO surveys made with the Columbia/CFA mini telescope and the SEST, as well as the [CII]158μm line observations have revealed that the molecular clouds in the Magellanic Clouds are different from those in our Galaxy. This can be understood as due to a stronger photodissociation of CO. Also, there should be little molecular hydrogen but we will have to wait for LYMAN to see it directly. In the mean time, high-resolution 21-cm line observations in progress with the Australia Telescope should tell us if, as expected, the "molecular" clouds detected in CO contain mainly atomic hydrogen! These observations will also provide us as a by-product with detailed maps of the Galactic background extinction which are badly needed by optical observers of the Clouds. Also, they should be able to give us an idea of the kinetic temperature of the atomic gas from 21-cm emission/absorption measurements. Future observations with ISO of the [CII]158μm and [OI]63μm lines especially in the diffuse medium will tell us about the thermal balance in the interstellar medium and the presence of shocks. Combining those observations, we will be able to obtain a complete picture of the interstellar gas in the Magellanic Clouds; will this help us to understand how they manage to form stars rather actively while there is so little molecular gas?

The properties of interstellar dust in the Magellanic Clouds, as revealed by interstellar reddening observations and observations with IRAS, are also different from those in the Galaxy. They are not really understood: in particular we do not understand why IRAS found so little emission in its 12μm band. This is probably a question that will be answered using ISO.

Much new is to be expected from the mere comparison between images of the Clouds obtained at different wavelengths. We have seen some beautiful results obtained in this way, e.g. on supernova remanants, but I am convinced that much more will come soon with the avaibility of data from ROSAT, ISO, near-infrared spectroscopy, etc. Here most of the problems that will be attacked are not really specific to the Magellanic Clouds, but the known distances make them easier to solve than in our Galaxy.

3 Stellar Populations

The recent years have been characterized by extensive surveys of various kinds of objects: carbon stars, planetary nebulae, X-ray sources (on which much more is coming from ROSAT). The Magellanic Clouds are ideal places to study objects whose distances are difficult or impossible to obtain in our Galaxy although

HIPPARCOS will change the situation to some extent: essentially the giants and the post-AGB objects. It is expected that the near-IR surveys presently contemplated both in the USA and in Europe as well as ISO will yield complete samples of such objects: these will be of prime importance in studies of the final stages of stellar evolution, which are still poorly understood.

The Magellanic Clouds also host a zoo of massive stars which are exceptional or difficult to study in our Galaxy, e.g. the Luminous Blue Variables. Considerable progress has been recently made in their observation and understanding. They are also very interesting places to study star formation, and particular that of the most massive stars. It appears that most luminous objects in the Clouds are indeed multiple, showing that they have been formed as small clusters of massive stars; more is to come from speckle interferometry and observations with SUSI at the 3.5-m NTT telescope and with the Hubble space telescope. Similarly, the Clouds are ideal places to look for star formation in clusters and to answer questions on the Initial Mass Function, the coevality or non-coevality of star formation, etc. Finally, the study of young stellar objects (pre-main sequence stars, embedded objects, bipolar flows etc.) is only starting in the Clouds (again with the advantage of known distances) and it is clear that much will come from the near-IR surveys and from near- and far-IR imaging from the ground and from ISO.

The Magellanic Clouds are ideal places to study the influence of metallicity on stellar evolution: there is a span in metallicities by a factor 20-30 between the Galactic bulge, the Solar neighborhood, the LMC and the SMC. The corresponding theoretical models have made considerable progresses, still we do not really understand the lack of bright carbon stars, the existence of faint carbon stars, the CH stars, the transition AGB to post-AGB in general, and even the evolution of the progenitor of SN 1987A! For the latter problem, the solution will probably be found through detailed photometric and spectroscopic observations of stars in Magellanic clusters with various metallicities, mapping the photospheric abundance anomalies on the HR diagram. Exotic objects like novae, symbiotic stars, B[e] stars etc. will be more rewarding to study in the Magellanic Clouds, due once again to their known distances.

4 Abundances and Chemical Evolution

The pattern of observed abundances of the various elements in stars and the intersellar matter of the Magellanic Clouds is a mess. Bernard Pagel has tried here to put some order in this mess and his conclusion was that the heavy-element abundances inside each Cloud might well be homogeneous after all. This is reassuring because i) chemical inhomogeneitie are not observed in HII regions and ii) they would be rather difficult to explain at least for young objects without invoking exotic processes like population interchanges between both Clouds and/or the Galaxy, or some miraculous preservation of low-metallicity "pockets" of interstellar gas against mixing and pollution. However I am not sure that everyone in the audience has accepted his conclusions. There is clearly still a lot of work to do. Particularly irritating is the bad situation on the abundance of carbon - notoriously difficult to determine- as this element plays a major role in sevral fields, in particular the abundance of interstellar molecules and dust.

Given the controversial situation about abundances, it is no surprise that the chemical evolution of the clouds is not well understood. Once we have said that the SMC is less evolved that the LMC as it contains more gas and has lower abundances, and the same for the LMC vs. our Galaxy, it's about all that can be told with any certainty. A troublesome problem is that of the yield in heavy elements. It is well-known that in the simplest model of chemical evolution, the closed-box model (also called the one-zone model), the evolution of the metallicity Z can be decribed by the simple expression:

$Z = y \ln(M_{tot}/M_{gas})$,

where y is the yield, M_{tot} the total mass (excluding dark matter without relation with stellar evolution) and M_{gas} the mass of interstellar matter. When plugging numbers for the Magellanic Clouds into this equation one obtains for both Clouds a yield of 0.003, while conventional models of stellar evolution and a normal initial mass function for stars predict a yield of about 0.01. Surprisingly, there are several possible explanations for this discrepancy between which it is not possible to choose. One can invoke infall of primordial gas, ejection out of the galaxy of some of the gas just enriched by massive stars, or an incorrect determination of M_{tot} due to the neglect of dark matter. Another possibility is the suggestion by Maeder that the stars more massive than 20-25 $M_\odot$ or so do not end as supernovae but rather collapse as black holes: this will decrease considerably the production of heavy elements by stars and then the yield. In any case, the yield is likely to depend strongly on metallicity due to the dependence of mass-loss by massive stars on Z, and this has never been taken into account in models of chemical evolution! There is still much work to do in this field, even if it is certainly a domain much less secure than others!

List of Participants

Andreani	Paola	Padova
Aparicio	Antonio	La Laguna
Appenzeller	Immo	Heidelberg
Azzopardi	Marc	Marseille
Baschek	Bodo	Heidelberg
Beaulieu	Sylvie	Paris
Bertelli	Gianpaolo	Padova
Bessell	Michael S.	Mt. Stromlo
Beuermann	Klaus	Göttingen
Bohannan	Bruce	Tucson
Bomans	Dominik J.	Bonn
Booth	Roy	Onsala
Bressan	Alessandro	Padova
Cheng	Kwang-Ping	Greenbelt
Clampin	Mark	Baltimore
Clegg	Robin E.S.	Cambridge UK
Danziger	I. John	Garching
De Boer	Klaas S.	Bonn
De Graauw	Thijs	Groningen
Dennefeld	Michel	Paris
Dennerl	Konrad	Garching
Dettmar	Ralf-Jürgen	Bonn
Domgörgen	Hildegard	Bonn
Dominik	Carsten	Berlin
Dubois	Pascal	Strasbourg
Duschl	Wolfgang J.	Heidelberg
Fagotto	Franco	Padova
Feast	Michael	Cambridge UK
Ferlet	Roger	Paris
Fleischer	Axel J.	Berlin
Gieren	Wolfgang P.	Santiago Chile
Gochermann	Josef	Bochum
Grison		Paris
Hatzidimitriou	Despina	Cambridge UK
Hauschildt	Peter H.	Tempe
Heydari-Malayeri	Mohammad	La Silla
Hintzen	Paul	Greenbelt
Israel	Frank P.	Leiden
Isserstedt	Jörg	Würzburg
Jüttner	Andreas	Heidelberg
Kahabka	Peter	Garching
Kegel	Wilhelm H.	Frankfurt

Kinkel	Ulrich	Heidelberg/Garching
Klare	Gerhard	Heidelberg
Klein	Ulrich	Bonn
Kneer	Reinhold	Heidelberg
Koornneef	Jan	Baltimore
Köppen	Joachim	Heidelberg
Kutner	Marc	Troy
Le Coarer	Etienne	Marseille
Lequeux	James	Meudon
Lortet	Marie-Claire	Meudon
Maeder	André	Sauverny
Magalhaes	Antonio Mario	Madison
Martin	Nicole	Marseille
Marx	Monika	Bonn
Maurice	Eric	Marseille
Mebold	Ulrich	Bonn
Meinert	Dieter	Bonn
Molaro	Paolo	Trieste
Morgan	David H.	Edinburgh
Muratorio	Gerard	Marseille
Nota	Antonella	Baltimore
Okumura	Koryo	Meudon
Pagel	Bernard E.J.	Copenhagen
Pakull	Manfred	Heidelberg
Pietsch	Wolfgang	Garching
Queloz	Didier	Sauverny
Rauter	Winfried	Berlin
Richtler	Tom	Bonn
Rohlfs	Kristen	Bochum
Rons	Nadine	Brussels
Rosa	Michael	Garching
Rosado	Margarita	Mexico
Roueff	Evelyne	Meudon
Rubio	Monica	Santiago
Russell	Stephen C.	Dublin
Schmidt-Kaler	Theodor	Bochum
Seggewiß	Wilhelm	Daun
Silvy	Jocelyne	Marseille
Spiller	Frank	Heidelberg
Spite	Francois	Meudon
Spite	Monique	Meudon
Stahl	Otmar	Heidelberg
Starrfield	Sumner	Tempe
Szeifert	Thomas	Heidelberg
Vallenari	Antonella	Bonn
Viallefond	Francois	Meudon
Vigroux	Laurent	Saclay

Walsh	Jeremy R.	Garching
Wamsteker	Willem	Villafranca
Wehrse	Rainer	Heidelberg
Weidemann	Volker	Kiel
Westerlund	Bengt E.	Uppsala
Will	Jean-Marie	Bonn
Winters	Jan Martin	Berlin
Woitke	Peter	Berlin
Wolf	Bernhard	Heidelberg
Xu	Cong	Bonn/Heidelberg
Ye	Taisheng	Sydney
Zickgraf	Franz-Josef	Heidelberg
Zinnecker	Hans	Würzburg

Author Index

Lecture Notes in Physics

For information about Vols. 1–374
please contact your bookseller or Springer-Verlag

Vol. 375: C. Bartocci, U. Bruzzo, R. Cianci (Eds.), Differential Geometric Methods in Theoretical Physics. Proceedings, 1990. XIX, 401 pages. 1991.

Vol. 376: D. Berényi, G. Hock (Eds.), High-Energy Ion-Atom Collisions. Proceedings, 1990. IX, 364 pages. 1991.

Vol. 377: W. J. Duschl, S.J. Wagner, M. Camenzind (Eds.), Variability of Active Galaxies. Proceedings, 1990. XII, 312 pages. 1991.

Vol. 378: C. Bendjaballah, 0. Hirota, S. Reynaud (Eds.), Quantum Aspects of Optical Communications. Proceedings 1990. VII, 389 pages. 1991.

Vol. 379: J. D. Hennig, W. Lücke, J. Tolar (Eds.), Differential Geometry, Group Representations, and Quantization. XI, 280 pages. 1991.

Vol. 380: I. Tuominen, D. Moss, G. Rüdiger (Eds.), The Sun and Cool Stars: activity, magnetism, dynamos. Proceedings, 1990. X, 530 pages. 1991.

Vol. 381: J. Casas-Vazquez, D. Jou (Eds.), Rheological Modelling: Thermodynamical and Statistical Approaches. Proceedings, 1990. VII, 378 pages. 1991.

Vol. 382: V.V. Dodonov, V. I. Man'ko (Eds.), Group Theoretical Methods in Physics. Proceedings, 1990. XVII, 601 pages. 1991.

Vol. 384: M. D. Smooke (Ed.), Reduced Kinetic Mechanisms and Asymptotic Approximations for Methane-Air Flames. V, 245 pages. 1991.

Vol. 385: A. Treves, G. C. Perola, L. Stella (Eds.), Iron Line Diagnostics in X-Ray Sources. Proceedings, Como, Italy 1990. IX, 312 pages. 1991.

Vol. 386: G. Pétré, A. Sanfeld (Eds.), Capillarity Today. Proceedings, Belgium 1990. XI, 384 pages. 1991.

Vol. 387: Y. Uchida, R. C. Canfield, T. Watanabe, E. Hiei (Eds.), Flare Physics in Solar Activity Maximum 22. Proceedings, 1990. X, 360 pages. 1991.

Vol. 388: D. Gough, J. Toomre (Eds.), Challenges to Theories of the Structure of Moderate-Mass Stars. Proceedings, 1990. VII, 414 pages. 1991.

Vol. 389: J. C. Miller, R. F. Haglund (Eds.), Laser Ablation-Mechanisms and Applications. Proceedings. IX, 362 pages, 1991.

Vol. 390: J. Heldmann, M. J. Klein (Eds.), Bioastronomy The Search for Extraterrestrial Life. Proceedings, 1990. XVII, 413 pages. 1991.

Vol. 391: A. Zdziarski, M. Sikora (Eds.), Ralativistic Hadrons in Cosmic Compact Objects. Proceedings, 1990. XII, 182 pages. 1991.

Vol. 392: J.-D. Fournier, P.-L. Sulem (Eds.), Large-Scale Structures in Nonlinear Physics. Proceedings. VIII, 353 pages. 1991.

Vol. 393: M. Remoissenet, M.Peyrard (Eds.), Nonlinear Coherent Structures in Physics and Biology. Proceedings. XII, 398 pages. 1991.

Vol. 394: M. R. J. Hoch, R. H. Lemmer (Eds.), Low Temperature Physics. Proceedings. X,374 pages. 1991.

Vol. 395: H. E. Trease, M. J. Fritts, W. P. Crowley (Eds.), Advances in the Free-Lagrange Method. Proceedings, 1990. XI, 327 pages. 1991.

Vol. 396: H. Mitter, H. Gausterer (Eds.), Recent Aspects of Quantum Fields. Proceedings. XIII, 332 pages. 1991.

Vol. 398: T. M. M. Verheggen (Ed.), Numerical Methods for the Simulation of Multi-Phase and Complex Flow. Proceedings, 1990. VI, 153 pages. 1992.

Vol. 399: Z. Švestka, B. V. Jackson, M. E. Machedo (Eds.), Eruptive Solar Flares. Proceedings, 1991. XIV, 409 pages. 1992.

Vol. 400: M. Dienes, M. Month, S. Turner (Eds.), Frontiers of Particle Beams: Intensity Limitations. Proceedings, 1990. IX, 610 pages. 1992.

Vol. 401: U. Heber, C. S. Jeffery (Eds.), The Atmospheres of Early-Type Stars. Proceedings, 1991. XIX, 450 pages. 1992.

Vol. 402: L. Boi, D. Flament, J.-M. Salanskis (Eds.), 1830-1930: A Century of Geometry. VIII, 304 pages. 1992.

Vol. 403: E. Balslev (Ed.), Schrödinger Operators. Proceedings, 1991. VIII, 264 pages. 1992.

Vol. 404: R. Schmidt, H. O. Lutz, R. Dreizler (Eds.), Nuclear Physics Concepts in the Study of Atomic Cluster Physics. Proceedings, 1991. XVIII, 363 pages. 1992.

Vol. 405: W. Hollik, R. Rückl, J. Wess (Eds.), Phenomenological Aspects of Supersymmetry. VII, 329 pages. 1992.

Vol. 406: R. Kayser, T. Schramm, L. Nieser (Eds.), Gravitational Lenses. Proceedings, 1991. XXII, 399 pages. 1992.

Vol. 407: P. L. Smith, W. L. Wiese (Eds.), Atomic and Molecular Data for Space Astronomy. VII, 158 pages. 1992.

Vol. 408: V. J. Martínez, M. Portilla, D. Sàez (Eds.), New Insights into the Universe. Proceedings, 1991. XI, 298 pages. 1992.

Vol. 409: H. Gausterer, C. B. Lang (Eds.), Computational Methods in Field Theory. Proceedings, 1992. XII, 274 pages. 1992.

Vol. 410: J. Ehlers, G. Schäfer (Eds.), Relativistic Gravity Research. Proceedings, VIII, 409 pages. 1992.

Vol. 411: W. Dieter Heiss (Ed.), Chaos and Quantum Chaos. Proceedings, XIV, 330 pages. 1992.

Vol. 412: A. W. Clegg, G. E. Nedoluha (Eds.), Astrophysical Masers. Proceedings, 1992. XX, 480 pages. 1993.

Vol. 413: Aa. Sandqvist, T. P. Ray (Eds.); Central Activity in Galaxies. From Observational Data to Astrophysical Diagnostics. XIII, 235 pages. 1993.

Vol. 414: M. Napolitano, F. Sabetta (Eds.), Thirteenth International Conference on Numerical Methods in Fluid Dynamics. Proceedings, 1992. XIV, 541 pages. 1993.

Vol. 415: L. Garrido (Ed.), Complex Fluids. Proceedings, 1992. XIII, 413 pages. 1993.

Vol. 416: B. Baschek, G. Klare, J. Lequeux (Eds.), New Aspects of Magellanic Cloud Research. Proceedings, 1992. XIII, 404 pages. 1993.

Vol. 417: K. Goeke P. Kroll, H.-R. Petry (Eds.), Quark Cluster Dynamics. Proceedings, 1992. XI, 297 pages. 1993.

New Series m: Monographs

Vol. m 1: H. Hora, Plasmas at High Temperature and Density. VIII, 442 pages. 1991.

Vol. m 2: P. Busch, P. J. Lahti, P. Mittelstaedt, The Quantum Theory of Measurement. XIII, 165 pages. 1991.

Vol. m 3: A. Heck, J. M. Perdang (Eds.), Applying Fractals in Astronomy. IX, 210 pages. 1991.

Vol. m 4: R. K. Zeytounian, Mécanique des fluides fondamentale. XV, 615 pages, 1991.

Vol. m 5: R. K. Zeytounian, Meteorological Fluid Dynamics. XI, 346 pages. 1991.

Vol. m 6: N. M. J. Woodhouse, Special Relativity. VIII, 86 pages. 1992.

Vol. m 7: G. Morandi, The Role of Topology in Classical and Quantum Physics. XIII, 239 pages. 1992.

Vol. m 8: D. Funaro, Polynomial Approximation of Differential Equations. X, 305 pages. 1992.

Vol. m 9: M. Namiki, Stochastic Quantization. X, 217 pages. 1992.

Vol. m 10: J. Hoppe, Lectures on Integrable Systems. VII, 111 pages. 1992.

Vol. m 11: A. D. Yaghjian, Relativistic Dynamics of a Charged Sphere. XII, 115 pages. 1992.

Vol. m 12: G. Esposito, Quantum Gravity, Quantum Cosmology and Lorentzian Geometries. XVI, 326 pages. 1992.

Vol. m 13: M. Klein, A. Knauf, Classical Planar Scattering by Coulombic Potentials. V, 142 pages. 1992.

Vol. m 14: A. Lerda, Anyons. XI, 138 pages. 1992.

Vol. m 15: N. Peters, B. Rogg (Eds.), Reduced Kinetic Mechanisms for Applications in Combustion Systems. X, 360 pages. 1993.

GPSR Compliance

The European Union's (EU) General Product Safety Regulation (GPSR) is a set of rules that requires consumer products to be safe and our obligations to ensure this.

If you have any concerns about our products, you can contact us on ProductSafety@springernature.com

In case Publisher is established outside the EU, the EU authorized representative is:

Springer Nature Customer Service Center GmbH
Europaplatz 3
69115 Heidelberg, Germany

Batch number: 10620854

Printed by Printforce, the Netherlands